I0028875

Numerical Analysis 2000, Volume 4

Optimization and Nonlinear Equations

Numerical Analysis 2000, Volume 4

Optimization and Nonlinear Equations

Edited by

L.T. Watson
Virginia Tech.,
Dept. of Computer Science,
660 McBryde Hall
Blacksburg, VA 24061-0106
USA

M. Bartholomew-Biggs
University of Hertfordshire,
Numerical Optimisation Centre,
Hatfield, Hertfordshire
UK

J.A. Ford
University of Essex,
Dept. of Computer Science,
Wivenhoe Park
Colchester, Essex CO4 3SQ
UK

N·H

2001
ELSEVIER
Amsterdam - London - New York - Oxford - Paris - Shannon - Tokyo

ELSEVIER SCIENCE B.V.
Sara Burgerhartstraat 25
P.O. Box 211, 1000 AE Amsterdam, The Netherlands

© 2001 Elsevier Science B.V. All rights reserved.

This work is protected under copyright by Elsevier Science, and the following terms and conditions apply to its use:

Photocopying
Single photocopies of single chapters may be made for personal use as allowed by national copyright laws. Permission of the Publisher and payment of a fee is required for all other photocopying, including multiple or systematic copying, copying for advertising or promotional purposes, resale, and all forms of document delivery. Special rates are available for educational institutions that wish to make photocopies for non-profit educational classroom use.

Permissions may be sought directly from Elsevier Science Global Rights Department, PO Box 800, Oxford OX5 1DX, UK; phone: (+44) 1865 843830, fax: (+44) 1865 853333, e-mail: permissions@elsevier.co.uk. You may also contact Global Rights directly through Elsevier's home page (http://www.elsevier.nl), by selecting 'Obtaining Permissions'.

In the USA, users may clear permissions and make payments through the Copyright Clearance Center, Inc., 222 Rosewood Drive, Danvers, MA 01923, USA; phone: (+1)(978) 7508400, fax: (+1)(978) 7504744, and in the UK through the Copyright Licensing Agency Rapid Clearance Service (CLARCS), 90 Tottenham Court Road, London W1P 0LP, UK; phone: (+44) 207 631 5555; fax: (+44) 207 631 5500. Other countries may have a local reprographic rights agency for payments.

Derivative Works
Tables of contents may be reproduced for internal circulation, but permission of Elsevier Science is required for external resale or distribution of such material. Permission of the Publisher is required for all other derivative works, including compilations and translations.

Electronic Storage or Usage
Permission of the Publisher is required to store or use electronically any material contained in this work, including any chapter or part of a chapter.

Except as outlined above, no part of this work may be reproduced, stored in a retrieval system or transmitted in any form or by any means, electronic, mechanical, photocopying, recording or otherwise, without prior written permission of the Publisher.
Address permissions requests to: Elsevier Global Rights Department, at the mail, fax and e-mail addresses noted above.

Notice
No responsibility is assumed by the Publisher for any injury and/or damage to persons or property as a matter of products liability, negligence or otherwise, or from any use or operation of any methods, products, instructions or ideas contained in the material herein. Because of rapid advances in the medical sciences, in particular, independent verification of diagnoses and drug dosages should be made.

First edition 2001

Library of Congress Cataloging in Publication Data
A catalog record from the Library of Congress has been applied for.

ISBN: 0 444 50599 7

♾ The paper used in this publication meets the requirements of ANSI/NISO Z39.48-1992 (Permanence of Paper).
Transferred to digital printing 2005

NH

ELSEVIER

Journal of Computational and Applied Mathematics 124 (2000) 373

JOURNAL OF
COMPUTATIONAL AND
APPLIED MATHEMATICS

www.elsevier.nl/locate/cam

Author Index Volume 124 (2000)

Allgower, E.L. and K. Georg, Piecewise linear methods for nonlinear equations and optimization   245–261

Bartholomew-Biggs, M., S. Brown, B. Christianson and L. Dixon, Automatic differentiation of algorithms   171–190
Billups, S.C. and K.G. Murty, Complementarity problems   303–318
Boggs, P.T. and J.W. Tolle, Sequential quadratic programming for large-scale nonlinear optimization   123–137
Brown, S., see Bartholomew-Biggs, M.   171–190

Christianson, B., see Bartholomew-Biggs, M.   171–190

Dixon, L., see Bartholomew-Biggs, M.   171–190
Driscoll, P.J., see Sherali, H.D.   319–340

Galántai, A., The theory of Newton's method   25–44
Georg, K., see Allgower, E.L.   245–261

Hoffman, K.L., Combinatorial optimization: Current successes and directions for the future   341–360

Lewis, R.M., V. Torczon and M.W. Trosset, Direct search methods: then and now   191–207
Lukšan, L. and E. Spedicato, Variable metric methods for unconstrained optimization and nonlinear least squares   61–95

Martínez, J.M., Practical quasi-Newton methods for solving nonlinear systems   97–121
Murty, K.G., see Billups, S.C.   303–318

Nash, S.G., A survey of truncated-Newton methods   45–59

Pardalos, P.M., H.E. Romeijn and H. Tuy, Recent developments and trends in global optimization   209–228
Potra, F.A. and S.J. Wright, Interior-point methods   281–302

Renaud, J.E., see Rodríguez, J.   139–154
Rheinboldt, W.C., Numerical continuation methods: a perspective   229–244
Rodríguez, J.F., J.E. Renaud, B.A. Wujek and R.V. Tappeta, Trust region model management in multidisciplinary design optimization   139–154
Romeijn, H.E., see Pardalos, P.M.   209–228

Sargent, R.W.H., Optimal control   361–371
Sherali, H.D. and P.J. Driscoll, Evolution and state-of-the-art in integer programming   319–340
Spedicato, E., see Lukšan, L.   61–95
Spedicato, E., Z. Xia and L. Zhang, ABS algorithms for linear equations and optimization   155–170

Tappeta, R.V., see Rodríguez, J.   139–154
Tolle, J.W., see Boggs, P.T.   123–137
Torczon, V., see Lewis, R.   191–207
Trosset, M.W., see Lewis, R.   191–207
Tuy, H., see Pardalos, P.M.   209–228

Wolfe, M.A., Interval mathematics, algebraic equations and optimization   263–280
Wright, S.J., see Potra, F.A.   281–302
Wujek, B.A., see Rodríguez, J.   139–154

Xia, Z., see Spedicato, E.   155–170

Yamamoto, T., Historical developments in convergence analysis for Newton's and Newton-like methods   1–23

Zhang, L., see Spedicato, E.   155–170

0377-0427/00/$-see front matter © 2000 Elsevier Science B.V. All rights reserved.
PII: S 0 3 7 7 - 0 4 2 7 (0 0) 0 0 6 0 6 - 3

JOURNAL OF COMPUTATIONAL AND APPLIED MATHEMATICS

Volume 124, Numbers 1–2, 1 December 2000

Contents

Preface: Numerical Analysis 2000. Vol. IV: Optimization and Nonlinear Equations   ix

T. Yamamoto
Historical developments in convergence analysis for Newton's and Newton-like methods   1

A. Galántai
The theory of Newton's method   25

S.G. Nash
A survey of truncated-Newton methods   45

L. Lukšan and E. Spedicato
Variable metric methods for unconstrained optimization and nonlinear least squares   61

J.M. Martínez
Practical quasi-Newton methods for solving nonlinear systems   97

P.T. Boggs and J.W. Tolle
Sequential quadratic programming for large-scale nonlinear optimization   123

J.F. Rodríguez, J.E. Renaud, B.A. Wujek and R.V. Tappeta
Trust region model management in multidisciplinary design optimization   139

E. Spedicato, Z. Xia and L. Zhang
ABS algorithms for linear equations and optimization   155

M. Bartholomew-Biggs, S. Brown, B. Christianson and L. Dixon
Automatic differentiation of algorithms   171

R.M. Lewis, V. Torczon and M.W. Trosset
Direct search methods: then and now   191

P.M. Pardalos, H.E. Romeijn and H. Tuy
Recent developments and trends in global optimization   209

W.C. Rheinboldt
Numerical continuation methods: a perspective   229

E.L. Allgower and K. Georg
Piecewise linear methods for nonlinear equations and optimization   245

M.A. Wolfe
Interval mathematics, algebraic equations and optimization   263

F.A. Potra and S.J. Wright
Interior-point methods   281

S.C. Billups and K.G. Murty
Complementarity problems   303

H.D. Sherali and P.J. Driscoll
Evolution and state-of-the-art in integer programming   319

K.L. Hoffman
Combinatorial optimization: Current successes and directions for the future   341

R.W.H. Sargent
Optimal control   361

Author Index   373

NH

ELSEVIER   Journal of Computational and Applied Mathematics 124 (2000) ix–x

JOURNAL OF
COMPUTATIONAL AND
APPLIED MATHEMATICS

www.elsevier.nl/locate/cam

Preface

Numerical Analysis 2000
Vol. IV: Optimization and Nonlinear Equations

In one of the papers in this collection, the remark that "nothing at all takes place in the universe in which some rule of maximum or minimum does not appear" is attributed to no less an authority than Euler. Simplifying the syntax a little, we might paraphrase this as *Everything is an optimization problem*. While this might be something of an overstatement, the element of exaggeration is certainly reduced if we consider the extended form: *Everything is an optimization problem or a system of equations*. This observation, even if only partly true, stands as a fitting testimonial to the importance of the work covered by this volume.

Since the 1960s, much effort has gone into the development and application of numerical algorithms for solving problems in the two areas of optimization and systems of equations. As a result, many different ideas have been proposed for dealing efficiently with (for example) severe nonlinearities and/or very large numbers of variables. Libraries of powerful software now embody the most successful of these ideas, and one objective of this volume is to assist potential users in choosing appropriate software for the problems they need to solve. More generally, however, these collected review articles are intended to provide both researchers and practitioners with snapshots of the 'state-of-the-art' with regard to algorithms for particular classes of problem. These snapshots are meant to have the virtues of immediacy through the inclusion of very recent ideas, but they also have sufficient depth of field to show how ideas have developed and how today's research questions have grown out of previous solution attempts.

The most efficient methods for *local optimization*, both unconstrained and constrained, are still derived from the classical Newton approach. The papers in this collection describe many interesting variations, particularly with regard to the organization of the linear algebra involved. The popular quasi-Newton techniques avoid the need to calculate second derivatives, but gradient-based methods have received more attention than their direct search (function-values only) counterparts. This volume does, however, include an up-to-date account of available direct search techniques, which are required, for instance, for problems where function-values are subject to uncertainty.

As well as dealing in depth with the various classical, or neo-classical, approaches, the selection of papers on optimization in this volume ensures that newer ideas are also well represented. Thus the reader will find an account of the impact that ABS methods for linear systems are beginning to make upon the subject. The potential of interval arithmetic for dealing with the global optimization problem is also discussed, as are the emerging methods and software tools of automatic differentiation for supplying the derivative information needed by most optimization techniques. No volume on optimization would be considered to have the necessary breadth unless the vital topics of linear programming (to quote from one of the papers: "undoubtedly the optimization problem solved most

0377-0427/00/$ - see front matter © 2000 Elsevier Science B.V. All rights reserved.
PII: S 0377-0427(00)00416-7

frequently in practice") and its related areas were given due attention. As a consequence, papers on interior point methods (covering both linear and nonlinear problems), complementarity, and both integer and combinatorial optimization may all be found here. A state-of-the-art review of numerical methods in the field of optimal control is also included.

Solving *nonlinear algebraic systems of equations* is closely related to optimization. The two are not completely equivalent, however, and usually something is lost in the translation. Reformulating an optimization problem as a nonlinear system often introduces spurious stationary point solutions that are not local optima of the original problem. Conversely, formulating a nonlinear system of equations as minimizing some merit function often introduces local minima of the merit function that are not roots of the original nonlinear system. Nevertheless, the algorithms and conceptual approaches for optimization and nonlinear systems of equations are closely related, and a new algorithmic trick for one generally suggests an analogous trick for the other.

Algorithms for nonlinear equations can be roughly classified as *locally convergent* or *globally convergent*. The characterization is not perfect. A globally convergent method applied to a problem for which no global convergence theory exists may only converge locally. Often strong claims of global convergence are made for locally convergent methods (e.g., trust region), but the convergence is to a stationary point of some merit function, which is not necessarily a true *solution* to the nonlinear system of equations. The suggested classification is useful, because some approaches use local models and local theory, while others use the global behavior of the function. Local and global analysis are qualitatively very different, and from this perspective so are the algorithms based on these respective theories.

Locally convergent algorithms include Newton's method, modern quasi-Newton variants of Newton's method, and trust region methods. All of these approaches are well represented in this volume. Globally convergent algorithms include pattern search, continuation, homotopy, probability-one homotopy, and interval methods. The homotopy approach has both discrete (simplicial) and continuous variants. All of these (except for the relatively new, globally convergent, probability-one homotopy algorithms) are described in this volume by researchers with a lifetime of experience in the topics, often from a uniquely personal perspective.

Finally, the editors wish to place on record their deep appreciation of the assistance given by many referees in the preparation of this volume. Without their generous and continued co-operation, it is doubtful whether this volume would have appeared on schedule. To all of them, we offer our thanks. We also gladly acknowledge the guidance and assistance we have received from Luc Wuytack throughout the process of preparing this volume.

Layne T. Watson
Departments of Computer Science and Mathematics
Virginia Polytechnic Institute & State University
Blacksburg, VA 24061-0106, USA
Michael Bartholomew-Biggs
Numerical Optimisation Centre, University of Hertfordshire
Hatfield, Hertfordshire AL10 9AB, UK
John Ford
Department of Computer Science, University of Essex
Wivenhoe Park, Colchester, Essex CO4 3SQ, UK
E-mail address: fordj@essex.ac.uk

ELSEVIER

Journal of Computational and Applied Mathematics 124 (2000) 1–23

JOURNAL OF
COMPUTATIONAL AND
APPLIED MATHEMATICS

www.elsevier.nl/locate/cam

Historical developments in convergence analysis for Newton's and Newton-like methods

Tetsuro Yamamoto

Department of Mathematical Sciences, Faculty of Science, Ehime University, Matsuyama 790-8577, Japan

Received 18 August 1999; received in revised form 22 November 1999

Abstract

Historical developments in convergence theory as well as error estimates for Newton's method and Newton-like methods for nonlinear equations are described, mainly for differentiable equations in Banach spaces. © 2000 Elsevier Science B.V. All rights reserved.

MSC: 47H17; 65J15

Keywords: Nonlinear equations; Convergence theorems; Error estimates; Newton's method; Newton-like methods; Halley's method

1. Introduction

More than three hundred years have passed since a procedure for solving an algebraic equation was proposed by Newton in 1669 and later by Raphson in 1690 [10]. The method is now called Newton's method or the Newton–Raphson method and is still a central technique for solving nonlinear equations. Many topics related to Newton's method still attract attention from researchers. For example, the construction of globally convergent effective iterative methods for solving nondifferentiable equations in $\mathbb{R}^n$ or $\mathbb{C}^n$ is an important research area in the fields of numerical analysis and optimization.

The purpose of this paper is to trace historical developments in Kantorovich-type convergence theory as well as error estimates for Newton's method and Newton-like methods, mainly for differentiable equations in Banach spaces.

First, in Section 2, we state fundamental results in the history of convergence study as well as error estimates for Newton's method based on the Newton–Kantorovich theorem. Next, in Section 3, we

---

E-mail address: yamamoto@math.sci.ehime-u.ac.jp (T. Yamamoto).

0377-0427/00/$ - see front matter © 2000 Elsevier Science B.V. All rights reserved.
PII: S 0377-0427(00)00417-9

state some convergence results for Newton-like methods which generalize the results for Newton's method.

In Section 4, we describe Dennis' result for convergence of an iterative method with his "recalculation sequence", which includes the forward and the backward secant methods as special cases. Furthermore, in Section 5, we trace historical developments in convergence analysis for Halley's method and for Chebyshev's method.

Convergence theorems for a class of iterative methods for not necessarily differentiable equations are summarized in Section 6. Finally, in Section 7, concluding remarks are given.

2. Newton's method

Let X and Y be Banach spaces and $F : D \subseteq X \to Y$ be an operator where D is a domain of F. If F is differentiable in an open convex set $D_0 \subseteq D$, then Newton's method for solving the equation

$$F(x) = 0 \qquad (2.1)$$

with a solution x^* is defined by the following:

(N1) Let x_k be an approximation to x^*;
(N2) Solve the linear equation

$$F(x_k) + F'(x_k)h = 0 \qquad (2.2)$$

with respect to h, provided that $F'(x_k)$ is nonsingular;
(N3) Set $x_{k+1} = x_k + h$, expecting for it to be an improvement to x_k, where $k = 0, 1, 2, \ldots$.

Since $0 = F(x^*) = F(x_k + h) = F(x_k) + F'(x_k)h + \xi$ with $\|\xi\| = \mathrm{o}(\|h\|)$ ($\|\xi\| = \mathrm{O}(\|h\|^2)$ if F' satisfies a Lipschitz condition), (2.2) is a linearization procedure for the operator F around x_k. The procedure first employed by Newton in 1669 for the cubic equation $3x^3 - 2x - 5 = 0$ is different from the (N1)–(N3) [10], but it is easily verified that both are mathematically equivalent. The procedure (N1)–(N3) can also be written

$$x_{k+1} = x_k - F'(x_k)^{-1}F(x_k), \quad k = 0, 1, 2, \ldots . \qquad (2.3)$$

Since Raphson described in 1690 the formula (2.3) for a general cubic equation $x^3 - bx = c$, the procedure (N1)–(N3) or (2.3) is also called the Newton–Raphson method.

Later, in 1818, Fourier [34] proved the quadratic convergence of the method for the case $X = \mathbb{R}$. In 1829, Cauchy first proved a convergence theorem which does not assume any existence of a solution. It asserts that if x_0 satisfies certain conditions, then Eq. (2.1) has a solution and iteration (2.3) starting from x_0 quadratically converges to the solution.

Cauchy's results are summarized as follows:

Theorem 2.1 (Cauchy [13, pp. 575–600]). *Let* $X = \mathbb{R}$, $F = f \in C^2$, $x_0 \in X$, $f'(x_0) \neq 0$, $\sigma_0 = -f(x_0)/f'(x_0)$, $\eta = |\sigma_0|$,

$$I = \langle x_0, x_0 + 2\sigma_0 \rangle \equiv \begin{cases} [x_0, x_0 + 2\sigma_0] & \text{if } \sigma_0 \geqslant 0, \\ [x_0 + 2\sigma_0, x_0] & \text{if } \sigma_0 < 0 \end{cases}$$

and $|f''(x)| \leqslant K$ *in I. Then the following results hold:*

(i) *If*

$$2K\eta < |f'(x_0)|, \tag{2.4}$$

then Eq. (2.1) has a unique solution x^* in I.

(ii) *If $|f'(x)| \geqslant m$ in I and*

$$2K\eta < m, \tag{2.5}$$

then the Newton sequence $\{x_k\}$ starting from x_0 satisfies the following:

$$|x_{k+1} - x_k| \leqslant \frac{K}{2m}|x_k - x_{k-1}|^2, \quad k \geqslant 1$$

and

$$x^* \in \langle x_k, x_k + 2\sigma_k \rangle,$$

where $\sigma_k = -f(x_k)/f'(x_k) = x_{k+1} - x_k$, so that

$$|x^* - x_k| \leqslant 2|x_{k+1} - x_k| \quad (k \geqslant 0)$$

$$\leqslant \frac{K}{m}|x_k - x_{k-1}|^2 \quad (k \geqslant 1)$$

$$\leqslant 2\eta \left(\frac{K\eta}{2m}\right)^{2^k - 1} \quad (k \geqslant 0).$$

A convergence theorem for an iterative method is called a local convergence theorem if it asserts convergence by assuming that a solution x^* exists and an initial approximation is chosen sufficiently close to x^*. On the other hand, a convergence theorem like Theorem 2.1, which does not assume the existence of any solution a priori, but assumes that some conditions hold at the initial point x_0, is called a semilocal convergence theorem.

Ostrowski [62] improved Theorem 2.1 by replacing (2.5) by $2K\eta \leqslant |f'(x_0)|$ and showing

$$|x_{k+1} - x_k| \leqslant \frac{K}{2|f'(x_0)|}|x_k - x_{k-1}|^2, \quad k \geqslant 1.$$

In 1916, Fine [33] first established a semilocal convergence theorem for (2.3) when $X = \mathbb{R}^n$ or $\mathbb{C}^n$, $n \geqslant 1$. He assumed each component of $F = (f_1, \ldots, f_n)^t$ has continuous first and second derivatives in the domain D and that

$$\bar{S} = \bar{S}(x_1, \eta) = \{x \in X \mid \|x - x_1\|_2 \leqslant \eta\} \subseteq D, \tag{2.6}$$

with

$$\eta = \|F'(x_0)^{-1}F(x_0)\|_2,$$

where $\|\cdot\|_2$ denotes the Euclidean norm. He then proved that if

$$|[F'(x)^{-1}]_{ij}| \leqslant \mu, \quad \left|\frac{\partial^2 f_i}{\partial x_j \partial x_k}\right| \leqslant v \quad \text{in } D$$

and

$$\|F(x_0)\|_2 < \frac{1}{n^{7/2}\mu^2\nu},\tag{2.7}$$

then Eq. (2.1) has one and only one solution in $\bar{S}$ and the solution is obtained by (2.3). Furthermore, if instead of x_0 any other point in $\bar{S}$ is chosen as the starting point, (2.3) will converge to the same solution as $k \to \infty$. For a comparison of Cauchy's condition with those of Fine and Ostrowski for the case of a single equation (i.e., $n = 1$), see a historical note in Ostrowski's book [65, pp. 400–401].

In 1922, Banach [7] introduced a notion of a space of the type (B), which is now called a Banach space, and developed the theory in his famous book [8]. As is widely recognized, results established for such an abstract space can usually be applied to different fields of mathematics including finite and infinite algebraic equations, differential equations, integral equations, etc.

In 1939, Kantorovich [43] published a paper on iterative methods for functional equations in a space of the type (B) and applied his theory developed there to derive a convergence theorem for Newton's method, on the basis of the contraction mapping principle of Banach.

Later, in 1948, he [44] established a semilocal convergence theorem for Newton's method in a Banach space, which is now called Kantorovich's theorem or the Newton–Kantorovich theorem. It may be considered as a generalization and an improvement of Theorem 2.1.

Theorem 2.2 (Kantorovich's theorem). *Let $F : D \subseteq X \to Y$ be twice differentiable in an open convex set $D_0 \subseteq D$ and suppose that, for some $x_0 \in D_0$, $F'(x_0)^{-1}$ exists, $\|F'(x_0)^{-1}\| \leqslant B$, $\|F''(x)\| \leqslant K$ in X. Assume that $F(x_0) \neq 0$ without loss of generality and that*

$$\|F'(x_0)^{-1}F(x_0)\| \leqslant \eta, \quad h = KB\eta \leqslant \tfrac{1}{2}$$

$$t^* = \frac{2\eta}{1 + \sqrt{1 - 2h}}, \quad \bar{S} = \bar{S}(x_0, t^*) \subseteq D_0.$$

Then:

(i) *The iterates (2.3) are well defined, $x_k \in S$ (interior of $\bar{S}$), $k \geqslant 0$ and $\{x_k\}$ converges to a solution $x^* \in \bar{S}$ of (2.1).*

(ii) *The solution is unique in*

$$\tilde{S} = \begin{cases} S(x_0, t^{**}) \cap D_0 & \text{if } 2h < 1, \\ \bar{S}(x_0, t^{**}) \cap D_0 & \text{if } 2h = 1, \end{cases}\tag{2.8}$$

where $t^{**} = (1 + \sqrt{1 - 2h})/KB$.

(iii) *Error estimates*

$$\|x^* - x_k\| \leqslant \frac{2\eta_k}{1 + \sqrt{1 - 2h_k}} \leqslant 2^{1-k}(2h)^{2^k-1}\eta, \quad k \geqslant 0,\tag{2.9}$$

hold, where $\{\eta_k\}$, $\{h_k\}$ are defined by the recurrence relations

$$B_0 = B, \quad \eta_0 = \eta, \quad h_0 = h,$$

$$B_k = \frac{B_{k-1}}{1 - h_{k-1}}, \quad \eta_k = \frac{h_{k-1}\eta_{k-1}}{2(1 - h_{k-1})}, \quad h_k = KB_k\eta_k, \quad k \geqslant 1.\tag{2.10}$$

Three years later, Kantorovich [45] introduced a majorant principle to give a new proof for Theorem 2.2. In fact, he showed that under the same conditions as in Theorem 2.2, Newton's sequence $\{x_k\}$ satisfies

$$\|x_{k+1} - x_k\| \leqslant t_{k+1} - t_k, \quad k \geqslant 0 \tag{2.11}$$

and

$$\|x^* - x_k\| \leqslant t^* - t_k, \quad k \geqslant 0, \tag{2.12}$$

with the scalar Newton sequence $\{t_k\}$ defined by

$$t_0 = 0, \quad t_{k+1} = t_k - \frac{f(t_k)}{f'(t_k)}, \quad k = 0, 1, 2, \ldots,$$

where

$$f(t) = \tfrac{1}{2} KB t^2 - t + \eta. \tag{2.13}$$

At first glance, the estimate (2.12) appears to be sharper than (2.9). However, we have (cf. [98])

$$t_{k+1} - t_k = \eta_k, \quad t^* - t_k = \frac{1 - \sqrt{1 - 2h_k}}{KB_k} = \frac{2\eta_k}{1 + \sqrt{1 - 2h_k}},$$

$$t^{**} - t_k = \frac{1 + \sqrt{1 - 2h_k}}{KB_k},$$

so that there is no difference between the estimates (2.9) and (2.12).

The condition $h = KB\eta \leqslant \frac{1}{2}$ in Theorem 2.2 is often called the Kantorovich condition.

Remark 2.3 (Updating the Newton–Kantorovich theorem). It is known that the conditions "$\|F''(x)\| \leqslant K$ in X" and "$\bar{S}(x_0, t^*) \subseteq D_0$" of Theorem 2.2 may be replaced by weaker ones

$$\|F'(x) - F'(y)\| \leqslant K\|x - y\|, \quad x, y \in D_0 \quad \text{(Fenyö [32])} \tag{2.14}$$

and

$$\bar{S}(x_1, t^* - \eta) \subseteq D_0 \quad \text{(Ostrowski [65])}, \tag{2.15}$$

respectively. Observe that $\bar{S}(x_1, t^* - \eta) \subseteq \bar{S}(x_0, t^*) \cap \bar{S}(x_1, \eta)$ and (2.15) improves Fine's condition (2.6), too. Furthermore, Deuflhard–Heindl [27] asserted that (2.14) should be replaced by an affine invariant condition

$$\|F'(x_0)^{-1}(F'(x) - F'(y))\| \leqslant K\|x - y\|, \quad x, y \in D_0, \tag{2.16}$$

since Newton's sequence is invariant under an affine transformation. If (2.16) replaces $\|F''(x)\| \leqslant K$ in X, then the conditions $h = KB\eta$, $t^{**} = (1 + \sqrt{1 - 2h})/(KB)$ and $B_0 = B$ in Theorem 2.2 should be replaced by $h = K\eta$, $t^{**} = (1 + \sqrt{1 - 2h})/K$ and $B_0 = 1$, respectively. Under these assumptions and notation, it can be shown that $x_k \in S$, $k \geqslant 1$, $x_k \to x^* \in \bar{S}$ as $k \to \infty$ and the assertions (ii) and (iii) hold [98].

Remark 2.4. If we replace the condition $\|F'(x_0)^{-1}\| \leqslant B$ by a stronger condition $\|F'(x)^{-1}\| \leqslant B$ in D_0 in Theorem 2.2, then we can prove convergence of Newton's method under the condition $h < 2$, which is weaker than the Kantorovich condition $2h \leqslant 1$. This is known as Mysovskii's theorem or the Newton–Mysovskii theorem (cf. Ortega–Rheinboldt [61] or Mysovskii [58]).

Remark 2.5. After the appearance of the Newton–Kantorovich theorem, sharp upper and lower bounds for the errors of Newton's method have been derived by many authors, for example, by Dennis [24], Döring [29], Rall–Tapia [75], Tapia [88], Ostrowski [64], Gragg–Tapia [37], Kornstaedt [49], Miel [53–55], Potra–Pták [69], Potra [68], Pták [70], Moret [57], etc.

In a series of papers [93–95,97,98], Yamamoto showed that their results follow from the Kantorovich theorem and that detailed comparison of the bounds can be made, which put a stop to the race for finding sharp error bounds for Newton's method. The following is a part of the chart for the upper error bounds arranged in an order reflecting the power of the result:

$$\|x^* - x_k\| \leqslant \frac{2d_k}{1 + \sqrt{1 - 2K(1 - K\varDelta_k)d_k}} \quad (k \geqslant 0) \qquad \text{(Moret [57])}$$

$$\leqslant \frac{2d_k}{1 + \sqrt{1 - 2KB_k d_k}} \quad (k \geqslant 0) \qquad \text{(Kantorovich [44])}$$

$$= \begin{cases} \dfrac{2d_k}{1 + \sqrt{1 - \frac{4}{\varDelta}\frac{1 - \theta^{2^k}}{1 + \theta^{2^k}} d_k}} & (2h < 1) \\[4mm] \dfrac{2d_k}{1 + \sqrt{1 - \frac{2^k}{\eta} d_k}} & (2h = 1) \end{cases} \quad (k \geqslant 0) \qquad \text{(Yamamoto [94])}$$

$$\leqslant \frac{t^* - t_k}{t_{k+1} - t_k} d_k \quad (k \geqslant 0) \qquad \text{(Yamamoto [96])}$$

$$= \frac{2d_k}{1 + \sqrt{1 - 2\bar{h}_k}} \quad (k \geqslant 0) \qquad \text{(Döring [29])}$$

$$\leqslant \frac{K d_{k-1}^2}{\sqrt{1 - 2h} + \sqrt{1 - 2h + (Kd_{k-1})^2}} \quad (k \geqslant 1)$$

$$= \sqrt{a^2 + d_{k-1}^2} - a \quad (k \geqslant 1) \qquad \text{(Potra–Pták [69])}$$

$$\leqslant t^* - t_k \quad (k \geqslant 0) \qquad \text{(Kantorovich [45])}$$

$$= \begin{cases} e^{-2^{k-1}\varphi} \dfrac{\sinh\varphi}{\sinh 2^{k-1}\varphi}\eta & (2h < 1) \\[3mm] 2^{1-k}\eta & (2h = 1) \end{cases} \qquad \text{(Ostrowski [64])}$$

$$= \begin{cases} \dfrac{\varDelta\theta^{2^k}}{1 - \theta^{2^k}} & (2h < 1) \\[3mm] 2^{1-k}\eta & (2h = 1) \end{cases} \quad (k \geqslant 0), \qquad \text{(Gragg–Tapia [37])}$$

where $\varphi \geqslant 0$, $(1 + \cosh \varphi)h = 1$, $\Delta_k = \|x_k - x_0\|$, $d_k = \|x_{k+1} - x_k\|$, $\Delta = t^{**} - t^*$, $\theta = t^*/t^{**}$, $a = \sqrt{1 - 2h}/KB_0$.
We remark that the famous Gragg–Tapia bound in the above chart is also obtained by Ostrowski
[63, p. 301]. Hence it should perhaps be known as the Ostrowski–Gragg–Tapia bound.

Similarly, we have the following chart for lower error bounds:

$$\|x^* - x_k\| \geqslant \frac{2d_k}{1 + \sqrt{1 + 2K(1 - K\Delta_k)^{-1}d_k}} \quad (k \geqslant 0) \qquad \text{(Yamamoto [98])}$$

$$\geqslant \frac{2d_k}{1 + \sqrt{1 + 2K(1 - Kt_k)^{-1}d_k}} \quad (k \geqslant 0) \qquad \text{(Schmidt [85])}$$

$$= \frac{2d_k}{1 + \sqrt{1 + 2KB_k d_k}} \quad (k \geqslant 0) \qquad \text{(Miel [54])}$$

$$= \begin{cases} \dfrac{2d_k}{1 + \sqrt{1 + \frac{4}{\Delta}\frac{1 - \theta^{2^k}}{1 + \theta^{2^k}}d_k}} & (2h < 1) \\[2em] \dfrac{2d_k}{1 + \sqrt{1 + \frac{2^k}{\eta}d_k}} & (2h = 1) \end{cases} \quad (k \geqslant 0) \qquad \text{(Miel [55])}$$

$$\geqslant \frac{2d_k}{1 + \sqrt{1 + \frac{2d_k}{d_k + \sqrt{a^2 + d_k^2}}}} \quad (k \geqslant 0)$$

$$= \frac{2d_k(d_k + \sqrt{a^2 + d_k^2})}{d_k + \sqrt{a^2 + d_k^2} + \sqrt{(d_k + \sqrt{a^2 + d_k^2})(3d_k + \sqrt{a^2 + d_k^2})}} \quad (k \geqslant 0).$$

$$\text{(Potra–Pták [69])}$$

3. Newton-like methods

The majorant principle due to Kantorovich is so powerful that many authors have applied it to establish convergence theorems for variants of Newton's method [86,87,77].

In particular, Rheinboldt applied his majorant theory to obtain a convergence theorem for Newton-like method

$$x_{k+1} = x_k - A(x_k)^{-1}F(x_k), \quad k = 0, 1, 2, \ldots, \tag{3.1}$$

for solving Eq. (2.1) in the Banach space, where $F : D \subseteq X \to Y$ is differentiable in an open convex set $D_0 \subseteq D$, $x_0 \in D_0$ and $A(x)$ denotes an invertible, bounded linear operator which may be considered as an approximation to $F'(x)$. Dennis [25] generalized Rheinboldt's result. Miel [54] improved their error bounds. Furthermore, Moret [57] gave a sharper error bound than Miel's, but under a rather stronger assumption on $A(x)$. Mysovskii-type theorems can be found in Dennis [23].

To state an updated version of Rheinboldt's and Dennis' results, we employ an affine-invariant formulation due to Deuflhard–Heindl [27] and assume

$$\|A(x_0)^{-1}(F'(x) - F'(y))\| \leqslant K\|x - y\|, \quad x, y \in D_0, \ K > 0$$

$$\|A(x_0)^{-1}(A(x) - A(x_0))\| \leqslant L\|x - x_0\| + l, \quad x \in D_0, \ L \geqslant 0, \ l \geqslant 0 \qquad (3.2)$$

$$\|A(x_0)^{-1}(F'(x) - A(x))\| \leqslant M\|x - x_0\| + m, \quad x \in D_0, \ M \geqslant 0, \ m \geqslant 0,$$

$$l + m < 1, \quad \sigma = \max\left(1, \frac{L + M}{K}\right), \quad F(x_0) \neq 0,$$

$$\|A(x_0)^{-1}F(x_0)\| \leqslant \eta, \quad h = \frac{\sigma K \eta}{(1 - l - m)^2} \leqslant \frac{1}{2},$$

$$t^* = \frac{(1 - l - m)(1 - \sqrt{1 - 2h})}{\sigma K}, \quad \tilde{t}^{**} = \frac{1 - m + \sqrt{(1 - m)^2 - 2K\eta}}{K},$$

$$\bar{S} = \bar{S}(x_1, t^* - \eta) \subseteq D_0.$$

Under these assumptions, define the sequence $\{t_k\}$ by

$$t_0 = 0, \quad t_{k+1} = t_k + \frac{f(t_k)}{g(t_k)}, \quad k = 0, 1, 2, \ldots,$$

where

$$f(t) = \tfrac{1}{2}\sigma K t^2 - (1 - l - m)t + \eta \quad \text{and} \quad g(t) = 1 - l - Lt,$$

and the sequence $\{p_k\}$, $\{q_k\}$, $\{B_k\}$, $\{\eta_k\}$ and $\{h_k\}$ by

$$p_0 = 1 - l, \quad q_0 = 1 - l - m, \quad B_0 = \frac{p_0}{q_0}, \quad \eta_0 = \frac{\eta}{1 - l}, \quad h_1 = \sigma K B_0 \eta_0,$$

$$p_k = 1 - l - L\sum_{j=0}^{k-1}\eta_j, \quad q_k = 1 - l - m - \sigma\sum_{j=0}^{k-1}\eta_j, \quad B_k = \frac{p_k}{q_k^2},$$

$$\eta_k = \{\tfrac{1}{2}\sigma K \eta_{k-1}^2 + (p_{k-1} - q_{k-1})\eta_{k-1}\}/p_k, \quad h_k = \sigma K B_k \eta_k, \quad k \geqslant 1.$$

Furthermore, put

$$\varphi(t) = 1 - l - m - (L + M)t, \quad \Delta_k = \|x_k - x_0\|, \quad d_k = \|x_{k+1} - x_k\|.$$

Then, we have the following result which is an updated version of the results due to Rheinboldt, Dennis, Miel, Moret and others.

Theorem 3.1 (Yamamoto [99]).

(i) *The iteration (3.1) is well defined for every $k \geqslant 0$, $x_k \in S$ (interior of $\bar{S}$) for $k \geqslant 1$ and $\{x_k\}$ converges to a solution $x^* \in \bar{S}$ of Eq. (2.1).*

(ii) *The solution is unique in*

$$\tilde{S} = \begin{cases} S(x_0, \tilde{t}^{**}) \cap D_0 & (\text{if } 2K\eta < (1-m)^2), \\ \bar{S}(x_0, \tilde{t}^{**}) \cap D_0 & (\text{if } 2K\eta = (1-m)^2). \end{cases}$$

(iii) *Let* $\bar{S}_0 = \bar{S}$, $\bar{S}_k = \bar{S}(x_k, t^* - t_k)$ $(k \geqslant 1)$, $K_0 = L_0 = K$,

$$K_k = \sup_{\substack{x,y \in \bar{S} \\ x \neq y}} \frac{\|A(x_k)^{-1}(F'(x) - F'(y))\|}{\|x - y\|} \quad (k \geqslant 1),$$

$$L_k = \sup_{\substack{x,y \in \bar{S} \\ x \neq y}} \frac{\|A(x_k)^{-1}(F'(x) - F'(y))\|}{\|x - y\|} \quad (k \geqslant 1). \tag{3.3}$$

Then

$$x^* \in \bar{S}_k \subseteq \bar{S}_{k-1} \subseteq \cdots \subseteq \bar{S}_0, \quad t_{k+1} - t_k = \eta_k, \quad 2h_k \leqslant 1$$

and

$$\|x^* - x_k\| \leqslant \alpha_k \equiv \frac{2g(\Delta_k)d_k}{\varphi(\Delta_k) + \sqrt{\varphi(\Delta_k)^2 - 2K_k g(\Delta_k)^2 d_k}} \quad (k \geqslant 0)$$

$$\leqslant \beta_k \equiv \frac{2g(\Delta_k)d_k}{\varphi(\Delta_k) + \sqrt{\varphi(\Delta_k)^2 - 2L_k g(\Delta_k)^2 d_k}} \quad (k \geqslant 0)$$

$$\leqslant \gamma_k \equiv \frac{2g(\Delta_k)d_k}{\varphi(\Delta_k) + \sqrt{\varphi(\Delta_k)^2 - 2K g(\Delta_k)d_k}} \quad (k \geqslant 0) \tag{3.4}$$

(Yamamoto [96], Moret [57])

$$\leqslant \delta_k \equiv \frac{2g(t_k)d_k}{\varphi(t_k) + \sqrt{\varphi(t_k)^2 - 2K g(t_k)d_k}} \quad (k \geqslant 0) \qquad \text{(Yamamoto [96])}$$

$$\leqslant \frac{t^* - t_k}{t_k - t_{k-1}} d_{k-1} \quad (k \geqslant 1)$$

$$= \frac{1}{\eta_{k-1}} \left(\frac{2(p_k/q_k)\eta_k}{1 + \sqrt{1 - 2h_k}} \right) d_{k-1} \quad (k \geqslant 1)$$

$$\leqslant t^* - t_k \quad (k \geqslant 0) \qquad \text{(Rheinboldt [77], Dennis [25])}$$

$$= \frac{2(p_k/q_k)\eta_k}{1 + \sqrt{1 - 2h_k}} \quad (k \geqslant 0).$$

Remark 3.2. Moret [57] obtained (3.4) under stronger assumptions that $K \geqslant L$, $M = K - L$ and replacing (3.2) by

$$\|A(x_0)^{-1}(A(x) - A(y))\| \leqslant L\|x - y\| \quad \text{for } x, y \in D_0.$$

Remark 3.3. If we put $A(x_k) = F'(x_k)$, then (3.1) reduces to the Newton method and Theorem 3.1 reduces to an updated version of Theorem 2.2.

Remark 3.4. If we put $A(x) = F'(x_0)$, then (3.1) is called the simplified (or modified) Newton method. Theorem 3.1 then reduces to:

Corollary 3.5. *Consider the simplified Newton method*

$$x_{k+1} = x_k - F'(x_0)^{-1}F(x_k), \quad k = 0, 1, 2, \ldots, \tag{3.5}$$

where we assume $x_0 \in D_0$ and $F'(x_0)^{-1}$ exists and the following conditions hold:

$$\|F'(x_0)^{-1}(F'(x) - F'(x_0))\| \leqslant K \|x - x_0\|, \quad x \in D_0$$

$$0 < \|F'(x_0)^{-1}F(x_0)\| \leqslant \eta, \quad h = K\eta \leqslant \tfrac{1}{2},$$

$$t^* = \frac{1 - \sqrt{1 - 2h}}{K}, \quad t^{**} = \frac{1 + \sqrt{1 - 2h}}{K},$$

$$\bar{S} = \bar{S}(x_1, t^* - \eta) \subseteq D_0.$$

Then:

(i) *The iteration (3.5) is well-defined for every $k \geqslant 0$, $x_k \in S$ for $k \geqslant 1$ and $\{x_k\}$ converges to a solution of (2.1).*

(ii) *The solution is unique in*

$$\tilde{S} = \begin{cases} S(x_0, t^{**}) \cap D_0 & (2h < 1), \\ \bar{S}(x_0, t^{**}) \cap D_0 & (2h = 1). \end{cases}$$

(iii) *Define the sequence $\{t_k\}$ by*

$$t_0 = 0, \quad t_{k+1} = \tfrac{1}{2}Kt_k^2 + \eta, \quad k = 0, 1, 2, \ldots .$$

Let $\bar{S}_0 = \bar{S}$, $\bar{S}_k = \bar{S}(x_k, t^* - t_k)$ $(k \geqslant 1)$ and

$$K_k = \sup_{\substack{x \in \bar{S}_k \\ x \neq x_0}} \frac{\|F'(x_0)^{-1}(F'(x) - F'(x_0))\|}{\|x - x_0\|} \quad (k \geqslant 0).$$

Then

$$\|x^* - x_k\| \leqslant \frac{2d_k}{1 - K_k \Delta_k + \sqrt{(1 - K_k \Delta_k)^2 - 2K_k d_k}} \quad (k \geqslant 0)$$

$$\leqslant \frac{2d_k}{1 - K\Delta_k + \sqrt{(1 - K_k \Delta_k)^2 - 2Kd_k}} \quad (k \geqslant 0)$$

$$\leqslant \frac{2d_k}{1 - Kt_k + \sqrt{(1 - Kt_k)^2 - 2Kd_k}} \quad (k \geqslant 0)$$

$$\leqslant \frac{t^* - t_k}{t_{k+1} - t_k} d_k \quad (k \geqslant 0)$$

$$\leqslant \frac{t^* - t_k}{t_k - t_{k-1}} d_{k-1} \quad (k \geqslant 1) \tag{3.6}$$

$$\leqslant t^* - t_k \quad (k \geqslant 0)$$

$$= \frac{2(t_{k+1} - t_k)}{1 - Kt_k + \sqrt{(1 - Kt_k)^2 - 2K(t_{k+1} - t_k)}} \quad (k \geqslant 0)$$

$$\leqslant \frac{1}{2K}(1 - \sqrt{1 - 2h})^{k+1} \quad (k \geqslant 1) \qquad \text{(Andrew [4])} \tag{3.7}$$

$$\leqslant \frac{1}{K}(1 - \sqrt{1 - 2h})^{k+1} \quad (k \geqslant 0), \qquad \text{(Kantorovich–Akilov [46])}$$

where the inequality in (3.7) is replaced by the strict inequality if $k > 1$.

4. Secant method

Consider a Newton-like method using a divided difference operator $\delta F(x, y) \in L(X, Y)$ (Banach space of bounded linear operators of X into Y) in place of $A(x)$ in (3.1) such that

$$\delta F(x, y)(x - y) = F(x) - F(y), \quad x, y \in D_0 \tag{4.1}$$

and

$$\|\delta F(x, y) - \delta F(y, u)\| \leqslant a\|x - u\| + b\|x - y\| + b\|y - u\|, \quad u \in D_0, \tag{4.2}$$

where $a \geqslant 0$ and $b \geqslant 0$ are constants independent of x, y, u. These conditions are due to Schmidt [83]. Later, Laarsonen (1969) put the condition

$$\|\delta F(x', x'') - \delta F(y', y'')\| \leqslant M(\|x' - y'\| + \|x'' - y''\|), \quad x, x', y, y' \in D_0 \tag{4.3}$$

in place of (4.2) (cf. [26]). Dennis [26] showed that (4.3) implies (4.2) and that the conditions (4.1) and (4.2) imply that F is Fréchet differentiable, $\delta F(x, x) = F'(x)$ and F' satisfies the Lipschitz condition in D_0 with the Lipschitz constant $2(a + b)$.

According to Dennis, we shall call the iteration

$$x_{k+1} = x_k - \delta F(x_k, x_{k-1})^{-1} F(x_k), \quad k = 0, 1, 2, \dots \tag{4.4}$$

(considered by Schmidt) the backward secant method and the iteration

$$x_{k+1} = x_k - \delta F(x_{k-1}, x_k)^{-1} F(x_k), \quad k = 0, 1, 2, \dots \tag{4.5}$$

the forward secant method, where x_0, x_{-1} are given.

Laarsonen considered the following iterations:

$$y_{k+1} = y_k - \delta F(y_k, \bar{y}_k)^{-1} F(y_k), \tag{4.6}$$

$$\bar{y}_{k+1} = y_{k+1} - \delta F(y_k, \bar{y}_k)^{-1} F(y_{k+1}), \quad k = 0, 1, 2, \dots \tag{4.7}$$

where y_0 and $\bar{y}_0$ are given.

Dennis gave a convergence theorem for the iteration

$$x_{k+1} = x_k - A_{\alpha_k}^{-1} F(x_k), \quad k = 0, 1, 2, \ldots, \tag{4.8}$$

which is the most general possible mixture of the forward and backward secant methods, where

$$A_{\alpha_k} = \delta F(x_{\alpha_k}, x_{\alpha_k - 1}) \quad \text{or} \quad A_{\alpha_k} = \delta F(x_{\alpha_k - 1}, x_{\alpha_k})$$

and $\{\alpha_k\}$ is a recalculation sequence defined by Dennis, that is, $\alpha_0 = 0$, $\alpha_k = k$ or $\alpha_k = \alpha_{k-1}$. Iterations (4.4), (4.5) and Laarsonen's (4.6)–(4.7) are included in (4.8) (cf. [26]). Dennis' result is stated as follows:

Theorem 4.1 (Dennis [26]). *Let* (4.1) *and* (4.2) *hold and let* $x_{-1}, x_0 \in D_0$ *have the following properties: for either* $A_0 = \delta F(x_0, x_{-1})$ *or for* $A_0 = \delta F(x_{-1}, x_0)$, $\|A_0^{-1}\| \leqslant \beta$, $\|x_{-1} - x_0\| \leqslant \eta_{-1}$,

$$\|A_0^{-1} F(x_0)\| \leqslant \eta, \quad \beta(a+b)\eta_{-1} < 1,$$

$$h \equiv \frac{(a+b)\beta\eta}{(1 - \beta(a+b)\eta_{-1})^2} < \frac{1}{4}$$

and $S(x_0, r_0) \subset D_0$, *where*

$$r_0 = \frac{1 - \sqrt{1 - 4h}}{2\beta(a+b)}(1 - \beta(a+b)\eta_{-1}).$$

Then, for an arbitrary recalculation sequence $\{\alpha_k\}$, (4.8) *converges to a solution* x^* *of* (2.1), *according to*

$$\|x^* - x_{k+1}\| \leqslant r_0 - t_{k+1}$$

$$\equiv r_0 - t_k - \frac{\beta(a+b)t_k^2 - (1 - \beta(a+b)\eta_{-1})t_k + \eta}{1 - \beta(a+b)(t_{\alpha_k} + t_{\alpha_{k-1}} + \eta_{-1})},$$

$$k = 0, 1, 2, \ldots,$$

$$t_{-1} = -\eta_{-1}, \quad t_0 = 0.$$

Independently of Theorem 4.1, Schmidt [84] established a semilocal convergence theorem for the backward secant iteration (4.4). (He called this iteration the regular-falsi iteration.)

We remark here that, in many cases, uniqueness assertions in Kantorovich-type convergence theorems follow from the Newton–Kantorovich theorem, which may simplify proofs of the theorems. This remark also applies to Schmidt's theorem for (4.4). See Yamamoto [99,102,103].

5. Halley's and Chebyshev's methods

Let $X = \mathbb{R}$ and $F = f$ be a single function of C^2-class with a zero x^*. Then, as was mentioned in Section 2, a linear approximation

$$f(x + h) \doteq f(x) + f'(x)h$$

leads to the Newton method whose convergence speed is of second order. If we use a second-order approximation

$$f(x+h) \doteq f(x) + f'(x)h + \tfrac{1}{2}f''(x)h^2,$$

then an iterative method

$$x_{k+1} = x_k + h, \quad k = 0,1,2,\ldots,$$

$$h = \cfrac{-\cfrac{2f(x_k)}{f'(x_k)}}{1 + \sqrt{1 - \cfrac{2f(x_k)f''(x_k)}{f'(x_k)^2}}}, \tag{5.1}$$

is obtained. Cauchy [13] first established a semilocal convergence result for (5.1) under some assumptions. As is shown there, the convergence speed of this method is cubic. A more detailed discussion on its convergence can be found in Hitotumatu [41].

To avoid the complexity of computation of the square root, we replace $\sqrt{1-x}$ by its approximation $1 - \tfrac{1}{2}x$ near $x = 0$. Then we obtain Halley's method (1694)

$$x_{k+1} = x_k - \cfrac{\cfrac{f(x_k)}{f'(x_k)}}{1 - \cfrac{1}{2}\cfrac{f(x_k)f''(x_k)}{f'(x_k)^2}}, \quad k = 0,1,2,\ldots, \tag{5.2}$$

which is also called the method of tangent hyperbolas. It is interesting to note that if we replace the denominator of (5.2) by

$$\sqrt{1 - \cfrac{f(x_k)f''(x_k)}{f'(x_k)^2}},$$

then we get the square root iteration due to Ostrowski [65]

$$x_{k+1} = x_k - K(x_k), \quad k = 0,1,2,\ldots,$$

$$K(x) = \cfrac{f(x)/f'(x)}{\sqrt{1 - f(x)f''(x)/f'(x)^2}}.$$

Convergence analysis of (5.2) can be found in Salehov [82], Brown [12], Alefeld [2], Gander [36], Hernandez [40], etc.

As in Newton's method, the procedure (5.2) is easily extended to nonlinear operators in Banach spaces: Let X and Y be Banach spaces and $F: D \subseteq X \to Y$ as before. If F is twice Fréchet differentiable in an open convex domain $D_0 \subseteq D$, then the Halley method applied to Eq. (2.1) in a Banach space is defined by

$$x_0 \in D_0, \quad x_{k+1} = x_k - A(x_k)^{-1}F(x_k), \quad k = 0,1,2,\ldots \tag{5.3}$$

with

$$A(x) = F'(x)[I - \tfrac{1}{2}F'(x)^{-1}F''(x)F'(x)^{-1}F(x)],$$

which is equivalent to the procedure:

(H1) Solve the linear equation $F(x_k) + F'(x_k)c_k = 0$ with respect to c_k.
(H2) Solve the linear equation $F(x_k) + F'(x_k)d_k + \frac{1}{2}F''(x_k)c_k d_k = 0$ with respect to d_k.
(H3) Set $x_{k+1} = x_k + d_k, \quad k = 0, 1, 2, \ldots$.

A convergence theorem for (5.3) as well as a uniqueness result was first given by Mertvecova [52], who extended Salehov's result (1952) for a single equation of a real or complex function. Since then, Kantorovich-type convergence theorems for (5.3) or its variants have been obtained by many authors (Altman [3], Safiev [80,81], Döring [30], Alefeld [1], Werner [91,92], Yamamoto [101], Chen–Argyros–Qian [14], Ezquerro–Hernandez [31], Hernandez [40], etc.)

According to Safiev, but slightly changing his notation, we assume

(I) The operator $\Gamma = F'(x_0)^{-1}$ exists. We put $\Phi(x) = \Gamma F(x)$;
(II) $\zeta = \|\Phi(x_0)\| > 0$, $M = \|\Phi''(x_0)\| > 0$;
(III) $\|\Phi''(x) - \Phi''(y)\| \leqslant N\|x - y\|$, $x, y \in D_0$, $N > 0$;
(IV) The equation $f(t) = \frac{1}{6}Nt^3 + \frac{1}{2}Mt^2 - t + \zeta = 0$ has one negative root and two positive roots t^*, t^{**} such that $t^* \leqslant t^{**}$. Equivalently,

$$\zeta \leqslant \frac{M^2 + 4N - M\sqrt{M^2 + 2N}}{3N(M + \sqrt{M^2 + 2N})}, \tag{5.4}$$

where the equality holds if and only if $t^* = t^{**}$.

Under these assumptions, define the scalar sequence $\{t_k\}$ by

$$t_0 = 0, \quad t_{k+1} = t_k - a(t_k)^{-1}f(t_k), \quad k = 0, 1, 2, \ldots,$$

where $a(t) = f'(t) - \frac{1}{2}f''(t)f'(t)^{-1}f(t)$. Then the following result holds, which improves Döring's:

Theorem 5.1 (Yamamoto [101]). *To the assumptions* (I)–(IV) *add*

$$\bar{S} = \bar{S}(x_1, t^* - t_1) \subseteq D_0.$$

Then:

(i) *The iteration* (5.3) *is well defined for every* $k \geqslant 0$, $\{x_{k+1}\}$ *lies in S (interior of* $\bar{S}$*) and converges to a solution* x^* *of* (2.1).
(ii) *The solution is unique in*

$$\tilde{S} = \begin{cases} S(x_0, t^{**}) \cap D_0 & \text{if } t^* < t^{**}, \\ \bar{S}(x_0, t^{**}) \cap D_0 & \text{if } t^* = t^{**}. \end{cases}$$

(iii) *Error estimates*

$$\sigma_k^* \leqslant \|x^* - x_k\| \leqslant \tau_k^* \leqslant d_k + (t^* - t_{k+1})\left(\frac{d_k}{t_{k+1} - t_k}\right)^3$$

$$\leqslant (t^* - t_k)\left(\frac{d_k}{t_{k+1} - t_k}\right) \leqslant t^* - t_k$$

and

$$\|x^* - x_k\| \leqslant (t^* - t_k)\left(\frac{d_{k-1}}{t_k - t_{k-1}}\right)^3$$

hold, where $d_k = \|x_{k+1} - x_k\|$, τ_k^* and σ_k^* are the smaller positive root of the equation

$$\frac{(t^* - t_{k+1})}{(t^* - t_k)^3}t^3 - t + d_k = 0,$$

and

$$\frac{(t^* - t_{k+1})}{(t^* - t_k)^3}t^3 + t - d_k = 0,$$

respectively.

Remark 5.2. Since

$$\lim_{N \to 0} \frac{M^2 + 4N - M\sqrt{M^2 + 2N}}{3N(M + \sqrt{M^2 + 2N})} = \frac{1}{2M},$$

condition (5.4) reduces to the Kantorovich condition $2M\zeta \leqslant 1$, if $N = 0$. Hence (III) admits the case $N = 0$.

Kanno [42] shows that Theorem 5.1 is better than Safiev's result. See also Cuyt [21], Cuyt–Rall [22] for computational implementation of the method.

Chebyshev's method (1951) (cf. [60]) is similarly defined by

$$x_{k+1} = x_k - F'(x_k)^{-1}F(x_k)$$
$$-\tfrac{1}{2}F'(x_k)^{-1}F''(x_k)[F'(x_k)^{-1}F(x_k)]^2, \quad k = 0, 1, 2, \ldots \tag{5.5}$$

or by the following procedure:

(C1) Solve $F(x_k) + F'(x_k)c_k = 0$ with respect to c_k.
(C2) Solve $F(x_k) + F'(x_k)d_k + \tfrac{1}{2}F''(x_k)c_k^2 = 0$ with respect to d_k.
(C3) Set $x_{k+1} = x_k + d_k, \quad k \geqslant 0$.

The method is also obtained if we replace

$$[I - \tfrac{1}{2}F'(x)^{-1}F''(x)F'(x)^{-1}F(x)]^{-1}$$

in the Halley method (5.3) by

$$I + \tfrac{1}{2}F'(x)F''(x)F'(x)^{-1}F(x).$$

Convergence theorems for Chebyshev's and Chebyshev-like methods have been obtained by Necepurenko [60], Safiev [81], Alefeld [2], Werner [92] and others. A numerical test by Alefeld shows that Halley's method gives a slightly sharper result than Chebyshev's, although convergence speed of both methods is of cubic order.

Remark 5.3. Two steps of an iterative method with cubic order correspond in general to three steps of a method with second order. However, computational cost depends on the form of F. Hence, it is difficult to give a general statement about the efficiency/computational cost of Halley's or Halley-like methods versus Newton's method. In Ostrowski's book [65] the efficiency index of a procedure is defined and discussed for some algorithms.

6. A class of iterative methods for not necessarily differentiable equations

Recently, much attention has been paid to iterative methods for solving (2.1) when F is not necessarily differentiable. In 1963, Zincenko [107] considered using a differentiable operator $f : D_f \subseteq X \to Y$, where X and Y are Banach spaces, and putting $g(x) = F(x) - f(x)$. Then $F(x) = f(x) + g(x)$ and he proved convergence theorems for the iterations

$$x_{k+1} = x_k - f'(x_k)^{-1}(f(x_k) + g(x_k)), \quad k = 0, 1, 2, \ldots \tag{6.1}$$

and

$$x_{k+1} = x_k - f'(x_0)^{-1}(f(x_k) + g(x_k)), \quad k = 0, 1, 2, \ldots. \tag{6.2}$$

In [107], Zincenko observed that (6.1) and (6.2) were suggested by Krasnoselskii. Rheinboldt used his majorant theory to prove Zincenko's results. His result for (6.1) is stated as follows:

Theorem 6.1. Let $g : D_g \subseteq X \to Y$ and suppose that on some convex set $D_0 \subseteq D_f \cap D_g$,

$$\|f'(x) - f'(y)\| \leqslant \gamma \|x - y\|,$$

$$\|g(x) - g(y)\| \leqslant \delta \|x - y\|, \quad x, y \in D_0.$$

Assume that for some $x_0 \in D_0$, $f'(x_0)^{-1} \in L(Y, X)$ exists and that $\|f'(x_0)^{-1}\| \leqslant \beta$, $\|f'(x_0)^{-1} f(x_0)\| \leqslant \alpha$ as well as $\beta\delta < 1$ and

$$h = \beta\gamma\alpha/(1 - \beta\delta)^2 \leqslant 1/2.$$

Define t^*, t^{**} by

$$t^* = \frac{2}{1 + \sqrt{1 - 2h}} \cdot \frac{\alpha}{1 - \beta\delta}, \quad t^{**} = \frac{1 + \sqrt{1 - 2h}}{h} \cdot \frac{\alpha}{1 - \beta\delta}.$$

If $\bar{S}(x_0, t^*) \subseteq D_0$, then the sequence (6.1) remains in $S(x_0, t^*)$ and converges to the only solution x^* of $F(x) = 0$ in $S(x_0, t^{**}) \cap D_0$.

(If $g = 0$, then Theorem 6.1 reduces to the Kantorovich theorem.)

Later, in 1982, Zabrejko–Zlepko [106] proved a semilocal convergence theorem for (6.1) and

$$x_{k+1} = x_k - C^{-1}(f(x_k) + g(x_k)), \quad k = 0, 1, 2, \ldots,$$

where $C \in L(X, Y)$. To state their result for (6.1), let f and g be defined in the closed ball $\bar{S}(x_0, R)$ of X and f be differentiable in the open ball $S(x_0, R)$. They considered the auxiliary scalar equation

$$r = \varphi(r), \tag{6.3}$$

where

$$\varphi(r) = a + b \left(\int_0^r \omega(\tau) \, d\tau + H(r) \right),$$

$a > 0$, $b > 0$, and $\omega(r)$, $H(r)$, $H(r+t)-H(r)$, $t \geqslant 0$ are nonnegative, monotonically increasing and continuous functions, vanishing for $r = 0$.

Assume that (6.3) has at least one positive solution. Denote by r^* the least of these solutions. If it is isolated and $\varphi(r) < r$ for r sufficiently close to r^* and larger than r^*, then we denote by R^* the least upper bound of the number ξ for which $\varphi(r) < r$ for $r \in (r^*, \xi)$; otherwise we put $R^* = r^*$. Then the iteration $r_{k+1} = \varphi(r_k)$, $k = 0, 1, 2, \ldots$, $r_0 = 0$ is monotonically increasing and converges to r^*, while the iteration $R_{k+1} = \varphi(R_k)$, $k = 0, 1, 2, \ldots$ starting from $R_0 \in [r^*, R^*)$ is monotonically decreasing and converges to r^*.

Similarly, the iteration

$$\rho_{k+1} = \rho_k - \frac{\rho_k - \varphi(\rho_k)}{1 - b\omega(\rho_k)}, \quad k = 0, 1, 2, \ldots, \quad \rho_0 = 0$$

is monotonically increasing and converges to r^*.

In particular, putting

$$a = \|f'(x_0)^{-1}(f(x_0) + g(x_0))\|, \quad b = \|f'(x_0)^{-1}\| \tag{6.4}$$

the following Theorem 6.2 can be proved by assuming

$$\|f'(x+h) - f'(x)\| \leqslant \omega(r + \|h\|) - \omega(r), \quad \|x - x_0\| \leqslant r, \quad r + \|h\| \leqslant R,$$

$$\|g(x+h) - g(x)\| \leqslant H(\|x - x_0\| + \|h\|) - H(\|x - x_0\|).$$

Theorem 6.2 (Zabrejko–Zlepko [106]). *Let $r^* \leqslant R < R^*$. Then*

$$f(x) + g(x) = 0 \tag{6.5}$$

has a solution $x^* \in \bar{S}(x_0, r^*)$, which is unique in $\bar{S}(x_0, R)$. The sequence $\{x_k\}$ from (6.1) satisfies $x_k \in \bar{S}(x_0, r_k)$ for $k \geqslant 0$ and $x_k \to x^*$ as $k \to \infty$; we have

$$\|x_{k+1} - x_k\| \leqslant \rho_{k+1} - \rho_k \quad and \quad \|x^* - x_k\| \leqslant r^* - \rho_k, \quad k \geqslant 0.$$

In 1987, Zabrejko–Nguen [105] reformulated Theorem 6.2 as follows: the operators f and g are defined in $\bar{S}(x_0, R)$, f is differentiable in $S(x_0, R)$ and

$$\|f'(x') - f'(x'')\| \leqslant \kappa(r)\|x' - x''\|, \quad x', x'' \in \bar{S}(x_0, r),$$

$$\|g(x') - g(x'')\| \leqslant \varepsilon(r)\|x' - x''\|, \quad x', x'' \in \bar{S}(x_0, r),$$

where $\kappa(r)$ and $\varepsilon(r)$ are nondecreasing functions on the interval $[0, R]$. Furthermore, assume that $f'(x_0)^{-1}$ exists and put

$$\omega(r) = \int_0^r \kappa(t)\,dt,$$

$$\varphi(r) = a + b\int_0^r \omega(t)\,dt - r = a + b\int_0^r (r - s)\kappa(s)\,ds - r,$$

$$\psi(r) = b\int_0^1 \varepsilon(t)\,dt.$$

Under these assumptions and notation, they proved the following:

Theorem 6.3 (Zabrejko–Nguen [105]). *Suppose that the function* $\chi(r) = \varphi(r) + \psi(r)$ *has a unique zero* ρ^* *in* $[0, R]$ *and* $\chi(R) \leqslant 0$. *Then Eq.* (6.5) *admits a solution* x^* *in* $\bar{S}(x_0, \rho)$, *which is unique in* $\bar{S}(x_0, R)$.

The iteration (6.1) *is defined for all* $k \geqslant 0$, $x_k \in \bar{S}(x_0, \rho^*)$, $k \geqslant 0$, *and* $\{x_k\}$ *converges to* x^* *as* $k \to \infty$. *Error estimates*

$$\|x_{k+1} - x_k\| \leqslant \rho_{k+1} - \rho_k, \quad k \geqslant 0$$

and

$$\|x^* - x_k\| \leqslant \rho^* - \rho_k, \quad k \geqslant 0$$

hold where ρ_k *is defined by*

$$\rho_{k+1} = \rho_k - \frac{\chi(\rho_k)}{\varphi'(\rho_k)}, \quad k \geqslant 0, \quad \rho_0 = 0$$

and

$$\rho_0 < \rho_1 < \cdots < \rho_k \to \rho^* \quad as \ k \to \infty.$$

They used this result to generalize Pták's error estimates for Newton's method obtained by the "nondiscrete induction technique" [70] and showed that his estimates are a simple consequence of the classical majorant method due to Kantorovich.

In the same paper, they mentioned the following error estimate without proof: Let $r_k = \|x_k - x_0\|$, $\kappa_k(r) = \kappa(r_k + r)$ and $\varepsilon_k(r) = \varepsilon(r_k + r)$ for $r \in [0, R - r_k]$ and put

$$a_k = \|x_{k+1} - x_k\|, \quad b_k = (1 - \omega(r_k))^{-1}.$$

Without loss of generality, we may assume that $a_k > 0$. Then the equation

$$r = a_k + b_k \int_0^r \{(r - t)\kappa_k(t) + \varepsilon_k(t)\} \, dt$$

has the unique positive zero ρ_k^* in $[0, R - r_k]$ and

$$\|x^* - x_k\| \leqslant \rho_k^*, \quad k = 0, 1, 2, \ldots, \tag{6.6}$$

where we understand $\rho_0^* = \rho^*$. A proof of (6.6) is given in Yamamoto [100] together with the following error estimates:

$$\|x^* - x_k\| \leqslant \rho_k^* \quad (k \geqslant 0)$$

$$\leqslant \frac{a_k}{\rho_{k+1} - \rho_k}(\rho^* - \rho_k) \quad (k \geqslant 0)$$

$$\leqslant \frac{a_{k-1}}{\rho_k - \rho_{k-1}}(\rho^* - \rho_k) \quad (k \geqslant 1)$$

$$\leqslant \rho^* - \rho_k \quad (k \geqslant 0).$$

Motivated by Zabrejko–Nguen's paper, Chen–Yamamoto [17] considered the iteration,

$$x_{k+1} = x_k - A(x_k)^{-1}(f(x_k) + g(x_k)), \quad k = 0, 1, 2, \ldots, \tag{6.7}$$

which includes (6.1) as a special case, where $A(x)$ is an approximation for $f'(x)$. Under the Zabrejko–Nguen type hypotheses, they determined a domain Ω such that the method (6.5) starting from any point of Ω converges to a solution of (6.5).

Furthermore, a study of the iteration (6.7) is found in Yamamoto–Chen [104] and Chen–Yamamoto [18], where ball convergence theorems as well as error estimates are given. The results generalize and deepen those of Kantorovich [44], Mysovskii [58], Rall [74], Rheinboldt [78], Dennis [25], Yamamoto [96,99,100], Zabrejko–Nguen [105], and others. See also the works of Argyros [5,6].

For the case $X = Y = \mathbb{R}^n$, Clark [20] proposed a generalized Newton method which uses an element in the generalized Jacobian $\partial F(x^k)$ in place of the Jacobian if F is locally Lipschitzian but not differentiable. Since then, there has been a growing interest in the study of nonsmooth equations, which is closely related to the study of Newton-like methods. Such equations arise, for examples, from (i) nonlinear complementarity problems, (ii) nonlinear constrained optimization problems, (iii) nonsmooth convex optimization problems, (iv) compact fixed point problems, (v) nonsmooth eigenvalue problem related to ideal MHD (magnetohydrodynamics), etc.

Reformulating problems (i)–(iii) to nonsmooth equations and global and superlinear convergence of algorithms for solving such nonsmooth equations can be found in the works of Chen [15], Qi [71], Qi–Sun [73], Pang [66], Pang–Qi [67], Qi–Chen [72], Robinson [79], Fukushima–Qi [35] and others. Heinkenschloss et al. [39] discuss (iv) and Rappaz [76] and Kikuchi [47,48] discuss (v).

Finally, we remark that the general Gauss–Newton method for solving singular or ill-posed equations is defined by

$$x_{k+1} = x_k - B(x_k)F(x_k), \quad k = 0, 1, 2 \ldots \tag{6.8}$$

where $F : D \subseteq X \to Y$ is differentiable and $B(x_k)$ is a linear operator which generalizes the Moore–Penrose pseudo-inverse. If $X = \mathbb{R}^n$, $Y = \mathbb{R}^m$ and $B(x_k) = F'(x_k)^+$, then (6.8) reduces to the ordinary Gauss–Newton method for solving the least square problem: Find $x \in D$ which minimizes $F(x)^t F(x)$. Convergence analysis for (6.8) is given in the works of Ben-Israel [11], Lawson and Hanson [50], Meyn [56], Häußler [38], Walker [89], Walker and Watson [90], Martínez [51], Deuflhard and Potra [28], Chen and Yamamoto [19], Nashed and Chen [59], Chen et al. [16] and others.

7. Concluding remarks

In this paper, we have traced historical developments in convergence theory and error estimates for Newton (and Newton-like) methods based on the Newton–Kantorovich theorem.

In spite of its simple principle, Newton's method is applicable to various types of equations such as systems of nonlinear algebraic equations including matrix eigenvalue problems, differential equations, integral equations, etc., and even to random operator equations [9]. Hence, the method fascinates many researchers.

However, as is well known, a disadvantage of the methods is that the initial approximation x_0 must be chosen sufficiently close to a true solution in order to guarantee their convergence. Finding a criterion for choosing x_0 is quite difficult and therefore effective and globally convergent algorithms are needed. This remark includes the important case of nonsmooth equations. It is expected that if such algorithms exist, then they too will be variants of Newton's method.

Acknowledgements

The author thanks Prof. Xiaojun Chen of Shimane University for her valuable comments during the preparation of this paper; and is also grateful to referees for their nice suggestions. By these suggestions this paper has been greatly improved.

References

[1] G. Alefeld, Zur Konvergenz des Verfahrens der tangierenden Hyperbeln und des Tschelyscheff-Verfahrens bei konvexen Abbildungen, Computing 11 (1973) 379–390.

[2] G. Alefeld, On the convergence of Halley's method, Amer. Math. Monthly 88 (1981) 530–536.

[3] M. Altman, Concerning the method of tangent hyperbolas for operator equations, Bull. Acad. Polon. Sci. Ser. Sci. Math. Astronom. Phys. 9 (1961) 633–637.

[4] A.L. Andrew, Error bounds for the modified Newton's method, Bull. Austral. Math. Soc. 14 (1976) 427–433.

[5] I.K. Argyros, The Theory and Applications of Iteration Methods, CRC Press, Boca Raton, 1993.

[6] I.K. Argyros, Improved á posteriori error bounds for Zincenko's iteration, Internat. J. Comput. Math. 51 (1994) 51–54.

[7] S. Banach, Sur les opérations dans les ensembles abstracts et leurs applications aux équations integrales, Fund. Math. 3 (1922) 133–181.

[8] S. Banach, Theorie des Opérations Linéaires, Warszawa, 1932. Reprinted by Chelsea, New York.

[9] A.T. Bharucha-Reid, R. Kannan, Newton's method for random operator equations, J. Nonlinear Anal. 4 (1980) 231–240.

[10] N. Bićanić, K.H. Johnson, Who was '-Raphson'?, Internat. J. Numer. Methods Eng. 14 (1979) 148–152.

[11] A. Ben-Israel, A Newton–Raphson method for the solution of systems of equations, J. Math. Anal. Appl. 15 (1966) 243–252.

[12] G.H. Brown, On Halley's variation of Newton's method, Amer. Math. Monthly 84 (1977) 726–728.

[13] A.L. Cauchy, Sur la détermination approximative des racines d'une équation algébrique ou transcendante, in: Leçons sur le Calcul Differentiel, Buré fréres, Paris, 1829, pp. 573–609.

[14] D. Chen, I.K. Argyros, Q.S. Qian, A note on the Halley method in Banach spaces, Appl. Math. Comp. 58 (1993) 215–224.

[15] X. Chen, Smoothing methods for complementarity problems and their applications: a survey, J. Oper. Res. Soc. Japan 43 (2000) 32–47.

[16] X. Chen, M.Z. Nashed, L. Qi, Convergence of Newton's method for singular smooth and nonsmooth equations using adaptive outer inverses, SIAM J. Optim. 7 (1997) 245–262.

[17] X. Chen, T. Yamamoto, Convergence domains of certain iterative methods for solving nonlinear equations, Numer. Funct. Anal. Optim. 10 (1989) 37–48.

[18] X. Chen, T. Yamamoto, A convergence ball for multistep simplified Newton-like methods, Numer. Funct. Anal. Optim. 14 (1993) 15–24.

[19] X. Chen, T. Yamamoto, Newton-like methods for solving undetermined nonlinear equations with nondifferentiable terms, J. Comput. Appl. Math. 55 (1994) 311–324.

[20] F.H. Clark, Optimization and Nonsmooth Analysis, Wiley, New York, 1983.

[21] A.A.M. Cuyt, Numerical stability of the Halley-iteration for the solution of a system of nonlinear equations, Math. Comp. 38 (1982) 171–179.

[22] A.A.M. Cuyt, L.B. Rall, Computational implementation of the multivariate Halley method for solving nonlinear systems of equations, ACM Trans. Math. Software 11 (1985) 20–36.

[23] J.E. Dennis Jr., On Newton-like methods, Numer. Math. 11 (1968) 324–330.

[24] J.E. Dennis Jr., On the Kantorovich hypothesis for Newton's method, SIAM J. Numer. Anal. 6 (1969) 493–507.

[25] J.E. Dennis Jr., On the convergence of Newton-like methods, in: P. Rabinowitz (Ed.), Numerical Methods for Nonlinear Algebraic Equations, Gordon & Breach, New York, 1970, pp. 163–181.

[26] J.E. Dennis Jr., Toward a unified convergence theory for Newton-like methods, in: L.B. Rall (Ed.), Nonlinear Functional Analysis and Applications, Academic Press, New York, 1971.

[27] P. Deuflhard, G. Heindl, Affine invariant convergence theorems for Newton's method and extensions to related methods, SIAM J. Numer. Anal. 16 (1979) 1–10.

[28] P. Deuflhard, F.A. Potra, A refined Gauss–Newton–Mysovskii theorem, ZIB SC 91-4 (1991) 1–9.

[29] B. Döring, Über das Newtonsche Näherungsverfahren, Math. Phys. Semi.-Ber. 16 (1969) 27–40.

[30] B. Döring, Einige Sätze über das Verfahren der tangierenden Hyperbeln in Banach-Räumen, Aplikace Mat. 15 (1970) 418–464.

[31] J.A. Ezquerro, M.A. Hernandez, Avoiding the computation of the second Fréchet-derivative in the convex acceleration of Newton's method, J. Comput. Appl. Math. 96 (1998) 1–12.

[32] I. Fenyö, Über die Lösung der in Banachschen Raume definierten nichtlinearen Gleichungen, Acta. Math. Acad. Sci. Hungar. 5 (1954) 85–93.

[33] H.B. Fine, On Newton's method of approximation, Proc. Nat. Acad. Sci. USA 2 (1916) 546–552.

[34] J.B.J. Fourier, Question d'analyse algébrique, in: Oeuvres Complétes (II), Gauthier-Villars, Paris, 1890, pp. 243–253.

[35] M. Fukushima, L. Qi, Reformulation: Nonsmooth, Piecewise Smooth, Semismooth and Smoothing Methods, Kluwer Academic Publishers, Dordrecht, 1989.

[36] W. Gander, On Halley's iteration method, Amer. Math. Monthly 92 (1985) 131–134.

[37] W.B. Gragg, R.A. Tapia, Optimal error bounds for the Newton–Kantorovich theorem, SIAM J. Numer. Anal. 11 (1974) 10–13.

[38] W.M. Häußler, A Kantorovich-type convergence analysis for the Gauss–Newton method, Numer. Math. 48 (1986) 119–125.

[39] M. Heinkenschloss, C.T. Kelley, H.T. Tran, Fast algorithms for nonsmooth compact fixed-point problems, SIAM J. Numer. Anal. 29 (1992) 1769–1792.

[40] M.A. Hernandez, A note on Halley's method, Numer. Math. 59 (1991) 273–276.

[41] S. Hitotumatu, A method of successive approximation based on the expansion of second order, Math. Japon. 7 (1962) 31–50.

[42] S. Kanno, Convergence theorems for the method of tangent hyperbolas, Math. Japon. 37 (1992) 711–722.

[43] L.V. Kantorovich, The method of successive approximations for functional analysis, Acta. Math. 71 (1939) 63–97.

[44] L.V. Kantorovich, On Newton's method for functional equations, Dokl Akad. Nauk SSSR 59 (1948) 1237–1240 (in Russian).

[45] L.V. Kantorovich, The majorant principle and Newton's method, Dokl. Akad. Nauk SSSR 76 (1951) 17–20 (in Russian).

[46] L.V. Kantorovich, G.P. Akilov, Functional Analysis, 2nd ed., Pergamon Press, Oxford, 1982.

[47] F. Kikuchi, An iteration scheme for a nonlinear eigenvalue problem, Theoret. Appl. Mech. 29 (1981) 319–333.

[48] F. Kikuchi, Finite element analysis of a nondifferentiable nonlinear problem related to MHD equilibria, J. Fac. Sci. Univ. Tokyo Sect. IA Math. 35 (1988) 77–101.

[49] H.J. Kornstaedt, Functional Ungleichungen und Iterations Verfahren, Aequationes Math. 13 (1975) 21–45.

[50] C.L. Lawson, R.J. Hanson, Solving Least Square Problems, Prentice-Hall, Englewood Cliffs, NJ, 1974.

[51] J.M. Martínez, Quasi-Newton methods for solving underdetermined nonlinear simultaneous equations, J. Comput. Appl. Math. 34 (1991) 171–190.

[52] M.A. Mertvecova, An analog of the process of tangent hyperbolas for general functional equations, Dokl. Akad. Nauk SSSR 88 (1953) 611–614 (in Russian).

[53] G.J. Miel, The Kantorovich theorem with optimal error bounds, Amer. Math. Monthly 86 (1979) 212–215.

[54] G.J. Miel, Majorizing sequences and error bounds for iterative methods, Math. Comp. 34 (1980) 185–202.

[55] G.J. Miel, An updated version of the Kantorovich theorem for Newton's method, Computing 27 (1981) 237–244.

[56] K.H. Meyn, Solution of underdetermined nonlinear equations by stationary iteration methods, Numer. Math. 42 (1983) 161–172.

[57] I. Moret, A note on Newton type iterative methods, Computing 33 (1984) 65–73.

[58] I. Mysovskii, On the convergence of L.V. Kantorovich's method of solution of functional equations and its applications, Dokl, Akad. Nauk SSSR 70 (1950) 565–585 (in Russian).

[59] M.Z. Nashed, X. Chen, Convergence of Newton-like methods for singular operator equations using outer inverses, Numer. Math. 66 (1993) 235–257.

[60] M.I. Necepurenko, On Chebyshev's method for functional equations, Uspehi Matem. Nauk 9 (1954) 163–170. English translation: L.B. Rall, MRC Technical Summary Report #648, 1966.

[61] J.M. Ortega, W.C. Rheinboldt, Iterative Solution of Nonlinear Equations in Several Variables, Academic Press, New York, 1970.

[62] A.M. Ostrowski, Sur la convergence et l'estimation des erreurs dans quelques procédés de résolution des équations numériques, in: Collection of Papers in Memory of D.A. Grave, Gosudarstv. Izdat. Tehn.-Teor. Lit., Moscow, 1940, pp. 213–234.

[63] A.M. Ostrowski, Solution of Equations and Systems of Equations, 2nd ed., Academic Press, New York, 1966.

[64] A.M. Ostrowski, La méthode de Newton dans les espaces de Banach, C.R. Acad. Sci. Paris Ser. A 272 (1971) 1251–1253.

[65] A.M. Ostrowski, Solution of Equations in Euclidean and Banach Space, Academic Press, New York, 1973.

[66] J.S. Pang, A B-differentiable equation-based, globally and locally quadratically convergent algorithm for nonlinear programs, complementarity and variational inequality problems, Math. Programming 51 (1991) 101–131.

[67] J.S. Pang, L. Qi, Nonsmooth equations: motivation and algorithms, SIAM J. Optim. 3 (1993) 443–465.

[68] F.A. Potra, On the a posteriori error estimates for Newton's method, Beiträge zur Numer. Math. 12 (1984) 125–138.

[69] F.A. Potra, V. Pták, Sharp error bounds for Newton's process, Numer. Math. 34 (1980) 63–72.

[70] V. Pták, The rate of convergence of Newton's process, Numer. Math. 25 (1976) 279–285.

[71] L. Qi, Convergence analysis of some algorithms for solving nonsmooth equations, Math. Oper. Res. 18 (1993) 227–244.

[72] L. Qi, X. Chen, A globally convergent successive approximation method for severely nonsmooth equations, SIAM J. Control. Optim. 33 (1995) 402–418.

[73] L. Qi, J. Sun, A nonsmooth version of Newton's method, Math. Programming 58 (1993) 353–367.

[74] L.B. Rall, A note on the convergence of Newton's method, SIAM J. Numer. Anal. 11 (1974) 34–36.

[75] L.B. Rall, R.A. Tapia, The Kantorovich theorem and error estimates for Newton's method, MRC Technical Summary Report No. 1043, Univ. of Wisconsin – Madison (1970).

[76] J. Rappaz, Approximation of a nondifferentiable nonlinear problem related to MHD equilibria, Numer. Math. 45 (1984) 117–133.

[77] W.C. Rheinboldt, A unified convergence theory for a class of iterative processes, SIAM J. Numer. Anal. 5 (1968) 42–63.

[78] W.C. Rheinboldt, An adaptive continuation process for solving systems of nonlinear equations, in: Mathematical Models and Numerical Methods, Banach Center Publications, Vol. 3, PWN–Polish Scientific Publishers, Warsaw, 1978, pp. 129–142.

[79] S.M. Robinson, Newton's method for a class of nonsmooth functions, Set-Valued Anal. 2 (1994) 291–305.

[80] R.A. Safiev, The method of tangent hyperbolas, Sov. Math. Dokl. 4 (1963) 482–485.

[81] R.A. Safiev, On some iterative processes, Ž . Vyčisl. Mat. i. Mat. Fiz. 4 (1964) 139–143 (in Russian).

[82] G.S. Salehov, On the convergence of the process of tangent hyperbolas, Dokl. A. N. SSSR 82 (1952) 525–528 (in Russian).

[83] J.W. Schmidt, Ein Übertragung der Regula falsi auf Gleichungen in Banachräumen, ZAMM 34 (1963) Part I, 1–8, Part II, 97–110.

[84] J.W. Schmidt, Regular-falsi-Verfahren mit konsistenter Steigung und Majorantenprinzip, Period. Math. Hungarica 5 (1974) 187–193.

[85] J.W. Schmidt, Untere Fehlerschranken für Regular-falsi-Verfahren, Period. Math. Hungarica 9 (1978) 241–247.

[86] J. Schröder, Nichtlineare Majoranten beim Verfahren der schrittweisen Näherung, Arch. Math. 7 (1956) 471–484.

[87] J. Schröder, Über das Newtonsche Verfahren, Arch. Rational Mech. Anal. 1 (1957) 154–180.

[88] R.A. Tapia, The Kantorovich theorem for Newton's method, Amer. Math. Monthly 78 (1971) 389–392.

[89] H.J. Walker, Newton-like methods for underdetermined systems, in: E.L. Allgower, K. Georg (Eds.), Computational Solution of Nonlinear Systems of Equations, Lectures in Appl. Math., Vol. 26, AMS, Providence, 1990.

[90] H.J. Walker, L.T. Watson, Least-change secant update methods for underdetermined systems, SIAM J. Numer. Anal. 27 (1990) 1227–1262.

[91] W. Werner, Some improvements of classical iterative methods for the solution of nonlinear equations, in: E.L. Allgower, K. Glashoft, H.O. Peitgen (Eds.), Numerical Solution of Nonlinear Equations, Lecture Notes in Math., Vol. 878, Springer, Heidelberg, 1981.

[92] W. Werner, Iterative solution of systems of nonlinear equations based upon quadratic approximations, Comput. Math. Appl. A 12 (1986) 331–343.

[93] T. Yamamoto, Error bounds for Newton's process derived from the Kantorovich theorem, Japan J. Appl. Math. 2 (1985) 258–292.

[94] T. Yamamoto, A unified derivation of several error bounds for Newton's process, J. Comput. Appl. Math. 12,13 (1985) 179–191.

[95] T. Yamamoto, Error bounds for Newton's iterates derived from the Kantorovich theorem, Numer. Math. 48 (1986) 91–98.

[96] T. Yamamoto, Error bounds for Newton-like methods under Kantorovich type assumptions, Japan J. Appl. Math. 3 (1986) 295–313.

[97] T. Yamamoto, A convergence theorem for Newton's method in Banach spaces, Japan J. Appl. Math. 3 (1986) 37–52.

[98] T. Yamamoto, A method for finding sharp error bounds for Newton's method under the Kantorovich assumptions, Numer. Math. 49 (1986) 203–220.

[99] T. Yamamoto, A convergence theorem for Newton-like methods in Banach spaces, Numer. Math. 51 (1987) 545–557.

[100] T. Yamamoto, A note on a posteriori error bound of Zabrejko and Nguen for Zincenko's iteration, Numer. Funct. Anal. Optim. 9 & 10 (1987) 987–994.

[101] T. Yamamoto, On the method of tangent hyerbolas in Banach spaces, J. Comput. Appl. Math. 21 (1988) 75–86.

[102] T. Yamamoto, Unified convergence analysis for a class of iterative methods for solving nonlinear equations, in: Th.M. Rassias (Ed.), Topics in Mathematical Analysis, World Scientific, Singapore, 1989, pp. 878–900.

[103] T. Yamamoto, Uniqueness of the solution in a Kantorovich-type theorem of Häußler for the Gauss–Newton method, Japan J. Appl. Math. 6 (1989) 77–81.

[104] T. Yamamoto, X. Chen, Ball convergence theorems and error estimates for certain iterative methods for nonlinear equations, Japan J. Appl. Math. 7 (1990) 131–143.

[105] P.P. Zabrejko, D.F. Nguen, The majorant method in the theory of Newton–Kantorovich approximations and the Pták error estimates, Numer. Funct. Anal. Optim. 9 (1987) 671–684.

[106] P.P. Zabrejko, P.P. Zlepko, On a generalization of the Newton–Kantorovich method for an equation with nondifferentiable operator, Ukr. Mat. Zhurn. 34 (1982) 365–369 (in Russian).

[107] A.I. Zincenko, Some approximate methods of solving equations with nondifferentiable operators, Dopovidi Akad. Nauk. Ukrain RSR (1963) 156–161 (in Ukrainian).

ELSEVIER

Journal of Computational and Applied Mathematics 124 (2000) 25–44

JOURNAL OF
COMPUTATIONAL AND
APPLIED MATHEMATICS

www.elsevier.nl/locate/cam

The theory of Newton's method

A. Galántai

Institute of Mathematics, The University of Miskolc, 3515 Miskolc-Egyetemvaros, Hungary

Received 18 June 1999; received in revised form 31 January 2000

Abstract

We review the most important theoretical results on Newton's method concerning the convergence properties, the error estimates, the numerical stability and the computational complexity of the algorithm. We deal with the convergence for smooth and nonsmooth equations, underdetermined equations, and equations with singular Jacobians. Only those extensions of the Newton method are investigated, where a generalized derivative and or a generalized inverse is used. © 2000 Elsevier Science B.V. All rights reserved.

1. Introduction

The Newton or Newton–Raphson method has the form

$$x_{k+1} = x_k - [F'(x_k)]^{-1} F(x_k), \quad k = 0, 1, \ldots \tag{1}$$

for the solution of the nonlinear equation

$$F(x) = 0 \quad (F : X \to Y), \tag{2}$$

where X and Y are Banach spaces and F' is the Fréchet derivative of F. The geometric interpretation of the Newton method is well known, if F is a real function. In such a case x_{k+1} is the point where the tangential line $y - F(x_k) = F'(x_k)(x - x_k)$ of function $F(x)$ at point $(x_k, F(x_k))$ intersects the x-axis. The geometric interpretation of the complex Newton method $(F : C \to C)$ is given by Yau and Ben-Israel [70]. In the general case $F(x)$ is approximated at point x_k as

$$F(x) \approx L_k(x) = F(x_k) + F'(x_k)(x - x_k). \tag{3}$$

The zero of $L_k(x) = 0$ defines the new approximation x_{k+1}.

Variants of the Newton method are the damped Newton method

$$x_{k+1} = x_k - t_k [F'(x_k)]^{-1} F(x_k) \quad (t_k > 0, \ k = 0, 1, 2, \ldots) \tag{4}$$

E-mail address: matgal@gold.uni-miskolc.hu (A. Galántai).

0377-0427/00/$ - see front matter © 2000 Elsevier Science B.V. All rights reserved.
PII: S 0377-0427(00)00435-0

and the modified Newton method

$$x_{k+1} = x_k - [F'(x_{\alpha_k})]^{-1}F(x_k) \quad (k = 0, 1, 2, \ldots),$$ (5)

where $\alpha_0 = 0$, $\alpha_{k-1} \leqslant \alpha_k \leqslant k$ $(k \geqslant 1)$. The latter formulation includes the Shamanskii–Newton method [42], where the Jacobian is always reevaluated after m consecutive iterations. The Newton-like methods are generally defined by the recursion

$$x_{k+1} = x_k - [M(x_k)]^{-1}F(x_k) \quad (k = 0, 1, 2, \ldots),$$

where $M(x)$ is usually an approximation to $F'(x^*)$, where x^* is a solution of Eq. (2). These methods formally include the quasi-Newton and inexact Newton methods, as well. In this paper we deal only with the theory of Newton's method. We concentrate on the convergence properties, error estimates, complexity and related issues. It is remarkable that much of these results were obtained in the last 30 years. Yet, the theory of Newton method is far from being complete. For the implementation of Newton's method we refer to Ortega–Rheinboldt [42], Dennis and Schnabel [13], Brown and Saad [8], and Kelley [29]. Kearfott [1, pp. 337–357] discusses the implementation of Newton's method in interval arithmetic. For other important results not quoted here we refer to the reference list of the paper.

2. Convergence results for smooth equations

Let X and Y be two Banach spaces. Let $S(x, r) = \{z \in X \mid \|z - x\| < r\}$ denote the open ball in X with center x and radius r and let $\overline{S(x, r)}$ be its closure. Let $F: S(x_0, R) \subset X \to Y$ be a given (nonlinear) mapping. Assume that a sphere $S(x_0, r)$ exists such that $\overline{S(x_0, r)} \subset S(x_0, R)$. Denote by $F'(x)$ and $F''(x)$ the first and second derivatives of F in the sense of Fréchet. Kantorovich proved the following classical result.

Theorem 1 (Kantorovich [27]). *Let* $F: S(x_0, R) \subset X \to Y$ *have a continuous second Fréchet derivative in* $\overline{S(x_0, r)}$. *Moreover, let* (i) *the linear operation* $\Gamma_0 = [F'(x_0)]^{-1}$ *exist;* (ii) $\|\Gamma_0 F(x_0)\| \leqslant \eta$; (iii) $\|\Gamma_0 F''(x)\| \leqslant K$ $(x \in \overline{S(x_0, r)})$. *Now, if*

$$h = K\eta \leqslant \tfrac{1}{2}$$

and

$$r \geqslant r_0 = \frac{1 - \sqrt{1 - 2h}}{h}\eta,$$

Eq. (2) *will have a solution* x^* *to which the Newton method is convergent. Here,*

$$\|x^* - x_0\| \leqslant r_0.$$

Furthermore, if for $h < \tfrac{1}{2}$

$$r < r_1 = \frac{1 + \sqrt{1 - 2h}}{h}\eta,$$

or for $h = \tfrac{1}{2}$

$$r \leqslant r_1,$$

the solution x^* will be unique in the sphere $\overline{S(x_0,r)}$. The speed of convergence is characterized by the inequality

$$\|x^* - x_k\| \leqslant \frac{1}{2^k}(2h)^{2^k}\frac{\eta}{h} \quad (k = 0,1,2,\ldots).$$

Remark 2. The conditions (ii) and (iii) of the theorem can be replaced by (i') $\|\Gamma_0\| \leqslant B'$; (ii') $\|F(x_0)\| \leqslant \eta'$; (iii') $\|F''(x)\| \leqslant K'$ $(x \in \overline{S(x_0,r)})$. In this case h, r_0 and r_1 are, respectively,

$$h = K'(B')^2\eta', \quad r_0 = \frac{1 - \sqrt{1 - 2h}}{h}B'\eta', \quad r_1 = \frac{1 + \sqrt{1 - 2h}}{h}B'\eta'.$$

Remark 3. The conditions $h \leqslant \frac{1}{2}$ and $r_0 \leqslant r$ are necessary for the existence of solution. The solution is not unique in the absence of condition $r < r_1$ or $r \leqslant r_1$.

Notice that $\overline{S(x_0,r)}$ gives an inclusion region for a zero of Eq. (2). If a bound is known for $\|[F'(x)]^{-1}\|$ in $\overline{S(x_0,r)}$, the condition imposed on h can be weakened by requiring $h < 2$ instead of $h \leqslant 1/2$.

Theorem 4 (Mysovskikh [38]). *Let the following conditions be satisfied:* (i) $\|F(x_0)\| \leqslant \eta$; (ii) *the linear operation* $\Gamma(x) = [F'(x)]^{-1}$ *exist for* $x \in \overline{S(x_0,r)}$, *where* $\|\Gamma(x)\| \leqslant B$ $(x \in \overline{S(x_0,r)})$; (iii) $\|F''(x)\| \leqslant K$ $(x \in \overline{S(x_0,r)})$. *Then, if*

$$h = B^2K\eta < 2$$

and

$$r > r' = B\eta\sum_{j=0}^{\infty}\left(\frac{h}{2}\right)^{2^j-1},$$

Eq. (2) has a solution $x^* \in \overline{S(x_0,r)}$ *to which the Newton method with initial point* x_0 *is convergent. The speed of convergence is given by*

$$\|x^* - x_k\| \leqslant B\eta\frac{(h/2)^{2^k-1}}{1 - (h/2)^{2^k}} \quad (k = 0,1,2,\ldots).$$

The Newton iterates x_k are invariant under any affine transformation $F \to G = AF$, where A denotes any bounded and bijective linear mapping from Y to any Banach space Z. This property is easily verified, since

$$[G'(x)]^{-1}G(x) = [F'(x)]^{-1}A^{-1}AF(x) = [F'(x)]^{-1}F(x).$$

The affine invariance property is clearly reflected in the Kantorovich theorem. For other affine invariant theorems we refer to Deuflhard and Heindl [14].

The Kantorovich theorem is a masterpiece not only by its sheer importance but by the original and powerful proof technique. The results of Kantorovich and his school initiated some very intensive research on the Newton and related methods. A great number of variants and extensions of his results emerged in the literature. Ortega–Rheinboldt [42] is a good survey for such developments until 1970 (see also [24,56] or [13]).

In the Newton–Kantorovich theorem the continuity conditions on $F''(x)$ can be easily replaced either by

$$\|F'(x) - F'(y)\| \leqslant K\|x - y\| \quad (x, y \in \overline{S(x_0, r)})$$ (6)

or

$$\|[F'(x_0)]^{-1}(F'(x) - F'(y))\| \leqslant K\|x - y\| \quad (x, y \in \overline{S(x_0, r)}).$$ (7)

This was done by several authors, the first of which was, perhaps, Fenyő [16]. A typical result of this kind is the following.

Theorem 5 (Tapia [64]). *Let X and Y be Banach spaces and $F : D \subset X \to Y$. Suppose that on an open convex set $D_0 \subset D$, F is Fréchet differentiable with*

$$\|F'(x) - F'(y)\| \leqslant K\|x - y\| \quad (x, y \in D_0).$$

Assume that $x_0 \in D_0$ is such that (i) $[F'(x_0)]^{-1}$ *exists,*

$$\|[F'(x_0)^{-1}]\| \leqslant B, \quad \|[F'(x_0)]^{-1}F(x_0)\| \leqslant \eta, \quad h = BK\eta \leqslant \tfrac{1}{2},$$

(ii) $\overline{S(x_0, t^*)} \subset D_0$ $(t^* = ((1 - \sqrt{1 - 2h})/h)\eta)$. *Then the Newton iterates x_k are well defined, remain in $\overline{S(x_0, t^*)}$, and converge to $x^* \in \overline{S(x_0, t^*)}$ such that $F(x^*) = 0$. In addition*

$$\|x^* - x_k\| \leqslant \frac{\eta}{h}\left(\frac{(1 - \sqrt{1 - 2h})^{2^k}}{2^k}\right), \quad k = 0, 1, 2, \dots .$$

This result, which is often referred to as the Kantorovich theorem, is an improvement of Ortega [41] and gives an optimum estimate for the rate of convergence.

If the hypotheses (i) and (ii) of the above Kantorovich theorem are satisfied, then not only the Newton sequence $\{x_k\}$ exists and converges to a solution x^* but $[F'(x^*)]^{-1}$ exists in this case. The following result of Rall [50] shows that the existence of $[F'(x^*)]^{-1}$ conversely guarantees that the hypotheses of the Kantorovich theorem with $h < \tfrac{1}{2}$ are satisfied at each point of an open ball S^* with center x^*. In such a case x^* is called a simple zero of F.

Theorem 6 (Rall [50]). *If x^* is a simple zero of F, $\|[F'(x^*)]^{-1}\| \leqslant B^*$, and*

$$S_* = \left\{ x \,\middle|\, \|x - x^*\| < \frac{1}{B^*K} \right\} \subset D_0,$$

then the hypotheses (i) *with $h < \tfrac{1}{2}$ and* (ii) *of the Kantorovich Theorem 5 are satisfied at each $x_0 \in S^*$, where*

$$S^* = \left\{ x \,\middle|\, \|x - x^*\| < \frac{2 - \sqrt{2}}{2B^*K} \right\}.$$

The value given for the radius of S^* is the best possible.

Vertgeim [71] was the first to weaken the C^2 condition of $F(x)$ to the Hölder condition

$$\|F'(x) - F'(y)\| \leqslant K\|x - y\|^\alpha \quad (x, y \in D_0),$$ (8)

where $0 < \alpha \leqslant 1$ is a constant. His result was improved or rediscovered by several authors ([42,24,28, 55,3,4,33] and others). A typical result of this kind is the following

Theorem 7 (Jankó-Coroian [24]). *Let $F : S(x_0, R) \subset X \to Y$ be a given nonlinear mapping and assume that* (i) $\Gamma_0 = [F'(x_0)]^{-1}$ *exists and* $\|\Gamma_0\| \leqslant B$; (ii) $\|\Gamma_0 F(x_0)\| \leqslant \eta$; (iii)

$$\|F'(x) - F'(y)\| \leqslant K\|x - y\|^{\alpha} \quad (x, y \in \overline{S(x_0, r)}, \ 0 < \alpha \leqslant 1),$$

where

$$r = \frac{(1+\alpha)^{1/\alpha}[(1+\alpha)^{1/\alpha}h]^{1-(1/\alpha)}}{(1+\alpha)^{(1/\alpha)} - 1}\eta,$$

(iv)

$$h = BK\eta^{\alpha} \leqslant \frac{\alpha}{1+\alpha}.$$

Then there is at least one zero x^* in $\overline{S(x_0, r)}$ and $\{x_k\}$ converges to x^* with the speed

$$\|x^* - x_k\| \leqslant \frac{[(1+\alpha)^{1/\alpha}h]^{(1+\alpha)^k - (1/\alpha)}}{[(1+\alpha)^{(1/\alpha)} - 1](1+\alpha)^{(k-1)/\alpha}}\eta.$$

Similar results hold for the Newton–Mysovskikh theorem (see [24]), one of which is the following.

Theorem 8 (Jankó [24]). *Let $F : X \to Y$ be a given nonlinear mapping and let the following conditions be satisfied:* (i) *the linear operator $\Gamma(x) = [F'(x)]^{-1}$ exists for all $x \in \overline{S(x_0, T\eta)}$, where*

$$T = \sum_{j=0}^{\infty}\left(\frac{h}{1+\alpha}\right)^{((1+\alpha)^j - 1)/\alpha}, \quad 0 < \alpha \leqslant 1,$$

(ii) $\|\Gamma(x_0)F(x_0)\| \leqslant \eta$;
(iii) $\|\Gamma(x)[F'(y) - F'(z)]\| \leqslant K\|x - y\|^{\alpha} \quad (x, y, z \in \overline{S(x_0, T\eta)})$;
(iv) $h = K\eta^{\alpha} < 1 + \alpha$.
Then Eq. (2) has a solution $x^* \in \overline{S(x_0, T\eta)}$ to which the Newton method with initial point x_0 is convergent. The speed of convergence is given by

$$\|x^* - x_k\| \leqslant T\eta\left(\frac{h}{1+\alpha}\right)^{((1+\alpha)^k - 1)/\alpha} \quad (k = 0, 1, 2, \ldots).$$

Assuming that operator F is analytic, Smale [59], Rheinboldt [53], and Wang and Han [73] gave convergence results which utilize only information at the starting point. Denote by $F^{(j)}(x)$ the jth Fréchet derivative of F at point x. A typical result of this type is given in

Theorem 9 (Rheinboldt [53]). *Let $F : X \to Y$ be analytic on some open set S of X and let*

$$\beta(x) = \|[F'(x)]^{-1}F(x)\|, \quad \gamma(x) = \sup_{j > 1}\left\|\frac{1}{j!}[F'(x)]^{-1}F^{(j)}(x)\right\|^{1/(j-1)}.$$

Consider a point x_0 of S where $F'(x_0)$ is invertible. Let ρ (≈ 0.16842669) be the positive root of the cubic $(\sqrt{2} - 1)(1 - \rho)^3 - \sqrt{2}\rho = 0$. If $\alpha(x_0) = \beta(x_0)\gamma(x_0) \leqslant \rho/\sqrt{2}$ (≈ 0.11909565) and the ball $S(x_0, r_0)$ with radius $r_0 = \rho/\gamma(x_0)$ is contained in S, then the Newton iterates x_k converge to a solution x^* of Eq. (2). Moreover, the convergence is at least R-quadratic with R_2-factor $\frac{1}{2}$.

Using the results of [73], Wang and Zhao [74] improved Smale's result by not assuming that $\gamma(x)$ is bounded.

3. Error estimates for the Newton method

There are several error estimates for the Newton method. The Kantorovich theorem is a basis for many of these, but there are others as well.

In the next four theorems we assume the conditions of Tapia's Theorem 5. Gragg and Tapia [20] added the following optimal error bounds to this theorem.

Theorem 10 (Gragg–Tapia [20]).

$$\|x^* - x_k\| \leqslant \begin{cases} \frac{2}{h}\sqrt{1 - 2h}\dfrac{\theta^{2^k}}{1 - \theta^{2^k}}\|x_1 - x_0\|, & \text{if } 2h < 1, \\ 2^{1-k}\|x_1 - x_0\|, & \text{if } 2h = 1 \end{cases}$$

and

$$\frac{2\|x_{k+1} - x_k\|}{1 + \sqrt{1 + 4\theta^{2^k}/(1 + \theta^{2^k})^2}} \leqslant \|x^* - x_k\| \leqslant \theta^{2^{k-1}}\|x_k - x_{k-1}\|, \quad k \geqslant 1,$$

where $\theta = (1 - \sqrt{1 - 2h})/(1 + \sqrt{1 + 2h})$.

Miel [34] gave a new proof for this theorem using the technique of Ortega [41]. An affine invariant version of the theorem was given by Deuflhard and Heindl [14]. Miel also constructed several error bounds for the Newton method [36,37].

Theorem 11 (Miel [36]). *Let us define the constants A_k, B_k and C_k recursively by*

$$A_1 = \frac{1}{\eta}B_1, \quad A_{k+1} = A_k(2 - \Delta A_k), \quad \Delta = \frac{2\eta\sqrt{1 - 2h}}{h},$$

$$B_1 = \theta, \quad B_{k+1} = B_k^2, \quad \theta = (1 - \sqrt{1 - 2h})/(1 + \sqrt{1 + 2h}),$$

$$C_1 = B_1, \quad C_{k+1} = \frac{C_k^2}{2C_k + \Delta/\eta}.$$

Then the error bounds

$$\|x^* - x_k\| \leqslant A_k\|x_k - x_{k-1}\|^2 \leqslant B_k\|x_k - x_{k-1}\| \leqslant C_k\|x_1 - x_0\|,$$

are valid and the best possible.

Theorem 12 (Miel [37]). *Let* $\Delta = (2\eta\sqrt{1-2h})/h$ *and*

$$\theta = (1 - \sqrt{1-2h})/(1 + \sqrt{1+2h}).$$

Then

$$\frac{2\|x_{k+1} - x_k\|}{1 + \sqrt{1 + (4/\Delta)(1 - \theta^{2^k})/(1 + \theta^{2^k})\|x_{k+1} - x_k\|}} \leqslant \|x^* - x_k\| \leqslant \frac{1 - \theta^{2^k}}{\Delta}\|x_k - x_{k-1}\|^2$$

if $2h < 1$, *and*

$$\frac{2\|x_{k+1} - x_k\|}{1 + \sqrt{1 + (2^k/\eta)\|x_{k+1} - x_k\|}} \leqslant \|x^* - x_k\| \leqslant \frac{2^{k-1}}{\eta}\|x_k - x_{k-1}\|^2$$

if $2h = 1$.

Other proofs of this result can be found in [68,69]. Using nondiscrete mathematical induction, Pták [46], Potra and Pták [45], and Lai and Wu [31] gave convergence results and error estimates for the Newton and Newton-like methods. Here we recall the following result.

Theorem 13 (Potra–Pták [45]). *Let* $a = (\eta\sqrt{1-2h})/h$ *and*

$$\gamma(r) = (a^2 + 4r^2 + 4r(a^2 + r^2)^{1/2})^{1/2} - (r + (a^2 + r^2)^{1/2}).$$

Then

$$\gamma(\|x_{k+1} - x_k\|) \leqslant \|x_k - x^*\| \leqslant (a^2 + \|x_k - x_{k-1}\|^2)^{1/2} - a.$$

Yamamoto [68] pointed out that the Gragg–Tapia estimates are derivations of the Kantorovich recurrence relations and the Potra–Pták and Miel estimates are improvements of the Gragg–Tapia theorem. He also showed that the latter two results also follow from the original Kantorovich theorem, and Miel's result (Theorem 12) improves on that of Potra and Pták. Yamamoto [69] gives a method for finding sharp posterior error bounds for Newton's method under the assumptions of Kantorovich's theorem. Yamamoto's paper [69], where a comparison of the best known bounds can also be found, is the best source for error estimates in the theory of the Newton method.

Other type of error estimate is given by Neumaier [40]. For any two vectors $x, y \in \mathbb{R}^n$ let $x \leqslant y$, if and only if $x_i \leqslant y_i$ for all i. Furthermore, let $|A| = [|a_{ij}|]_{i,j=1}^{m,n}$ for any $A \in \mathbb{R}^{m \times n}$.

Theorem 14 (Neumaier [40]). *Let* $F: D \subset \mathbb{R}^n \to \mathbb{R}^n$ *be continuous,* $x_0 \in D$, $A \in \mathbb{R}^{n \times n}$ *be nonsingular and* $\delta_0 = A^{-1}F(x_0)$. *Suppose further that there is a constant* $\kappa > 1$ *such that*

$$\overline{S(x_0, \kappa\|\delta_0\|)} \subset D$$

and a nonnegative vector $c \in \mathbb{R}^n$ *such that, for some monotone norm* $\|.\|$

$$|F(x) - F(x_0) - A(x - x_0)| \leqslant \|\delta_0\|c \quad (x \in \overline{S(x_0, \kappa\|\delta_0\|)}).$$

If the vector $b = |A^{-1}|c$ satisfies the condition $\|b\| \leqslant \kappa - 1$, then $F(x)$ has at least one zero in $\overline{S(x_0, \kappa\|\delta_0\|)}$, and any such zero $\hat{x}$ satisfies

$$(2 - \kappa)\|\delta_0\| \leqslant \|\hat{x} - x_0\| \leqslant \kappa\|\delta_0\|.$$

The inclusion region $\overline{S(x_0, r)}$ of the Kantorovich theorem easily follows for the choice of scaled l_∞-norm and $A = F'(x_0)$.

4. Monotone convergence

The Newton method exhibits monotone convergence under partial ordering. We use the natural partial ordering for vectors and matrices; that is, $A \leqslant B$ ($A, B \in \mathbb{R}^{n \times n}$) if and only if $a_{ij} \leqslant b_{ij}$ ($i, j = 1, \ldots, n$). The function $F : \mathbb{R}^n \to \mathbb{R}^n$ is said to be *convex* on a convex set $D \subseteq \mathbb{R}^n$ if

$$F(\lambda x + (1 - \lambda)y) \leqslant \lambda F(x) + (1 - \lambda)F(y) \tag{9}$$

holds for all $x, y \in D$ and $\lambda \in [0, 1]$. Assume that $F : \mathbb{R}^n \to \mathbb{R}^n$ is differentiable on the convex set D. Then F is convex on D if and only if

$$F(y) - F(x) \geqslant F'(x)(y - x) \tag{10}$$

holds for all $x, y \in D$. The following basic result of Baluev is a special case of Theorem 13.3.7 of [42].

Theorem 15 (Baluev [42]). *Assume that $F : \mathbb{R}^n \to \mathbb{R}^n$ is continuously differentiable and convex on all of $\mathbb{R}^n$, that $F'(x)$ is nonsingular and $[F'(x)]^{-1} \geqslant 0$ for all $x \in \mathbb{R}^n$, and that $F(x) = 0$ has a solution x^*. Then x^* is unique and the Newton iterates x_k converge to x^* for any x_0. Moreover $x^* \leqslant x_{k+1} \leqslant x_k$ ($k = 1, 2, \ldots$).*

Note that Baluev's theorem guarantees global convergence for the given function class. Characterizations of such functions and related results can be found in [42], or [18]. We show later that Newton's method has optimal complexity under special circumstances. It is not the case however for this kind of monotone convergence. Frommer [17] proved that Brown's method is faster than Newton's method under partial ordering. A similar result was proved for the ABS methods, as well [19].

5. The Newton method for underdetermined equations

For underdetermined equations of the form

$$F(x) = 0 \quad (F : \mathbb{R}^n \to \mathbb{R}^m, \ n \geqslant m) \tag{11}$$

an extension of the Newton iteration requires the solution of

$$F(x) + F'(x)(x_+ - x) = 0. \tag{12}$$

If $n > m$, then the new approximation x_+ is not uniquely determined. There are various ways to define x_+. The first result of this kind is due to Ben-Israel [6] who used the Moore–Penrose inverse to define $x_+ = x - F'(x)^+ F(x)$.

Theorem 16 (Ben-Israel [6]). *Let* $F: \mathbb{R}^n \rightarrow \mathbb{R}^m$ *be a function,* $x_0 \in \mathbb{R}^n$*, and* $r > 0$ *be such that* $F \in C^1(S(x_0, r))$*. Let* M, N *be positive constants such that for all* $x, y \in S(x_0, r)$ *with* $x - y \in \mathbb{R} \, (F'(y)^\mathrm{T})$:

$$\|F(x) - F(y) - F'(y)(x - y)\| \leqslant M \|x - y\|,$$

$$\|(F'(x)^+ - F'(y)^+)F(y)\| \leqslant N \|x - y\|$$

and

$$M \|F'(x)^+\| + N = \gamma < 1 \quad (x \in S(x_0, r)),$$

$$\|F'(x_0)^+\| \|F(x_0)\| < (1 - \gamma)r.$$

Then the sequence

$$x_{k+1} = x_k - F'(x_k)^+ F(x_k) \quad (k = 0, 1, 2, \ldots) \tag{13}$$

converges to a solution of $F'(x)^\mathrm{T} F(x) = 0$ *which lies in* $S(x_0, r)$*.*

Condition $F'(x)^\mathrm{T} F(x) = 0$ is equivalent to $F'(x)^+ F(x) = 0$. Algorithm (13) is called the normal flow algorithm. The name comes from the $n = m + 1$ case, in which the iteration steps $-F'(x_k)^+ F(x_k)$ are asymptotically normal to the Davidenko flow. For any n, the step $-F'(x_k)^+ F(x_k)$ is normal to the manifold $\{y \in \mathbb{R}^n \mid F(y) = F(x_k)\}$. For the special case $n = m + 1$ we refer to [2] and [10].

Walker [72] investigates the normal flow algorithm and the augmented Jacobian algorithm which is defined as follows. For a specified $V \in \mathbb{R}^{k \times n}$ and a given approximate solution x_k, determine x_{k+1} by

$$x_{k+1} = x_k + s, \quad \text{where} \quad F'(x_k)s = -F(x_k) \quad \text{and} \quad Vs = 0. \tag{14}$$

Walker gives two local convergence theorems under the following hypotheses.

Normal flow hypothesis. *F is differentiable, F' is of full rank m in an open convex set D, and the following hold:*
   (i) *there exists $K \geqslant 0$ and $\alpha \in (0, 1]$ such that $\|F'(x) - F'(y)\| \leqslant K \|x - y\|^\alpha$ for all $x, y \in D$;*
   (ii) *there is a constant B for which $\|F'(x)^+\| \leqslant B$ for all $x \in D$.*

Augmented Jacobian hypothesis. *F is differentiable and*

$$\begin{bmatrix} F'(x) \\ V \end{bmatrix}$$

is nonsingular in an open convex set D, and the following hold:
   (i) *there exists $K \geqslant 0$ and $\alpha \in (0, 1]$ such that $\|F'(x) - F'(y)\| \leqslant K \|x - y\|^\alpha$ for all $x, y \in D$;*
   (ii) *there is a constant B for which*

$$\left\| \begin{bmatrix} F'(x) \\ V \end{bmatrix}^{-1} \right\| \leqslant B \quad \text{for all } x \in D.$$

Furthermore for $\eta > 0$ let

$$D_\eta = \{x \in D \mid \|x - y\| < \eta \Rightarrow y \in D\}.$$

Theorem 17 (Walker [72]). *Let F satisfy the normal flow hypothesis and suppose D_η is given for some $\eta > 0$. Then there is an $\varepsilon > 0$ depending only on K, α, B, and η such that if $x_0 \in D_\eta$ and $\|F(x_0)\| < \varepsilon$, then the iterates $\{x_k\}_{k=0}^\infty$ determined by the normal flow algorithm (13) are well defined and converge to a point $x^* \in D$ such that $F(x^*) = 0$. Furthermore, there is a constant β for which*

$$\|x_{k+1} - x^*\| \leqslant \beta \|x_k - x^*\|^{1+\alpha}, \quad k = 0, 1, 2, \dots .$$

Theorem 18 (Walker [72]). *Let F satisfy the augmented Jacobian hypothesis and suppose D_η is given for some $\eta > 0$. Then there is an $\varepsilon > 0$ depending only on K, α, B, and η such that if $x_0 \in D_\eta$ and $\|F(x_0)\| < \varepsilon$, then the iterates $\{x_k\}_{k=0}^\infty$ determined by the augmented Jacobian algorithm (14) are well defined and converge to a point $x^* \in D$ such that $F(x^*) = 0$. Furthermore, there is a constant β for which*

$$\|x_{k+1} - x^*\| \leqslant \beta \|x_k - x^*\|^{1+\alpha}, \quad k = 0, 1, 2, \dots .$$

In fact, the latter theorem is a consequence of the previous one. The augmented Jacobian algorithm applied to F is equivalent to the normal flow algorithm applied to

$$\bar{F}(x) = \begin{bmatrix} F(x) \\ V(x - x_0) \end{bmatrix}.$$

The Moore–Penrose inverse is not the only possibility to define a Newton step in the underdetermined case. Nashed and Chen [39] suggested the use of outer inverses in a more general setting. Let X and Y be Banach spaces and let $L(X, Y)$ denote the set of all bounded linear operators on X into Y. Let $A \in L(X, Y)$. A linear operator $B : Y \to X$ is said to be an outer inverse, if $BAB = B$. The outer inverse of A will be denoted by $A^\#$. So the Newton method is given in the form

$$x_{k+1} = x_k - F'(x_k)^\# F(x_k), \quad k = 0, 1, 2, \dots . \tag{15}$$

The following result is true.

Theorem 19 (Nashed–Chen [39]). *Let $F : D \subset X \to Y$ be Fréchet differentiable. Assume that there exist an open convex subset D_0 of D, $x_0 \in D_0$, a bounded outer inverse $F'(x_0)^\#$ of $F'(x_0)$ and constants $\eta, K > 0$, such that for all $x, y \in D_0$ the following conditions hold:*

$$\|F'(x_0)^\# F(x_0)\| \leqslant \eta,$$

$$\|F'(x_0)^\# (F'(x) - F'(y))\| \leqslant K \|x - y\|,$$

$$h := K\eta \leqslant \tfrac{1}{2}, \quad S(x_0, t^*) \subset D_0,$$

where $t^* = (1 - \sqrt{1 - 2h})/K$. Then (i) the sequence $\{x_k\}$ defined by (15) with

$$F'(x_k)^\# = [I + F'(x_0)^\# (F'(x_k) - F'(x_0))]^{-1} F'(x_0)^\# \tag{16}$$

lies in $S(x_0, t^*)$ and converges to a solution $x^* \in \overline{S(x_0, t^*)}$ of $F'(x_0)^\# F(x) = 0$; (ii) the equation $F'(x_0)^\# F(x) = 0$ has a unique solution in

$$\tilde{S} \cap \{R(F'(x_0)^\#) + x_0\},$$

where

$$\tilde{S} = \begin{cases} \overline{S(x_0, t^*)} \cap D_0 & \text{if } h = \frac{1}{2} \\ S(x_0, t^{**}) \cap D_0 & \text{if } h < \frac{1}{2} \end{cases}$$

$$R(F'(x_0)^\#) + x_0 = \{x + x_0 \mid x \in R(F'(x_0)^\#)\}$$

and

$$t^{**} = (1 + \sqrt{1 - 2h})/K;$$

(iii) *the speed of convergence is quadratic:*

$$\|x^* - x_{k+1}\| \leqslant \frac{1}{1 - Kt^*} \frac{K}{2} \|x^* - x_k\|^2 \quad (k = 0, 1, 2, \ldots).$$

For related results, we also refer to [9] where a generalization of Rall's Theorem 6 can be found.

Finally we mention that Tapia [63] proved the convergence of the Newton method when the left inverse of $F'(x)$ is used. By extending the Gragg–Tapia results [20] Paardekooper [43] gave a Kantorovich-type inclusion region for the zero of $F(x) = 0$ ($F : X \to Y$) when X and Y are Hilbert spaces, and the right inverse of $F'(x)$ is used.

6. Newton's method at singular points

Let X be a Banach space and $F : X \to X$. Assume that $F(x^*) = 0$ and the Jacobian $F'(x^*)$ is singular. The solution x^* is then called multiple or singular or non-isolated. Such situations may occur, for example, in the Bairstow method (see Blish–Curry [1, pp. 47–60]). The case of multiple zeros was first investigated by Rall [49]. Later Reddien [51] found a basic result which initiated an intensive research into singularities (see [22]). Here we recall only Reddien's result.

Assume that F is C^3 and $F'(x^*)$ has finite-dimensional null space N and closed range R so that $X = N \oplus R$. Let P_N denote a projection onto N parallel to R, and let $P_R = I - P_N$. The singular set of $F'(x)$ near x^* may range from a single point to a codimension one smooth manifold through x^*. Hence the nonsingularity of F' can be expected only in carefully selected regions about x^*. An added difficulty is that the Newton iterates must remain in the chosen region of invertability of F'. The following set satisfies both requirements:

$$W_{\rho,\theta} = \{x \in X \mid 0 < \|x - x^*\| \leqslant \rho, \|P_R(x - x^*)\| \leqslant \theta \|P_N(x - x^*)\|\}. \tag{17}$$

Theorem 20 (Reddien [52]). *Assume that*
(i) $\dim(N) = 1$;
(ii) $F''(x^*)(N, N) \cap R = \{0\}$;
(iii) *there is $c > 0$ so that for all $\phi \in N$, $x \in X$, $\|F''(x^*)(\phi, x)\| \geqslant c\|\phi\| \|x\|$.*

Then for ρ and θ sufficiently small, $F'(x)^{-1}$ exists for $x \in W_{\rho,\theta}$, the map $G(x)=x-F'(x)^{-1}F(x)$ takes $W_{\rho,\theta}$ onto itself, and there is $c_1 > 0$ such that $\|F'(x)^{-1}\| \leqslant c_1\|x - x^*\|^{-1}$ for all $x \in W_{\rho,\theta}$. Moreover if $x_0 \in W_{\rho,\theta}$ and $x_k = G(x_{k-1})$ for $k \geqslant 1$, the sequence x_k converges to x^* and the following hold:

$$\|P_R(x_k - x^*)\| \leqslant c_2\|x_{k-1} - x^*\|^2,$$

$$\lim_{k\to\infty} \|P_N(x_k - x^*)\|/\|P_N(x_{k-1} - x^*)\| = \tfrac{1}{2}.$$

Also, x^* is the only solution to equation $F(x)=0$ in the ball $\overline{S(x^*,\rho)}$.

Reddien's result indicates that the convergence region around x^* must have quite a special structure. Griewank [21] constructed an open star-like domain of initial points, from which the Newton method converges linearly to x^*. Griewank [22] provides a comprehensive survey of the singularity results.

7. The continuous Newton method

Gavurin [42] was the first to consider the continuous analogue of the Newton method

$$x'(t) = -[F'(x)]^{-1}F(x), \quad x(0) = x_0. \tag{18}$$

Let $x(t,x_0)$ denote the solution of (18) such that $x(0,x_0) = x_0$. We assume that $x(t,x_0)$ is defined on the maximum interval $[0,M)$. This solution satisfies the first integral

$$F(x(t,x_0)) = \exp(-t)F(x_0).$$

Hence, the image of the trajectory moves in the direction $F(x_0)$ towards the origin as time proceeds. Along this line the magnitude of $F(x)$ is reduced exponentially. If a solution exists for the interval $[0,\infty)$, then

$$\lim_{t\to\infty} F(x(t,x_0)) = 0.$$

Therefore we may expect that the solution approaches the set $V = \{x \mid F(x) = 0\}$. We also expect this behavior from the numerical solution of (18). If the explicit Euler method is applied on the grid

$$\{t_{k+1} \mid t_{k+1} = t_k + h_k, \ h_k > 0, \ k = 0, 1, 2,\ldots, \ t_0 = 0\},$$

we have the recursion

$$x_{k+1} = x_k - h_k[F'(x_k)]^{-1}F(x_k) \quad (k = 0, 1, 2,\ldots),$$

which becomes the "discrete" Newton method for $h_k = 1$ ($k \geqslant 0$). This is why the differential equation (18) is called the continuous or global Newton method. There are two questions:
(a) Under what conditions does $x(t,x_0)$ tend to a solution x^* of Eq. (2) as $t \to \infty$?
(b) What discretization method follows the solution path $x(t,x_0)$ to infinity?

Concerning question (a) we present the following results of Tanabe [61]. Let $F : \mathbb{R}^n \to \mathbb{R}^m$ be twice continuously differentiable, $n \geqslant m$ and consider the continuous analogue of the Newton–Ben-Israel method

$$x'(t) = -[F'(x)]^+F(x), \quad x(0) = x_0, \tag{19}$$

where "+" stands for the Moore–Penrose inverse. Let $V_F = \{x \in \mathbb{R}^n \mid F(x) = 0\}$ and $S_F = \{x \in \mathbb{R}^n \mid \operatorname{rank}(F'(x)) = m\}$.

Theorem 21 (Tanabe [61]). *If* $\mathrm{rank}(F'(x^*)) = m$ *for a solution* $x^* \in V_F$ *then there exists a neighborhood* U^* *of* x^* *such that for each* $x_0 \in U^*$ *there exists a solution* $x(t, x_0)$, $0 \leqslant t < \infty$, *of* (19) *with* $x(0, x_0) = x_0$, *and as* t *tends to infinity it always converges to a point in* V_F *which may be different from* x^* *in the case where* $m < n$.

Theorem 22 (Tanabe [61]). *For a given* $x_0 \in S_F$, *there exists a solution* $x(t, x_0)$, $0 \leqslant t < M$, *of* (19) *with* $x(0, x_0) = x_0$. *As* t *tends to* M, *its trajectory either* (i) *converges to a solution* $x^* \in V_F \cap S_F$, *in which case we have* $M = \infty$ *and*

$$\|x(t, x_0) - x^*\| \leqslant k\|F(x_0)\|\exp(-t), \quad 0 \leqslant t < \infty$$

for some positive number k, *or* (ii) *approaches the set*

$$S = \{x \in \mathbb{R}^n \mid \mathrm{rank}(F'(x)) < m\}$$

of singular points of (19), *or* (iii) *diverges.*

The case $m > n$ is investigated in Tanabe [62]. We also mention that for $F : \mathbb{R}^n \to \mathbb{R}^n$ and a certain open bounded region $\Omega \subset \mathbb{R}^n$, Smale gave boundary conditions on $\partial\Omega$ under which the solution of (18) leads to a zero point x^* of F in Ω (see [2,23]).

Concerning question (b) we only note that it is not at all easy although the application of any sophisticated ODE solver seems straightforward. A list of papers dealing with the numerical implementation of the continuous Newton method is given in [62] (see also [2]). For derivation and theory of the continuous Newton method we refer to [2,11]. A quantitative analysis of the solution flow $x(t)$ of (18) in the presence of parameters is given in [25].

8. The Newton method for nonsmooth equations

Nonsmooth equations arise concerning various problems such as the nonlinear complementarity problem, variational inequalities and the Karush–Kuhn–Tucker system (see, e.g. [44,26]). The nonsmooth Newton methods are defined for functions $F : \mathbb{R}^n \to \mathbb{R}^m$ which are locally Lipschitz continuous. In such cases F is almost everywhere differentiable by Rademacher's theorem [15,54] and it is possible to use various kinds of generalized derivatives.

Suppose $F : \mathbb{R}^n \to \mathbb{R}^m$ is a locally Lipschitzian function and let D_F denote the set of points at which F is differentiable. Let

$$\partial_B F(x) = \left\{ \lim_{x_i \to x, \; x_i \in D_F} F'(x_i) \right\}.$$

Let ∂F be the generalized Jacobian of F in the sense of Clarke [12]. Then $\partial F(x)$ is the convex hull of $\partial_B F(x)$,

$$\partial F(x) = \mathrm{conv}\, \partial_B F(x).$$

Let us denote

$$\partial_b F(x) = \partial_B F_1(x) \times \partial_B F_2(x) \times \cdots \times \partial_B F_m(x).$$

The classical directional derivative of F is defined by

$$F'(x; h) = \lim_{t \downarrow 0} \frac{F(x + th) - F(x)}{t}.$$

The function F is called semismooth at x if F is locally Lipschitzian at x and

$$\lim_{V \in \partial F(x+th'), h' \rightarrow h, t \downarrow 0} \{Vh'\}$$

exists for any $h \in \mathbb{R}^n$.

Suppose now that $F : \mathbb{R}^n \rightarrow \mathbb{R}^n$. The first nonsmooth Newton method is defined by

$$x_{k+1} = x_k - V_k^{-1} F(x_k) \quad (V_k \in \partial F(x_k), \ k = 0, 1, 2, \ldots). \tag{20}$$

An extension of the classical Newton–Kantorovich theorem is the following.

Theorem 23 (Qi–Sun [47]). *Suppose that F is locally Lipschitzian and semismooth on $\overline{S(x_0, r)}$. Also suppose that for any $V \in \partial F(x)$, $x, y \in \overline{S(x_0, r)}$, V is nonsingular,*

$$\|V^{-1}\| \leqslant \beta, \quad \|V(y - x) - F'(x; y - x)\| \leqslant K\|y - x\|,$$

$$\|F(y) - F(x) - F'(x; y - x)\| \leqslant \delta\|y - x\|,$$

where $q = \beta(\gamma + \delta) < 1$ and $\beta\|F(x_0)\| \leqslant r(1 - q)$. Then the iterates (20) remain in $\overline{S(x_0, r)}$ and converge to the unique solution x^* of $F(x)$ in $\overline{S(x_0, r)}$. Moreover, the error estimate

$$\|x_k - x^*\| \leqslant \frac{q}{1 - q}\|x_k - x_{k-1}\|$$

holds for $k = 1, 2, \ldots$.

A modified nonsmooth Newton method is defined by

$$x_{k+1} = x_k - V_k^{-1} F(x_k) \quad (V_k \in \partial_B F(x_k), \ k = 0, 1, 2, \ldots). \tag{21}$$

Qi [48] proved the local superlinear convergence of this method. For the case $F : \mathbb{R}^n \rightarrow \mathbb{R}^m$ Chen et al. [9] suggested the following nonsmooth algorithm:

$$x_{k+1} = x_k - V_k^{\#} F(x_k) \quad (V_k \in \partial_B F(x_k), \ k = 0, 1, 2, \ldots), \tag{22}$$

where $V_k^{\#}$ denotes the outer inverse of V_k. Convergence results and numerical experiments can be found in [9].

Papers [44,26,9], contain further references to nonsmooth Newton papers.

9. The convergence and divergence of the Newton method

Under the standard assumptions the Newton method is locally convergent in a suitable sphere centered at the solution. We may ask however for the set of all points x_0 from which the Newton method is converging to a solution. The continuous Newton methods provide a possibility to characterize the set of convergence points. The case $\mathbb{R}^n$ is investigated in [2] (see also [62]). Braess [7] studies the case of complex polynomials. Another possibility is to use the results and techniques of iteration theory (see e.g. [65]). The best results in this direction are obtained for real and complex polynomials. The following observation indicates the difficulty of the convergence problem.

Theorem 24 (Rényi [52]). *Let $f : \mathbb{R} \to \mathbb{R}$ be defined on $(-\infty, +\infty)$. Let us suppose that $f''(x)$ is monotone increasing for all $x \in R$ and that $f(x) = 0$ has exactly three real roots A_i $(i = 1, 2, 3)$. The sequence $x_{k+1} = x_k - f(x_k)/f'(x_k)$ converges to one of the roots for every choice of x_0 except for x_0 belonging to an enumerable set E of singular points, which can be explicitly given. For any $\varepsilon > 0$ there exists an interval $(t, t + \varepsilon)$ and in this interval three points a_i, $(t < a_i < t + \varepsilon, \, i = 1, 2, 3)$ having the property that if $x_0 = a_i$, $\{x_k\}$ converges to A_i $(i = 1, 2, 3)$.*

The possibility that a small change in x_0 can cause a drastic change in convergence indicates the nasty nature of the convergence problem. The set of divergence points of the Newton method is best described for real polynomials.

Theorem 25 (Barna [5]). *If f is a real polynomial having all real roots and at least four distinct ones, then the set of initial values for which Newton's method does not yield a root of f is homeomorphic to a Cantor set. The set of exceptional initial values is of Lebesgue measure zero.*

Smale [58] gives a survey of results and related problems for complex polynomials. A geometric interpretation of the complex Newton method and its use for the convergence problem is given in [70] where a list of relevant publications is also given. The probability that the damped Newton method is converging to a zero is investigated in Smale [57] for complex polynomials.

10. Error analysis

Lancaster [32], Rokne [55] and Miel [35] investigated the following error propagation model of the Newton method:

$$\xi_{k+1} = \xi_k - [F'(\xi_k) + E_k]^{-1}(F(\xi_k) + e_k) + g_k \quad (k = 0, 1, 2, \ldots),$$

where E_k, e_k and g_k are perturbations and ξ_k is the computed Newton iterate instead of x_k. Under certain assumptions it is shown that the error sequence $\{\|x_k - \xi_k\|\}$ is bounded. If for some $k = p$ and some $l \geqslant 1$, $\xi_p = \xi_{p+l}$ and $x_k \to x^*$, then $\|\xi_k - x^*\| \leqslant \delta_0$ holds for $k \geqslant p$.

Wozniakowski [75] investigates the Newton method on the parametrized nonlinear system

$$F(x) = F(x; d) = 0 \quad (F, x \in C^n, \, d \in C^m), \tag{23}$$

where vector d is the parameter. It is assumed that a simple zero x^* of (23) exists and F is sufficiently smooth in x and d. Let $\{x_k\}$ be a computed sequence of the successive approximations of x^* by an iteration ϕ. Let ζ be the relative computer precision in fl arithmetics. An iteration ϕ is called *numerically stable*, if

$$\overline{\lim_{k}} \|x_k - x^*\| \leqslant \zeta(k_1 \|x^*\| + k_2 \|F'_x(x^*; d)^{-1} F'_d(x^*; d)\| \, \|d\|) + \mathrm{O}(\zeta^2).$$

An iteration ϕ called *well behaved* if there exist $\{\delta x_k\}$ and $\{\delta d_k\}$ such that

$$\overline{\lim_{k}} \|F(x_k + \delta x_k; d + \delta d_k)\| = \mathrm{O}(\zeta^2)$$

and

$$\|\delta x_k\| \leqslant k_3 \zeta \|x_k\|, \quad \|\delta d_k\| \leqslant k_4 \zeta \|d\|.$$

for large k. The k_i values can only depend on n and m $(i = 1, 2, 3, 4)$. If ϕ is well behaved then it is also numerically stable. An algorithm of one Newton step in fl arithmetics is given by

  (i) compute $F(x_k)$, $F'(x_k)$,
 (ii) solve a linear system $F'(x_k)z_k = F(x_k)$,
(iii) set $x_{k+1} = x_k - z_k$.

Let us assume that F is computed by a well-behaved algorithm, that is

$$fl(F(x_k; d)) = (I + \Delta F_k)F(x_k + \Delta x_k; d + \Delta d_k) = F(x_k) + \delta F_k, \tag{24}$$

where $\|\Delta F_k\| \leqslant \zeta K_F$, $\|\Delta x_k\| \leqslant K_x \|x_k\|$, $\|\Delta d_k\| \leqslant K_d \|d\|$ and

$$\delta F_k = \Delta F_k F(x_k) + F'_x(x_k)\Delta x_k + F'_d(x_k)\Delta d_k + O(\zeta^2). \tag{25}$$

Further, let us assume that

$$fl(F'(x_k; d)) = F'(x_k) + \delta F'_k, \quad \delta F'_k = O(\zeta). \tag{26}$$

This means that we do not need a well behaved algorithm for the evaluation of $F'(x_k)$. Finally, let us assume that a computed solution of the linear system $F'(x_k)z_k = F(x_k)$ satisfies

$$(F'(x_k) + \delta F'_k + E_k)z_k = F(x_k) + \delta F_k, \tag{27}$$

where $E_k = O(\zeta)$. Then a computed approximation x_{k+1} from $x_{k+1} = x_k - z_k$ satisfies

$$x_{k+1} = (I + \delta I_k)(x_k - z_k), \tag{28}$$

where δI_k is a diagonal matrix and $\|\delta I_k\| \leqslant C_1 \zeta$, C_1 depends on the norm.

Theorem 26 (Wozniakowski [75]). *If (24), (26) and (27) hold, then the Newton iteration is well behaved. Specifically it produces a sequence $\{x_k\}$ such that*

$$\varlimsup_k \|F(x_{k+1} + \Delta x_k - \delta I_k x_k; d + \Delta d_k)\| = O(\zeta^2),$$

where Δx_k, δI_k and Δd_k are defined by (24) and (28).

A different error model is given by Spellucci [60].

11. Complexity results

The computational complexity of the Newton method was investigated by Kung and Traub [30], Traub and Wozniakowski [66,67]. Kung and Traub investigated real functions f and a class of rational two-evaluation iterations $x_{k+1} = \phi(f)(x_k)$ without memory given in the following form. Let U_0, U_1, U_2 and nonnegative integers h, t, independent of f such that $U_0 : \mathbb{R} \to \mathbb{R}$ is a rational function,

$$U_1(x, y) = \sum_0^l a_i(x)y^i, \tag{29}$$

where $a_i : \mathbb{R} \to \mathbb{R}$ is a rational function, and

$$U_2(x, y, z) = x + \frac{\sum_0^m b_{i,j}(x)y^i z^j}{\sum_0^q c_{i,j}(x)y^i z^j}, \tag{30}$$

where $b_{i,j}, c_{i,j} : \mathbb{R} \to \mathbb{R}$ are rational functions. Then

$$\phi(f)(x) = U_2(x, f^{(h)}(z_0), f^{(t)}(z_1)), \tag{31}$$

where

$$z_0 = U_0(x), \quad z_1 = U_1(x, f^{(h)}(z_0)). \tag{32}$$

Without loss of generality $U_0(x) = x$ and $h = 0$ can be assumed and thus the iteration function ϕ can be written in the form

$$\phi(f)(x) = x + \frac{\sum_0^m b_{i,j}(x) f^i(x)(f^{(t)}(z_1))^j}{\sum_0^q c_{i,j}(x) f^i(x)(f^{(t)}(z_1))^j}. \tag{33}$$

Kung and Traub defined the efficiency measure of an iteration ϕ by

$$e(\phi, f) = \frac{\log_2 p(\phi)}{v(\phi, f) + a(\phi)},$$

where $p(\phi)$ is the order of convergence of ϕ, $v(\phi, f)$ is the evaluation cost and $a(\phi)$ is the combinatory cost. The cost is the number of arithmetic operations. Let $E_2(f)$ denote the optimal efficiency achievable by a rational two-evaluation iteration without memory. Kung and Traub [30] showed that

$$E_2(f) = \max\left(\frac{1}{c(f) + c(f') + 2}, \frac{1}{2c(f) + 5}\right),$$

where $c(f)$ and $c(f')$ are the cost of evaluating f and f', respectively. Depending on the relative cost of evaluation f or f', the optimal efficiency $E_2(f)$ is achieved by either the Newton iteration

$$\gamma(f)(x) = x - \frac{f(x)}{f'(x)},$$

or the Steffensen iteration ψ,

$$\psi(f)(x) = x - \frac{f^2(x)}{f(x + f(x)) - f(x)}.$$

The result of Kung and Traub shed new lights on the intrinsic values of the Newton method. For Banach spaces the complexity of the Newton method is investigated in [66,67].

Acknowledgements

The author wishes to thank Prof. T. Yamamoto for pointing out references [73,74], and Prof. E. Spedicato for his comments. The author is also indebted to the unknown referees for their several useful remarks which helped him to improve the paper.

References

[1] E.L. Allgower, K. Georg (Eds.), Computational Solution of Nonlinear Systems of Equations, Lecture Notes in Applied Mathematics, Vol. 26, AMS, Providence, RI, 1990.

[2] E.L. Allgower, K. Georg, Numerical Continuation Methods, Springer, Berlin, 1990.

[3] I.K. Argyros, On Newton's method and nondiscrete mathematical induction, Bull. Austral. Math. Soc. 38 (1988) 131–140.

[4] I.K. Argyros, On Newton's method under mild differentiability conditions and applications, Appl. Math. Comput. 102 (1999) 177–183.

[5] B. Barna, Über die Divergenzpunkte des Newtonschen Verfahrens zur Bestimmung von Wurzeln algebraischer Gleichungen, I, Publ. Math. Debrecen 3 (1953) 109–118; II, 4 (1956) 384–397; III, 8 (1961) 193–207; IV, 14 (1967) 91–97.

[6] A. Ben-Israel, A Newton–Raphson method for the solution of systems of equations, J. Math. Anal. Appl. 15 (1966) 243–252.

[7] D. Braess, Über die Einzugsbereiche der Nullstellen von Polynomen beim Newton–Verfahren, Numer. Math. 29 (1977) 123–132.

[8] P.N. Brown, Y. Saad, Hybrid Krylov methods for nonlinear systems of equations, SIAM J. Sci. Statist. Comput. 11 (1990) 450–481.

[9] X. Chen, Z. Nashed, L. Qi, Convergence of Newton's method for singular smooth and nonsmooth equations using adaptive outer inverses, SIAM J. Optim. 7 (1997) 445–462.

[10] M.T. Chu, On a numerical treatment for the curve-tracing of the homotopy method, Numer. Math. 42 (1983) 323–329.

[11] M.T. Chu, On the continuous realization of iterative processes, SIAM Rev. 30 (1988) 375–387.

[12] F.H. Clarke, Optimization and Nonsmooth Analysis, Wiley, New York, 1983.

[13] J.R. Dennis, R.B. Schnabel, Numerical Methods for Unconstrained Optimization and Nonlinear Equations, Prentice-Hall, Englewood Cliffs, NJ, 1983.

[14] P. Deuflhard, G. Heindl, Affine invariant convergence theorems for Newton's method and extensions to related methods, SIAM J. Numer. Anal. 16 (1979) 1–10.

[15] H. Federer, Geometric Measure Theory, Springer, Berlin, 1969.

[16] I. Fenyő, Über die Lösung der im Banachsen Raume definierten nichtlinearen Gleichungen, Acta Math. Hungar. 5 (1954) 85–93.

[17] A. Frommer, Comparison of Brown's and Newton's method in the monotone case, Numer. Math. 52 (1988) 511–521.

[18] A. Frommer, Verallgemeinerte Diagonaldominanz bei nichtlinearer Funktionen I–II, Z. Angew. Math. Mech. 72 (1992) 431–444, 591–605.

[19] A. Galántai, The global convergence of the ABS methods for a class of nonlinear problems, Optim. Methods Software 4 (1995) 283–295.

[20] W.B. Gragg, R.A. Tapia, Optimal error bounds for the Newton–Kantorovich theorem, SIAM J. Numer. Anal. 11 (1974) 10–13.

[21] A.O. Griewank, Starlike domains of convergence for Newton's method at singularities, Numer. Math. 35 (1980) 95–111.

[22] A.O. Griewank, On solving nonlinear equations with simple singularities or nearly singular solutions, SIAM Rev. 27 (1985) 537–563.

[23] M.W. Hirsch, S. Smale, On algorithms for solving $f(x) = 0$, Commun. Pure Appl. Math. 32 (1979) 281–312.

[24] B. Jankó, Rezolvarea Ecuatiilor Operationale Nelineare in Spatii Banach (Romanian), Editura Academiei RSP, Bucuresti, 1969.

[25] V. Janovský, V. Seige, Qualitative analysis of Newton's flow, SIAM J. Numer. Anal. 33 (1996) 2068–2097.

[26] H. Jiang, L. Qi, X. Chen, D. Sun, Semismoothness and superlinear convergence in nonsmooth optimization and nonsmooth equations, in: G. Di Pillo, F. Gianessi (Eds.), Nonlinear Optimization and Applications, Plenum Press, New York, 1996.

[27] L.V. Kantorovich, G.P. Akilov, Functional Analysis in Normed Spaces, Pergamon, Oxford, 1964.

[28] H.B. Keller, Newton's method under mild differentiability conditions, J. Comput. Systems Sci. 4 (1970) 15–28.

[29] C.T. Kelley, Iterative Methods for Linear and Nonlinear Equations, SIAM, Philadelphia, 1995.

[30] K.T. Kung, J.F. Traub, Optimal order and efficiency for iterations with two evaluations, SIAM J. Numer. Anal. 13 (1976) 84–99.

[31] Lai, Hang-Chin, Wu, Pou-Yah, Error bounds of Newton type process on Banach spaces, Numer. Math. 39 (1982) 175–193.

[32] P. Lancaster, Error analysis for the Newton–Raphson method, Numer. Math. 9 (1966) 55–68.

[33] Yu.V. Lisenko, The convergence conditions of the Newton–Kantorovich method for nonlinear equations with Hölderian linearization (Russian), Dokl. Akad. Nauk Belarusi 38 (1994) 20–24.

[34] G.J. Miel, The Kantorovich theorem with optimal error bounds, Amer. Math. Monthly 86 (1979) 212–215.

[35] G.J. Miel, Unified error analysis for Newton-type methods, Numer. Math. 33 (1979) 391–396.

[36] G.J. Miel, Majorizing sequences and error bounds for iterative methods, Math. Comput. 34 (1980) 185–202.

[37] G.J. Miel, An updated version of the Kantorovich theorem for Newton's method, Computing 27 (1981) 237–244.

[38] I.P. Mysovskikh, On the convergence of L.V. Kantorovich's method of solution of functional equations and its applications (Russian), Dokl. Akad. Nauk SSSR 70 (1950) 565–568.

[39] M.Z. Nashed, X. Chen, Convergence of Newton-like methods for singular operator equations using outer inverses, Numer. Math. 66 (1993) 235–257.

[40] A. Neumaier, Simple bounds for zeros of systems of equations, in: R. Ansorge, Th. Meis, W. Törnig (Eds.), Iterative Solution of Nonlinear Systems of Equations, Lecture Notes in Mathematics, Vol. 953, Springer, Berlin, 1982, pp. 88–105.

[41] J.M. Ortega, The Newton–Kantorovich theorem, Amer. Math. Monthly 75 (1968) 658–660.

[42] J.M. Ortega, W.C. Rheinboldt, Iterative Solution of Nonlinear Equations in Several Variables, Academic Press, New York, 1970.

[43] M.H.C. Paardekooper, An upper and a lower bound for the distance of a manifold to a nearby point, J. Math. Anal. Appl. 150 (1990) 237–245.

[44] J. Pang, L. Qi, Nonsmooth equations: motivation and algorithms, SIAM J. Optim. 3 (1993) 443–465.

[45] F.A. Potra, V. Pták, Sharp error bounds for Newton's process, Numer. Math. 34 (1980) 63–72.

[46] V. Pták, The rate of convergence of Newton's process, Numer. Math. 25 (1976) 279–285.

[47] L. Qi, J. Sun, A nonsmooth version of Newton's method, Math. Programming 58 (1993) 353–367.

[48] L. Qi, Convergence analysis of some algorithms for solving nonsmooth equations, Math. Oper. Res. 18 (1993) 227–244.

[49] L.B. Rall, Convergence of the Newton process to multiple solutions, Numer. Math. 9 (1966) 23–37.

[50] L.B. Rall, A note on the convergence of Newton's method, SIAM J. Numer. Anal. 11 (1974) 34–36.

[51] G.W. Reddien, On Newton's method for singular problems, SIAM J. Numer. Anal. 15 (1978) 993–996.

[52] A. Rényi, On Newton's method of approximation (Hungarian), Mat. Lapok 1 (1950) 278–293.

[53] W.C. Rheinboldt, On a theorem of S. Smale about Newton's method for analytic mappings, Appl. Math. Lett. 1 (1988) 69–72.

[54] R.T. Rockafellar, R.J.-B. Wets, Variational Analysis, Springer, Berlin, 1998.

[55] J. Rokne, Newton's method under mild differentiability conditions with error analysis, Numer. Math. 18 (1972) 401–412.

[56] H. Schwetlick, Numerische Lösung nichtlinearer Gleichungen, VEB DVW, Berlin, 1979.

[57] S. Smale, The fundamental theorem of algebra and complexity theory, Bull. Amer. Math. Soc. 4 (1981) 1–36.

[58] S. Smale, On the efficiency of algorithms of analysis, Bull. Amer. Math. Soc. 13 (1985) 87–121.

[59] S. Smale, Newton's method estimates from data at one point, in: R.E. Ewing, K.I. Gross, C.F. Clyde (Eds.), The Merging Disciplines: New Directions in Pure, Applied, and Computational Mathematics, Springer, New York, 1986, pp. 185–196.

[60] P. Spellucci, An approach to backward analysis for linear and nonlinear iterative methods, Computing 25 (1980) 269–282.

[61] K. Tanabe, Continuous Newton–Raphson method for solving an underdetermined systems of nonlinear equations, Nonlinear Anal. Theory Methods Appl. 3 (1979) 495–503.

[62] K. Tanabe, Global analysis of continuous analogues of the Levenberg–Marquardt and Newton–Raphson methods for solving nonlinear equations, Ann. Inst. Statist Math. Part B 37 (1985) 189–203.

[63] R.A. Tapia, The weak Newton method and boundary value problems, SIAM J. Numer. Anal. 6 (1969) 539–550.

[64] R.A. Tapia, The Kantorovich theorem for Newton's method, Amer. Math. Monthly 78 (1971) 389–392.

[65] G. Targonski, Topics in Iteration Theory, Vandenhoeck & Rupprecht, Göttingen, 1981.

[66] J.F. Traub, H. Wozniakowski, Convergence and complexity of Newton iteration for operator equations, J. Assoc. Comput. Mach. 26 (1979) 250–258.

[67] J.F. Traub, H. Wozniakowski, Convergence and complexity of interpolatory-Newton iteration in a Banach space, Comput. Math. Appl. 6 (1980) 385–400.

[68] T. Yamamoto, Error bounds for Newton's iterates derived from the Kantorovich theorem, Numer. Math. 48 (1986) 91–98.

[69] T. Yamamoto, A method for finding sharp error bounds for Newton's method under the Kantorovich assumptions, Numer. Math. 49 (1986) 203–220.

[70] L. Yau, A. Ben-Israel, The Newton and Halley methods for complex roots, Amer. Math. Monthly 105 (1998) 806–818.

[71] B.A. Vertgeim, On the conditions of Newton's approximation method, Dokl. Akad. Nauk SSSR 110 (1956) 719–722 (in Russian).

[72] H.F. Walker, Newton-like methods for underdetermined systems, in: E.L. Allgower, K. Georg (Eds.), Computational Solution of Nonlinear Systems of Equations, Lecture Notes in Applied Mathematics, Vol. 26, AMS, Providence, RI, 1990, pp. 679–699.

[73] X. Wang, D. Han, On dominating sequence method in the point estimate and Smale's theorem, Sci. Sinica Ser. A (1989) 905–913.

[74] D. Wang, F. Zhao, The theory of Smale's point estimation and its applications, J. Comput. Appl. Math. 60 (1995) 253–269.

[75] H. Wozniakowski, Numerical stability for solving nonlinear equations, Numer. Math. 27 (1977) 373–390.

Journal of Computational and Applied Mathematics 124 (2000) 45–59

JOURNAL OF
COMPUTATIONAL AND
APPLIED MATHEMATICS

www.elsevier.nl/locate/cam

A survey of truncated-Newton methods

Stephen G. Nash[1]

Systems Engineering and Operations Research Department, George Mason University, Fairfax, VA 22030, USA

Received 18 June 1999; received in revised form 8 December 1999

Abstract

Truncated-Newton methods are a family of methods for solving large optimization problems. Over the past two decades, a solid convergence theory has been derived for the methods. In addition, many algorithmic enhancements have been developed and studied, resulting in a number of publicly available software packages. The result has been a collection of powerful, flexible, and adaptable tools for large-scale nonlinear optimization. © 2000 Elsevier Science B.V. All rights reserved.

1. Introduction

Truncated-Newton methods are a family of methods suitable for solving large nonlinear optimization problems. At each iteration, the current estimate of the solution is updated (i.e., a step is computed) by approximately solving the Newton equations using an iterative algorithm. This results in a doubly iterative method: an outer iteration for the nonlinear optimization problem, and an inner iteration for the Newton equations. The inner iteration is typically stopped or "truncated" before the solution to the Newton equations is obtained.

More generally, an "inexact" Newton method computes a step by approximately solving the Newton equations. This need not be done using an iterative method. These definitions, however, are not universal. In some papers, "inexact" Newton methods refer to methods for solving systems of nonlinear equations, and "truncated" Newton methods refer to methods for solving optimization problems. I focus here on truncated-Newton methods and optimization problems.

A truncated-Newton method will be effective if

- a small number of inner iterations is sufficient to produce a "good" step,
- each inner iteration can be performed efficiently,

[1] Partially supported by National Science Foundation grant DMI-9800544.

0377-0427/00/$ - see front matter © 2000 Elsevier Science B.V. All rights reserved.
PII: S 0377-0427(00)00426-X

- the overall method is implemented with appropriate safeguards (a "globalization" strategy) to guarantee convergence to a stationary point or local optimum, in cases where the optimization problem satisfies appropriate assumptions.

These issues motivate much of my discussion.

Choices are available for the components of a truncated-Newton method:

- the globalization procedure (some form of line search or trust region strategy),
- the inner iterative algorithm,
- the preconditioner for the inner algorithm,
- the truncation rule for the inner algorithm,
- the technique for computing or estimating second-derivative information.

These choices provide a great deal of flexibility, and allow the method to be adapted to the optimization problem and the computing environment. "Black-box" software is available, but a sophisticated practitioner can enhance the basic method in many ways when faced with a difficult optimization problem.

In much of this paper I focus on the unconstrained problem

$$\min f(x) \tag{1}$$

since the ideas can be explained more simply in this setting, and many of the ideas carry over directly to the constrained case. The first-order optimality condition for this problem is

$$\nabla f(x) = 0,$$

which is a system of nonlinear equations. For this reason, results for nonlinear equations provide insight in the optimization setting.

Given some guess x_k of a solution x_*, Newton's method computes a step p_k as the solution to the linear system

$$\nabla^2 f(x_k)p = -\nabla f(x_k) \tag{2}$$

and then sets $x_{k+1} \leftarrow x_k + p_k$. In this simple form, Newton's method is not guaranteed to converge.

In a truncated-Newton method, an iterative method is applied to (2), and an approximate solution accepted. In [3], the rate of convergence of the outer iteration is proven to be related to the accuracy with which (2) is solved. The paper [3] focuses on nonlinear equations, but the results apply (with minor modification) to optimization problems. These results clarify the local convergence behavior of a truncated-Newton method (i.e., the behavior of the method when x_k is close to the solution x_*).

If the problem (1) satisfies appropriate assumptions, then global convergence (to a local solution) can be guaranteed in either a line search or a trust region framework by making adjustments to the inner algorithm (see Section 3). (In this paper, "global convergence" for an unconstrained problem means that the limit of the gradient norms is zero.) Building upon this foundation, many practical enhancements can be made to the overall method.

A basic question in a truncated-Newton method is the choice of an inner iterative algorithm for solving (2). Some variant of the linear conjugate-gradient method is almost always used. The conjugate-gradient method is an optimal iterative method for solving a positive-definite linear system $Ap = b$, in the sense that the ith iterate p_i minimizes the associated quadratic function $Q(p) = \frac{1}{2}p^{\mathrm{T}}Ap - p^{\mathrm{T}}b$ over the Krylov subspace spanned by $\{b, Ab, \ldots, A^{i-1}b\}$.

The Hessian matrix $\nabla^2 f(x_k)$ need not be positive definite, so the assumptions underlying the conjugate-gradient method may not be satisfied. However, the Hessian matrix is always symmetric. At a local minimizer of (1), the Hessian is guaranteed to be positive semi-definite; in nondegenerate cases it will be positive definite. Thus, as the solution is approached (and the Newton model for (1) is more accurate and appropriate) we can anticipate that the requirements for the conjugate-gradient method will be satisfied. If the Hessian matrix is not positive definite, then the techniques discussed in Section 5 should be used.

A truncated-Newton method will only be competitive if further enhancements are used. For example, a preconditioner for the linear system will be needed, and the stopping rule for the inner algorithm will have to be chosen so that it is effective both close to and far from the solution. With these enhancements, truncated-Newton methods are a powerful tool for large-scale optimization.

Because of all the choices that can be made in designing truncated-Newton methods, they form a flexible class of algorithms. For this reason, the method can be adapted to the problem being solved. Thus, if "black box" software is not able to solve a problem successfully, it is possible to modify the inner algorithm, the preconditioner, the stopping rule for the inner iteration, or a number of other details to enhance performance.

A constrained optimization problem

$$\begin{aligned} \min \quad & f(x) \\ \text{subject to} \quad & g(x) = 0, \\ & h(x) \geqslant 0, \end{aligned}$$

can be solved using a penalty-barrier method, in which one solves a sequence of unconstrained problems of the form

$$\min_x f(x) + \rho_j \sum g_i(x)^2 - \frac{1}{\rho_j} \sum \log(h_i(x))$$

for an increasing sequence of values of $\rho_j \to \infty$ [19]. Each of the unconstrained problems can be solved using a truncated-Newton method, and so all of the above comments apply in this case. (There are also some new issues; see Section 10.) This is not the only possible approach to constrained problems, but it does indicate the relevance of unconstrained optimization techniques in this setting.

Some applications where truncated-Newton methods have been effective include:

- weather modeling,
- potential-energy minimization,
- molecular geometry,
- multicommodity flow,
- medical imaging,
- molecular conformation.

Truncated-Newton methods have been extended to the infinite-dimensional case, at least in the setting of nonlinear equations. See, for example, [23].

Many of the above ideas are discussed in greater detail in the remainder of the paper. Here is an outline of the topics covered:

- controlling the convergence rate (Section 2),
- guaranteeing convergence (Section 3),

- computing second-derivative information (Section 4),
- handling nonconvex problems (Section 5),
- preconditioning (Section 6),
- parallel algorithms (Section 7),
- practical behavior (Section 8),
- software (Section 9),
- constrained problems (Section 10).

A version of this paper containing an expanded reference list can be obtained from `http://iris.gmu.edu/~snash/` under "New Papers", or by contacting the author.

1.1. Basics

The default norm $|| \cdot ||$ used in this paper is the 2-norm: for a vector $x = (x_1, \ldots, x_n)$, $||x|| = \sqrt{x_1^2 + \cdots + x_n^2}$. All vectors are column vectors.

The conjugate-gradient method for solving a linear system $Ap = b$ is initialized with $p_0 = 0$, $r_0 = b$ (r_i is the ith residual $b - Ap_i$), $v_{-1} = 0$, and $\beta_0 = 0$. Then

For $i = 0, 1, \ldots$

    If stopping rule satisfied, stop

    If $i > 0$ set $\beta_i = r_i^T r_i / r_{i-1}^T r_{i-1}$

    Set $v_i = r_i + \beta_i v_{i-1}$

    Set $\alpha_i = r_i^T r_i / v_i^T A v_i$

    Set $p_{i+1} = p_i + \alpha_i v_i$

    Set $r_{i+1} = r_i - \alpha_i A v_i$

Stopping rules are discussed in Section 2. The algorithm requires the computation of the matrix–vector product Av_i, but other information about A need not be supplied.

A line-search method for solving (1) has the following basic form: Specify some initial guess of the solution x_0. Then

For $k = 0, 1, \ldots$

    If stopping rule satisfied, stop

    Compute a search direction p_k

    Determine an improved estimate of the solution $x_{k+1} = x_k + \alpha_k p_k$

        [line search]

"Improvement" is often measured in terms of the function value $f(x_{k+1})$. For example, the new estimate x_{k+1} might be required to satisfy a "sufficient decrease" condition of the form

$$f(x_{k+1}) \leqslant f(x_k) + \mu \alpha p_k^T \nabla f(x_k)$$

for some $0 < \mu < 1$ [19]. That is, there must be a decrease in the function value that is a fraction of the decrease predicted by the first-order Taylor series approximation to $f(x_k + \alpha p_k)$.

A trust-region method for solving (1) has the following basic form: Specify some initial guess of the solution x_0, and specify Δ_0, the bound on the size of the "trust region", i.e., the bound on the

length of the allowable step at the current iteration. Then
> For $k = 0, 1, \ldots$
>> If stopping rule satisfied, stop
>> Choose p_k so as to minimize some approximation $\psi_k(p) \approx f(x_k + p)$,
>>> subject to the constraint $||p|| \leqslant \Delta_k$
>> Compute x_{k+1} and Δ_{k+1} using p_k.

Algorithms for constrained problems can also be imbedded inside line search or trust region approaches, but the details are more complicated (both practically and theoretically). For more information on these topics, see [19].

2. Controlling the convergence rate

The basic local convergence theorem appeared in [3], in the context of nonlinear equations. Here is an adaptation of that theorem to unconstrained optimization. The definition of a q [strong] rate of convergence can be found in [3].

Theorem 1. *Assume that ∇f is continuously differentiable in a neighborhood of a local solution x_* of (1). In addition, assume that $\nabla^2 f(x_*)$ is nonsingular and that $\nabla^2 f$ is Lipschitz continuous at x_*. Assume that iteration k of the truncated-Newton method computes a step p_k that satisfies*

$$||\nabla f(x_k) + \nabla^2 f(x_k) p_k|| \leqslant \eta_k ||\nabla f(x_k)||$$

for a specified value of η_k; the new estimate of the solution is computed using $x_{k+1} \leftarrow x_k + p_k$. If x_0 is sufficiently close to x_* and $0 \leqslant \eta_k \leqslant \eta_{\max} < 1$ then $\{x_k\}$ converges to x_* q-linearly in the norm $|| \cdot ||_*$ defined by $||v||_* \equiv ||\nabla^2 f(x_*) v||$, with asymptotic rate constant no greater than $\eta_{\max}$. If $\lim_{k \to \infty} \eta_k = 0$, then the convergence is q-superlinear. If $\eta_k = O(||\nabla f(x_k)||^r)$ for $0 < r \leqslant 1$, then the convergence is of order at least $(1 + r)$.

The sequence $\{\eta_k\}$ is referred to as a "forcing" sequence. The theorem shows that there is a direct relationship between the forcing sequence and the rate of convergence of the truncated-Newton method for (1). In [3] the authors suggest using

$$\eta_k = \min\{\tfrac{1}{2}, c||\nabla f(x_k)||^r\}$$

as a practical forcing sequence, where c is a positive constant, and $0 < r \leqslant 1$. This sequence leads to a method with a fast asymptotic convergence rate. However, it is not scale invariant, i.e., the behavior of the truncated-Newton method will change if the objective function $f(x)$ is multiplied by a positive constant.

If the conjugate-gradient method is used for the inner iteration, then the ith inner iteration finds a minimizer of the quadratic model

$$f(x_k + p) \approx f(x_k) + p^{\mathrm{T}} \nabla f(x_k) + \tfrac{1}{2} p^{\mathrm{T}} \nabla^2 f(x_k) p \equiv Q_k(p) \tag{3}$$

over the Krylov subspace spanned by $\{\nabla f(x_k), \ldots, [\nabla^2 f(x_k)]^{i-1} \nabla f(x_k)\}$. The model (3) has a global minimum when the residual of the Newton equations is zero. If p is not the minimizer of the quadratic model, however, the magnitudes of the residual and the quadratic model can be dramatically different [15], and the residual can be a deceptive measure of the quality of the search direction. For this reason, it may be preferable to base a stopping rule on the value of the quadratic model.

Let p_i be the search direction computed at the ith inner iteration, and let $Q_i = Q(p_i)$. The stopping rule suggested in [15] is to accept a search direction if

$$i(Q_i - Q_{i-1})/Q_i \leqslant \eta_k.$$

Quoting from [15]: "This criterion ... compares the reduction in the quadratic model at the current iteration ($Q_i - Q_{i-1}$) with the average reduction per iteration (Q_i/i). If the current reduction is small relative to the average reduction (with 'small' measured by η_k), then the inner iteration is terminated."

Newton's method is based on the Taylor series approximation (3). If this approximation is inaccurate then it may not be sensible to solve the Newton equations accurately. ("Over-solving" the Newton equations will not produce a better search direction.) In this circumstance, the inner algorithm should be truncated after a small number of iterations.

The conjugate-gradient method minimizes the quadratic model (3); in particular, it computes the value of the quadratic model. The quadratic model predicts the amount of decrease that will be obtained in the objective value. The outer iteration will typically compute $f(x_k + p)$, and hence determines the actual decrease in the objective value. By comparing these two quantities, the algorithm can determine if the quadratic model is accurate. If not, an alternative value of the forcing term η_{k+1} can be used at the next outer iteration. A simple rule of this type is used in [16]; more sophisticated approaches are analyzed in [5].

The paper [5] identifies two successful forcing sequences (in the context of solving nonlinear equations, but adapted here for optimization). Let p_k be the search direction at the kth outer iteration, and let α_k be the step length. The first stopping sequence uses $\eta_0 \in [0, 1)$, and then for $k = 0, 1, \ldots$

$$\eta_{k+1} = \frac{|\,\|\nabla f(x_{k+1})\| - \|\nabla f(x_k) + \alpha_k \nabla^2 f(x_k) p_k\|\,|}{\|\nabla f(x_k)\|}.$$

The second uses $\eta_0 \in [0, 1)$, $\gamma \in [0, 1]$, $\phi \in (1, 2)$, and then for $k = 0, 1, \ldots$

$$\eta_{k+1} = \gamma \left(\frac{\|\nabla f(x_{k+1})\|}{\|\nabla f(x_k)\|} \right)^{\phi}.$$

Both are designed to provide good asymptotic performance while at the same time preventing over-solving.

All of these results can be applied directly to the sequence of unconstrained problems that arise when a penalty-barrier method is used to solve a constrained problem. The convergence of the overall penalty-barrier method is discussed in [19]. Convergence results for a gradient-projection method for linearly-constrained problems can be found in [8].

3. Guaranteeing convergence

Convergence can be guaranteed by imbedding a truncated-Newton method in either a line-search or a trust-region framework. This is straightforward, although minor adjustments to the inner algorithm must be made. Basic convergence theorems for line-search and trust-region frameworks can be found (for example) in [19].

Global convergence results for the more general setting of inexact-Newton methods are developed in [4]. I assume here that a truncated-Newton method is used, with the conjugate-gradient method as the inner algorithm.

A variety of convergence results are available for line-search methods. In one such (from [19]), the line search method can be guaranteed to converge (in the sense that the limit of the gradient norms is zero) if the following assumptions are satisfied:

- the level set $S = \{x: f(x) \leqslant f(x_0)\}$ is bounded,
- ∇f is Lipschitz continuous for all $x \in S$,
- the search directions p_k satisfy a sufficient-descent condition

$$-\frac{p_k^{\mathrm{T}} \nabla f(x_k)}{\|p_k\| \cdot \|\nabla f(x_k)\|} \geqslant \varepsilon > 0$$

for some fixed ε,
- the search directions are gradient related: $\|p_k\| \geqslant m\|\nabla f(x_k)\|$ for some fixed $m > 0$,
- the search directions are bounded: $\|p_k\| \leqslant M$ for some fixed M,
- an "appropriate" line search is used.

The first two conditions are assumptions on the optimization problem, and the final condition is independent of the inner algorithm.

Before discussing the other three conditions, it is useful to discuss the Lanczos method. The Lanczos method can be applied to any symmetric matrix A. It determines a sequence of orthogonal matrices V_i and tridiagonal matrices T_i such that

$$V_i^{\mathrm{T}} A V_i = T_i.$$

The Lanczos method is equivalent to the conjugate-gradient method. If $A = \nabla^2 f(x_k)$ and p_i is the result of the ith iteration of the conjugate-gradient method applied to (2), then

$$p_i = -V_i T_i(x_k)^{(-1)} V_i^{\mathrm{T}} \nabla f(x_k).$$

See [21] for further details. The other three conditions for convergence will be satisfied if the eigenvalues of the matrices T_i are uniformly bounded for all x_k:

$$0 < c_1 \leqslant \lambda_{\min}[T_i(x_k)] \leqslant \lambda_{\max}[T_i(x_k)] \leqslant c_2.$$

The upper bound can be guaranteed if the level set S is bounded, and if the Hessian is continuous on S. The lower bound can be guaranteed by making adjustments to the conjugate-gradient method. These ideas are discussed further in Section 5.

These convergence results are based on a "traditional" line search, i.e., the new estimate of the solution is obtained as $x_{k+1} \leftarrow x_k + \alpha p_k$, where α is chosen to ensure that the objective function decreases at every iteration. Convergence can also be proved for algorithms that use a curvilinear line search [9]; the new estimate of the solution is of the form

$$x_{k+1} = x_k + \alpha^2 p_k + \alpha d_k$$

where d_k is a direction of negative curvature (see Section 5). In addition, convergence can be proved for algorithms that use a non-monotone line search [7], where decrease in $f(x_k)$ is not required at every iteration.

When using a trust-region method, fewer adjustments need be made to the conjugate-gradient method. One significant issue, though, is to ensure that the output vector p_k satisfies the trust-region constraint:

$$\|p_k\| \leqslant \Delta_k$$

where Δ_k is a parameter in the trust-region method. A technique for this was suggested by [25], and is outlined here. It is straightforward to prove that the iterates from the conjugate-gradient method increase monotonically in norm as long as the tridiagonal matrix T_i is positive definite:

$$\|p_0\| < \|p_1\| < \|p_2\| < \cdots.$$

Thus, it is easy to determine at which iteration the trust-region constraint is violated, and to choose p_k as the point between p_i and p_{i+1} which exactly satisfies the constraint. Of course, if the termination rule is satisfied before the trust-region constraint is encountered, then the inner algorithm terminates before this occurs. If T_i becomes indefinite, then the quadratic model is unbounded below, and the next step in the inner iteration will cause the trust-region constraint to be violated. In [6] the authors examine more closely what happens in a truncated-Newton method when the trust-region boundary is encountered, and propose alternatives to simply truncating the inner iteration in this case.

Just as in Section 2, all of these results can be applied when a penalty-barrier method is used to solve a constrained problem. Global convergence results for a trust-region method for linearly constrained problems can be found in [8].

4. Computing second-derivative information

The conjugate-gradient method requires the computation or estimation of matrix-vector products involving the Hessian of the objective function

$$w = \nabla^2 f(x_k)v \tag{4}$$

for any vector v. This can be accomplished in a variety of ways.

If the Hessian $\nabla^2 f$ is explicitly available then (4) can be computed directly. This can be especially efficient if the Hessian is sparse. The user must be able to derive and program the formulas for the Hessian to use this technique. The remaining techniques require less effort on the part of the user.

An estimate of (4) can be obtained using finite differencing:

$$w \approx \frac{\nabla f(x_k + hv) - \nabla f(x_k)}{h}$$

for some "small" h. Each matrix–vector product requires one gradient evaluation, since $\nabla f(x_k)$ is already available as the right-hand side of (2). The choice of h is discussed in [19], as are alternative finite-difference formulas. This approach is widely used in practical truncated-Newton methods.

If it is possible to use complex arithmetic, then a more accurate finite-difference approximation to w can be obtained using

$$v_h = x_k + \sqrt{-1}hv, \quad g_h = \nabla f(v_h), \quad w \approx \mathrm{Im}(g_h)/h.$$

With this technique it is possible to choose a very small value of h (e.g., $h = 10^{-16}$) and obtain an estimate of w that is accurate to $O(h)$.

A third alternative is to use automatic differentiation [19] to compute (4). This is an exact calculation (up to the limits of computer arithmetic). The computational cost is comparable to a gradient evaluation, and thus comparable to the finite-difference technique.

5. Nonconvex problems

As was mentioned in Section 3, the conjugate-gradient method is equivalent to the Lanczos method in the sense that

$$p_i = -V_i T_i(x_k)^{(-1)} V_i^{\mathrm{T}} \nabla f(x_k),$$

where $A = \nabla^2 f(x_k)$ and

$$V_i^{\mathrm{T}} A V_i = T_i.$$

Here $\{V_i\}$ is a sequence of orthogonal matrices and $\{T_i\}$ is a sequence of tridiagonal matrices.

If A is positive definite, then the formulas for the conjugate-gradient method correspond to computing the factorization

$$T_i = L_i D_i L_i^{\mathrm{T}}$$

where D_i is diagonal (with positive diagonal entries), and L_i is lower triangular (with ones along the diagonal). This factorization exists if and only if T_i is positive definite.

If T_i is not positive definite then this factorization cannot be computed. The algorithm will break down if a diagonal entry of D_i is zero, and will be numerically unstable if a diagonal entry of D_i is negative. To guarantee convergence (see Section 3) the diagonal entries of D_i must be positive and bounded away from zero.

The same situation occurs for certain implementations of Newton's method. In that setting a variety of proposals have been made that correspond to "modifying" the Hessian (or, equivalently, the factorization) to obtain a new, positive definite matrix that then replaces the Hessian in (2).

Any of these techniques could, in principle, be applied to the factorization of the tridiagonal matrix T_i. This is not usually done, however, because the components of the matrix T_i are generated iteratively, and the matrices T_i and V_i are not stored.

An alternative approach that uses information from two successive iterations of the conjugate-gradient method is developed in [12]. This "modified" conjugate-gradient method is iterative (like the regular conjugate-gradient method), and has many of the same theoretical and practical properties as modified-Newton methods.

It is possible to use a simpler technique, and develop a "modified" method using only information from the current iteration of the conjugate-gradient method. This approach is mentioned in [12]. The drawback to this approach is that the modification to the Hessian can be very large in norm, much larger than if information from two successive iterations is used.

If the tridiagonal matrix T_i is not positive semi-definite, then the matrix D_i must have a negative diagonal entry. This corresponds to a direction of negative curvature, i.e., a vector d satisfying

$$d^{\mathrm{T}}[\nabla^2 f(x_k)]d < 0.$$

Such a direction can be used as part of a search direction, since either d or $-d$ is a direction of nonascent. This idea is discussed in [9].

The trust-region techniques discussed in Section 3 provide an alternative way of handling non-convex problems. If a diagonal entry of D_i is negative, then the quadratic model can be decreased to $-\infty$ by following this direction of negative curvature. Thus, a sufficiently long step along such a direction (or any direction) is guaranteed to violate the trust-region constraint.

The application of these ideas to constrained problems is discussed in Section 10.

6. Preconditioning

The convergence of the conjugate-gradient method is strongly influenced by the condition number of the Hessian (i.e., its extreme eigenvalues), and by the number of distinct eigenvalues of the Hessian. Reducing either of these accelerates the convergence of the method.

Ideally, a preconditioner will be chosen based on the problem being solved. This can require considerable analysis and programming to accomplish, however, and is not suitable for routine cases.

If the Hessian matrix is available, a good "generic" choice of a preconditioner is an incomplete Cholesky factorization. The preconditioner is formed by factoring the Hessian, and ignoring some or all of the fill-in that occurs during Gaussian elimination. It may be necessary to modify the factorization (as discussed in Section 5) so that the preconditioner is positive definite. This idea is discussed in [24].

It is also possible to develop preconditioners based on partial separability in the objective function [2]. (A function $f(x)$ is partially separable if it can be written as the sum of functions $f_i(x)$, each of which has a large invariant subspace.)

If neither of these is possible, "automatic" preconditioners can be developed that do not require Hessian information. These preconditioners are based on quasi-Newton approximation to the Hessian. A quasi-Newton approximation is computed based on vector pairs (s_i, y_i). Traditionally, $s_i = x - \hat{x}$ for some pair of variable values, and $y_i = \nabla f(x) - \nabla f(\hat{x})$, the corresponding difference of gradient values. In the context of a truncated-Newton method, these might be $x = x_k$ and $\hat{x} = x_{k+1}$, i.e., two successive iterates.

It is also possible to use an arbitrary vector s_i with $y_i = \nabla^2 f(x_k)s_i$. At each iteration of the conjugate-gradient method, a matrix–vector product of this form is computed or estimated, and each of these matrix–vector products can be used to help construct a Hessian approximation.

In [13], both these ideas are combined to form a preconditioner. The matrix–vector products from the inner iteration are used to construct a diagonal approximation to the Hessian, using a BFGS update formula in which only the diagonal matrix entries are computed. This is in turn used to initialize a two-step limited-memory BFGS update formula which is the actual preconditioner. The limited-memory update is constructed using pairs in which s_i is the difference between a pair of x-vectors. Precise information is given in [14]. This preconditioner is implemented in the TN/TNBC software discussed in Section 9.

A more elaborate preconditioner is described in [11], based on an m-step limited memory BFGS update, with the (s_i, y_i) pairs chosen as a subset of the matrix–vector products in the inner iteration. Experiments are conducted with various choices of m. The authors propose an algorithm that "dynamically stores the correction pairs so that they are as evenly distributed as possible" among the set of pairs for a complete inner iteration.

The application of these ideas to constrained problems is discussed in Section 10.

7. Parallel algorithms

A parallel algorithm could be obtained by executing each of the steps of the truncated-Newton method in parallel. This would require converting the line search and the conjugate-gradient method so that they execute in parallel. By itself, this is not likely to be an effective strategy, since the steps in these algorithms consist of

- scalar operations,
- vector operations,
- function and gradient evaluations.

The scalar operations cannot be made parallel, and the vector operations do not offer much potential for speed-up on a parallel machine (since communication and synchronization delays could easily wipe out any computational savings obtained).

The function and gradient evaluations offer more hope, but this requires that the person solving the optimization problem be willing and able to compute these values effectively in parallel. For very large and difficult problems, however, this may be essential.

An alternative is to replace the line search and the inner algorithm with alternatives that are better able to exploit parallelism. Ideally, it should be possible to take advantage of both parallel linear algebra computations as well as parallel function and gradient evaluations (that is, simultaneous evaluations of the function and/or gradient on separate processors).

An approach of this type for unconstrained problems is discussed in [16]. In this work, the block conjugate-gradient method is used as the inner algorithm; this is a generalization of the conjugate-gradient method in which a block of vectors (rather than a single vector) is updated at every inner iteration. A simple parallel line search is used to compute $x_{k+1} \leftarrow x_k + \alpha p_k$. If the block size in the block conjugate-gradient method is equal to m, then each inner iteration requires the computation of m independent matrix–vector products (which can be approximated by m independent gradient evaluations). There is also considerable opportunity for parallel linear algebra computations. Each iteration of the line search requires m independent function evaluations.

A hybrid approach (combining parallelism in the algorithm with parallelism in the individual function evaluations) is also possible within the block conjugate-gradient method. The block size m need not be equal to the number of processors. This can be an advantage if the individual function and gradient evaluations can be performed in parallel. For example, suppose that a computer with 32 processors were available, and that each function or gradient evaluation could be spread over 4 processors. Then, if the block-size were chosen as $m = 8$, an inner iteration would require 8 gradient evaluations, each of which would require 4 processors. Thus a total of $4 \times 8 = 32$ processors would be used.

This algorithm is implemented in the software package BTN; see Section 9.

8. Practical behavior

Truncated-Newton methods use an inner iteration to compute a search direction, and thus expend considerable computational effort at each outer iteration. In contrast, nonlinear conjugate-gradient methods and limited-memory quasi-Newton methods use relatively few computations to obtain each search direction. (Precise operation counts can be found in [14].) A basic question is whether the effort per iteration for a truncated-Newton method can be worthwhile.

The tests in [14] compare the truncated-Newton method TN against the limited-memory quasi-Newton method L–BFGS. The tests imply that L–BFGS becomes more effective as the optimization problem (1) becomes more nonlinear. In a sense, the truncated-Newton method is more effective when the quadratic model (3) is more effective.

Attempts have been made to combine the best properties of both these methods. This consisted in

- monitoring the effectiveness of the quadratic model to avoid "over-solving" in cases where the quadratic model is poor,
- using limited-memory quasi-Newton formulas as preconditioners,
- combining both techniques in a single algorithm.

Testing of specific features of truncated-Newton software can be found in [12,13] and, for the parallel case, in [16]. The results of these tests have influenced the development of the software packages mentioned in Section 9. The paper [22] describes software for nonlinear equations, but many of the comments are also applicable to optimization. Tests of truncated-Newton methods for bound-constrained problems can be found in [8].

9. Software

Truncated-Newton software is available for unconstrained and bound constrained problems. The first three packages are available from the Netlib collection (www.netlib.org).

The package TN/TNBC solves both classes of problems. It requires that the user provide a subroutine to evaluate the function value and gradient of the objective function. The algorithm is described in [14]. This software is designed to be easy to use, and does not require or expect customization by the user.

The package TNPACK solves unconstrained minimization problems. In addition to function and gradient information, the user must provide formulas for the Hessian matrix and a user-supplied preconditioner. This software expects the user to supply information about the Hessian so that a preconditioner can be constructed. This can require considerable effort, but with the promise of improved performance.

The package BTN solves unconstrained problems on parallel computers (both shared and distributed memory). It requires that the user provide a (scalar) subroutine to evaluate the function value and gradient of the objective function. In addition BTN can take advantage of a parallel subroutine for the function and gradient if one is provided. The algorithm is described in [16]. This software comes with both easy-to-use and customizable top-level subroutines.

The TRON software [8] solves bound-constrained problems. It requires that the user supply function, gradient, and Hessian information. It uses an incomplete Cholesky factorization as a preconditioner. The software can be obtained from

    www.mcs.anl.gov/~more/tron/

The Lancelot software [1] is a more general package, but a variety of truncated Newton algorithms can be used within it by appropriately selecting software parameters. Considerable customization is possible. Information about this software can be obtained from

www.cse.clrc.ac.uk/Activity/LANCELOT

A variety of software packages for solving nonlinear systems of equations are mentioned in [22]. Software for the quasi-Newton preconditioner in [11] is available from

www.ece.nwu.edu/~nocedal/preqn.html

10. Constrained problems

Many algorithms for constrained optimization problems are built upon algorithms, techniques, or principles from unconstrained optimization. It should not be surprising that truncated-Newton methods can be used in this setting.

One approach is to use a penalty-barrier method to solve the constrained problem [19]. The constrained problem is replaced by a sequence of unconstrained problems, where violations in the constraints are included as penalty terms in the objective function (see Section 1). A truncated-Newton method can then be applied to the sequence of unconstrained problems. Under appropriate assumptions, it is possible to derive complexity results for an algorithm of this type [20].

It has been known for decades that the unconstrained problems become increasingly ill-conditioned as the barrier parameter is increased (i.e., as the solution is approached). This ill-conditioning causes the behavior of the inner algorithm to deteriorate. This ill-conditioning is not inherent, however.

In [17], an approximation to the inverse of the Hessian matrix is derived that can be used within the conjugate-gradient method. With this approximation, the conjugate-gradient method is applied to a linear system whose conditioning reflects that of the underlying optimization problem, and not that of the penalty-barrier problem. The approximation formula requires, though, that an active set be identified (a prediction of the set of constraints that are binding at the solution to the optimization problem).

Similar techniques can be applied within augmented Lagrangian and modified barrier methods [1].

It is also possible to adapt truncated-Newton techniques to constrained methods based on sequential quadratic programming. At each iteration of such a method, the nonlinear constrained problem is approximated by a quadratic program. An inner iterative method can then be applied to the quadratic program. There are a number of choices in how this is done; for example, an interior-point method could be used to solve the quadratic program.

If the quadratic program is solved using a null-space approach, then the conjugate-gradient method would be applied to a linear system with a matrix of the form

$$Z^{\mathrm{T}}HZ$$

where H is an approximation to the Hessian, and Z is a null-space matrix for the Jacobian of the constraints. Optimality conditions for the constrained optimization problem imply that this matrix will be positive semi-definite at the solution of the optimization problem. Indefiniteness can be dealt with as in the unconstrained case.

The matrix $Z^{\mathrm{T}}HZ$ may be a dense matrix even if H and the constraint Jacobian are sparse. Matrix–vector products involving this matrix should be computed in stages:

$$w_1 = Zv, \quad w_2 = Hw_1, \quad w = Z^{\mathrm{T}}w_2.$$

The convergence of the conjugate-gradient method is enhanced if a preconditioner for $Z^\mathrm{T}HZ$ is available. This is more challenging than in the unconstrained case. Since H is an approximation to the Hessian matrix, preconditioners for H will depend on properties of the optimization problem (i.e., this requires input from the user of the software). The null-space matrix Z depends on algorithmic details (i.e., this requires input from the developer of the optimization software). The paper [18] develops preconditioners for $Z^\mathrm{T}HZ$ that combine preconditioning information from these two sources.

The quadratic program might be solved by looking at the combined linear system for the variables and multipliers of the quadratic program, which has a matrix of the form

$$\begin{pmatrix} H & A^\mathrm{T} \\ A & 0 \end{pmatrix}. \tag{5}$$

H is an approximation to the Hessian of the Lagrangian, and A is the Jacobian matrix of the constraints. The matrix (5) is symmetric but indefinite. An iterative method can be applied to this system, derived from the Lanczos method (see Section 3). One possibility is to use SYMMLQ [21], which is designed for symmetric indefinite systems of equations.

It may be difficult to guarantee that the search direction that results is a descent direction. Straight-forward tests require factorizations of $Z^\mathrm{T}HZ$, where Z is a null-space matrix for A. In an iterative method, where matrices are not stored and matrix factorizations are not available, this is not usually possible.

It is also possible to derive preconditioners based on the structure of the system (5) [10].

11. Conclusions, recommendations

Truncated-Newton methods are a flexible set of methods for solving large optimization problems. They are built upon a sound theoretical foundation. They can be adjusted to achieve a desired asymptotic convergence rate, and they can be designed to limit the waste of over-solving at points far from the solution. They can also be customized to the problem being solved.

The easiest way to use a truncated-Newton method is via the software packages discussed in Section 9. If these are not adequate, then perhaps use of parallel software will help. Customization of the method may also be necessary. The greatest improvements in performance can be obtained by improving the preconditioner. In addition, the forcing sequence can be modified, as can the technique used to compute the matrix–vector product. The references in this paper provide much guidance in these areas.

Acknowledgements

I would like to thank an anonymous referee for many helpful and perceptive suggestions.

References

[1] A.R. Conn, N.M. Gould, P.L. Toint, LANCELOT: A Fortran Package for Large-Scale Nonlinear Optimization (Release A), Springer, Berlin, 1992.

[2] M.J. Daydé, J.-Y. L'Excellent, N.I.M. Gould, Element-by-element preconditioners for large partially separable optimization problems, SIAM J. Sci. Comput. 18 (1997) 1767–1787.

[3] R.S. Dembo, S.C. Eisenstat, T. Steihaug, Inexact Newton methods, SIAM J. Numer. Anal. 19 (1982) 400–408.

[4] S.C. Eisenstat, H.F. Walker, Globally convergent inexact Newton methods, SIAM J. Optim. 4 (1994) 393–422.

[5] S.C. Eisenstat, H.F. Walker, Choosing the forcing terms in an inexact Newton method, SIAM J. Sci. Comput. 17 (1996) 16–32.

[6] N.I.M. Gould, S. Lucidi, M. Roma, P.L. Toint, Solving the trust-region subproblem using the Lanczos method, SIAM J. Optim. 9 (1999) 504–525.

[7] L. Grippo, F. Lampariello, S. Lucidi, A truncated Newton method with nonmonotone line search for unconstrained optimization, J. Optim. Theory Appl. 609 (1989) 401–419.

[8] C.-J. Lin, J.J. Moré, Newton's method for large bound-constrained optimization problems, SIAM J. Optim. 9 (1999) 1100–1127.

[9] S. Lucidi, F. Rochetich, M. Roma, Curvilinear stabilization techniques for truncated Newton methods in large scale unconstrained optimization, SIAM J. Optim. 8 (1998) 916–939.

[10] L. Lukšan, J. Vlček, Indefinitely preconditioned inexact Newton method for large sparse equality constrained non-linear programming problems, Numer. Linear Algebra Appl. 5 (1998) 219–247.

[11] J.L. Morales, J. Nocedal, Automatic preconditioning by limited memory quasi-Newton updating, Rep. OTC 97/08, Opt. Tech. Ctr., Northwestern Univ., Evanston, IL, 1997.

[12] S.G. Nash, Newton-type minimization via the Lanczos method, SIAM J. Numer. Anal. 21 (1984) 770–788.

[13] S.G. Nash, Preconditioning of truncated-Newton methods, SIAM J. Sci. Statist. Comput. 6 (1985) 599–616.

[14] S.G. Nash, J. Nocedal, A numerical study of the limited memory BFGS method and the truncated-Newton method for large scale optimization, SIAM J. Optim. 1 (1991) 358–372.

[15] S.G. Nash, A. Sofer, Assessing a search direction within a truncated-Newton method, Oper. Res. Lett. 9 (1990) 219–221.

[16] S.G. Nash, A. Sofer, A general-purpose parallel algorithm for unconstrained optimization, SIAM J. Optim. 1 (1991) 530–547.

[17] S.G. Nash, A. Sofer, A barrier method for large-scale constrained optimization, ORSA J. Comput. 5 (1993) 40–53.

[18] S.G. Nash, A. Sofer, Preconditioning reduced matrices, SIAM J. Mat. Anal. Appl. 17 (1996) 47–68.

[19] S.G. Nash, A. Sofer, Linear and Nonlinear Programming, McGraw-Hill, New York, 1996.

[20] S.G. Nash, A. Sofer, On the complexity of a practical interior-point method, SIAM J. Optim. 8 (1998) 833–849.

[21] C.C. Paige, M.A. Saunders, Solution of sparse indefinite systems of linear equations, SIAM J. Numer. Anal. 12 (1975) 617–629.

[22] M. Pernice, H.F. Walker, NITSOL: A Newton iterative solver for nonlinear systems, SIAM J. Sci. Comput. 19 (1998) 302–318.

[23] E. Sachs, Rates of convergence for adaptive Newton methods, J. Optim. Theory Appl. 48 (1986) 175–190.

[24] T. Schlick, Modified Cholesky factorizations for sparse preconditioners, SIAM J. Sci. Comput. 14 (1993) 424–445.

[25] T. Steihaug, The conjugate gradient method and trust regions in large scale optimization, SIAM J. Numer. Anal. 20 (1983) 626–637.

ELSEVIER

Journal of Computational and Applied Mathematics 124 (2000) 61–95

JOURNAL OF
COMPUTATIONAL AND
APPLIED MATHEMATICS

www.elsevier.nl/locate/cam

Variable metric methods for unconstrained optimization and nonlinear least squares ☆

Ladislav Lukšan[a], Emilio Spedicato[b], *

[a]*Institute of Computer Science, Academy of Sciences of the Czech Republic, Pod vodárenskou věží 2, 182 07 Prague 8, Czech Republic*
[b]*Department of Mathematics, University of Bergamo, 24129 Bergamo, Italy*

Received 20 May 1999; received in revised form 13 November 1999

Abstract

Variable metric or quasi-Newton methods are well known and commonly used in connection with unconstrained optimization, since they have good theoretical and practical convergence properties. Although these methods were originally developed for small- and moderate-size dense problems, their modifications based either on sparse, partitioned or limited-memory updates are very efficient on large-scale sparse problems. Very significant applications of these methods also appear in nonlinear least-squares approximation and nonsmooth optimization. In this contribution, we give an extensive review of variable metric methods and their use in various optimization fields. © 2000 Elsevier Science B.V. All rights reserved.

Keywords: Quasi-Newton methods; Variable metric methods; Unconstrained optimization; Nonlinear least squares; Sparse problems; Partially separable problems; Limited-memory methods

1. Introduction

This paper reviews the efficient class of methods known as variable metric methods or quasi-Newton methods for local unconstrained minimization, i.e., for finding a point $x_* \in \mathbb{R}^n$ such that $F(x_*) = \min_{x \in \mathbb{R}^n} F(x)$ (we consider only local minima). Here $F : \mathbb{R}^n \to \mathbb{R}$ is a twice continuously differentiable objective function and $\mathbb{R}^n$ is an n-dimensional vector space.

☆ Work partly supported by grant GA ČR No. 201/00/0080, by CNR-GNIM visiting professorship program and by MURST Programma di Cofinanziamento.
* Corresponding author.
E-mail addresses: luksan@uivt.cas.cz (L. Lukšan), emilio@unibg.it (E. Spedicato).

0377-0427/00/$ - see front matter © 2000 Elsevier Science B.V. All rights reserved.
PII: S 0377-0427(00)00420-9

Methods for unconstrained minimization are iterative. Starting with an initial point $x_1 \in \mathbb{R}^n$, they generate a sequence $x_i \in \mathbb{R}^n$, $i \in \mathcal{N}$, by the simple process

$$x_{i+1} = x_i + \alpha_i d_i, \tag{1.1}$$

where $d_i \in \mathbb{R}^n$ is a direction vector and $\alpha_i \geqslant 0$ is a scalar, the stepsize ($\mathcal{N}$ is the set of natural numbers). The most efficient optimization methods belong to three classes: the modified Newton, variable metric and conjugate gradient methods. We mention basic properties of these classes here in order to clarify the application of variable metric methods in particular cases.

Modified Newton methods are based on a local quadratic model

$$Q_i(d) = \tfrac{1}{2} d^{\mathrm{T}} G_i d + g_i^{\mathrm{T}} d, \tag{1.2}$$

where $G_i = G(x_i)$ and $g_i = g(x_i)$ are, respectively, the Hessian matrix and the gradient of the objective function $F : \mathbb{R}^N \to \mathbb{R}$ at the point $x_i \in \mathbb{R}^n$. The direction vector $d_i \in \mathbb{R}^n$, $i \in \mathcal{N}$, is chosen to minimize $Q_i(d)$ (approximately) on $\mathbb{R}^n$ or on some subset of $\mathbb{R}^n$. Modified Newton methods converge fast, if they converge, but they have some disadvantages. Minimization of $Q_i(d)$ requires $\mathrm{O}(n^3)$ operations and computation of second-order derivatives can be difficult and time consuming. Moreover, if the Hessian matrices are not positive definite, then simple implementations of modified Newton methods need not be globally convergent. Nevertheless, modified Newton methods can be very efficient for large-scale problems. If $Q_i(d)$ is minimized iteratively, then the matrix–vector products involving G_i can be replaced by numerical differentiation. This leads to truncated Newton methods which do not require computation of second-order derivatives. Moreover, if G_i is sparse, then we need substantially less than $\mathrm{O}(n^3)$ operations for minimization of $Q_i(d)$.

Variable metric methods are based on the local quadratic model

$$Q_i(d) = \tfrac{1}{2} d^{\mathrm{T}} B_i d + g_i^{\mathrm{T}} d, \tag{1.3}$$

where B_i is some positive-definite approximation of G_i. Matrices B_i, $i \in \mathcal{N}$, are constructed iteratively so that B_1 is an arbitrary positive-definite matrix and B_{i+1} is determined from B_i in such a way that it is positive definite, is as close as possible to B_i and satisfies the quasi-Newton condition

$$B_{i+1} s_i = y_i,$$

where $s_i = x_{i+1} - x_i$ and $y_i = g_{i+1} - g_i$. The BFGS formula

$$B_{i+1} = B_i + \frac{y_i y_i^{\mathrm{T}}}{y_i^{\mathrm{T}} s_i} - \frac{B_i s_i (B_i s_i)^{\mathrm{T}}}{s_i^{\mathrm{T}} B_i s_i}$$

is widely used (cf. (2.13) and (2.17)). Variable metric methods have some advantages over modified Newton methods. The matrices B_i are positive definite and so variable metric methods can be forced to be globally convergent. Moreover, we can update the inverse $H_i = B_i^{-1}$ or the Cholesky decomposition $L_i D_i L_i^{\mathrm{T}} = B_i$, instead of B_i itself, using only $\mathrm{O}(n^2)$ operations per iteration. Even when variable metric methods require more iterations than modified Newton methods, they are usually more efficient for small- and moderate-size dense problems.

Conjugate gradient methods, see [51,37,73], use only n-dimensional vectors. Direction vectors $d_i \in \mathbb{R}^n$, $i \in \mathcal{N}$ are generated so that $d_1 = -g_1$ and

$$d_{i+1} = -g_{i+1} + \beta_i d_i, \tag{1.4}$$

where $g_{i+1} = g(x_{i+1})$ is the gradient of the objective function $F : \mathbb{R}^N \to \mathbb{R}$ at the point x_{i+1} and β_i is a suitably defined scalar parameter. Conjugate gradient methods require only $\mathrm{O}(n)$ storage elements and $O(n)$ operations per iteration, but they use more iterations than variable metric methods. Of course, these iterations are less expensive. Conjugate gradient methods are intended for large-scale problems.

In this paper, we review variable metric methods for basic unconstrained optimization problems. Our approach is mainly devoted to the computational aspects, i.e., to the derivation of efficient methods and their implementation; therefore, while we quote a number of fundamental convergence results in the field, the difficult and partly still open field of analysis of convergence is not dealt with at great length. Section 2 is devoted to variable metric methods for dense (small- and moderate-size) problems. In Section 3, we describe various modifications of variable metric methods for large-scale problems. Section 4 concerns the use of variable metric updates for improving the efficiency of methods for nonlinear least squares.

In this paper, properties of variable metric methods are sometimes demonstrated by computational experiments. For this purpose, we used FORTRAN codes TEST14 (22 test problems for general unconstrained optimization), TEST15 (22 test problems for nonlinear least squares) and TEST18 (30 test problems for systems of nonlinear equations) which are described in [62] and can be downloaded from the web homepage `http://www.uivt.cas.cz/~luksan#software`. Computational experiments were realized by using the optimization system UFO [61] (see also the above web homepage).

Optimization methods can be realized in various ways which differ in direction determination and stepsize selection. Line-search and trust-region realizations are the most popular, especially for variable metric methods. A basic framework for these methods is given in the following subsection. (Readers already familiar with this material may wish to skip it.)

1.1. Line-search methods

Line-search methods require the vectors $d_i \in \mathbb{R}^n$, $i \in \mathcal{N}$, to be descent directions, i.e.,

$$c_i \triangleq -g_i^T d_i / \|g_i\| \|d_i\| > 0. \tag{1.5}$$

Then the stepsizes α_i, $i \in \mathcal{N}$, can be chosen in such a way that $\alpha_i > 0$ and

$$F_{i+1} - F_i \leqslant \varepsilon_1 \alpha_i g_i^T d_i, \tag{1.6}$$

$$g_{i+1}^T d_i \geqslant \varepsilon_2 g_i^T d_i, \tag{1.7}$$

where $0 < \varepsilon_1 < \frac{1}{2}$ and $\varepsilon_1 < \varepsilon_2 < 1$ (here $F_{i+1} = F(x_{i+1})$, $g_{i+1} = g(x_{i+1})$, where x_{i+1} is defined by (1.1)). The following theorem, see [32], characterizes the global convergence of line-search methods.

Theorem 1.1. *Let the objective function $F : \mathbb{R}^N \to \mathbb{R}$ be bounded from below and have bounded second-order derivatives. Consider the line-search method* (1.1) *with d_i and α_i satisfying* (1.5)–(1.7). *If*

$$\sum_{i \in \mathcal{N}} c_i^2 = \infty, \tag{1.8}$$

then $\liminf_{i \to \infty} \|g_i\| = 0$.

If d_i is determined by minimizing (1.3), i.e., $d_i = B_i^{-1} g_i$ with B_i positive definite, then (1.8) can be replaced by

$$\sum_{i \in \mathcal{N}} \frac{1}{\kappa_i} = \infty, \tag{1.9}$$

where $\kappa_i = \kappa(B_i)$ is the spectral condition number of the matrix B_i. Note that (1.8) (or (1.9)) is satisfied if a constant $\underline{c} > 0$ (or $\bar{c} > 0$) and an infinite set $\mathcal{M} \subset \mathcal{N}$ exist so that $c_i \geqslant \underline{c}$ (or $\kappa_i < \bar{c}$) $\forall i \in \mathcal{M}$.

Variable metric methods in a line-search realization require the direction vectors to satisfy condition (1.5) and $\|B_i d_i + g_i\| \leqslant \omega_i \|g_i\|$, where $0 \leqslant \omega_i \leqslant \bar{\omega} < 1$ is a prescribed precision (the additional condition $\omega_i \to 0$ is required for obtaining a superlinear rate of convergence). Such vectors can be obtained in two basic ways. If the original problem is of small or moderate size or if it has a suitable sparsity pattern, we can set

$$d_i = -H_i g_i, \tag{1.10}$$

where $H_i = B_i^{-1}$, or use back substitution to solve

$$L_i D_i L_i^{\mathrm{T}} d_i = -g_i \tag{1.11}$$

after Cholesky decomposition of B_i. Otherwise, an iterative method may be preferable. The preconditioned conjugate gradient method is especially suitable. It starts with the vectors $s_1 = 0$, $r_1 = -g_i$, $p_1 = C_i^{-1} r_1$ and uses the recurrence relations

$$\begin{aligned}
q_j &= B_i p_j, \\
\alpha_j &= r_j^{\mathrm{T}} C_i^{-1} r_j / p_j^{\mathrm{T}} q_j, \\
s_{j+1} &= s_j + \alpha_j p_j, \\
r_{j+1} &= r_j - \alpha_j q_j, \\
\beta_j &= r_{j+1}^{\mathrm{T}} C_i^{-1} r_{j+1} / r_j^{\mathrm{T}} C_i^{-1} r_j, \\
p_{j+1} &= C_i^{-1} r_{j+1} + \beta_j p_j
\end{aligned} \tag{1.12}$$

for $j \in \mathcal{N}$. This process is terminated if either $\|r_j\| \leqslant \omega_i \|g_i\|$ (sufficient precision) or $p_j^{\mathrm{T}} q_j \leqslant 0$ (non-positive curvature). In both cases we set $d_i = s_j$. The matrix C_i is a preconditioner which should be chosen to make $B_i C_i$ as well conditioned as possible. Very efficient preconditioners can be based on incomplete Cholesky decomposition, see [5].

1.2. Trust-region methods

Trust-region methods use direction vectors $d_i \in \mathbb{R}^n$, $i \in \mathcal{N}$, which satisfy

$$\|d_i\| \leqslant \Delta_i, \tag{1.13}$$

$$\|d_i\| < \Delta_i \quad \Rightarrow \quad \|B_i d_i + g_i\| \leqslant \omega_i \|g_i\|, \tag{1.14}$$

$$-Q_i(d_i) \geqslant \underline{\sigma} \|g_i\| \min(\|d_i\|, \|g_i\| / \|B_i\|), \tag{1.15}$$

where $0 \leqslant \omega_i \leqslant \bar{\omega} < 1$ and $0 < \underline{\sigma} < 1$ (we consider spectral norms here, but $\|d_i\|$ can be an arbitrary norm). Steplengths $\alpha_i \geqslant 0$, $i \in \mathcal{N}$, in (1.1)) are chosen so that

$$\rho_i(d_i) \leqslant 0 \quad \Rightarrow \quad \alpha_i = 0, \tag{1.16}$$

$$\rho_i(d_i) > 0 \quad \Rightarrow \quad \alpha_i = 1, \tag{1.17}$$

where $\rho_i(d_i) = (F(x_i + d_i) - F(x_i))/Q_i(d_i)$. Trust-region radii $0 < \Delta_i \leqslant \bar{\Delta}$, $i \in \mathcal{N}$, are chosen so that $0 < \Delta_1 \leqslant \bar{\Delta}$ is arbitrary and

$$\rho_i(d_i) < \underline{\rho} \quad \Rightarrow \quad \underline{\beta}\|d_i\| \leqslant \Delta_{i+1} \leqslant \bar{\beta}\|d_i\|, \tag{1.18}$$

$$\rho_i(d_i) \geqslant \underline{\rho} \quad \Rightarrow \quad \Delta_i \leqslant \Delta_{i+1} \leqslant \bar{\Delta}, \tag{1.19}$$

where $0 < \underline{\beta} \leqslant \bar{\beta} < 1$ and $0 < \underline{\rho} < 1$. The following theorem, see [75], characterizes the global convergence of trust-region methods.

Theorem 1.2. *Let the objective function* $F : \mathbb{R}^N \to \mathbb{R}$ *be bounded from below and have bounded second-order derivatives. Consider the trust-region method* (1.13)–(1.19) *and denote* $M_i = \max(\|B_1\|, \ldots, \|B_i\|)$, $i \in \mathcal{N}$. *If*

$$\sum_{i \in \mathcal{N}} \frac{1}{M_i} = \infty, \tag{1.20}$$

then $\liminf_{i \to \infty} \|g_i\| = 0$.

Note that (1.20) is satisfied if a constant $\bar{B}$ and an infinite set $\mathcal{M} \subset \mathcal{N}$ exist, so that $\|B_i\| \leqslant \bar{B}$, $\forall i \in \mathcal{M}$.

Trust-region methods require the direction vectors to satisfy conditions (1.13)–(1.15). Such vectors can be obtained in three basic ways. The most sophisticated way consists in solving the constrained minimization subproblem

$$d_i = \underset{\|d\| \leqslant \Delta_i}{\arg\min} \, Q_i(d), \tag{1.21}$$

where $Q_i(d)$ is given by (1.2) or (1.3). This approach, which leads to the repeated solution of the equation $(B_i + \lambda I)d_i(\lambda) + g_i = 0$ for selected values of λ, see [66], is time consuming since it requires, on average, 2 or 3 Cholesky decompositions per iteration. Moreover, an additional matrix has to be used. Therefore, easier approaches have been looked for.

One such approach consists in replacing (1.21) by the two-dimensional subproblem

$$d_i = \underset{\|d(\alpha,\beta)\| \leqslant \Delta_i}{\arg\min} \, Q_i(d(\alpha,\beta)), \tag{1.22}$$

where $d(\alpha,\beta) = \alpha g_i + \beta B_i^{-1} g_i$. Subproblem (1.22) is usually solved approximately by the so-called dog-leg methods [25,74].

If the original problem is large then the inexact trust-region method, [90], can be used. This method is based on the fact that the vectors s_j, $j \in \mathcal{N}$, determined by the preconditioned conjugate gradient method (1.12), satisfy the recurrence inequalities

$$s_{j+1}^T C s_{j+1}^T > s_j^T C s_j,$$

$$Q(s_{j+1}) < Q(s_j),$$

where Q is the quadratic function (1.2) or (1.3). Thus a suitable path is generated in the trust region. If $\|s_j\| \leqslant \Delta_i$ and $\|r_j\| \leqslant \omega_i \|g_i\|$, then we set $d_i = s_j$. If $\|s_j\| \leqslant \Delta_i$ and $p_j^T q_j \leqslant 0$, then we set $d_i = s_j + \lambda_j p_j$, where λ_j is chosen in such a way that $\|d_i\| = \Delta$. If $\|d_j\| \leqslant \Delta$ and $\|d_{j+1}\| > \Delta$, then we set $d = d_j + \lambda_j(d_{j+1} - d_j)$, where λ_j is chosen in such a way that $d = \Delta$. Otherwise we continue the conjugate gradient process.

2. Variable metric methods for dense problems

2.1. Derivation of variable metric methods

Variable metric methods were originally developed for general unconstrained minimization of objective functions with dense Hessian matrices. As mentioned above, these methods use positive-definite matrices B_i, $i \in \mathcal{N}$, which are generally constructed iteratively using a least-change update satisfying the quasi-Newton condition $B_i s_i = y_i$, where $s_i = x_{i+1} - x_i$ and $y_i = g_{i+1} - g_i$. This condition is fulfilled by the matrix

$$\tilde{G}_i = \int_0^1 G(x_i + t s_i)\, \mathrm{d}t \tag{2.1}$$

which can be considered as a good approximation of the matrix $G_{i+1} = G(x_{i+1})$. Roughly speaking, the least-change principle guarantees that as much information from previous iterations as possible is saved while the quasi-Newton condition brings new information because it is satisfied by matrix (2.1). Notice that there are many least-change principles based on various potential functions and also that it is not necessary to satisfy the quasi-Newton equation accurately (see Theorem 3.1 and [98]).

More sophisticated quasi-Newton conditions are sometimes exploited, based on the fact that the matrix $G(x_{i+1})$ satisfies the condition

$$G(x_{i+1}) \frac{\mathrm{d}x(t)}{\mathrm{d}t}\bigg|_{t=1} = \frac{\mathrm{d}g(t)}{\mathrm{d}t}\bigg|_{t=1}, \tag{2.2}$$

where $x(t)$ is a smooth curve such that $x(0) = x_i$ and $x(1) = x_{i+1}$, say, and $g(t) = g(x(t))$. Starting from (2.2), Ford and Moghrabi [40] used a polynomial curve $x(t)$ interpolating the most recent iterates together with the gradient curve $g(t)$ determined by using the same interpolation coefficients. In the quadratic case when $x(t_{i-1}) = x_{i-1}$, $x(0) = x_i$ and $x(1) = x_{i+1}$, this approach gives the quasi-Newton equation

$$B_{i+1}\left(s_i + \frac{1}{t_{i-1}(t_{i-1} - 2)} s_{i-1}\right) = y_i + \frac{1}{t_{i-1}(t_{i-1} - 2)} y_{i-1},$$

where $s_{i-1} = x_i - x_{i-1}$ and $y_{i-1} = g_i - g_{i-1}$. The efficiency of this approach strongly depends on the value $t_{i-1} < 0$. Some ways of choosing this value are described in [39,41].

Another approach based on (2.2) was used in [99]. In this case, $x(t) = x_i + ts_i$ and $g(t)$ is a quadratic polynomial interpolating $g(0) = g_i$, $g(1) = g_{i+1}$ and satisfying the condition

$$F_{i+1} - F_i = \int_0^1 s_i^T g(t) \, dt.$$

This approach leads to the quasi-Newton equation

$$B_{i+1} s_i = y_i + \gamma_i \frac{s_i}{\|s_i\|},$$

where $\gamma_i = 3(g_{i+1} + g_i)^T s_i - 6(F_{i+1} - F_i)$.

The simplest way to incorporate function values into the quasi-Newton equation, known as the nonquadratic correction, was introduced in [7]. Consider the function $\phi(t) = F(x_i + ts)$. Using the backward Taylor expansion, we can write $\phi(0) = \phi(1) - \phi'(1) + (1/2)\phi''(\tilde{t})$, where $0 \leqslant \tilde{t} \leqslant 1$. On the other hand, if we write the quasi-Newton condition as

$$B_{i+1} s_i = \frac{1}{\rho_i} y_i, \tag{2.3}$$

then $s_i^T B_{i+1} s_i = s_i^T y_i / \rho_i$. Approximating $s_i^T B_{i+1} s_i$ by $\phi''(\tilde{t})$ obtained from the backward Taylor expansion, we get

$$\rho_i = \frac{s_i^T y_i}{2(F_i - F_{i+1} + s_i^T g_{i+1})}. \tag{2.4}$$

Formula (2.4) was derived in [84]. Similar formulas are also proposed in [7,8]. Alternatively instead of matrices B_i, $i \in \mathcal{N}$, we can construct matrices $H_i = B_i^{-1}$, since the equation $B_i d_i = -g_i$ can easily be solved in this case by setting

$$d_i = -H_i g_i \tag{2.5}$$

To simplify the notation, we now omit the index i and replace the index $i + 1$ by $+$ so that (2.3) can be rewritten in the form

$$H_+ y = \rho s. \tag{2.6}$$

Moreover, we define the scalars a, b, c by

$$a = y^T H y, \quad b = y^T s, \quad c = s^T H^{-1} s. \tag{2.7}$$

In what follows, we will take the nonquadratic correction (2.6) into account, together with a suitable scaling.

Scaling of the matrix H was first introduced in [69]. A simple heuristic idea for scaling is the replacement of H by γH before updating to make the difference $H_+ - \gamma H$ as small as possible. One possibility is to derive γ from (2.6) after premultiplying it by a vector and replacing H_+ by γH. Using the vector y, we obtain

$$\gamma/\rho = b/a. \tag{2.8}$$

Similarly, using the vector $H^{-1} s$, we obtain

$$\gamma/\rho = c/b. \tag{2.9}$$

Another useful value is the geometric mean

$$\gamma/\rho = \sqrt{c/a}. \tag{2.10}$$

It is interesting that these simple values often considerably improve the efficiency of variable metric methods, while more sophisticated formulae, derived by minimization of certain potential functions, usually give worse results, see [57]. Scaling applied in every iteration is inefficient in general, see [78], but can be very useful on very difficult functions, see [81]. Therefore, some selective scaling strategies have been developed. The simplest possibility, scaling only in the first iteration (or preliminary scaling, PS), is proposed in [78]. In [18], it is recommended to use the scaling parameter $\gamma = \max(1, \min(\tilde{\gamma}, \bar{\gamma}))$ in every iteration, where $\tilde{\gamma}$ is a theoretically computed value (e.g. (2.8)–(2.10)) and $\bar{\gamma}$ is a suitable upper bound. This choice follows from the fact that global convergence can be proved in this case (cf. Theorem 2.2). A slightly modified strategy, interval scaling IS, is proposed in [58]. Here the value $\gamma = \tilde{\gamma}$ is used, if $\underline{\gamma} \leqslant \tilde{\gamma} \leqslant \bar{\gamma}$. Otherwise we set $\gamma = 1$. Recommended values $0 < \underline{\gamma} < 1 < \bar{\gamma}$, corresponding to individual formulae (2.8)–(2.10), are also given in [58].

Now, we are in a position to derive a class of scaled variable metric methods satisfying the generalized quasi-Newton condition (2.6). Our problem can be formulated as finding a symmetric least-change update $\Delta H = H_+ - \gamma H$, satisfying the condition $\Delta Hy = \rho s - \gamma Hy$. We can intuitively suppose that the rank of this update should be as small as possible. Since two vectors s and Hy appear in the generalized quasi-Newton condition (2.6), we restrict our attention to rank two updates of the form $\Delta H = \gamma UMU^T$, where $U = [s, Hy]$ and M is a symmetric 2×2 matrix. Substituting this expression into the quasi-Newton condition and comparing the coefficients, we obtain, with η a free parameter

$$\frac{1}{\gamma}H_+ = H + \frac{\rho}{\gamma}\frac{1}{b}ss^T - \frac{1}{a}Hy(Hy)^T + \frac{\eta}{a}\left(\frac{a}{b}s - Hy\right)\left(\frac{a}{b}s - Hy\right)^T. \tag{2.11}$$

Formula (2.11) defines a three-parameter class, the so-called Huang–Oren class of variable metric updates, see [53,69,84]. If we assume ρ and γ to be fixed or computed by (2.4) and (2.8)–(2.10), we get a one-parameter class, the so-called scaled Broyden class (the original Broyden class corresponds to the values $\rho = 1$ and $\gamma = 1$). Three classic values of the parameter η are very popular. Setting $\eta = 0$, we get the scaled DFP [19,36] update

$$\frac{1}{\gamma}H_+ = H + \frac{\rho}{\gamma}\frac{1}{b}ss^T - \frac{1}{a}Hy(Hy)^T. \tag{2.12}$$

Setting $\eta = 1$, we get the scaled BFGS [11,31,46,77] update

$$\frac{1}{\gamma}H_+ = H + \left(\frac{\rho}{\gamma} + \frac{a}{b}\right)\frac{1}{b}ss^T - \frac{1}{b}(Hys^T + s(Hy)^T). \tag{2.13}$$

Setting $\eta = (\rho/\gamma)/(\rho/\gamma - a/b)$, we get the scaled symmetric rank-one (SR1) update

$$\frac{1}{\gamma}H_+ = H + \left(\frac{\rho}{\gamma} - \frac{a}{b}\right)^{-1}\frac{1}{b}\left(\frac{\rho}{\gamma}s - Hy\right)\left(\frac{\rho}{\gamma}s - Hy\right)^T. \tag{2.14}$$

Formula (2.11) gives another idea for scaling. It can be proved, see [69], that if $0 \leqslant \eta \leqslant 1$ and $b/c \leqslant \rho/\gamma \leqslant a/b$, then $\kappa(\tilde{G}H_+) \leqslant \kappa(\tilde{G}H)$, where $\tilde{G}$ is the matrix defined by (2.1) (κ denotes the spectral condition number). It is clear that for (2.8)–(2.10) the inequality $b/c \leqslant \rho/\gamma \leqslant a/b$ holds ($b/c \leqslant a/b$

follows from the Schwartz inequality). A more sophisticated reason for scaling, based on optimal conditioning of the matrix $H^{-1}H_+$, will be mentioned later (see (2.24)).

Writing $\Delta B = B_+ - (1/\gamma)B$, we can write (2.6) in the form $\Delta Bs = (1/\rho)y - (1/\gamma)Bs$. Proceeding as above, we obtain

$$\gamma B_+ = B + \frac{\gamma}{\rho}\frac{1}{b}yy^{\mathrm{T}} - \frac{1}{c}Bs(Bs)^{\mathrm{T}} + \frac{\beta}{c}\left(\frac{c}{b}y - Bs\right)\left(\frac{c}{b}y - Bs\right)^{\mathrm{T}}, \tag{2.15}$$

see (2.11), if we replace H, s, y, η, ρ, γ by B, y, s, β, $1/\rho$, $1/\gamma$, respectively. Using the Woodbury formula, we can prove that $B = H^{-1}$ implies $B_+ = H_+^{-1}$ if and only if the parameters η and β are related by the following *duality* relation:

$$\beta\eta(ac - b^2) + (\beta + \eta)b^2 = b^2. \tag{2.16}$$

For example, setting $\beta = 0$, we get the scaled BFGS update

$$\gamma B_+ = B + \frac{\gamma}{\rho}\frac{1}{b}yy^{\mathrm{T}} - \frac{1}{c}Bs(Bs)^{\mathrm{T}}. \tag{2.17}$$

Variable metric methods for general unconstrained problems are usually realized in the form (2.11), but form (2.15) is also possible. In the second case, the Cholesky decomposition LDL^{T} of the matrix B is updated using $O(n^2)$ operations by the numerically stable method described in [45]. This possibility is very attractive, since positive definiteness can be controlled. However, numerical experiments indicate that the form (2.11) is more efficient, measured by computational time, since cheaper operations are used and stability is not lost. Nevertheless, form (2.15) is the only possible one for sparse problems and for improving the Gauss–Newton method for nonlinear least squares.

2.2. Theoretical properties of variable metric methods

From now on we shall assume that the vectors s and Hy are linearly independent. Otherwise, the generalized quasi-Newton condition (2.6) can be fulfilled by simple scaling. Assuming γ and ρ to be fixed, we have one degree of freedom in the choice of the parameter η (or β). We introduce the critical values

$$\eta^{\mathrm{c}} = \beta^{\mathrm{c}} = \frac{b^2}{b^2 - ac} < 0. \tag{2.18}$$

We can then deduce from (2.16) that $\eta < \eta^{\mathrm{c}}$, $\eta^{\mathrm{c}} < \eta < 0$, $0 \leqslant \eta \leqslant 1$, $1 < \eta$, if and only if $\beta < \beta^{\mathrm{c}}$, $1 < \beta$, $0 \leqslant \beta \leqslant 1$, $\beta^{\mathrm{c}} < \beta < 0$, respectively. Moreover, one can prove, see [80], that the matrix H_+ (or B_+) is positive definite if and only if $b > 0$ and $\eta > \eta^{\mathrm{c}}$ (or $\beta > \beta^{\mathrm{c}}$). Value (2.18) is negative by the Schwartz inequality, since H is assumed to be positive definite and the vectors s and Hy are assumed to be linearly independent. The interval given by $0 \leqslant \eta \leqslant 1$ (or $0 \leqslant \beta \leqslant 1$) defines the so-called *restricted Broyden subclass*, whose updates can be written as convex combinations of the DFP and the BFGS update.

First, we introduce some basic results concerning the scaled Broyden class of variable metric methods. We begin with the quadratic termination property, see [11].

Theorem 2.1. *Let the objective function $F : \mathbb{R}^N \to \mathbb{R}$ be quadratic with positive-definite Hessian matrix G. Consider the variable metric method (1.1) with stepsizes chosen so that $g_{i+1}^{\mathrm{T}}d_i = 0$*

(perfect line search) and direction vectors determined by (2.5) and (2.11). Then there exists an index i, $1 \leq i \leq n$, such that the direction vectors d_j, $1 \leq j \leq i$, are mutually G-conjugate (i.e. $d_j^{\mathrm{T}} G d_k = 0$ whenever $j \neq k$ and $1 \leq j \leq i, 1 \leq k \leq i$) and, moreover, $g_{i+1} = 0$ and $x_{i+1} = x_*$.

In general, the quadratic termination property requires perfect line searches. Since this property seemed essential in the past, many authors proposed variable metric methods keeping this property even without perfect line searches (see [20]). These methods are not used presently since they require expensive computations while quadratic termination was shown to be unnecessary for obtaining a superlinear rate of convergence (cf. Theorem 2.3). Time-consuming perfect line searches are also not used even if they have nice theoretical implications: Dixon [30] proved that all variable metric methods from the Broyden class generate identical points when perfect line searches are used.

Very general global-convergence results for imperfect line searches can be found in [16]. We summarize and generalize them in the following theorem, see [60].

Theorem 2.2. *Consider the variable metric method (1.1) with $B_i d_i = -g_i$, (1.6), (1.7) and (2.15) with $0 < \underline{\gamma} \leq \gamma_i \leq \bar{\gamma}$, $0 < \underline{\rho} \leq \rho_i \leq \bar{\rho}$ and $(1 - \underline{\delta})\beta_i^c \leq \beta_i \leq 1 - \underline{\delta}$, where $0 < \underline{\delta} < 1$. Let the initial point $x_1 \in \mathbb{R}^n$ be chosen so that the objective function $F : \mathbb{R}^n \to \mathbb{R}$ is uniformly convex and has bounded second-order derivatives on the convex hull of the level set $\mathscr{L}_1 = \{x \in \mathbb{R}^n : F(x) \leq F(x_1)\}$. If there exist $k \in \mathcal{N}$ such that $\gamma_i \geq 1 \; \forall i \geq k$, then $\liminf_{i \to \infty} \|g_i\| = 0$.*

The above theorem has some important consequences. First, it cannot be proved when $\beta \geq 1$, which may be related to the bad properties of the DFP method. Secondly, it confirms that values $\beta^c < \beta < 0$ (or $1 < \eta$) are permissible (computational experiments have shown that some particular methods from this subclass are very efficient in practice). Third, the restriction $\gamma \geq 1$ has also a practical consequence and it was used in [18] as an efficient strategy for scaling.

The above theorem has a weakness, namely the fact that it requires uniform convexity of the objective function. Fortunately, global convergence of the line-search method can be controlled by using restarts of the iterative process. If the value c_i, defined by (1.5), is not sufficiently positive, we can replace the unsuitable matrix H_i by an arbitrary well-conditioned positive-definite matrix ($H_i = I$, say). Theorem 2.2 shows that restarting eventually does not occur if the objective function is uniformly convex in a neighborhood of the minimizer.

Another way to guarantee global convergence of the line-search method consists in turning the search direction towards the negative gradient when necessary, i.e., when (1.5) is not satisfied. This idea is realized, e.g., if (2.5) is replaced by the formula $d = -\bar{H}g$ with

$$\bar{H} = H + \sigma\|Hg\|I \quad \text{or} \quad \bar{H} = H + \sigma\|Hg\|\frac{gg^{\mathrm{T}}}{g^{\mathrm{T}}g}, \tag{2.19}$$

where H is a matrix obtained by update (2.11) and $\sigma > 0$ is a small number. Theoretical investigation of such modifications of variable metric methods is given in [76].

An important property of variable metric methods belonging to the Broyden class is their superlinear rate of convergence. Very general results concerning superlinear rate of convergence are given in [14]. We summarize them in the following theorem.

Theorem 2.3. *Let the assumptions of Theorem 2.2 be satisfied with $\rho_i = 1$ and $\gamma_i = 1$ and the line search be implemented in such a way that it always tries the steplength $\alpha_i = 1$ first. Let $x_i \to x_*$ and $G(x)$ be Lipschitz continuous at x_* (i.e. $\|G(x) - G(x_*)\| \leqslant \bar{L}\|x - x_*\|$ for all x from some neighborhood of x_*). Then a value $\underline{\beta} < 0$ exists such that if $\beta_i \geqslant \underline{\beta} \; \forall i \in \mathcal{N}$, then $\lim_{i \to \infty} \|x_{i+1} - x^*\|/\|x_i - x^*\| = 0$.*

This theorem generalizes results proposed in [49], where a superlinear rate of convergence was proved for the restricted Broyden subclass corresponding to the values $0 \leqslant \beta \leqslant 1$ in (2.15) (it also generalizes results given in [26], where only DFP and BFGS symmetric updates are considered). The fact that a superlinear rate of convergence can be obtained for suitable negative values of the parameter β is very useful, since negative values positively influence the global convergence of variable metric methods, see [14].

The statement of Theorem 2.3 is true only if $\rho_i = 1$ and $\gamma_i = 1$. The influence of nonunit values of these parameters on the superlinear rate of convergence of the BFGS method was studied in [68], where it was shown that scaling applied in every iteration eventually requires nonunit values of the stepsize α_i (unless ρ_i and γ_i tend to one). This effect again increases the number of function evaluations.

2.3. Selected variable metric updates

Now we focus our attention to the choice of the value η (or β). Motivated by the above theoretical results, we will assume that $\beta^c < \beta \leqslant 1$ (or $0 \leqslant \eta$), defining the *perfect Broyden subclass*. Among all classic updates (2.12)–(2.14), only the BFGS method can be used in the basic unscaled form. The DFP method requires either accurate line search or scaling in every iteration, otherwise it need not converge. The problem of the unscaled SR1 formula consists in the fact that it does not guarantee positive definiteness of the generated matrices, so that the line search can fail. Therefore, either suitable scaling or a trust-region realization are necessary. Another simple choice

$$\eta = \frac{\rho/\gamma}{\rho/\gamma + a/b} \tag{2.20}$$

is proposed in [52]. This value is self-dual, lies in the restricted Broyden subclass and interpolates properties of both the DFP and the BFGS methods.

Particular variable metric methods are usually obtained by minimizing some potential functions. The most popular, used first in [82], see also [70], is a condition number

$$\kappa(H^{-1}H_+) = \bar{\lambda}(H^{-1}H_+)/\underline{\lambda}(H^{-1}H_+),$$

where H_+ is given by (2.7) and $\bar{\lambda}$ and $\underline{\lambda}$ are the maximum and the minimum eigenvalues, respectively. Writing $\tilde{\eta} = 1 - \eta/\eta^c$ and $\tilde{\omega} = (\rho/\gamma)(c/b)$, we can see that the matrix $H^{-1}H_+$ has $n - 2$ unit eigenvalues and the remaining two eigenvalues $0 < \lambda_1 \leqslant \lambda_2$ are solutions of the quadratic equation $\lambda^2 - (\tilde{\eta} + \tilde{\omega})\lambda + \tilde{\eta}\tilde{\omega}b^2/(ac) = 0$. This fact implies that the ratio λ_2/λ_1 reaches its minimum if $\tilde{\eta} = \tilde{\omega}$ or

$$\eta(ac - b^2) = b^2 \left(\frac{\rho}{\gamma} \frac{c}{b} - 1 \right). \tag{2.21}$$

Taking into account the unit eigenvalues, we can see that the optimal value of η is given by

$$\eta = \frac{bc(\rho/\gamma - b/c)}{ac - b^2} \quad \text{if } b \leqslant \frac{2(\rho/\gamma)ac}{a + (\rho/\gamma)^2 c}, \tag{2.22}$$

$$\eta = \frac{\rho/\gamma}{\rho/\gamma - a/b} \quad \text{if } b > \frac{2(\rho/\gamma)ac}{a + (\rho/\gamma)^2 c} \tag{2.23}$$

(notice that (2.23) corresponds to the SR1 update). This optimally conditioned update was introduced in [20], although formula (2.22) was independently derived in [70].

Formula (2.21) can also be used for deriving the optimal ratio γ/ρ for a given value η, since we can write

$$\frac{\gamma}{\rho} = \frac{bc}{\eta(ac - b^2) + b^2}. \tag{2.24}$$

For $\eta = 1$ (BFGS) we obtain (2.8). For $\eta = 0$ (DFP) we obtain (2.9). For η given by (2.20) (Hoshino) we obtain (2.10). Substituting (2.10) back into (2.20) (or into (2.22)) we get the Oren–Spedicato update $\eta = b/(b + \sqrt{ac})$. Both the Hoshino and the Oren–Spedicato updates lie in the restricted Broyden subclass and, therefore, they are usually less efficient than the BFGS method in the unscaled case. The last case shows us a simple way for obtaining new variable metric updates. By finding the optimal ratio γ/ρ for a given value of η and substituting it back into the expression for η, we get a new update which differs from the original one if γ/ρ is not optimal.

This approach can also be used for the SR1 update. The analysis of update (2.14) shows that the matrix H keeps positive definiteness for b positive if and only if the ratio γ/ρ lies in the union of two disjoint open intervals $0 < \gamma/\rho < b/a$ and $c/b < \gamma/\rho < \infty$, see [85]. Inside each of these two intervals, exactly one value of the ratio γ/ρ exists which satisfies the Oren–Spedicato criterion. We consider only the interval $0 < \gamma/\rho < b/a$, since ratios $c/b < \gamma/\rho < \infty$ lead to unsuitable values $\eta < 0$. The optimal ratio $0 < \gamma/\rho < b/a$ for the SR1 update, derived from (2.23) to (2.24), can be expressed in the form

$$\frac{\gamma}{\rho} = \frac{c}{b}(1 - \sqrt{1 - b^2/(ac)}) = \frac{b}{a} \Big/ (1 + \sqrt{1 - b^2/(ac)}), \tag{2.25}$$

which is the value proposed in [71]. The important property of this optimally scaled SR1 update is the fact that it generates positive-definite matrices. Unfortunately, this update leads to scaling applied in every iteration, which has a negative influence on the superlinear rate of convergence, as mentioned above. Substituting (2.25) in (2.23) (or (2.22)) we get

$$\eta = 1 + 1/\sqrt{1 - b^2/(ac)}. \tag{2.26}$$

This choice lies outside the restricted Broyden subclass and usually gives better results than the BFGS update in the unscaled case (see [56]). Another very efficient modification of the SR1 method is proposed in [3,56]. This is a combination of the SR1 and the BFGS updates which can be written in the form

$$\eta = 1 \quad \text{if } \rho/\gamma \leqslant a/b, \tag{2.27}$$

$$\eta = \frac{\rho/\gamma}{\rho/\gamma - a/b} \quad \text{if } \rho/\gamma > a/b, \tag{2.28}$$

i.e., $\eta = \max(1,(\rho/\gamma)/(\rho/\gamma - a/b))$. In other words, the SR1 update is chosen if and only if it lies in the perfect Broyden subclass.

Another potential function, which has frequently been used for deriving variable metric updates, is the weighted Frobenius norm $\|W^{-1}(\gamma B_+ - B)\|$ with W symmmetric and positive-definite. It was proved, see [42], that this Frobenius norm reaches its minimum on the set of matrices satisfying the generalized quasi-Newton condition (2.6), if and only if

$$\gamma B_+ = B + \frac{wv^{\mathrm{T}} + vw^{\mathrm{T}}}{s^{\mathrm{T}}v} - \frac{w^{\mathrm{T}}s}{s^{\mathrm{T}}v}\frac{vv^{\mathrm{T}}}{s^{\mathrm{T}}v}, \tag{2.29}$$

where $w = (\gamma/\rho)y - Bs$ and $v = Ws$. If the matrix W is chosen so that $v = Ws$ lies in the subspace generated by the vectors y and Bs (i.e., if $v = y + \lambda Bs$, say) we obtain a portion of the scaled Broyden class (2.15). This portion contains variable metric methods for which $p \geqslant 0$, where p is defined by (2.35) below. The relation between λ and β is given by

$$\beta = \frac{b(b - \lambda^2(\gamma/\rho)c)}{(b - \lambda c)^2}. \tag{2.30}$$

For $\beta = 0$ (BFGS) we get $\lambda = \sqrt{(\rho/\gamma)(b/c)}$. For $\beta = 1$ (DFP) we get $\lambda = 0$. For $\beta = (\gamma/\rho)/(\gamma/\rho - c/b)$ (SR1) we get $\lambda = \rho/\gamma$.

If we set $W = I$ in (2.29), we get the Powell symmetric Broyden (PSB) update

$$\gamma B_+ = B + \frac{sw^{\mathrm{T}} + ws^{\mathrm{T}}}{s^{\mathrm{T}}s} - \frac{w^{\mathrm{T}}s}{s^{\mathrm{T}}s}\frac{ss^{\mathrm{T}}}{s^{\mathrm{T}}s}. \tag{2.31}$$

The PSB method does not guarantee positive definiteness of the generated matrices, so that the line search can fail. Therefore, a trust-region realization is necessary. Generally, this method is highly inefficient even if it is superlinearly convergent (and the proof of its superlinear rate of convergence, cf. Theorem 3.1, is much easier than the proof of Theorem 2.3).

Other potential functions have been used for deriving variable metric methods. If $X = H^{-1}H_+$ then [34] shows that the DFP update minimizes the function

$$\psi(X) = \mathrm{trace}(X) - \log(\det(X)), \tag{2.32}$$

on the set of positive-definite matrices H_+ satisfying the quasi-Newton condition $H_+y = d$. Similarly, the BFGS method minimizes (2.32), where $X = B^{-1}B_+$, on the set of positive-definite matrices B_+ satisfying the quasi-Newton condition $B_+d = y$. The functions

$$\sigma(X) = \bar{\lambda}(X)/\sqrt{\det(X)},$$

$$\tau(X) = \mathrm{trace}(X)/(n\underline{\lambda}(X))$$

are both minimized (either for $X = H^{-1}H_+$ or for $X = B^{-1}B_+$) by the optimally scaled SR1 updates, see [95,96] ($\bar{\lambda}$ and $\underline{\lambda}$ are maximum and minimum eigenvalues).

Besides the above potential functions, other principles have been used for the derivation of free parameters in the Broyden class of variable metric methods. Byrd et al. [14] recommend a theoretical value $\beta = \beta^c + (1/c)(1/v^{\mathrm{T}}G^{-1}v)$, where $v = (1/b)y - (1/c)Bd$ and G is the exact Hessian matrix. Unfortunately, the exact Hessian matrix is usually unknown and so must be approximated. In [57], a simple approximation $G \approx (1/\gamma)B$ is used with γ given by (2.10) (with $\rho = 1$). Using the expression

Table 1

η	NS with $\rho = 1$	PS with $\rho = 1$	IS with $\rho = 1$
BFGS	7042–10 409	7182–8008	4162–5059
DFP	26 failures	36 failures	6301–7642
(2.20)	8288–10 701	9538–10 118	4316–4892
(2.22)–(2.23)	7038–9290	6821–7557	4522–5052
(2.26)	5940–9979	5358–6543	4065–5340
(2.27)–(2.28)	5888–9596	5022– 6085	4173–5095
(2.33)	6044–9047	5663–6538	4152–4913
η	NS with (2.4)	PS with (2.4)	IS with (2.4)
BFGS	6800–10 120	6742–7430	4127–5049
DFP	24 failures	36 failures	5027–6102
(2.20)	8648–11 003	8720–9356	4218–4883
(2.22)–(2.23)	7444–9542	6130–6684	4324–4821
(2.26)	6112–10 203	5402–6559	3962–5230
(2.27)–(2.28)	5882–9645	4881–6075	4106–5066
(2.33)	5787–8538	5315–6042	3927–4589

for v, we can write $v^{\mathrm{T}}G^{-1}v \approx \gamma v^{\mathrm{T}}Hv = (\gamma/c)(ac/b^2 - 1)$, which together with (2.18) and (2.16) gives $\eta = (ac\sqrt{c/a} - b^2)/(ac - b^2)$. Keeping the numerator nonnegative, we obtain the formula

$$\eta = \frac{\max(0, \sqrt{c/a} - b^2/(ac))}{1 - b^2/(ac)}. \tag{2.33}$$

Note that the denominator in (2.33) and the same expression in (2.26) are usually replaced by $\max(\varepsilon, 1 - b^2/(ac))$ with ε a small number (10^{-60}, say). This is a safeguard against division by zero caused by round-off errors.

Finally, we notice that the rank-two update classes we have considered so far, namely updates (2.11) and (2.29), are only special cases of the set of solutions of the quasi-Newton equation. Since the quasi-Newton equation can be viewed as a set of n linear systems, each consisting of a single equation and all differing only in the right hand side, the general solution can easily be obtained using the techniques offered by the ABS class of algorithms for linear equations, see [1]. The general formula obtained contains two parameter matrices, see [87], and is equivalent to a formula previously obtained in [2], using the theory of generalized inverses. No new updates in this general class have yet been developed.

Table 1 compares several variable metric methods of the form (2.11) with standard line-search. They are either unscaled (NS) or use preliminary scaling (PS) or interval scaling (IS). Both the value $\rho = 1$ and the nonquadratic correction (2.4) were used. Values of scaling parameter γ have been selected from (2.8) to (2.10) to give the best results for individual methods — i.e., (2.9) for the DFP update and (2.10) for all other updates. Total numbers of iterations and function evaluations for 74 problems (22 from TEST14, 22 from TEST15, 30 from TEST18, [62]) with 20 variables are presented.

Table 1 implies recommendations for the choice of suitable variable metric methods. First, a reasonable scaling strategy, e.g., IS, should be used since it improves efficiency of all investigated updates. Furthermore, if interval scaling is used, then the easily implementable Hoshino method (2.20) is very efficient. Also, update (2.33) is excellent but more complicated (it must be safeguarded against division by zero as shown above). The nonquadratic correction (2.4) improves this update significantly.

An interesting realization of variable metric methods is based on product-form updates. Suppose that $H = ZZ^T$, where Z is a nonsingular square matrix. Then the direction vector $d = -Hg$ can be obtained using three substitutions

$$d = Z\tilde{d}, \quad \tilde{d} = -\tilde{g}, \quad \tilde{g} = Z^T g. \tag{2.34}$$

We write $\tilde{s} = Z^{-1}s = \alpha\tilde{d}$ and $\tilde{y} = Z^T y$ so that $a = \tilde{y}^T\tilde{y}$, $b = \tilde{y}^T\tilde{s}$, $c = \tilde{s}^T\tilde{s}$. If

$$p = \frac{1}{ab}\left(\eta\left(\frac{a}{b} - \frac{\rho}{\gamma}\right) + \frac{\rho}{\gamma}\right) \geq 0, \tag{2.35}$$

$$q = \frac{\rho}{\gamma}\frac{1}{ab}(\eta(ac - b^2) + b^2) \geq 0, \tag{2.36}$$

then the matrix H_+ can be expressed in the form $H_+ = Z_+Z_+^T$, where Z_+ is obtained from Z by a rank one formula. The general update, derived in [20], is rather complicated, but it contains special cases, which have acceptable complexity. Setting $\eta = 0$ (DFP), we get $p = \rho/(\gamma ab)$ and $q = \rho b/(\gamma a)$, so that

$$\frac{1}{\sqrt{\gamma}}Z_+ = Z + \frac{1}{a}Z\left(\sqrt{\frac{\rho a}{\gamma b}}\tilde{s} - \tilde{y}\right)\tilde{y}^T. \tag{2.37}$$

Setting $\eta = 1$ (BFGS), we get $p = 1/b^2$ and $q = \rho c/(\gamma b)$, so that

$$\frac{1}{\sqrt{\gamma}}Z_+ = Z + \frac{1}{b}Z\tilde{s}\left(\sqrt{\frac{\rho b}{\gamma c}}\tilde{s} - \tilde{y}\right)^T. \tag{2.38}$$

Setting $\eta = (\rho/\gamma)(\rho/\gamma - a/b)$ (SR1), we get $p = 0$ and $q = ((\rho/\gamma)c - b)/(b - (\gamma/\rho)a)$, so that

$$\frac{1}{\sqrt{\gamma}}Z_+ = Z + \frac{\sqrt{q} - 1}{(\rho/\gamma)^2 c - 2(\rho/\gamma)b + a}Z\left(\frac{\rho}{\gamma}\tilde{s} - \tilde{y}\right)\left(\frac{\rho}{\gamma}\tilde{s} - \tilde{y}\right)^T. \tag{2.39}$$

Theoretically, it would be possible to invert the above formulas to obtain similar expressions for the matrix $A_+ = Z_+^{-1}$. Unfortunately, the vector $\tilde{y} = Z^T y = (A^T)^{-1}y$, required in that case, cannot be determined without inversion of the matrix A. The BFGS update, obtained by inversion of (2.38), is the only one that allows us to overcome this difficulty by using the following transformation:

$$\sqrt{\gamma}A_+ = A + \frac{1}{c}\tilde{s}\left(\sqrt{\frac{\gamma c}{\rho b}}\tilde{y} - \tilde{s}\right)^T A = A + \frac{1}{c}As\left(\sqrt{\frac{\gamma c}{\rho b}}y - A^T As\right)^T. \tag{2.40}$$

Formulae (2.37)–(2.39) are very advantageous for seeking minima on linear manifolds, when the matrix H is singular and the matrix Z is rectangular. Formula (2.40) is useful for nonlinear least squares.

3. Variable metric methods for large-scale problems

Basic variable metric methods cannot be used for large-scale optimization, since they utilize dense matrices. Therefore, new principles have to be found, which take into account the sparsity pattern of the Hessian matrix. There are three basic approaches: preserving the sparsity pattern by special updates; using classic updates applied to submatrices of lower dimension; and reconstruction of matrices from vectors by limited memory methods. The first approach was initiated in [91], the second was proposed in [48] and the third was introduced in [67].

3.1. Sparse variable metric updates

Preserving a sparsity pattern is a strong restriction, which eliminates some important properties of variable metric methods. In general, updates cannot have a low rank. For instance, a diagonal update of a diagonal matrix, which changes it to satisfy the quasi-Newton condition, can have rank n. Moreover, positive definiteness of the updated matrix can be lost for an arbitrary sparse update, which can again be demonstrated on a diagonal matrix. From this point of view, it is interesting that a superlinear rate of convergence can be obtained even if the quadratic termination property does not hold.

Sparse variable metric updates should satisfy the quasi-Newton condition, not violate symmetry and preserve sparsity. Let us write

$$\mathscr{V}_Q = \{B \in \mathbb{R}^{n \times n}: Bs = y\},$$
$$\mathscr{V}_S = \{B \in \mathbb{R}^{n \times n}: B^T = B\},$$
$$\mathscr{V}_G = \{B \in \mathbb{R}^{n \times n}: G_{ij} = 0 \Rightarrow B_{ij} = 0\}$$

(we assume, that $G_{ii} \neq 0 \; \forall 1 \leqslant i \leqslant n$). Clearly, $\mathscr{V}_Q$, $\mathscr{V}_S$, $\mathscr{V}_G$ are linear manifolds ($\mathscr{V}_S$ and $\mathscr{V}_G$ are subspaces) in $\mathbb{R}^{n \times n}$. We can define orthogonal projections $\mathscr{P}_Q$, $\mathscr{P}_S$, $\mathscr{P}_G$ into $\mathscr{V}_Q$, $\mathscr{V}_S$, $\mathscr{V}_G$ as matrices B_+ minimizing the Frobenius norm $\|B_+ - B\|_F$ on $\mathscr{V}_Q$, $\mathscr{V}_S$, $\mathscr{V}_G$, respectively. Similarly, we can define orthogonal projections $\mathscr{P}_{QS}$, $\mathscr{P}_{QG}$, $\mathscr{P}_{SG}$ and $\mathscr{P}_{QSG}$ into $\mathscr{V}_Q \cap \mathscr{V}_S$, $\mathscr{V}_Q \cap \mathscr{V}_G$, $\mathscr{V}_S \cap \mathscr{V}_G$ and $\mathscr{V}_Q \cap \mathscr{V}_S \cap \mathscr{V}_G$, respectively. It is clear that the requirements laid down on a sparse update are satisfied by the matrix $B^+ = \mathscr{P}_{QSG} B$.

To eliminate the zero elements from the quasi-Newton condition, we define vectors $\mathscr{P}_i s \in \mathbb{R}^n$, $1 \leqslant i \leqslant n$, in such a way that

$$e_j^T \mathscr{P}_i s = e_j^T s, \quad G_{ij} \neq 0,$$
$$e_j^T \mathscr{P}_i s = 0, \quad G_{ij} = 0$$

and we rewrite the quasi-Newton condition in the form

$$e_i^T (B_+ - B) \mathscr{P}_i s = e_i^T (y - Bs), \quad 1 \leqslant i \leqslant n.$$

It can be proved, [27], that the orthogonal projections considered can be expressed as

$$\mathscr{P}_Q B = B + \frac{(y - Bs)s^T}{s^T s},$$
$$\mathscr{P}_S B = \tfrac{1}{2}(B + B^T),$$
$$(\mathscr{P}_G B)_{ij} = B_{ij}, \quad G_{ij} \neq 0,$$

$$(\mathscr{P}_G B)_{ij} = 0, \quad G_{ij} = 0,$$

$$\mathscr{P}_{QS} B = B + \frac{(y - Bs)s^\mathsf{T} + s(y - Bs)^\mathsf{T}}{s^\mathsf{T} s} - \frac{(y - Bs)^\mathsf{T} s}{s^\mathsf{T} s} \frac{ss^\mathsf{T}}{s^\mathsf{T} s},$$

$$\mathscr{P}_{QG} B = B + \mathscr{P}_G(us^\mathsf{T}),$$

$$\mathscr{P}_{SG} B = \mathscr{P}_S \mathscr{P}_G B = \mathscr{P}_G \mathscr{P}_S B,$$

$$\mathscr{P}_{QSG} B = B + \mathscr{P}_G(vs^\mathsf{T} + sv^\mathsf{T}),$$

where $u \in \mathbb{R}^n$ solves the linear system $Du = y - Bs$ with positive-semidefinite diagonal matrix

$$D = \sum_{i=1}^n \|\mathscr{P}_i s\|^2 e_i e_i^\mathsf{T}$$

and $v \in \mathbb{R}^n$ solves the linear system $Qv = y - Bs$ with positive-semidefinite matrix

$$Q = \mathscr{P}_G(ss^\mathsf{T}) + \sum_{i=1}^n \|\mathscr{P}_i s\|^2 e_i e_i^\mathsf{T},$$

which has the same sparsity pattern as the matrix B.

The variable metric method which uses the update

$$B_+ = \mathscr{P}_{QSG} B, \tag{3.1}$$

was proposed in [91]. Realization of this method is time consuming, since an additional linear system has to be solved. Moreover, its convergence properties are not very good, since its variational derivation is similar to the derivation of the inefficient PSB method. Therefore, easier methods with better convergence properties have been looked for. Steihaug [89] has shown that the updates based on the composite projections

$$B^+ = \mathscr{P}_S \mathscr{P}_{QG} B, \tag{3.2}$$

$$B^+ = \mathscr{P}_G \mathscr{P}_{QS} B, \tag{3.3}$$

$$B^+ = \mathscr{P}_{SG} \mathscr{P}_Q B \tag{3.4}$$

and realized in the trust-region framework, lead to methods which are globally and superlinearly convergent. We summarize his results in the following theorem.

Theorem 3.1. *Consider the trust-region method* (1.13)–(1.19), *where* $B_{i+1} = B_i$, *if* (1.16) *holds, or updates of the form* (3.1)–(3.4) *are used, if* (1.17) *holds. Let the objective function* $F : \mathbb{R}^N \to \mathbb{R}$ *be bounded from below and have bounded and Lipschitz continuous second-order derivatives. Then* $\liminf_{i \to \infty} \|g_i\| = 0$. *If, in addition,* $x_i \to x_*$ *and* $\omega_i \to 0$, *see* (1.14), *then* $\lim_{i \to \infty} \|x_{i+1} - x^*\| / \|x_i - x^*\| = 0$.

Unfortunately, a similar result cannot be obtained for a line-search realization, since the hereditary positive definiteness of generated matrices is not guaranteed. Nevertheless, our unpublished experiments indicate that a line-search realization usually outperforms a trust-region implementation. These experiments also imply that update (3.2) is the most efficient one among all composite projections. This fact is also mentioned in [29,94].

Table 2

Method	Iterations	f. eval.	g. eval.	CG steps	CPU time	Failures
LVVM	26 739	27 901	27 901	—	1 : 23	—
LMVM	27 282	31 723	31 723	—	1 : 35	—
LRVM	28 027	30 061	30 061	—	1 : 32	—
SCVM	13 145	27 292	27 292	51 0773	4 : 10	1
SFVM	5308	16 543	41 732	—	1 : 54	1
SPVM	3769	5190	5190	—	0 : 30	—
SDNM	1958	2000	10 238	—	0 : 34	—
STNM	2203	2980	60 420	57 195	1 : 14	—
NCGM	19 974	39 854	39 854	—	1 : 29	—

To eliminate difficulties arising in connection with update (3.1), Tůma has proposed sparse fractioned updates [94]. Let $\mathcal{G} = (V, E)$, $V = \{v_1, \ldots, v_n\}$, $E \in V \times V$, be the adjacency graph of the matrix G so that $(v_i, v_j) \in E$ if and only if $G_{ij} \neq 0$ (structurally). Let $c : V \rightarrow \{1, \ldots, r\}$, $r \leqslant n$ be a colouring of the graph $\mathcal{G}$ so that $c(v_i) \neq c(v_j)$ if and only if $(v_i, v_j) \in E$ (the minimum possible r is the so-called chromatic number of the graph $\mathcal{G}$). This colouring induces a partition $V = \bigcup_{i=1}^r C_i$ where $C_i = \{v \in V : c(v) = i\}$. Assume now that $s = \sum_{i=1}^r s^i$ where $s^i = \sum_{j \in C_i} e_j e_j^T s$ and set

$$B_+ = B^r, \tag{3.5}$$

where

$$x^0 = x, \quad g^0 = g, \quad B^0 = B$$

and

$$x^i = x^{i-1} + s^i, \quad g^i = g(x^i), \quad y^i = g^i - g^{i-1},$$

$$B^i = \mathcal{P}_{Q^i SG} B^{i-1}, \quad \mathcal{V}_{Q^i} = \{B \in \mathbb{R}^{n \times n} : Bs^i = y^i\}$$

for $1 \leqslant i \leqslant r$. As has been already shown, $\mathcal{P}_{Q^i SG} B^{i-1} = B^{i-1} + \mathcal{P}_G(v^i (s^i)^T + s^i (v^i)^T)$, where $Q^i v^i = y^i - B^{i-1} s^i$ and where

$$Q^i = \mathcal{P}_G(s^i (s^i)^T) + \sum_{j=1}^n \|\mathcal{P}_j s^i\|^2 e_j e_j^T = \sum_{j \in C_i} e_j e_j^T s s^T e_j e_j^T + \sum_{j=1}^n \|\mathcal{P}_j s^i\|^2 e_j e_j^T$$

is now a diagonal matrix. Since the matrices Q^i, $1 \leqslant i \leqslant r$ are diagonal, the partial updates $B^i = \mathcal{P}_{Q^i SG} B^{i-1}$, are very simple and can be realized in an efficient way. Notice that this simplicity is compensated by evaluation of intermediate gradients $g^1, \ldots, g^{r-1}$. This is a common feature with the method of approximating sparse Hessian matrices proposed by Coleman and Moré [17]. However, the number of groups induced by colouring c given above can be much smaller than the number of groups induced by the symmetric or lower triangular colouring used by Coleman and Moré. Computational experiments confirm that sparse fractioned updates are more efficient than update (3.1) and than composite projections (3.2)–(3.4) (see Table 2).

Another way of obtaining sparse quasi-Newton updates is described in [35]. This method is based on the minimization of the potential function (2.32), where $X = HB_+$, on the linear manifold $\mathcal{V}_Q \cap \mathcal{V}_S \cap \mathcal{V}_G$. Function (2.32) has two advantages. First, its minimization leads to the efficient BFGS formula in the dense case and, secondly, it serves as a barrier function against losing positive definiteness. Fletcher [35] proved that if the minimum of (2.32) on $\mathscr{V}_Q \cap \mathscr{V}_S \cap \mathscr{V}_G$ exists, it is characterized by the existence of $\lambda \in \mathbb{R}^n$ such that

$$\mathscr{P}_G H_+ = \mathscr{P}_G(H + \lambda s^{\mathrm{T}} + s\lambda^{\mathrm{T}}). \tag{3.6}$$

The vector λ cannot be obtained explicitly in the sparse case. Instead, the nonlinear system of equations $B_+(\lambda)s - y = 0$ must be solved using the Newton method, where $B_+(\lambda)$ is a matrix determined from (3.6). This approach has two difficulties. Firstly, the determination of $B_+(\lambda)$ from $\mathscr{P}_G H_+(\lambda)$ is rather complicated and it requires a sparsity pattern which is not changed during the Cholesky decomposition. Secondly, the nonlinear equations have to be solved with the Jacobian matrix M, say, which has the same pattern as B in general. Therefore, the whole process is time consuming and moreover three sparse matrices B, $\mathscr{P}_G H$ and M are necessary. Nevertheless, numerical experiments in [35] indicate robustness and good convergence properties of this method.

Finally, we observe that the approach based upon use of the ABS algorithm can also provide the general solution of the quasi-Newton equation with sparsity and symmetry conditions, since they are just additional linear equations, see [85,88]. The sparse symmetric update is given in explicit form, while in the approach of, e.g., [91], a sparse linear system has to be solved. By requiring that the diagonal element be sufficiently large, extra linear conditions are given which in general allow us to obtain symmetric sparse quasi-positive-definite updates (i.e., updates where the $(n-1)$th principal submatrix is SPD) and quasi-diagonally dominant updates, see [88,86]. The last result can be used to produce full SPD sparse updates by imbedding the minimization of the function $F(x)$ in a suitable equivalent $(n+1)$-dimensional problem. No particular algorithms or numerical experiments are yet available based upon this approach.

3.2. Partitioned variable metric updates

A quite different approach to large-scale optimization, leading to partitioned updating methods, is proposed in [48]. It is based on properties of partially separable functions of the form

$$F(x) = \sum_{k=1}^{m} f_k(x), \tag{3.7}$$

where each of the element function $f_k(x)$ depends only on n_k variables and n_k is much less than n, the size of the original problem. In this case, we can define packed element-gradients $\hat{g}_k(x) \in \mathbb{R}^{n_k}$ and packed element-Hessian matrices $\hat{G}_k(x) \in \mathbb{R}^{n_k \times n_k}$, $1 \leqslant k \leqslant m$, as dense but small-size vectors and matrices. Such a formulation is highly practical since, e.g., sparse nonlinear least-square problems (see (4.1) below) have this structure.

Partitioned updating methods consider each element function separately and update approximations $\hat{B}_k$, $1 \leqslant k \leqslant m$, of the packed element-Hessian matrices $\hat{G}_k(x)$ using the quasi-Newton conditions $\hat{B}_k^+ \hat{s}_k = \hat{y}_k$, where $\hat{s}_k$ is a part of the vector s consisting of components corresponding to variables of f_k and $\hat{y}_k = \hat{g}_k^+ - \hat{g}_k$ (we use $+$ as the upper index in the partitioned case). Therefore, a variable metric update of the form (2.15) can be used for each of the element functions. However, there are some differences between the classic and the partitioned approach. First, the main reason for partitioned update is an approximation of the element Hessian matrix, so that scaling and nonquadratic corrections do not usually improve efficiency. Secondly, denoting $\hat{b}_k = \hat{y}_k^{\mathrm{T}} \hat{s}_k$, $\hat{c}_k = \hat{s}_k^{\mathrm{T}} \hat{B}_k \hat{s}_k$, we can observe that $\hat{b}_k \geqslant 0$ does not have to be guaranteed for all $1 \leqslant k \leqslant m$. This difficulty is unavoidable and an efficient algorithm has to handle this situation. Therefore, the following partitioned BFGS method is recommended:

$$\hat{B}_k^+ = \hat{B}_k + \frac{1}{\hat{b}_k}\hat{y}_k\hat{y}_k^{\mathrm{T}} - \frac{1}{\hat{c}_k}\hat{B}_k\hat{s}_k(\hat{B}_k\hat{s}_k)^{\mathrm{T}}, \quad \hat{b}_k > 0,$$

$$\hat{B}_k^+ = \hat{B}_k, \quad \hat{b}_k \leqslant 0. \tag{3.8}$$

Another possibility is the partitioned rank one method

$$\hat{B}_k^+ = \hat{B}_k + \frac{1}{\hat{b}_k - \hat{c}_k}(\hat{y}_k - \hat{B}_k\hat{s}_k)(\hat{y}_k - \hat{B}_k\hat{s}_k)^{\mathrm{T}}, \quad |\hat{b}_k - \hat{c}_k| \neq 0,$$

$$\hat{B}_k^+ = \hat{B}_k, \quad |\hat{b}_k - \hat{c}_k| = 0. \tag{3.9}$$

which can be used for indefinite matrices. Usually, the latter method works worse but can be useful in some pathological cases. Therefore, combined methods are welcome. One such combination is proposed in [50]. It starts with the partitioned BFGS update (3.8). When a negative curvature $\hat{b}_k < 0$ appears in some iteration then (3.8) is switched to (3.9) for $\hat{B}_k$ and is kept in all subsequent iterations. We suggest another strategy, which was used in our experiments reported in Table 2. This is based on the observation that (3.8) usually fails in the case when too many elements have indefinite Hessian matrices. Therefore, we start with the partitioned BFGS update (3.8). If $m_{\mathrm{neg}} \geqslant \theta m$, where m_{neg} is a number of elements with a negative curvature and θ is a threshold value, then (3.9) is used for all elements in all subsequent iterations (we recommend $\theta = \frac{1}{2}$).

Partitioned variable metric methods are very efficient for solving real-world problems, but their convergence properties have not yet been satisfactorily investigated. Griewank and Toint [49] have proved a superlinear rate of convergence of partitioned variable metric methods belonging to the restricted Broyden class. Unfortunately, a general global-convergence theory, which would include the most efficient algorithms, e.g., the partitioned BFGS method given above, is not known. Some partial results are given in [92], where global convergence is proved under complicated and restrictive conditions. Some globally convergent modifications of partitioned variable metric methods are also given in [47]. Unfortunately, we have experimentally found that these modifications are computationally less efficient and cannot be competitive with the best strategies given above.

A disadvantage of partitioned variable metric methods is that approximations of packed element-Hessian matrices have to be stored. Therefore, the number of stored elements can be much greater than the number of nonzero elements in the standard sparse pattern. For this reason, it is suitable to construct the standard sparse Hessian approximation before solving a linear system, since a multiplication by a sparse matrix is more efficient than the use of the partitioned structure.

3.3. Variable metric methods with limited memory

Variable metric methods with limited memory are based on the application of a limited number of BFGS updates, which are computed recursively using previous differences s_j, y_j, $i - n \leqslant j \leqslant i - 1$ (i is the iteration number). Their development started by the observation that an application of the BFGS update is equivalent to a conjugate gradient step in the case of perfect line search, see [72], and is more efficient in other cases. In [12,13] a limited number of BFGS steps was used for construction of a suitable preconditioner to the conjugate gradient method; and a similar approach has been used for the approximation of the Hessian matrix, see [67,55,43]. Such applications have been made possible by a special form of the BFGS update

$$H_+ = \gamma V^{\mathrm{T}} H V + \frac{\rho}{b} s s^{\mathrm{T}},$$

$$V = I - \frac{1}{b} y s^{\mathrm{T}}$$

We define the m-step BFGS method with limited memory as the iterative process (1.1) and (2.5), where $H_i = H_i^i$ and the matrix H_i^i is generated by the recurrence formula

$$H_{j+1}^i = \gamma_j^i V_j^{\mathrm{T}} H_j^i V_j + \frac{\rho_j}{b_j} s_j s_j^{\mathrm{T}} \tag{3.10}$$

for $i - m \leqslant j \leqslant i - 1$, where $H_{i-m}^i = I$. At the same time $\gamma_{i-m}^i = b_{i-1}/a_{i-1}$ and $\gamma_j^i = 1$ for $i - m < j \leqslant i - 1$. Using induction, we can rewrite (3.10) in the form

$$H_{j+1}^i = \frac{b_{i-1}}{a_{i-1}} \left(\prod_{k=i-m}^{j} V_k \right)^{\mathrm{T}} \left(\prod_{k=i-m}^{j} V_k \right) + \sum_{l=i-m}^{j} \frac{\rho_l}{b_l} \left(\prod_{k=l+1}^{j} V_k \right)^{\mathrm{T}} s_l s_l^{\mathrm{T}} \left(\prod_{k=l+1}^{j} V_k \right) \tag{3.11}$$

for $i - m \leqslant j \leqslant i - 1$. From (3.11), we can deduce that the matrix H_i^i can be determined using $2m$ vectors $s_j, y_j, i - m \leqslant j \leqslant i - 1$, without storing the matrices H_j^i, $i - m \leqslant j \leqslant i - 1$. This matrix need not be constructed explicitly since we need only the vector $s_i = -H_i^i g_i$, which can be computed using two recurrences (the Strang formula [67]). First, the vectors

$$u_j = -\left(\prod_{k=j}^{i-1} V_k \right) g_i,$$

where $i - 1 \geqslant j \geqslant i - m$, are computed using the backward recurrence

$$\sigma_j = s_j^{\mathrm{T}} u_{j+1}/b_j,$$

$$u_j = u_{j+1} - \sigma_j y_j$$

for $i - 1 \geqslant j \geqslant i - m$, where $u_i = -g_i$. Then the vectors

$$v_{j+1} = \frac{b_{i-1}}{a_{i-1}} \left(\prod_{k=i-m}^{j} V_k \right)^{\mathrm{T}} u_{i-m} + \sum_{l=i-m}^{j} \frac{\rho_l}{b_l} \left(\prod_{k=l+1}^{j} V_k \right)^{\mathrm{T}} s_l s_l^{\mathrm{T}} u_{l+1},$$

where $i - m \leqslant j \leqslant i - 1$, are computed using the forward recurrence

$$v_{i-m} = (b_{i-1}/a_{i-1}) u_{i-m},$$

$$v_{j+1} = v_j + (\rho_j \sigma_j - y_j^{\mathrm{T}} v_j) s_j$$

for $i - m \leqslant j \leqslant i - 1$, where $v_{i-m} = (b_{i-1}/a_{i-1}) u_{i-m}$. Finally we set $s_i = v_i$.

Recently, a new approach to variable metric methods with limited memory, based on explicit expression of the matrix $H_i = H_i^i$ using low-order matrices, was proposed in [15]. Let $H_i = H_i^i$ be the matrix obtained after m steps of the form

$$H_{j+1}^i = H_j^i + [s_j, H_j^i y_j] M_j [s_j, H_j^i y_j]^{\mathrm{T}},$$

$i - m \leqslant j \leqslant i - 1$, where M_j, $i - m \leqslant j \leqslant i - 1$, are symmetric 2×2 matrices which realize a particular variable metric method (2.11) with $\rho_j = \gamma_j = 1$. We need an expression

$$H_i = H_{i-m}^i - [S_i, H_{i-m}^i Y_i] N_i^{-1} [S_i, H_{i-m}^i Y_i]^{\mathrm{T}}, \tag{3.12}$$

where $S_i = [s_{i-m}, \ldots, s_{i-1}]$, $Y_i = [y_{i-m}, \ldots, y_{i-1}]$ and N_i is a symmetric matrix of order $2m$. Formula (3.12) was obtained for classical variable metric methods (DFP, BFGS, SR1), since the matrices M_j^{-1}, $i - m \leqslant j \leqslant i - 1$, have a relatively simple form in these cases. Derivations, which can be found in [15], are formally rather complicated. Therefore, we introduce only the final results. For this purpose, we denote by R_i the upper triangular matrix of order m, such that $(R_i)_{kl} = s_k^{\mathrm{T}} y_l$, for $k \leqslant l$, and $(R_i)_{kl} = 0$, otherwise. Furthermore, we denote by C_i the diagonal matrix of order m, such that $(C_i)_{kk} = s_k^{\mathrm{T}} y_k$. Taking

$$N_i = \begin{bmatrix} -C_i & R_i - C_i \\ (R_i - C_i)^{\mathrm{T}} & Y_i^{\mathrm{T}} H_{i-m}^i Y_i \end{bmatrix} \tag{3.13}$$

in (3.12), we get the m-step DFP update. Taking

$$N_i = \begin{bmatrix} 0 & R_i \\ R_i^{\mathrm{T}} & C_i + Y_i^{\mathrm{T}} H_{i-m}^i Y_i \end{bmatrix} \tag{3.14}$$

in (3.12), we get the m-step BFGS update. The m-step SR1 update can be written in the following slightly simpler form:

$$H_i = H_{i-m}^i + (S_i - H_{i-m}^i Y_i)(R_i + R_i^{\mathrm{T}} - C_i - Y_i^{\mathrm{T}} H_{i-m}^i Y_i)^{-1} (S_i - H_{i-m}^i Y_i)^{\mathrm{T}}. \tag{3.15}$$

In the sequel, we restrict our attention to the BFGS method. If we choose $H_{i-m}^i = \gamma_{i-m}^i I$, where $\gamma_{i-m}^i = b_{i-1}/a_{i-1}$, and if we explicitly invert matrix (3.14), we can write

$$H_i = \gamma_{i-m} I + [S_i, \gamma_{i-m} Y_i] \begin{bmatrix} (R_i^{-1})^{\mathrm{T}} (C_i + \gamma_{i-m} Y_i^{\mathrm{T}} Y_i) R_i^{-1} & -(R_i^{-1})^{\mathrm{T}} \\ -R_i^{-1} & 0 \end{bmatrix} [S_i, \gamma_{i-m} Y_i]^{\mathrm{T}}. \tag{3.16}$$

This formula has the advantage that no inversion or matrix decomposition is used.

Similar explicit expressions can be obtained for the matrices $B_i = H_i^{-1}$ using duality relations. Since we replace S_i and Y_i by Y_i and S_i, respectively, we have to replace the upper part of $S_i^{\mathrm{T}} Y_i$ by the upper part of $Y_i^{\mathrm{T}} S_i$ (or by the transposed lower part of $S_i^{\mathrm{T}} Y_i$). Therefore, we define the lower triangular matrix L_i, such that $(L_i)_{kl} = s_k^{\mathrm{T}} y_l$, $k \geqslant l$ and $(L_i)_{kl} = 0$, otherwise. Then the m-step BFGS update can be written in the form

$$B_i = B_{i-m}^i - [Y_i, B_{i-m}^i S_i] \begin{bmatrix} -C_i & (L_i - C_i)^{\mathrm{T}} \\ L_i - C_i & S_i^{\mathrm{T}} B_{i-m}^i S_i \end{bmatrix}^{-1} [Y_i, B_{i-m}^i S_i]^{\mathrm{T}}. \tag{3.17}$$

The limited-memory variable metric methods described above require a double set of difference vectors. Fletcher [33] has proposed a method that requires only a single set of these vectors. The same property is possessed by the limited-memory reduced-Hessian variable metric methods introduced in [44] and based on product form updates investigated in [79]. Consider variable metric methods of the form (2.15) with $B_1 = I$ (the unit matrix). Let $\mathscr{G}_i$ and $\mathscr{D}_i$ be linear subspaces spanned by the columns of matrices $G_i = [g_1, \ldots, g_i]$ and $D_i = [d_1, \ldots, d_i]$, respectively. In [79] it is proved that $\mathscr{D}_i = \mathscr{G}_i$ and that $B_i v \in \mathscr{G}_i$ and $B_i w = \lambda_i w$, whenever $v \in \mathscr{G}_i$ and $w \in \mathscr{G}_i^\perp$ (a possible nonunit value λ_i is a consequence of nonquadratic correction and scaling). Let Z_i be a matrix whose columns form an orthonormal basis in $\mathscr{G}_i$ and let $Q_i = [Z_i, W_i]$ be a square orthogonal matrix. Then the above result implies

$$Q_i^T B_i Q_i = \begin{bmatrix} Z_i^T B_i Z_i & 0 \\ 0 & \lambda_i I \end{bmatrix}, \quad Q_i^T g_i = \begin{bmatrix} Z_i^T g_i \\ 0 \end{bmatrix}, \tag{3.18}$$

so that

$$d_i = Z_i \tilde{d}_i, \quad Z_i^T B_i Z_i \tilde{d}_i = -\tilde{g}_i, \quad \tilde{g}_i = Z_i^T g_i. \tag{3.19}$$

In other words, all information concerning variable metric updates is contained in the reduced Hessian approximation $Z_i^T B_i Z_i$ so that the reduced system (3.19) is sufficient for obtaining the direction vector.

This idea can be used for developing limited-memory reduced Hessian variable metric methods. These methods use limited-dimension subspaces $\mathscr{G}_i = \text{span}[g_{i-m+1}, \ldots, g_i]$ and $\mathscr{D}_i = \text{span}[d_{i-m+1}, \ldots, d_i]$ which change on every iteration. Now $\mathscr{D}_i = \mathscr{G}_i$ does not hold in the limited-dimension case, but the quadratic termination property requires columns of Z_i to form a basis in $\mathscr{D}_i$ instead of $\mathscr{G}_i$. Hence the above process has to be slightly reformulated. Instead of Z_i we use an upper triangular matrix T_i such that $D_i = Z_i T_i$ and the reduced Hessian approximation is given in the factorized form $Z_i^T B_i Z_i = R_i^T R_i$ with R_i again upper triangular. Using a scaling parameter γ_1, we can set

$$D_1 = [g_1], \quad T_1 = [\|g_1\|], \quad R_1 = [\sqrt{1/\gamma_1}], \quad \tilde{g}_1 = [\|g_1\|].$$

On every iteration, we first solve two equations $R_i^T R_i \tilde{d}_i = -\tilde{g}_i$, $T_i v_i = \tilde{d}_i$ and set $d_i = D_i v_i$. After determining the direction vector d_i, the line search is performed to obtain a new point $x_{i+1} = x_i + \alpha_i d_i$. Moreover, the matrices D_i and T_i have to be changed to correspond to the subspace $\mathscr{D}_i$. For this purpose, we replace the last column of D_i by d_i and the last column of T_i by $\tilde{d}_i$. Now a representation of the subspace $\mathscr{D}_{i+1}$ has to be formed. First, we project the new gradient $g_{i+1} = g(x_{i+1})$ into the subspace $\mathscr{D}_i$ by solving the equation $T_i^T r_{i+1} = D_i^T g_{i+1}$. Then we determine the quantity $\rho_{i+1} = \|g_{i+1}\| - \|r_{i+1}\|$, set $D_{i+1} = [D_i, g_{i+1}]$ and

$$T_{i+1} = \begin{bmatrix} T_i & r_{i+1} \\ 0 & \rho_{i+1} \end{bmatrix}, \quad \tilde{g}_{i+1} = \begin{bmatrix} r_{i+1} \\ \rho_{i+1} \end{bmatrix}.$$

Using the scaling parameter γ_{i+1}, we obtain a temporary representation of the reduced Hessian approximation in the form $Z_{i+1}^T B_i Z_{i+1} = R_{i+1}^T R_{i+1}$, where

$$R_{i+1} = \begin{bmatrix} R_i & 0 \\ 0 & \sqrt{1/\gamma_{i+1}} \end{bmatrix}, \quad \tilde{g}_{i+1} = \begin{bmatrix} r_{i+1} \\ \rho_{i+1} \end{bmatrix}.$$

This factorization is updated to satisfy the quasi-Newton condition $R_{i+1}^T R_{i+1} \tilde{s}_i = \tilde{y}_i$, where

$$\tilde{s}_i = \alpha_i \begin{bmatrix} \tilde{d}_i \\ 0 \end{bmatrix}, \quad \tilde{y}_i = \tilde{g}_{i+1} - \begin{bmatrix} \tilde{g}_i \\ 0 \end{bmatrix}.$$

Numerically stable methods described in [45] can be used for this purpose. If the subspace $\mathscr{D}_{i+1}$ has dimension $m + 1$, then it must be reduced before the new iteration is started. Denote the matrices after such reduction by $\bar{D}_{i+1}, \bar{T}_{i+1}, \bar{R}_{i+1}$. Then $\bar{D}_{i+1}$ is obtained from D_{i+1} by deleting its first column and matrices $\bar{T}_{i+1}, \bar{R}_{i+1}$ are constructed using elementary Givens rotations (see [44]). Notice that the scaling parameters used above have a similar meaning to those in (2.15). Values $\gamma_1 = 1$ and $\gamma_{i+1} = \tilde{s}_i^T \tilde{y}_i / \tilde{y}_i^T \tilde{y}_i$ are recommended.

3.4. Computational experiments

Now, we can present computational experiments with various variable metric methods for large-scale unconstrained optimization. Table 2 compares the sparse-composite update SCVM (3.2), the sparse-fractioned update SFVM (3.5), the sparse-partitioned BFGS update SPVM (3.8), the limited-memory BFGS update in vector form LVVM (3.11), the limited-memory BFGS update in matrix form LMVM (3.16) and the limited memory BFGS update in reduced-Hessian form LRVM (3.19). The limited-memory updates LVVM and LMVM were constructed from 5 previous steps ($m = 5$) and LRVM was constructed from 10 previous steps ($m = 10$) . For further comparison, we introduce results for the sparse discrete Newton method SDNM [17], the truncated Newton method STNM [21] and the nonlinear conjugate gradient method NCGM [37]. Most of the tested methods were implemented in a line-search framework with direct computation of direction vectors (limited-memory methods in the form (1.10), SFVM and SPVM using sparse Cholesky decomposition (1.11)). The sparse composite method SCVM and the truncated Newton method STNM were implemented by using the unpreconditioned inexact conjugate gradient method (1.12) (again with standard line search). The sparse discrete Newton method SDNM was implemented in a trust-region framework by using the optimal procedure (1.21). We have chosen the most suitable implementations for individual methods. Computational experiments were performed on a DIGITAL UNIX workstation using 22 sparse test problems from TEST14 [62] with 1000 variables. The CPU times in Table 2 represent total time for all 22 test problems and are measured in minutes.

From Table 2, it appears that only the SPVM and SDNM methods are worth considering and other variable metric methods are unsuitable for large-scale problems. Indeed, SPVM and SDNM are excellent for general partially separable problems or general problems with sufficiently sparse Hessian matrices (they can be inefficient for ill-conditioned sum of squares as shown in Table 4 below). On the other hand, variable metric methods with limited memory LVVM, LMVM, LRVM, the truncated Newton method STNM and the nonlinear conjugate gradient method NCGM also work well for problems with dense Hessian matrices. Such problems frequently appear in practice. For instance, a product of functions or a squared sum of functions have the same complexity as a sum of functions (3.7) but their Hessian matrices can be completely full. The sparse composite update SCVM is not robust in general. It sometimes fails for difficult problems and generates matrices which are not suitable for sparse Cholesky decomposition (an iterative solution is then required). We review SCVM here, since it gives an excellent tool for improving methods for large sparse sum of squares as demonstrated in Section 4.

4. Variable metric methods for nonlinear least squares

4.1. Basic ideas for using variable metric updates

Suppose that the objective function $F : \mathbb{R}^N \to \mathbb{R}$ has the form

$$F(x) = \tfrac{1}{2} f^{\mathrm{T}}(x) f(x) = \frac{1}{2} \sum_{k=1}^{m} f_k^2(x), \tag{4.1}$$

where $f_k : \mathbb{R}^n \to \mathbb{R}$, $1 \leqslant k \leqslant m$, are twice continuously differentiable functions. This objective function is frequently used for nonlinear regression and for solving systems of nonlinear equations. We can express the gradient and Hessian matrix of (4.1) in the form

$$g(x) = J^{\mathrm{T}}(x)f(x) = \sum_{k=1}^{m} f_k(x)g_k(x), \tag{4.2}$$

$$G(x) = J^{\mathrm{T}}(x)J(x) + C(x) = \sum_{k=1}^{m} g_k(x)g_k^{\mathrm{T}}(x) + \sum_{k=1}^{m} f_k(x)G_k(x), \tag{4.3}$$

where $g_k(x)$ and $G_k(x)$ are the gradients and the Hessian matrices of the functions $f_k : \mathbb{R}^n \to R$, $1 \leqslant k \leqslant m$ and $f^{\mathrm{T}}(x) = [f_1(x), \ldots f_m(x)]$, $J^{\mathrm{T}}(x) = [g_1(x), \ldots g_m(x)]$. $J(x)$ is the Jacobian matrix of the mapping f at the point x.

The most popular method for nonlinear least squares is the Gauss–Newton method, which uses the first part of (4.3) as an approximation of the Hessian matrix, i.e., $B_i = J_i^{\mathrm{T}} J_i$, $\forall i \in \mathcal{N}$. This method is very efficient for zero-residual problems with $F(x_*) = 0$. In this case, $x_i \to x_*$ implies $F(x_i) \to F(x_*) = 0$ and, therefore, $f_k(x_i) \to 0 \; \forall k, 1 \leqslant k \leqslant m$. If $\|G_k(x)\| \leqslant \bar{G}$, $\forall k, 1 \leqslant k \leqslant m$, then also

$$\|C(x_i)\| = \left\| \sum_{k=1}^{m} f_k(x_i)G_k(x_i) \right\| \leqslant \bar{G} \sum_{k=1}^{m} |f_k(x_i)| \to 0$$

and, therefore, $\|G(x_i) - B_i\| = \|C(x_i)\| \to 0$, which implies a superlinear rate of convergence, see [26]. Since the Jacobian matrices J_i, $i \in \mathcal{N}$, are usually ill-conditioned, even singular, the Gauss–Newton method is most frequently implemented in a trust-region framework.

The Gauss–Newton method is very efficient when applied to a zero-residual problem. It usually outperforms variable metric methods in this case. On the other hand, the rapid convergence can be lost if $F(x_*)$ is large, since $B_i = J_i^{\mathrm{T}} J_i$ can be a bad approximation of G_i. For these reasons, combinations of the Gauss–Newton method with special variable metric updates may be advantageous. Such combined methods exist and can be very efficient, but three problems have to be carefully solved. Firstly, suitable variable metric updates have to be found, together with corresponding quasi-Newton conditions. Secondly, a way for combining these updates with the Gauss–Newton method has to be chosen. Thirdly, a strategy for suppressing the influence of variable metric updates, in case the Gauss–Newton method converges rapidly, has to be proposed. We will investigate these problems in reverse order.

The main idea for suppressing the influence of variable metric updates consists in using the Gauss–Newton method, if it converges rapidly, and variable metric corrections otherwise. The choice of a suitable switching criterion is very important. The most general and, at the same time, most efficient strategy is proposed in [38]. It uses the condition

$$F - F_+ \leqslant \bar{\theta}_1 F, \tag{4.4}$$

where $0 < \bar{\theta}_1 < 1$. If (4.4) holds, then a variable metric correction is applied in the subsequent iteration. Otherwise, the Gauss–Newton method is used. This strategy is based on the fact that $F_{i+1}/F_i \to 0$, if $F_i \to F_* = 0$ superlinearly, and $F_{i+1}/F_i \to 1$, if $F_i \to F_* > 0$.

Now, we describe techniques for combining variable metric updates with the Gauss–Newton method. We consider the following techniques: simple correction, cumulative correction and successive approximation of the second-order term in (4.3). We shall use A to denote a matrix such that $A^{\mathrm{T}}A$ approximates the Hessian matrix $J^{\mathrm{T}}J + C$ (see (2.40)).

A simple correction is useful in the sparse case, when a cumulative correction cannot be realized. On non-Gauss–Newton iterations we compute the matrix B_+ (or A_+) from $J_+^T J_+$ (or J_+) using a variable metric update. Otherwise, we set $B_+ = J_+^T J_+$ (or $A_+ = J_+$).

A cumulative correction is proposed in [38]. On non-Gauss-Newton iterations we compute the matrix B_+ (or A_+) from B (or A) using a variable metric update. Otherwise, we set $B_+ = J_+^T J_+$ (or $A_+ = J_+$).

A successive approximation of the second-order term is based on the model $B = J^T J + C$. The matrix C_+ is computed from the matrix C using variable metric updates. If the Gauss–Newton method should not be used, we set $B_+ = J_+^T J_+ + C_+$. Otherwise, we set $B_+ = J_+^T J_+$. While simple and cumulative corrections use the standard updates described in previous sections, the successive approximation of the second-order term requires special updates (known as structured updates) which we now describe. We will suppose that $\rho = 1$ and $\gamma = 1$ in (2.15). Later we will consider a special scaling technique.

4.2. Structured variable metric updates

There are two possibilities for construction of structured variable metric updates. The first method is based on the transformed quasi-Newton condition $C_+ s = z = J_+^T f_+ - J^T f - J_+^T J_+ s$. Therefore, the general update has the form (2.15) with B and y replaced by C and z, respectively. The SR1 update, derived in this way, can be written in the form

$$C_+ = C + \frac{(z - Cs)(z - Cs)^T}{s^T(z - Cs)}. \tag{4.5}$$

This SR1 update is very efficient and usually outperforms other structured variable metric updates. Notice that the BFGS method cannot be realized in this approach since positivity of $s^T z$ is not guaranteed.

The second possibility involves updating $\bar{B} = J_+^T J_+ + C$ to obtain $B_+ = J_+^T J_+ + C_+$ satisfying the quasi-Newton condition $B_+ s = y = J_+^T f_+ - J^T f$. The resulting general update has the form (2.15) with B replaced by $\bar{B}$. Since $y - \bar{B}s = z - Cs$, it is advantageous to use formula (2.29). Then

$$
\begin{aligned}
C_+ &= C + \frac{(y - \bar{B}s)v^T + v(y - \bar{B}s)^T}{s^T v} - \frac{(y - \bar{B}s)^T s}{s^T v}\frac{vv^T}{s^T v} \\
&= C + \frac{(z - Cs)v^T + v(z - Cs)^T}{s^T v} - \frac{(z - Cs)^T s}{s^T v}\frac{vv^T}{s^T v}
\end{aligned} \tag{4.6}
$$

with $v=s$ for the structured PSB update, $v=y$ for the structured DFP update and $v=y+(y^T s/s^T \bar{B}s)^{1/2}\bar{B}s$ for the structured BFGS update. Methods based on formula (4.6) have been investigated in [24], where superlinear convergence of the structured BFGS method was proved.

The vectors y and z, used in formulae (4.5)–(4.6), can be defined in various ways, but always based on $z = y - J_+^T J_+ s$. The standard choice

$$z = J_+^T f_+ - J^T f - J_+^T J_+ s, \tag{4.7}$$

corresponding to the quasi-Newton condition $(J_+^{\mathrm{T}}J_+ + C_+)s = J_+^{\mathrm{T}}f_+ - J^{\mathrm{T}}f$, is introduced in [22]. In [6], a similar choice

$$z = J_+^{\mathrm{T}}f_+ - J^{\mathrm{T}}f - J^{\mathrm{T}}Js \tag{4.8}$$

corresponding to the quasi-Newton condition $(J^{\mathrm{T}}J + C_+)s = J_+^{\mathrm{T}}f_+ - J^{\mathrm{T}}f$, is given. Another choice [4,83]) is based on the objective function $\tilde{F}(x) = (1/2)(f^{\mathrm{T}}(x)f(x) - x^{\mathrm{T}}J^{\mathrm{T}}Jx)$, whose Hessian matrix is just the matrix $J^{\mathrm{T}}J$ that we want to approximate. Applying the standard variable metric method to the function $\tilde{F}$, we obtain the quasi-Newton condition $C_+s = \tilde{g}_+ - \tilde{g} = z$, where

$$z = J_+^{\mathrm{T}}f_+ - J_+^{\mathrm{T}}J_+x_+ - J^{\mathrm{T}}f + J^{\mathrm{T}}Jx. \tag{4.9}$$

A popular choice, proposed in [9], is based on the explicit form of the second-order term in (4.3). Suppose that the approximations B_k^+ of the Hessian matrices G_k satisfy the quasi-Newton conditions $B_k^+s = g_k^+ - g_k$, $1 \leqslant k \leqslant m$. Then we can write

$$z = C_+s \triangleq \sum_{k=1}^{m} f_k^+ B_k^+ s = \sum_{k=1}^{m} f_k^+ (g_k^+ - g_k) = (J_+ - J)^{\mathrm{T}}f_+. \tag{4.10}$$

Interesting methods for nonlinear least squares have been obtained from the product-form BFGS update (2.40) (other product-form updates are less suitable since they require the inversion of the matrix $A^{\mathrm{T}}A$). A generalization of (2.40) (with $\rho = 1$ and $\gamma = 1$), related to structured update (4.6), is described in [97]. Here A is replaced by the matrix $J + L$, where J is the Jacobian matrix and L plays a similar role to C in (4.6). Thus $B = (J + L)^{\mathrm{T}}(J + L)$, $B_+ = (J_+ + L_+)^{\mathrm{T}}(J_+ + L_+)$ and if we set $\bar{B} = (J_+ + L)^{\mathrm{T}}(J_+ + L)$, we can express (4.6) as

$$L_+ = L + \frac{(J_+ + L)s}{s^{\mathrm{T}}\bar{B}s}\left(\sqrt{\frac{s^{\mathrm{T}}\bar{B}s}{s^{\mathrm{T}}y}}\,y - \bar{B}s\right)^{\mathrm{T}}, \tag{4.11}$$

which is similar to (2.40).

Structured variable metric updates can be improved by a suitable scaling technique. The main reason for scaling is controlling the size of the matrix C. Therefore, the quasi-Newton condition $C_+s = z$ is preferred. The scaling parameter γ is chosen in such a way that $(1/\gamma)Cs$ is close to z in some sense. In analogy with (2.9), we can choose $\gamma = s^{\mathrm{T}}Cs/s^{\mathrm{T}}z$ or $\gamma = \max(s^{\mathrm{T}}Cs/s^{\mathrm{T}}z, 1)$, which is the value proposed in [23]. Biggs [9] recommends the value $\gamma = f^{\mathrm{T}}f/f_+^{\mathrm{T}}f$ based on a quadratic model. If we choose the scaling parameter γ, then we replace C by $(1/\gamma)C$ in (4.5)–(4.6) to obtain a scaled structured update. A more complicated process, described in [97], is used in connection with product form update (4.11). All the above methods can be realized efficiently using switching strategy (4.4). Structured variable metric updates can also be used permanently (without switching), as follows from the theory given in [24], but such a realization is usually less efficient.

Interesting variable metric updates are based on an approximation of the term

$$T(x) = \sum_{k=1}^{m} \frac{f_k(x)}{\|f(x)\|}G_k(x).$$

Table 3

Line-search realization	Iterations	f. eval.	g. eval.	CPU time	Failures
Scaled BFGS	4229	5301	5301	1.43	1
Standard GN	4809	8748	13 555	3.46	7
GN with (4.5) and (4.4)	1447	2546	3993	1.37	—
GN with (4.13) and (4.4)	1594	2807	4400	1.32	—
GN with (2.17) and (4.4)	1658	2805	4461	1.15	—
Trust-region realization	Iterations	f. eval.	g. eval.	CPU time	Failures
Standard GN	2114	2512	2194	1.31	—
GN with (4.5) and (4.4)	1497	1777	1579	1.05	—
GN with (4.13) and (4.4)	1480	1753	1562	1.04	—
GN with (2.17) and (4.4)	1476	1846	1555	0.99	—

Thus we have the model $B = J^T J + \|f\| T$. By analogy with structured variable metric methods, Huschens [54] proposed totally structured variable metric methods which consist in updating the matrix $\bar{B} = J_+^T J_+ + \|f\| T$ to get the matrix $\tilde{B}_+ = J_+^T J_+ + \|f\| T_+$, satisfying the quasi-Newton condition $\tilde{B}_+ s = y$. Finally, the matrix $B_+ = J_+^T J_+ + \|f_+\| T_+$ is chosen. Using expression (2.27), we can write

$$T_+ = T + \frac{1}{\|f\|} \left(\frac{(y - \bar{B}s)v^T + v(y - \bar{B}s)^T}{s^T v} - \frac{(y - \bar{B}s)^T s}{s^T v} \frac{vv^T}{s^T v} \right)$$

$$= T + \frac{(\tilde{z} - Ts)v^T + v(\tilde{z} - Ts)^T}{s^T v} - \frac{(\tilde{z} - Ts)^T s}{s^T v} \frac{vv^T}{s^T v}, \tag{4.12}$$

where $\tilde{z} = z/\|f\| = (y - J_+^T J_+ s)/\|f\|$. Setting $v = s$, we get the totally structured PSB method. Setting $v = y$, we get the totally structured DFP method. Setting $v = y + (y^T s/s^T \bar{B}s)^{1/2} \bar{B}s$, we get the totally structured BFGS method. The totally structured SR1 method has the form

$$T_+ = T + \frac{(\tilde{z} - Ts)(\tilde{z} - Ts)^T}{s^T(\tilde{z} - Ts)}. \tag{4.13}$$

The use of $\|f\|$ instead of $\|f_+\|$ in the quasi-Newton condition $(J_+^T J_+ + \|f\| T_+)s = y$ leads to methods which have a quadratic rate of convergence in the case of zero-residual problems and a superlinear rate of convergence otherwise, see [54]. This is the most significant theoretical result concerning permanent realization of structured variable metric updates.

We now present numerical experiments with various methods for nonlinear least squares. Table 3 compares the BFGS method with interval scaling (2.10) and nonquadratic correction (2.4), the standard Gauss–Newton method, the Gauss–Newton method with structured SR1 update (4.5) and switching strategy (4.4), the Gauss–Newton method with totally structured SR1 update (4.13) and switching strategy (4.4) and the Gauss–Newton method with the cumulative BFGS correction (2.17) and switching strategy (4.4). The first part of Table 3 refers to the standard line-search implementation and the second part refers to the trust-region implementation (1.22). Structured updates (4.5) and (4.13) were scaled in each iteration as in [23]. The cumulative BFGS update was scaled only on the first iteration. Computational experiments have been performed on a PENTIUM PC computer using 82 test problems (30 from [65], 22 from TEST15, 30 from TEST18, [62]) with 20 variables (62 of them have zero residual at the solution). The CPU times in Table 3 represent total time for all 82 test problems and are measured in seconds.

Results in Table 3 suggest that trust region realizations are preferable whenever the matrix $B_i = J_i^T J_i$ is used (this matrix is usually ill-conditioned). Furthermore, they show the efficiency of switching strategy (4.4). Structured updates were also tested without switching but results obtained were much worse. The efficiency of scaled BFGS method with line-search confirms its robustness for nonlinear least squares (CPU time is low since $O(n^2)$ operations per iteration are used).

4.3. Variable metric updates for sparse least squares

The Gauss–Newton method can also be combined with variable metric updates in the sparse case. We will now describe some such possibilities. One is a combination of the Gauss–Newton method with the composite update (3.2), so that

$$B_+ = \begin{cases} \mathscr{P}_S \mathscr{P}_{QG}(J_+^T J_+) & \text{if } F - F_+ \leq \bar{\theta}_1 F, \\ J_+^T J_+ & \text{if } F - F_+ > \bar{\theta}_1 F. \end{cases} \tag{4.14}$$

Computational efficiency of this hybrid method was studied in [59].

An interesting approach, based on the partitioned SR1 update, was proposed in [93] and also studied in [59]. The partitioned SR1 update is applied to the approximations $\hat{T}_k$ of the packed element-Hessian matrices $\hat{G}_k(x)$ of the functions $f_k : \mathbb{R}^n \to \mathbb{R}$, $1 \leq k \leq m$, contained in (4.1). These matrices are updated in such a way that

$$\hat{T}_k^+ = \begin{cases} \hat{T}_k + \dfrac{(\hat{y}_k - \hat{T}_k \hat{s}_k)(\hat{y}_k - \hat{T}_k \hat{s}_k)^T}{\hat{s}_k^T(\hat{y}_k - \hat{T}_k \hat{s}_k)} & \text{if } |\hat{s}_k^T(\hat{y}_k - \hat{T}_k \hat{s}_k)| > \bar{\theta}_0, \\ \hat{T}_k & \text{if } |\hat{s}_k^T(\hat{y}_k - \hat{T}_k \hat{s}_k)| \leq \bar{\theta}_0 \end{cases} \tag{4.15}$$

and are used for construction of approximations $\hat{B}_k$ of the packed element-Hessian matrices $\hat{g}_k \hat{g}_k^T + f_k \hat{G}_k$. Using (4.4), we can write

$$\hat{B}_k^+ = \hat{g}_k^+ (\hat{g}_k^+)^T + f_k^+ \hat{T}_k^+ \quad \text{if } F - F_+ \leq \bar{\theta}_1 F, \tag{4.16}$$

$$\hat{B}_k^+ = \hat{g}_k^+ (\hat{g}_k^+)^T \quad \text{if } F - F_+ > \bar{\theta}_1 F. \tag{4.17}$$

In the first iteration we set $\hat{T}_k = I$, $1 \leq k \leq m$. Notice that the matrices $\hat{T}_k^+$, $1 \leq k \leq m$, have to be stored simultaneously, which is a disadvantage of this method.

Another interesting way for improving the sparse Gauss–Newton method is based on the factorized formula (4.11), which is used as a simple update so that $L = 0$. Taking $L = 0$ in (5.11), we can express $A_+ = J_+ + L_+$ in the form

$$\begin{aligned} A_+ &= J_+ + \frac{J_+ s}{s^T J_+^T J_+ s} \left(\sqrt{\frac{s^T J_+^T J_+ s}{s^T y}} \, y - J_+^T J_+ s \right)^T \\ &= J_+ + \frac{J^+ s}{\|J^+ s\|} \left(\frac{y}{\sqrt{s^T y}} - J_+^T \frac{J^+ s}{\|J^+ s\|} \right)^T. \end{aligned} \tag{4.18}$$

Table 4

Method	Iterations	f. eval.	g. eval.	CPU time	Failures
GN	11 350	11 760	11 402	3 : 49	2
GNCVM	7264	7688	7316	2 : 36	—
GNPVM	8562	9588	8614	3 : 48	1
GNDNM	7012	7604	9286	2 : 35	—
SPVM	14 009	29 161	29 161	4 : 59	3
SDNM	12 588	84 484	84 337	8 : 38	4

Then we can use the matrix (4.18) if $F - F_+ \leqslant \bar{\theta}_1 F$, and set $A_+ = J_+$, otherwise (see [59] for more detail).

An interesting sparse hybrid method is based on the SR1 update. Consider the augmented linear least-squares problem $\tilde{J}_+ d_+ \approx -\tilde{f}_+$ where

$$\tilde{J}_+ = \begin{bmatrix} J_+ \\ w \end{bmatrix}, \qquad \tilde{f}_+ = \begin{bmatrix} f_+ \\ 0 \end{bmatrix}. \tag{4.19}$$

The normal equations for this problem have the form $B_+ d_+ = -J_+^T f_+$, where

$$B^+ = \tilde{J}_+^T \tilde{J}_+ = J_+^T J_+ + w w^T. \tag{4.20}$$

If we choose

$$w = (y - J_+^T J_+ s)/\sqrt{s^T(y - J_+^T J_+ s)}, \tag{4.21}$$

then (4.20) gives exactly the SR1 update (with B replaced by $J_+^T J_+$). Note that (4.21) can be used only if $s^T(y - J_+^T J_+ s) > 0$, which slightly restricts the use of update (4.19). We use the augmented linear least-squares problem $\tilde{J}_+ d_+ \approx -\tilde{f}_+$ (with w given by (4.21)), if $F - F_+ \leqslant \bar{\theta}_1 F$ and $s^T(y - J_+^T J_+ s) > \bar{\theta}_0$ hold simultaneously, and the standard linear least-squares problem $J_+ d \approx -f_+$, otherwise.

Table 4 compares the standard Gauss–Newton method GN, the Gauss–Newton method with composite update GNCVM (4.14) and the Gauss–Newton method with partitioned update GNPVM (4.15)–(4.16). For further comparison, we quote results for the combined Gauss–Newton and discrete Newton method GNDNM, utilizing switching strategy (4.4) and also for the partitioned BFGS method SPVM (3.8) and the sparse discrete Newton method SDNM. All these methods have been implemented within a trust-region strategy (1.21), see [66]. Computational experiments were performed on a DIGITAL UNIX workstation using 52 sparse test problems (22 from TEST15, 30 from TEST18, [62]) with 1000 variables (38 of them have zero residual at the solution). The CPU times in Table 4 represent the total for all 52 test problems and are quoted in minutes. Sparse and limited-memory variable metric methods have not been efficient for solving these problems.

Table 4 implies that special methods for least-squares problems are usually more efficient than methods for general problems. This conclusion also holds for other classes of problems. For instance, the last 30 problems used in Table 4 are solutions to systems of nonlinear equations, which can also be solved more efficiently by special methods. Inefficiency of SDNM was mainly caused by four failures (3000 iterations or 5000 function evaluations did not suffice). But SDNM did not outperform combined methods even if difficult problems were excluded.

5. Conclusion

In this paper, we have given a review of variable metric or quasi-Newton methods for unconstrained optimization, paying particular attention to the derivation of formulas and their efficient implementation (we have tried to quote all relevant literature). Quasi-Newton methods can be also used for solving systems of nonlinear equations, see, e.g., [10,28,64], but theoretical investigation and practical realization require a slightly different point of view. Another field for application of variable metric methods is general constrained optimization. Nevertheless, problems connected with potential functions, constraint handling or interior point approach are dominant in this case and go beyond the scope of this contribution.

Numerical experience, partially reported in this paper, gives implications for the choice of a suitable optimization method. We would like to give few recommendations for potential users. Standard variable metric methods described in Section 2 are mostly suitable for dense small or moderate-size general problems (up to 100–200 variables, say). Reasonable scaling and nonquadratic correction can improve the efficiency of these methods.

If we have a large-scale problem, then the choice of method depends on the problem structure. General problems with sparse Hessian matrices are successfully solved by the discrete Newton method. Partially separable problems can be efficiently solved by partitioned variable metric updates. If the Hessian matrix has no structure, then limited memory variable metric methods as well as the truncated Newton method and the nonlinear conjugate gradient method are suitable.

If the objective function is a sum of squares, then special methods for least squares should be used. Trust region realizations are most suitable in this case. We recommend the Gauss–Newton method with variable metric corrections. The switching strategy (4.4) is very efficient. If the problem is dense then the cumulative BFGS update is of a primary interest. The simple composite update (4.14) is suitable in the sparse case.

Variable metric methods can be successfully adapted to solve nondifferentiable problems. An efficient variable metric method for nonsmooth optimization is proposed in [63].

References

[1] J. Abaffy, E. Spedicato, ABS Projection Algorithms, Mathematical Techniques for Linear and Nonlinear Equations, Ellis Horwood, Chichester, 1989.

[2] N. Adachi, On variable metric algorithm, J. Optim. Theory Appl. 7 (1971) 391–409.

[3] M. Al-Baali, Highly efficient Broyden methods of minimization with variable parameter, Optim. Methods Software 1 (1992) 301–310.

[4] M. Al-Baali, R. Fletcher, Variational methods for nonlinear least squares, J. Optim. Theory Appl. 36 (1985) 405–421.

[5] O. Axelsson, Iterative Solution Methods, Cambridge University Press, Cambridge, 1996.

[6] J.T. Betts, Solving the nonlinear least squares problem: application of a general method, J. Optim. Theory Appl. 18 (1976) 469–483.

[7] M.C. Biggs, Minimization algorithms making use of nonquadratic properties of the objective function, J. Inst. Math. Appl. 8 (1971) 315–327.

[8] M.C. Biggs, A note on minimization algorithms which make use of non-quadratic properties of the objective function, J. Inst. Math. Appl. 12 (1973) 337–338.

[9] M.C. Biggs, The estimation of the Hessian matrix in nonlinear least squares problems with nonzero residuals, Math. Programming 12 (1977) 67–80.

[10] C.G. Broyden, A class of methods for solving nonlinear simultaneous equations, Math. Comp. 19 (1965) 577–593.

[11] C.G. Broyden, The convergence of a class of double rank minimization algorithms, Part 1 – general considerations, Part 2 – the new algorithm, J. Inst. Math. Appl. 6 (1970) 76–90, 222–231.

[12] A. Buckley, A combined conjugate-gradient quasi-Newton minimization algorithm, Math. Programming 15 (1978) 200–210.

[13] A. Buckley, A. LeNir, QN-like variable storage conjugate gradients, Math. Programming 27 (1983) 155–175.

[14] R.H. Byrd, D.C. Liu, J. Nocedal, On the behavior of Broyden's class of quasi-Newton methods, Report No. NAM 01, Dept. of Electrical Engn. and Computer Science, Northwestern University, Evanston, 1990.

[15] R.H. Byrd, J. Nocedal, R.B. Schnabel, Representation of quasi-Newton matrices and their use in limited memory methods, Math. Programming 63 (1994) 129–156.

[16] R.H. Byrd, J. Nocedal, Y.X. Yuan, Global convergence of a class of quasi-Newton methods on convex problems, SIAM J. Numer. Anal. 24 (1987) 1171–1190.

[17] M. Coleman, J.J. Moré, Estimation of sparse Hessian matrices and graph coloring problems, Math. Programming 42 (1988) 245–270.

[18] M. Contreras, R.A. Tapia, Sizing the BFGS and DFP updates: a numerical study, J. Optim. Theory Appl. 78 (1993) 93–108.

[19] W.C. Davidon, Variable metric method for minimisation, A.E.C. Research and Development Report ANL-5990, 1959.

[20] W.C. Davidon, Optimally conditioned optimization algorithms without line searches, Math. Programming 9 (1975) 1–30.

[21] R.S. Dembo, T. Steihaug, Truncated-Newton algorithms for large-scale unconstrained minimization, Math. Programming 26 (1983) 190–212.

[22] J.E. Dennis, Some computational techniques for the nonlinear least squares problem, in: G.D. Byrne, C.A. Hall (Eds.), Numerical Solution of Nonlinear Algebraic Equations, Academic Press, London, 1974.

[23] J.E. Dennis, D. Gay, R.E. Welsch, An adaptive nonlinear least squares algorithm, ACM Trans. Math. Software 7 (1981) 348–368.

[24] J.E. Dennis, H.J. Martinez, R.A. Tapia, Convergence theory for the structured BFGS secant method with application to nonlinear least squares, J. Optim. Theory Appl. 61 (1989) 161–177.

[25] J.E. Dennis, H.H.W. Mei, An unconstrained optimization algorithm which uses function and gradient values, Report No. TR-75-246, Dept. of Computer Science, Cornell University, 1975.

[26] J.E. Dennis, J.J. Moré, A characterization of superlinear convergence and its application to quasi-Newton methods, Math. Comp. 28 (1974) 549–560.

[27] J.E. Dennis, R.B. Schnabel, Least change secant updates for quasi-Newton methods, Report No. TR78-344, Dept. of Computer Sci., Cornell University, Ithaca, 1978.

[28] J.E. Dennis, R.B. Schnabel, Numerical Methods for Unconstrained Optimization and Nonlinear Equations, Prentice-Hall, Englewood Cliffs, NJ, 1983.

[29] J.E. Dennis, R.B. Schnabel, A view of unconstrained optimization, in: G.L. Nemhauser, A.H.G. Rinnooy Kan, M.J. Todd (Eds.), Optimization, North-Holland, Amsterdam, 1989.

[30] L.C.W. Dixon, Quasi-Newton algorithms generate identical points, Math. Programming 2 (1972) 383–387.

[31] R. Fletcher, A new approach to variable metric algorithms, Comput. J. 13 (1970) 317–322.

[32] R. Fletcher, Practical Methods of Optimization, Wiley, New York, 1987.

[33] R. Fletcher, Low storage methods for unconstrained optimization, in: E.L. Algower, K. Georg (Eds.), Computational Solution of Nonlinear Systems of Equations, Lectures in Applied Mathematics, Vol. 26, AMS Publications, Providence, RI, 1990.

[34] R. Fletcher, A new variational result for quasi-Newton formulae, SIAM J. Optim. 1 (1991) 18–21.

[35] R. Fletcher, An optimal positive definite update for sparse Hessian matrices, SIAM J. Optim. 5 (1995) 192–218.

[36] R. Fletcher, M.J.D. Powell, A rapidly convergent descent method for minimization, Comput. J. 6 (1963) 163–168.

[37] R. Fletcher, C.M. Reeves, Function minimization by conjugate gradients, Comput. J. 7 (1964) 149–154.

[38] R. Fletcher, C. Xu, Hybrid methods for nonlinear least squares, IMA J. Numer. Anal. 7 (1987) 371–389.

[39] J.A. Ford, I.A. Moghrabi, Alternative parameter choices for multi-step quasi-Newton methods, Optim. Methods Software 2 (1993) 357–370.

[40] J.A. Ford, I.A. Moghrabi, Multi-step quasi-Newton methods for optimization, J. Comput. Appl. Math. 50 (1994) 305–323.

[41] J.A. Ford, I.A. Moghrabi, Minimum curvature multi-step quasi-Newton methods for unconstrained optimization, Report No. CSM-201, Department of Computer Science, University of Essex, Colchester, 1995.

[42] J. Greenstadt, Variations on variable metric methods, Math. Comput. 24 (1970) 1–18.

[43] J.C. Gilbert, C. Lemarechal, Some numerical experiments with variable-storage quasi-Newton algorithms, Math. Programming 45 (1989) 407–435.

[44] P.E. Gill, M.W. Leonard, Limited-memory reduced-Hessian methods for large-scale unconstrained optimization, Report NA 97-1, Dept. of Mathematics, University of California, San Diego, La Jolla, 1997.

[45] P.E. Gill, W. Murray, M.A. Saunders, Methods for computing and modifying LDV factors of a matrix, Math. Comput. 29 (1975) 1051–1077.

[46] D. Goldfarb, A family of variable metric algorithms derived by variational means, Math. Comput. 24 (1970) 23–26.

[47] A. Griewank, The global convergence of partitioned BFGS on problems with convex decompositions and Lipschitzian gradients, Math. Programming 50 (1991) 141–175.

[48] A. Griewank, P.L. Toint, Partitioned variable metric updates for large-scale structured optimization problems, Numer. Math. 39 (1982) 119–137.

[49] A. Griewank, P.L. Toint, Local convergence analysis for partitioned quasi-Newton updates, Numer. Math. 39 (1982) 429–448.

[50] A. Griewank, P.L. Toint, Numerical experiments with partially separable optimization problems, in: D.F. Griffits (Ed.), Numerical Analysis, Proc. Dundee 1983, Lecture Notes in Mathematics, Vol. 1066, Springer, Berlin, 1984, pp. 203–220.

[51] M.R. Hestenes, C.M. Stiefel, Methods of conjugate gradient for solving linear systems, J. Res. NBS 49 (1964) 409–436.

[52] S. Hoshino, A formulation of variable metric methods, J. Inst. Math. Appl. 10 (1972) 394–403.

[53] H.Y. Huang, Unified approach to quadratically convergent algorithms for function minimization, J. Optim. Theory Appl. 5 (1970) 405–423.

[54] J. Huschens, On the use of product structure in secant methods for nonlinear least squares, SIAM J. Optim. 4 (1994) 108–129.

[55] D.C. Liu, J. Nocedal, On the limited memory BFGS method for large-scale optimization, Math. Programming 45 (1989) 503–528.

[56] L. Lukšan, Computational experience with improved variable metric methods for unconstrained minimization, Kybernetika 26 (1990) 415–431.

[57] L. Lukšan, Computational experience with known variable metric updates, J. Optim. Theory Appl. 83 (1994) 27–47.

[58] L. Lukšan, J. Vlček, Simple scaling for variable metric updates, Report No. 611, Institute of Computer Science, Academy of Sciences of the Czech Republic, Prague 1995.

[59] L. Lukšan, Hybrid methods for large sparse nonlinear least squares, J. Optim. Theory Appl. 89 (1996) 575–595.

[60] L. Lukšan, Numerical methods for unconstrained optimization, Report DMSIA 12/97, University of Bergamo, 1997.

[61] L. Lukšan, M. Tůma, M. Šiška, J. Vlček, N. Ramešová, Interactive system for universal functional optimization (UFO) — Version 1998, Report No. 766, Institute of Computer Science, Academy of Sciences of the Czech Republic, Prague, 1998.

[62] L. Lukšan, J.Vlček, Subroutines for testing large sparse and partially separable unconstrained and equality constrained optimization problems, Report No. 767, Institute of Computer Science, Academy of Sciences of the Czech Republic, Prague, 1999.

[63] L. Lukšan, J. Vlček, Globally convergent variable metric method for convex nonsmooth unconstrained minimization, J. Optim. Theory Appl. 102 (1999) 593–613.

[64] J.M. Martinez, A quasi-Newton method with modification of one column per iteration, Computing 33 (1984) 353–362.

[65] J.J. Moré, B.S. Garbow, K.E. Hillstrom, Testing unconstrained optimization software, ACM Trans. Math. Software 7 (1981) 17–41.

[66] J.J. Moré, D.C. Sorensen, Computing a trust region step, Report ANL-81-83, Argonne National Laboratory, 1981.

[67] J. Nocedal, Updating quasi-Newton matrices with limited storage, Math. Comp. 35 (1980) 773–782.

[68] J. Nocedal, Y. Yuan, Analysis of a self-scaling quasi-Newton method, Math. Programming 61 (1993) 19–37.

[69] S.S. Oren, D.G. Luenberger, Self scaling variable metric (SSVM) algorithms, Part 1 — criteria and sufficient condition for scaling a class of algorithms, Part 2 — implementation and experiments, Management Sci. 20 (1974) 845–862, 863–874.

[70] S.S. Oren, E. Spedicato, Optimal conditioning of self scaling variable metric algorithms, Math. Programming 10 (1976) 70–90.

[71] M.R. Osborne, L.P. Sun, A new approach to the symmetric rank-one updating algorithm. Report No. NMO/01, School of Mathematics, Australian National University, Canberra, 1988.

[72] A. Perry, A modified conjugate gradient algorithm, Oper. Res. 26 (1978) 1073–1078.

[73] E. Polak, G. Ribiére, Note sur la convergence des méthodes de directions conjugées, Rev. Française Inform. Mech. Oper. 16-R1 (1969) 35–43.

[74] M.J.D. Powell, A new algorithm for unconstrained optimization, in: J.B. Rosen, O.L. Mangasarian, K. Ritter (Eds.), Nonlinear Programming, Academic Press, London, 1970.

[75] M.J.D. Powell, On the global convergence of trust region algorithms for unconstrained minimization, Math. Programming 29 (1984) 297–303.

[76] D.G. Pu, W.W. Tian, A class of modified Broyden algorithms, J. Comput. Math. 12 (1994) 366–379.

[77] D.F. Shanno, Conditioning of quasi-Newton methods for function minimization, Math. Comp. 24 (1970) 647–656.

[78] D.F. Shanno, K.J. Phua, Matrix conditioning and nonlinear optimization, Math. Programming 14 (1978) 144–160.

[79] D. Siegel, Implementing and modifying Broyden class updates for large scale optimization, Report DAMTP NA12, Department of Applied Mathematics and Theoretical Physics, University of Cambridge, 1992.

[80] E. Spedicato, Stability of Huang's update for the conjugate gradient method, J. Optim. Theory Appl. 11 (1973) 469–479.

[81] E. Spedicato, On condition numbers of matrices in rank-two minimization algorithms, in: L.C.W. Dixon, G.P. Szego (Eds.), Towards Global Optimization, North-Holland, Amsterdam, 1975.

[82] E. Spedicato, A bound on the condition number of rank-two corrections and applications to the variable metric method, Calcolo 12 (1975) 185–199.

[83] E. Spedicato, Parameter estimation and least squares, in: F. Archetti, M. Cugiani (Eds.), Numerical Techniques for Stochastic Systems, North-Holland, Amsterdam, 1980.

[84] E. Spedicato, A class of rank-one positive definite quasi-Newton updates for unconstrained minimization, Math. Operationsforsch. Statist. Ser. Optim. 14 (1983) 61–70.

[85] E. Spedicato, A class of sparse symmetric quasi-Newton updates, Ricerca Operativa 22 (1992) 63–70.

[86] E. Spedicato, N.Y. Deng, Z. Li, On sparse quasi-Newton quasi-diagonally dominant updates, Report DMSIA 1/96, University of Bergamo, 1996.

[87] E. Spedicato, Z. Xia, Finding general solutions of the quasi-Newton equation via the ABS approach, Optim. Methods Software 1 (1992) 243–252.

[88] E. Spedicato, J. Zhao, Explicit general solution of the Quasi-Newton equation with sparsity and symmetry, Optim. Methods Software 2 (1993) 311–319.

[89] T. Steihaug, Local and superlinear convergence for truncated iterated projections methods, Math. Programming 27 (1983) 176–190.

[90] T. Steihaug, The conjugate gradient method and trust regions in large-scale optimization, SIAM J. Numer. Anal. 20 (1983) 626–637.

[91] P.L. Toint, On sparse and symmetric matrix updating subject to a linear equation, Math. Comp. 31 (1977) 954–961.

[92] P.L. Toint, Global convergence of the partitioned BFGS algorithm for convex partially separable optimization, Math. Programming 36 (1986) 290–306.

[93] P.L. Toint, On large scale nonlinear least squares calculations, SIAM J. Sci. Statis. Comput. 8 (1987) 416–435.

[94] M. Tůma, Sparse fractioned variable metric updates, Report No. 497, Institute of Computer and Information Sciences, Czechoslovak Academy of Sciences, Prague 1991.

[95] H. Wolkowicz, Measures for rank-one updates, Math. Oper. Res. 19 (1994) 815–830.

[96] H. Wolkowicz, An all-inclusive efficient region of updates for least change secant methods, SIAM J. Optim. 5 (1996) 172–191.

[97] H. Yabe, T. Takahashi, Factorized quasi-Newton methods for nonlinear least squares problems, Math. Programming 51 (1991) 75–100.

[98] Y. Yuan, Non-quasi-Newton updates for unconstrained optimization, J. Comput. Math. 13 (1995) 95–107.

[99] J. Zhang, N.Y. Deng, L. Chen, A new quasi-Newton equation and related methods for unconstrained optimization, Report MA-96-05, City University of Hong Kong, 1996.

ELSEVIER

Journal of Computational and Applied Mathematics 124 (2000) 97–121

JOURNAL OF
COMPUTATIONAL AND
APPLIED MATHEMATICS

www.elsevier.nl/locate/cam

Practical quasi-Newton methods for solving nonlinear systems [☆]

José Mario Martínez

Institute of Mathematics, University of Campinas, CP 6065, 13081-970 Campinas SP, Brazil

Received 28 April 1999; received in revised form 27 January 2000

Abstract

Practical quasi-Newton methods for solving nonlinear systems are surveyed. The definition of quasi-Newton methods that includes Newton's method as a particular case is adopted. However, especial emphasis is given to the methods that satisfy the secant equation at every iteration, which are called here, as usually, secant methods. The least-change secant update (LCSU) theory is revisited and convergence results of methods that do not belong to the LCSU family are discussed. The family of methods reviewed in this survey includes Broyden's methods, structured quasi-Newton methods, methods with direct updates of factorizations, row-scaling methods and column-updating methods. Some implementation features are commented. The survey includes a discussion on global convergence tools and linear-system implementations of Broyden's methods. In the final section, practical and theoretical perspectives of this area are discussed. © 2000 Elsevier Science B.V. All rights reserved.

Keywords: Single equations; Systems of equations

1. Introduction

In this survey we consider nonlinear systems of equations

$$F(x) = 0, \tag{1}$$

where $F : \mathbb{R}^n \to \mathbb{R}^n$ has continuous first partial derivatives. We denote $F = (f_1, \ldots, f_n)^{\mathrm{T}}$ and $J(x) = F'(x)$ for all $x \in \mathbb{R}^n$.

Problem (1) is a particular case of the problem of minimizing $\|F(x)\|_2^2$. However, special methods are far more efficient than minimization and nonlinear least-squares methods for solving this problem, especially when n is large.

All practical algorithms for solving (1) are iterative. Given an initial approximation $x_0 \in \mathbb{R}^n$, a sequence of iterates x_k, $k = 0, 1, 2, \ldots$, is generated in such a way that, hopefully, the approximation to

☆ This work was supported by PRONEX-Optimization Project, FAPESP (grant 90-3724-6), FINEP, CNPq, FAEP-UNICAMP.

0377-0427/00/$ - see front matter © 2000 Elsevier Science B.V. All rights reserved.
PII: S 0377-0427(00)00434-9

some solution is progressively improved. Newton's is the most widely used method in applications. See [27,55,81,94]. The Newtonian iteration is defined whenever $J(x_k)$ is nonsingular. In this case, the iterate that follows x_k is given by

$$x_{k+1} = x_k - J(x_k)^{-1}F(x_k). \qquad (2)$$

The Jacobian inverse $J(x_k)^{-1}$ does not need to be calculated. Instead, $s_k \in \mathbb{R}^n$ results from solving

$$J(x_k)s_k = -F(x_k) \qquad (3)$$

and the new iterate is defined by

$$x_{k+1} = x_k + s_k. \qquad (4)$$

Newton's method has very attractive theoretical and practical properties: if x_* is a solution of (1) at which $J(x_*)$ is nonsingular and x_0 is close enough to x_*, then x_k converges superlinearly to x_*. This means that, given an arbitrary norm $\|\cdot\|$ in $\mathbb{R}^n$,

$$\lim_{k \to \infty} \frac{\|x_{k+1} - x_*\|}{\|x_k - x_*\|} = 0. \qquad (5)$$

Moreover, if $J(x)$ satisfies the Lipschitz condition

$$\|J(x) - J(x_*)\| \leqslant L\|x - x_*\| \qquad (6)$$

for all x close enough to x_*, the convergence is quadratic, so the error at iteration $k+1$ is proportional to the square of the error at iteration k. In other words, the number of correct digits of the approximation x_{k+1} tends to double the number of correct digits of x_k.

Another remarkable property of Newton's method is its invariancy with respect to linear transformations both in the range-space and in the domain space. Invariancy in the range space means that, given any nonsingular matrix A, the iterates of the method applied to

$$AF(x) = 0$$

coincide with the iterates of the method applied to (1). Domain space invariancy means that the iterates of the method applied to

$$F(Ay) = 0$$

are given by $A^{-1}x_k$, provided that $y_0 = A^{-1}x_0$, where $\{x_k\}$ is the sequence generated by (2). The main consequence of invariancy is that bad scaling of the variables or the components of the system cannot affect the performance of the method, if rounding errors (which can affect the quality of the solution of (3)) are disregarded.

The Newton iteration can be costly, since partial derivatives must be computed and the linear system (3) must be solved at every iteration. This fact motivated the development of quasi-Newton methods, which are defined as the generalizations of (2) given by

$$x_{k+1} = x_k - B_k^{-1}F(x_k). \qquad (7)$$

In quasi-Newton methods, the matrices B_k are intended to be approximations of $J(x_k)$. In many methods, the computation of (7) does not involve computing derivatives at all. Moreover, in many particular methods, B_{k+1}^{-1} is obtained from B_k^{-1} using simple procedures thanks to which the linear algebra cost involved in (7) is much less than the one involved in (3).

According to definition (7), Newton's method is a quasi-Newton method. So is the *stationary Newton* method, where $B_k = J(x_0)$ for all $k = 0, 1, 2, \ldots$ and Newton's method "with p refinements", in which $B_k = J(x_k)$ when k is a multiple of $p + 1$, whereas $B_k = B_{k-1}$ otherwise. The "discrete Newton" method is a quasi-Newton method too. It consists in defining

$$B_k = \left(\frac{F(x_k + h_{k,1}e_1) - F(x_k)}{h_{k,1}}, \ldots, \frac{F(x_k + h_{k,n}e_n) - F(x_k)}{h_{k,n}} \right) \tag{8}$$

where $\{e_1, \ldots, e_n\}$ is the canonical basis of $\mathbb{R}^n$ and $h_{k,j} \neq 0$ is a discretization parameter. This parameter must be small enough so that the difference approximation to the derivatives is reliable but large enough so that rounding errors in the differences (8) are not important.

In many problems, $J(x)$ is a sparse matrix, whose sparsity pattern is known. In this case, a procedure given in [20] and refined in [18] (see also [17,58]) allows one to compute a finite difference approximation to $J(x)$ using less than n auxiliary functional evaluations. When the Jacobian matrix is dense, the discrete Newton method is not competitive with the cheap linear algebra versions of (7). But, in many large sparse problems, discrete Newton implementations are quite effective. In these cases, the finite difference technique allows one to compute the approximate Jacobian using a small number of functional evaluations and the matrix structure is such that factorization is not expensive.

In the sixties it was common to justify the existence of most quasi-Newton methods saying that the task of computing derivatives is prone to human errors. However, automatic differentiation techniques have been developed in the last 20 yr that, in practice, eliminates the possibility of error. See [31,45,50,54,87–89] and many others. Moreover, in most cases, the computation of derivatives using automatic differentiation is not expensive. This implies that, in modern practice, the most interesting quasi-Newton methods are those in which the Jacobian approximations are defined in such a way that much linear algebra is saved per iteration. It must be warned that there are many *minimization* problems in which automatic differentiation techniques cannot be applied to compute gradients [19,86] but this is not frequent in *nonlinear systems* coming from practical applications.

Usually, in large and sparse problems, the resolution of (3) using direct methods [32,37,104] is expensive but not prohibitive. (When it is prohibitive it is probably better to use inexact-Newton methods [7,22,55].) In these cases, to use $B_0 = J(x_0)$ generating B_k, $k \geqslant 1$, using cheap linear algebra quasi-Newton techniques is worthwhile.

The name "quasi-Newton" was used after 1965 to describe methods of the form (7) such that the equation

$$B_{k+1}s_k = y_k \equiv F(x_{k+1}) - F(x_k) \tag{9}$$

was satisfied for all $k = 0, 1, 2, \ldots$ See [9]. Eq. (9) was called "the fundamental equation of quasi-Newton methods". Following the Dennis–Schnabel book [27], most authors call quasi-Newton to all the methods of the form (7), whereas the class of methods that satisfy (9) are called "secant methods". Accordingly, (9) is called "secant equation".

The iteration (7) admits an interesting and pedagogical interpretation. Assume that, for all $k = 0, 1, 2, \ldots$ we approximate $F(x)$ by a "linear model"

$$F(x) \approx L_k(x) \equiv F(x_k) + B_k(x - x_k). \tag{10}$$

Then, x_{k+1} is the unique solution of the simpler problem $L_k(x) = 0$. By (10) we also have that

$$L_k(x_k) = F(x_k) \quad \text{for all } k = 0, 1, 2, \ldots \tag{11}$$

It is easy to see that (9) implies that

$$L_k(x_{k-1}) = F(x_{k-1}) \quad \text{for all } k = 1, 2, \dots \tag{12}$$

Therefore, the affine function $L_k(x)$ interpolates $F(x)$ at x_k and x_{k-1}. "Multipoint" secant methods can be defined satisfying

$$L_k(x_j) = F(x_j) \quad \text{for all } j \in I_k, \tag{13}$$

where $\{k-1, k\} \subset I_k$ for all $k = 1, 2, \dots$ See [4,5,12,13,36,42,51,59,66,67,81,92,94,103].

This survey is organized as follows. In Section 2 we sketch a local convergence theory that applies to most secant methods introduced after 1965. In Section 3 we give the most used examples of least-change secant-update methods. In Section 4 we introduce interesting quasi-Newton methods that cannot be justified by the theory of Section 2. In Section 5 we discuss large-scale implementations. In Section 6 we show how to deal with possible singularity of the matrices B_k. In Section 7 we discuss procedures used for obtaining global convergence. In Section 8 we study the behavior of some quasi-Newton methods for linear systems. In Section 9 we survey a few numerical studies on large-scale problems. Finally, in Section 10, we discuss the prospective of the area and we formulate some open problems.

2. Least-change update theory

Most practical quasi-Newton methods can be analyzed under the framework of a general theory introduced in [72]. See, also, [73,75]. This framework can be useful to understand practical methods. However, this section can be skipped at a first reading of this paper, without risk of missing the main algorithmic ideas presented in the remaining sections.

In our analysis, we will use a finite dimensional linear space $\mathscr{E}$ with a scalar product $\langle \cdot, \cdot \rangle_{x,z}$ determined by each pair $x, z \in \mathbb{R}^n$. Denote $|E|^2_{x,z} = \langle E, E \rangle_{x,z}$, where $E \in \mathscr{E}$. Let $V(x,z) \subset \mathscr{E}$ denote an affine subspace determined by any fixed pair $x, z \in \mathbb{R}^n$.

The general algorithm analyzed in this section is defined by (7), where

$$B_k = \varphi(x_k, E_k), \tag{14}$$

where $\varphi : \mathbb{R}^n \times \mathscr{E} \to \mathbb{R}^{n \times n}$. The initial approximation $x_0 \in \mathbb{R}^n$ and the initial parameter $E_0 \in \mathscr{E}$ are arbitrary. Moreover, the parameters are generated by

$$E_{k+1} = P_k(E_k), \tag{15}$$

where $P_k \equiv P_{x_k x_{k+1}}$ is the projection operator on $V(x_k, x_{k+1})$, with respect to the norm $|\cdot|_{x_k x_{k+1}}$. Therefore, E_{k+1} is the parameter in $V(x_k, x_{k+1})$ which is closest to E_k. This justifies the term "least-change" in the definition of these methods.

The most simple example of (14), (15) is Broyden's "good" method (BGM) [8], which is defined by

$$\mathscr{E} = \mathbb{R}^{n \times n}, \tag{16}$$

$$|\cdot|_{x,z} = \|\cdot\|_F = \text{the Frobenius norm for all}, z \in \mathbb{R}^n, \tag{17}$$

$$\varphi(x, E) = E \quad \text{for all } x \in \mathbb{R}^n, E \in \mathscr{E} \tag{18}$$

and

$$V(x,z) = \{B \in \mathbb{R}^{n \times n} \mid B(z-x) = F(z) - F(x)\}. \tag{19}$$

Broyden's sparse (or Schubert's) method [10,93] is defined by (7), (16)–(18) and

$$V(x,z) = \{B \in \mathscr{S} \mid B(z-x) = F(z) - F(x)\}, \tag{20}$$

where $\mathscr{S} \subset \mathbb{R}^{n \times n}$ is the set of matrices that have the sparsity pattern of $J(x)$. See [6] for a variation of this method.

Broyden's "bad" method (BBM) is defined by (7), (16) and (17),

$$\varphi(x,E) = E^{-1} \quad \text{for all } x \in \mathbb{R}^n, \ E \in \mathscr{E}, E \text{ nonsingular} \tag{21}$$

and

$$V(x,z) = \{H \in \mathbb{R}^{n \times n} \mid H[F(z) - F(x)] = z - x\}. \tag{22}$$

Many other examples are given in [72,73]. In most cases $|\cdot|_{x,z}$ does not depend on x and z. However, situations where $|\cdot|_{x,z}$ changes appear when one analyzes quasi-Newton methods with symmetric Jacobian. This is the case of function minimization. The analysis of the popular DFP and BFGS methods for unconstrained optimization require explicit dependence of the norm with respect to x,z. See [27]. In Section 3.3 we will define least-change methods where φ explicitly depends of x.

Under standard assumptions, which we will consider below, methods defined by (7), (14) and (15) are locally (and "quickly") convergent. The first two are assumptions on the functional F and the remaining ones are assumptions on the method. A convergence analysis for Broyden's method in a situation where the first assumption is violated can be found in [21]. In the rest of this section, $\|\cdot\|$ denotes an arbitrary norm in $\mathbb{R}^n$ as well as its subordinate norm in $\mathbb{R}^{n \times n}$. Moreover, $\langle \cdot, \cdot \rangle$ will denote a scalar product in $\mathscr{E}$ and $|\cdot|$ will be the associated norm. (So, $|E|^2 = \langle E, E \rangle$ for all $E \in \mathscr{E}$.)

Assumption 1. There exists $x_* \in \mathbb{R}^n$ such that $F(x_*) = 0$ and $J(x_*)$ is nonsingular.

Assumption 2. There exists $L > 0$ such that

$$\|J(x) - J(x_*)\| \leq L\|x - x_*\| \tag{23}$$

if x belongs to some neighborhood of x_*.

The following assumption says that there exists an ideal parameter E_* which is associated to the solution x_* in the sense that $\varphi(x_*, E_*)^{-1} J(x_*)$ is close to the identity matrix. From now on, we write $B_* = \varphi(x_*, E_*)$. In many algorithms, $B_* = J(x_*)$.

Assumption 3. There exist $E_* \in \mathscr{E}$ and $r_* \in [0,1)$ such that φ is well defined and continuous in a neighborhood of (x_*, E_*). Moreover, $\varphi(x_*, E_*)$ is nonsingular and

$$\|I - \varphi(x_*, E_*)^{-1} J(x_*)\| \leq r_*. \tag{24}$$

Assumption 3 implies that we could define an ideal iteration, given by

$$x_{k+1} = x_k - B_*^{-1} F(x_k), \tag{25}$$

satisfying

$$\lim_{k\to\infty} x_k = x_* \quad \text{and} \quad \limsup_{k\to\infty} \frac{||x_{k+1} - x_*||}{||x_k - x_*||} \leqslant r_* \tag{26}$$

if x_0 is close enough to x_*. Of course, the ideal method defined by (25) cannot be implemented in practice because we do not know the solution x_*. However, the least-change update theory consists in showing that some implementable methods enjoy the property (26). Observe that, in the case $r_* = 0$, (26) means superlinear convergence.

Let us define, for all $x, z \in \mathbb{R}^n$,

$$\sigma(x, z) = \max\{||x - x_*||, ||z - x_*||\}.$$

Assumption 4. In addition to Assumption 3, for all x, z close enough to x_* there exists $E \in V(x, z)$, $c_1 > 0$ such that

$$|E - E_*| \leqslant c_1 \sigma(x, z). \tag{27}$$

In the description of the algorithm, we saw that E_{k+1} is a projection of E_k on $V(x_k, x_{k+1})$. Assumption 4 says that the distance between E_* and this affine subspace is of the same order as the maximum distance between $\{x_k, x_{k+1}\}$ and x_*. In other words, we are projecting on manifolds that are not far from the ideal parameter E_*. An algorithm where projections are performed on the intersection of manifolds with boxes can be found in [15]. The relation between the different norms used in the projections is given by Assumption 5.

Assumption 5. There exists $c_2 > 0$ such that, for all x, z close enough to x_*, $E \in \mathscr{E}$,

$$|E|_{x,z} \leqslant [1 + c_2 \sigma(x, z)]|E| \quad \text{and} \quad |E| \leqslant [1 + c_2 \sigma(x, z)]|E|_{x,z}. \tag{28}$$

Assumption 5 says that the different norms tend to be the same when x and z are close to x_*. Assumptions 4 and 5 do not guarantee that the approximation of E_k to E_* improves through consecutive iterations. (This improvement certainly occurs if $E_* \in V(x_k, x_{k+1})$.) In fact, E_{k+1} might be a worse approximation to E_* than E_k. However, using these assumptions one can prove that the deterioration of E_{k+1} as an approximation to E_* is bounded in such a way that the "error" $|E_{k+1} - E_*|$ is less than the error $|E_k - E_*|$ plus a term which is proportional to the error $||x_k - x_*||$. This is a typical "bounded deterioration principle", as introduced in [11]. See, also, [25,26,29,91] and many other papers. By bounded deterioration, the parameters E_k cannot escape from a neighborhood of E_* for which it can be guaranteed that local convergence holds. Therefore, Assumptions 1–5 are sufficient to prove the following theorem.

Theorem 6. *Suppose that Assumptions 1–5 hold and let* $r \in (r_*, 1)$. *If* $\{x_k\}$ *is generated by* (7), (14) *and* (15), *there exist* $\varepsilon > 0$, $\delta > 0$ *such that, if* $||x_0 - x_*|| \leqslant \varepsilon$ *and* $|E_0 - E_*| \leqslant \delta$, *the sequence is well-defined, converges to* x_* *and satisfies*

$$||x_{k+1} - x_*|| \leqslant r||x_k - x_*|| \tag{29}$$

for all $k = 0, 1, 2, \dots$

Moreover,

$$\lim_{k \to \infty} |E_{k+1} - E_k| = \lim_{k \to \infty} ||B_{k+1} - B_k|| = 0. \tag{30}$$

At a first sight, result (29) is disappointing because the same result can be obtained (with $r_* = 0$) if one uses (7) with $B_k = J(x_0)$ for all $k = 0, 1, 2, \ldots$ It could be argued that there is no reason for modifying B_k at every iteration if one can obtain the same result not modifying this Jacobian approximation at all. Obviously, (30) also holds for this stationary-Newton choice of B_k.

Fortunately, some additional results help us to prove that, under some conditions, the ideal speed of convergence (26) can be reached. From a well-known theorem of Dennis and Walker [29] the following result can be obtained.

Theorem 7. *In addition to the hypotheses of Theorem 1, suppose that*

$$\lim_{k \to \infty} \frac{||[B_k - B_*](x_{k+1} - x_k)||}{||x_{k+1} - x_k||} = 0. \tag{31}$$

Then, (26) *holds.*

Theorem 7 corresponds, in the case $r_* = 0$, to the well-known Dennis–Moré condition [24], which characterizes the superlinear convergence of sequences generated by (7). Now, by (30), Theorem 2 implies the following more practical result.

Theorem 8. *Assume the hypotheses of Theorem 1, and*

$$\lim_{k \to \infty} \frac{||[B_{k+1} - B_*](x_{k+1} - x_k)||}{||x_{k+1} - x_k||} = 0. \tag{32}$$

Then, (26) *holds.*

Theorem 7 says that (26) holds if $B_k v_k \approx B_* v_k$, where v_k is the normalized increment. Since the increment is computed *after* B_k, it is not evident that many methods satisfy this condition. On the other hand, Theorem 3 says that (26) holds if $B_{k+1} v_k \approx B_* v_k$. Observe that the increment is computed *before* B_{k+1}. Since, in general, we know how to approximate $B_*(x_{k+1} - x_k)$ (for example, if $B_* = J(x_*)$, we have that $B_*(x_{k+1} - x_k) \approx F(x_{k+1}) - F(x_k)$) the task of computing B_{k+1} satisfying (32) is not so difficult. The most popular situation corresponds to the case $B_* = J(x_*)$ and consists in defining $V(x, z)$ in such a way that B_{k+1} satisfies the secant equation (9). In this case, (32) is equivalent to

$$\lim_{k \to \infty} \frac{||F(x_{k+1}) - F(x_k) - J(x_*)(x_{k+1} - x_k)||}{||x_{k+1} - x_k||} = 0$$

and this identity holds, if $x_k \to x_*$, due to the assumption (23).

The most important consequence of Theorem 3 is that superlinear convergence of the sequence $\{x_k\}$ takes place when $B_* = J(x_*)$.

None of the theorems above imply that, even when $r_* = 0$, E_k converges to E_*. Simple counterexamples can be shown where this is not true. Moreover, nothing guarantees that E_k is convergent at all. Even in the case of BGM, the best studied least-change secant-update method, it is not known if, under the conditions that are sufficient to prove local-superlinear convergence, the sequence of matrices B_k is convergent.

3. Some least-change secant-update methods

3.1. Broyden's methods

Broyden's "good" method is defined by (7) and (14)–(19). A simple quadratic programming exercise shows that, for this method,

$$B_{k+1} = B_k + \frac{(y_k - B_k s_k)s_k^T}{s_k^T s_k}. \tag{33}$$

Moreover, the relation between the inverses of B_k and B_{k+1} is, in this case,

$$B_{k+1}^{-1} = B_k^{-1} + \frac{(s_k - B_k^{-1} y_k)s_k^T B_k^{-1}}{s_k^T B_k^{-1} y_k}. \tag{34}$$

This formula shows that iteration (7) can be computed without solving a linear system at each iteration. For computing B_{k+1}^{-1} we only need to perform $O(n^2)$ operations, whereas $O(n^3)$ operations are necessary for solving a (dense) linear system. It is generally believed that the most stable way in which BGM can be implemented (when the number of variables is small) requires to store the QR factorization of B_k. Since B_{k+1} differs from B_k by a rank-one matrix, the factorization of B_{k+1} can be obtained using $O(n)$ plane rotations. See [80].

Broyden's "bad" method is given by (7), (16), (17), (21) and (22). As in the case of BGM, after some linear algebra the calculations can be organized so that the definition of the method becomes

$$B_{k+1}^1 = B_k^{-1} + \frac{(s_k - B_k^{-1} y_k)y_k^T}{y_k^T y_k} \tag{35}$$

for all $k = 0, 1, 2, \ldots$ Moreover, according to (35) we have

$$B_{k+1} = B_k + \frac{(y_k - B_k s_k)y_k^T B_k}{y_k^T B_k s_k}. \tag{36}$$

From (34) it is easy to deduce that, if the proper choices are made on the initial point and the initial Jacobian approximation, Broyden's "good" method is invariant under linear transformations in the range space. From (35) we see that Broyden's "bad" has the same property in the domain space. Therefore, if rounding errors are not considered and the behavior of Broyden's "good" for $F(x) = 0$ is satisfactory, it must also be satisfactory for solving $AF(x) = 0$. On the other hand, if Broyden's "bad" method works well on $F(x) = 0$, it will also work on $F(Ax) = 0$.

The reasons why BGM is good and BBM is bad are not well understood. Moreover, it is not clear that, in practice, BGM is really better than BBM. In [76] it was observed that, for BGM, since $B_k s_{k-1} = y_{k-1}$, we have, if $k \geqslant 1$,

$$B_{k+1} s_{k-1} - y_{k-1} = \frac{(y_k - B_k s_k)s_k^T s_{k-1}}{s_k^T s_k}.$$

Analogously, for BBM,

$$B_{k+1}s_{k-1} - y_{k-1} = \frac{(y_k - B_k s_k)y_k^T y_{k-1}}{y_k^T B_k s_k}.$$

Therefore, the "secant error" $B_{k+1}s_{k-1} - y_{k-1}$ is, in both cases, a multiple of $y_k - B_k s_k$. It is natural to conjecture that the BGM iteration will be better than the BBM iteration when

$$\frac{|s_k^T s_{k-1}|}{s_k^T s_k} < \frac{|y_k^T y_{k-1}|}{|y_k^T B_k s_k|}. \tag{37}$$

An analogous reasoning involving $B_{k+1}^{-1} y_{k-1} - s_{k-1}$ leads to conjecture that BGM is better than BBM when

$$\frac{|s_k^T s_{k-1}|}{|s_k^T (B_k)^{-1} y_k|} < \frac{|y_k^T y_{k-1}|}{y_k^T y_k}. \tag{38}$$

In [76] a combined method was implemented that chooses BGM or BBM according to the test (38). This method was tested using a set of small problems and turned out to be superior to both BGM and BBM. By (37)–(38) BGM tends to be better than BBM if B_k underestimates the true Jacobian. This means that, if B_0 is arbitrarily chosen and the true Jacobian is "larger than B_0", Broyden's "good" method tends to be better than Broyden's "bad". This is also confirmed by small numerical experiments.

3.2. Direct updates of factorizations

Suppose that, for all $x \in \mathbb{R}^n$, $J(x)$ can be factorized in the form

$$J(x) = M(x)^{-1}N(x), \tag{39}$$

where $N(x) \in \mathscr{S}_1$, $M(x) \in \mathscr{S}_2$ for all $x \in \mathbb{R}^n$, and $\mathscr{S}_1$, $\mathscr{S}_2$ are affine subspaces of $\mathbb{R}^{n \times n}$. A least-change secant update method associated to the factorization (39) can be defined by

$$x_{k+1} = x_k - N_k^{-1} M_k F(x_k). \tag{40}$$

In this method, (N_{k+1}, M_{k+1}) is the row-by-row orthogonal projection of (N_k, M_k) on the affine subspace of $\mathbb{R}^{n \times n} \times \mathbb{R}^{n \times n}$ defined by

$$V = \{(N,M) \in \mathscr{S}_1 \times \mathscr{S}_2 \mid Ns_k = My_k\}. \tag{41}$$

If, in a neighborhood of a solution x_*, $M(x)$ and $N(x)$ are continuous, the theory of Section 2 can be applied to this family of methods to prove that they are locally and superlinearly convergent. See [71]. If (39) represents the LU factorization, we obtain the method introduced in [52]. If we take into account possible sparsity of L^{-1} and U we obtain a method introduced in [16]. Orthogonal factorizations and structured situations were considered in [71]. In this paper it was also shown that the Dennis–Marwil method [23] is a limit method in the family (40)–(41). By this we mean that, although Dennis–Marwil is not a least-change superlinear convergent method, each Dennis–Marwil iteration can be arbitrarily approximated by iterations of the least-change family. Finally, it is easy to show that Broyden's "good" and "bad" methods are also particular cases of (40)–(41). Nontrivial methods based on (40)–(41) can be useful when the system

$$N_k s_k = -M_k F(x_k) \tag{42}$$

is easy to solve.

3.3. Structured methods

Suppose that $J(x) = C(x) + D(x)$ for all $x \in \mathbb{R}^n$, where $C(x)$ is easy to compute whereas $D(x)$ is not. In this case, it is natural to introduce the quasi-Newton iteration:

$$x_{k+1} = x_k - [C(x_k) + D_k]^{-1} F(x_k), \tag{43}$$

where, for each $k = 0, 1, 2, \ldots$, D_{k+1} is a projection of D_k on the affine subspace

$$V_{\text{full}} = \{ D \in \mathbb{R}^{n \times n} \mid D s_k = y_k - C(x_{k+1}) s_k \}. \tag{44}$$

Writing $\bar{y}_k = y_k - C(x_{k+1}) s_k$ and considering the Frobenius projection, we see that

$$D_{k+1} = D_k + \frac{(\bar{y}_k - D_k s_k) s_k^{\mathrm{T}}}{s_k^{\mathrm{T}} s_k}. \tag{45}$$

If $C(x_k)^{-1}$ is easy to compute (perhaps because $C(x)$ has a nice sparsity structure) and k is small, some linear algebra can be saved in the computation of $[C(x_k) + D_k]^{-1} F(x_k)$ using the techniques that will be explained in Section 5.

Sometimes one also knows that $D(x)$ belongs to some fixed affine subspace $\mathscr{S}$ for all $x \in \mathbb{R}^n$. In this case, we can define D_{k+1} as the projection of D_k on

$$V_{\text{structured}} = \{ D \in \mathscr{S} \mid D s_k = \bar{y}_k \equiv y_k - C(x_{k+1}) s_k \}, \tag{46}$$

but formula (45) is not valid anymore, even for Frobenius projections. Moreover, the affine subspace given by (46) can be empty so that the method only makes sense if this definition is conveniently modified. Let us redefine:

$$V_{\text{minimizers}} = \{ \text{Minimizers of } \| D s_k - \bar{y}_k \|_2 \text{ subject to } D \in \mathscr{S} \}. \tag{47}$$

The affine subspace given by (47) is obviously nonempty and, so, it is possible to project on it. Algorithms for computing this projection were given in [26]. Defining $s = z - x$, $y = F(z) - F(x)$, $\bar{y} = y - C(z) s$ and

$$V(x, z) = \{ \text{Minimizers of } \| D s - \bar{y} \|_2 \text{ subject to } D \in \mathscr{S} \} \tag{48}$$

we can apply the theory of Section 2 so that the resulting method turns out to be locally and superlinearly convergent. In principle, Assumption 4 is necessary for proving superlinear convergence. See, also, [29]. However, it can be conjectured that this assumption can be deduced, in this case, from the definitions (46) and (47).

Examples of applied structured quasi-Newton methods can be found, among others, in [3,46,47,56, 61,62].

4. Other secant methods

The Column-Updating method (COLUM) was introduced in [69] with the aim of reducing the computational cost of BGM. The idea is that, at each iteration, only the jth column of B_k is changed, where j is defined by $||s_k||_\infty = |[s_k]_j|$. So, COLUM is defined by (7) and

$$B_{k+1} = B_k + \frac{(y_k - B_k s_k)e_{j_k}^T}{e_{j_k}^T s_k} \tag{49}$$

where $\{e_1, \ldots, e_n\}$ is the canonical basis of $\mathbb{R}^n$ and $|e_{j_k}^T s_k| = ||s_k||_\infty$. The QR and the LU factorizations of B_{k+1} can be obtained from the corresponding factorizations of B_k using classical linear-programming updating techniques. By (49), we have that

$$B_{k+1}^{-1} = B_k^{-1} + \frac{(s_k - B_k^{-1} y_k)e_{j_k}^T B_k^{-1}}{e_{j_k}^T B_k^{-1} y_k}. \tag{50}$$

Partial convergence results for COLUM were given in [39,69,74]. It has been proved that COLUM enjoys local and superlinear convergence if the method is restarted (taking $B_k = J(x_k)$) every m iterations, where m is an arbitrary positive integer. Moreover, when the method (with or without restarts) converges, the convergence is r-superlinear and quadratic every $2n$ iterations. Finally, COLUM (without restarts) is superlinearly convergent if $n = 2$.

The Inverse Column-Updating method (ICUM), introduced in [78], is given by (7) and

$$B_{k+1}^{-1} = B_k^{-1} + \frac{(s_k - B_k^{-1} y_k)e_{j_k}^T}{e_{j_k}^T y_k}, \tag{51}$$

where $|e_{j_k}^T y_k| = ||y_k||_\infty$. Therefore, B_{k+1}^{-1} is identical to B_k^{-1} except on the j_kth column. So,

$$B_{k+1} = B_k + \frac{(y_k - B_k s_k)e_{j_k}^T B_k}{e_{j_k}^T B_k s_k}. \tag{52}$$

Similar local convergence results to those of COLUM were given in [60,78].

It is easy to see that COLUM and ICUM have the invariancy properties of BGM and BBM respectively. Probably, combined methods in the sense of [76] can also be efficient. See the rationale preceding formula (38) in Section 3.1 of this survey.

The discussion that leads to (38) suggests the introduction of quasi-Newton methods of the form

$$B_{k+1} = B_k + \frac{(y_k - B_k s_k)v_k^T}{v_k^T s_k}, \tag{53}$$

where $v_k \perp s_{k-1}$, or

$$B_{k+1}^{-1} = B_k^{-1} + \frac{(s_k - B_k^{-1} y_k)w_k^T}{w_k^T y_k}, \tag{54}$$

where $w_k \perp y_{k-1}$. These methods are close to the multipoint secant methods studied in [4,5,12,13,36, 42,51,59,66,67,81,92,94,103] in the sense that they satisfy an additional interpolatory condition. Their convergence analysis using the techniques of the above cited papers must be easy, but their practical efficiency does not seem to have been studied. Some authors [49,98] choose the parameter w_k in (54) with the aim of maintaining well-conditioning properties of the matrix B_k.

The first quasi-Newton method with direct updates of factorizations was introduced by Dennis and Marwil in [23]. We did not talk about this method in Section 3.2 because this is not a least-change method in the sense of Section 2. The Dennis–Marwil algorithm modifies the upper-triangular factor of the LU factorization of B_k at each iteration, so that the secant equation is always satisfied (with some stability safeguards). See, also, [83,99]. Convergence results for the Dennis–Marwil method are even weaker than the ones that can be proved for the Stationary Newton method commented in Section 1. The work [23] inspired the introduction of other methods with direct updates of factorizations with stronger convergence results. We have already mentioned the least-change secant update methods introduced in [71], which enjoy local and superlinear convergence. Other methods, having the same theoretical convergence properties as the Stationary Newton method, were introduced in [41,68,70]. The Row-Scaling method (see [41]) is particularly simple and, sometimes, quite effective. It consists in the updating $B_{k+1} = D_k B_k$, where D_k is diagonal and it is chosen so that the secant equation is satisfied, when this is possible. The good numerical properties of the Row-Scaling method are quite surprising. Unfortunately, this updating technique cannot be used in function minimization because it does not preserve possible symmetry of the Jacobian approximations.

We finish this section mentioning the quasi-Newton method introduced by Thomas [100], which is given by

$$B_{k+1}^{-1} = B_k^{-1} + \frac{(s_k - B_k^{-1} y_k) d_k^{\mathrm{T}} B_k^{-1}}{d_k^{\mathrm{T}} B_k^{-1} y_k}, \tag{55}$$

where

$$d_k = [R_k + (\|s_k\|_2/2)I]s_k$$

and

$$R_{k+1} = (1 + \|s_k\|_2)\left(\|s_k\|_2 I + R_k - \frac{d_k d_k^{\mathrm{T}}}{d_k^{\mathrm{T}} s_k}\right).$$

The properties of this method are not yet well understood. However, in spite of its larger cost per iteration, very good numerical results have been reported in several works. See, for example [48].

5. Large-scale implementations

The best known general-purpose modern implementations of quasi-Newton methods for solving large nonlinear systems are based on rank-one correction formulae like BGM, BBM, COLUM and ICUM. See [39,41,64]. Unfortunately, the methods based on direct updates of factorizations which have pleasant convergence properties [71] need sparsity of the L^{-1} factor in the LU decomposition of the true Jacobian, a property that holds only in very structured problems.

A crucial decision involves the choice of the initial Jacobian approximation B_0. The most favorable situation occurs when one is able to compute a good approximation of $J(x_0)$ (perhaps using automatic differentiation or the techniques given in [18,20]) and the LU factorization of this approximation is sparse. In this case, after possible permutations of rows and columns, we compute

$$B_0 = LU \tag{56}$$

and we use this sparse factorization throughout the calculations.

If a sparse factorization of a suitable approximation of $J(x_0)$ is not available, it is sensible to use

$$B_0^{-1} = \left(\frac{\nabla f_1(x_0)}{||\nabla f_1(x_0)||_2^2}, \ldots, \frac{\nabla f_n(x_0)}{||\nabla f_n(x_0)||_2^2} \right). \tag{57}$$

In (57), B_0 approximates $J(x_0)$ in the sense that the $J(x_0)B_0^{-1}$ has only 1's on the diagonal. With this choice the initial iteration is scale-invariant and the vector s_0 is a descent direction for $||F(x)||_2^2$ (see Section 7). Of course, some alternative choice must be employed for a column of B_0^{-1} if the involved gradient is null.

In BGM and COLUM, we have

$$B_{k+1} = B_k + \frac{(y_k - B_k s_k)v_k^T}{v_k^T s_k}, \tag{58}$$

and, consequently,

$$B_{k+1}^{-1} = B_k^{-1} + \frac{(s_k - B_k^{-1} y_k)v_k^T}{v_k^T B_k^{-1} y_k} B_k^{-1}. \tag{59}$$

Therefore,

$$B_{k+1}^{-1} = (I + u_k v_k^T)B_k^{-1}, \tag{60}$$

where $u_k = (s_k - B_k^{-1} y_k)/v_k^T B_k^{-1} y_k$. Thus,

$$B_k^{-1} = (I + u_{k-1}v_{k-1}^T)\ldots(I + u_0 v_0^T)B_0^{-1}. \tag{61}$$

Formula (61) shows that methods of the form (58) can be implemented associated to (56) or (57) adding O(n) operations and storage positions per iteration. By (61), for computing $B_k^{-1}F(x_k)$ one needs to store u_j, v_j, $j = 0, 1, \ldots, k - 1$ and the factorization (or the inverse) of B_0. Moreover, this computation involves the solution of a linear system whose matrix is B_0 plus a sequence of k operations consisting in a scalar product, a scalar-vector product and the sum of two vectors. The whole procedure can be quite economic if k is small but becomes prohibitive if k is large. In the case of COLUM, it is obvious that only one additional vector is needed per iteration. For BGM, a clever trick given in [30] allows one to implement (7,61) storing only one additional vector per iteration. See [39,41,43,79].

In BBM and ICUM we have

$$B_{k+1}^{-1} = B_k^{-1} + \frac{(s_k - B_k^{-1} y_k)w_k^T}{w_k^T y_k}. \tag{62}$$

Therefore, defining $u_k = (s_k - B_k^{-1} y_k)/w_k^T y_k$, we obtain

$$B_{k+1}^{-1} = B_0^{-1} + u_0 w_0^T + \cdots + u_k w_k^T. \tag{63}$$

This formula suggests straightforward associations of (62) with (56) or (57). A recent numerical study by Lukšan and Vlček [64] indicates that ICUM could be the most effective secant method for large-scale problems with the initial choice (56).

6. Dealing with singularity

The quasi-Newton iteration (7) is well defined only if B_k is nonsingular. Local convergence theories usually assume that $J(x_*)$ is nonsingular and that B_0 is close to $J(x_*)$. Under these conditions it can usually be proved that B_k is nonsingular for all k. However, in practice, the initial choice of B_0 could be singular and, moreover, B_{k+1} could be singular even when B_k is not.

Singularity of B_0 might occur when one chooses $B_0 = J(x_0)$ (or some very good approximation of the Jacobian). Since the (nonsingular) Newton step minimizes $\|J(x_0)s + F(x_0)\|$, it is natural, in the singular case, to choose s_0 as any minimizer of $\|J(x_0)s + F(x_0)\|_2^2$. Choosing the minimum-norm minimizer, we obtain

$$s_0^\dagger = -J(x_0)^\dagger F(x_0), \tag{64}$$

where $J(x_0)^\dagger$ is the Moore–Penrose pseudoinverse of the initial Jacobian (see [38]). Using a well-known approximation of the pseudoinverse, we can also compute, for some $\mu > 0$,

$$s_0(\mu) = -(J(x_0)^\mathrm{T} J(x_0) + \mu I)^{-1} J(x_0)^\mathrm{T} F(x_0). \tag{65}$$

When $\mu \to 0$, $s_0(\mu)$ tends to $s_0^\dagger$. The step $s_0(\mu)$ can be interpreted as the minimizer of $\|J(x_0)s + F(x_0)\|_2$ on a ball whose radius is smaller than $\|s_0^\dagger\|_2$.

In practical computations, singularity of $J(x_0)$ is detected during the LU factorization of this matrix: at some stage of the LU algorithm it is impossible to choose a safe nonnull pivot. When the problem is large, and $J(x_0)$ is possibly sparse, computing (65) is expensive and, so, this device is seldom used. It is usually preferred to continue the LU factorization replacing the null or very small pivot by some suitable nonnull quantity that takes into account the scaling of the matrix. See [41]. There is no strong justification for this procedure except that, perhaps, it is not necessary to choose carefully B_0 when x_0 is far from the solution. (Even this statement can be argued.)

On the other hand, a singular B_{k+1} can appear even if B_k is nonsingular.

When B_{k+1} is obtained from B_k by means of a secant rank-one correction,

$$B_{k+1} = B_k + \frac{(y_k - B_k s_k) v_k^\mathrm{T}}{v_k^\mathrm{T} s_k}, \tag{66}$$

as in the case of BGM and COLUM, we have

$$\det(B_{k+1}) = \frac{v_k^\mathrm{T} B_k^{-1} y_k}{v_k^\mathrm{T} s_k} \det(B_k). \tag{67}$$

If

$$\left| \frac{v_k^\mathrm{T} B_k^{-1} y_k}{v_k^\mathrm{T} s_k} \right|$$

is very small or very large then, either the scaling of B_{k+1} is very different from that of B_k or their stability characteristics are very different. Conservative small-variation arguments recommend us to impose

$$\sigma |\det(B_k)| \leqslant |\det(B_{k+1})| \leqslant \frac{1}{\sigma} |\det(B_k)|, \tag{68}$$

where $\sigma \in (0, 1)$ is small (say, $\sigma \approx 0.1$). By (67), if

$$\left| \frac{v_k^\mathrm{T} B_k^{-1} y_k}{v_k^\mathrm{T} s_k} \right| \notin [\sigma, 1/\sigma],$$

the inequalities (68) do not hold and B_{k+1} must be modified. Following [80], we can replace (66) by

$$B_{k+1} = B_k + \eta_k \frac{(y_k - B_k s_k)v_k^{\mathrm{T}}}{v_k^{\mathrm{T}} s_k}, \tag{69}$$

where $\eta_k \in [0,1]$. Clearly, (68) is satisfied if $\eta_k = 0$, but $\eta_k = 1$ is the best choice in the sense that B_{k+1} satisfies the secant equation. Therefore, it is natural to choose η_k as the maximum $\eta \in [0,1]$ such that (68) is satisfied. This motivates the definition

$$\eta_k = \max \left\{ \eta \in [0,1] \,\Big|\, \sigma \leqslant \left| (1-\eta) + \eta \frac{v_k^{\mathrm{T}} B_k^{-1} y_k}{v_k^{\mathrm{T}} s_k} \right| \leqslant \frac{1}{\sigma} \right\}. \tag{70}$$

In "inverse" rank-one correction methods like BBM and ICUM, it is easier to write directly

$$B_{k+1}^{-1} = B_k^{-1} + \frac{(s_k - B_k^{-1} y_k)w_k^{\mathrm{T}}}{w_k^{\mathrm{T}} y_k}. \tag{71}$$

An analogous reasoning to the one used to choose (70) leads us to the modification

$$B_{k+1}^{-1} = B_k^{-1} + \eta_k \frac{(s_k - B_k^{-1} y_k)w_k^{\mathrm{T}}}{w_k^{\mathrm{T}} y_k} \tag{72}$$

and, consequently, to the choice

$$\eta_k = \max \left\{ \eta \in [0,1] \,\Big|\, \sigma \leqslant \left| (1-\eta) + \eta \frac{w_k^{\mathrm{T}} B_k s_k}{w_k^{\mathrm{T}} y_k} \right| \leqslant \frac{1}{\sigma} \right\}. \tag{73}$$

Both in the initial iteration as in the updated ones a close-to-singular matrix B_k usually generates a very large increment s_k. Very simple step-control procedures are always associated to the implementation of quasi-Newton methods. In practical problems, it has been verified that opportunistic ways of controlling the step-length may prevent many divergence situations.

7. Global convergence tools

The results presented in this paper are local, in the sense that convergence to a solution can be guaranteed if the solution is assumed to exist and both the initial point and the initial Jacobian approximation are close enough to the solution and its Jacobian, respectively.

It is of maximal practical importance to analyze what happens with sequences generated by quasi-Newton methods when no restrictions are made on the initial approximations. Unfortunately, almost nothing positive can be said about sequences generated by pure formulae like (7), unless strong assumptions are made on F. Newtonian sequences can oscillate between neighborhoods of two or more nonsolutions or tend to infinity, even in problems where a unique solution exists. So, if we want to devise algorithms with global convergence properties, the basic iteration (7) must be modified.

Usually, modifications of the basic iteration make use of some merit function. Almost always, some norm of $F(x)$ is used. The squared 2-norm $\|F(x)\|_2^2$ is frequently preferred because of its differentiability properties. We will call $f(x)$ the (continuous and nonnegative) merit function, whose main property is that $f(x) = 0$ if, and only if, $F(x) = 0$. Therefore, the problem of solving $F(x) = 0$

turns out to be equivalent to the problem of finding a global minimizer of $f(x)$. If, at a global minimizer, $f(x)$ does not vanish, the original system has no solution at all. For simplicity, assume that

$$f(x) = \tfrac{1}{2}\|F(x)\|_2^2, \tag{74}$$

so

$$\nabla f(x) = J(x)^{\mathrm{T}} F(x). \tag{75}$$

From (75), we see that the Newton direction fits well with the necessity of decreasing $f(x)$. Computing the directional derivative, we obtain

$$\langle -J(x)^{-1} F(x), \nabla f(x) \rangle = -2f(x) < 0,$$

So, it is always possible to decrease $f(x)$ along the Newton direction, if this direction is well defined and $f(x) \neq 0$. Many algorithms can be interpreted as adaptations of unconstrained optimization techniques (see [27]) to the minimization of $f(x)$. In particular, the iteration

$$x_{k+1} = x_k - \alpha_k J(x_k)^{-1} F(x_k) \tag{76}$$

has been exhaustively analyzed. See [33,65] and references therein. If $\alpha_k > 0$ is chosen is such a way that $f(x_{k+1})$ is sufficiently smaller than $f(x_k)$, then every limit point of the sequence generated by (76) either is a solution or a point where the Jacobian is singular. So, if the Jacobian is nonsingular for all $x \in \mathbb{R}^n$ and $f(x)$ has bounded level sets, (76) necessarily finds a solution. Finally, in a vicinity of such a solution it can be proved that $\alpha_k \equiv 1$ satisfies the sufficient decrease requirements, therefore the method (76) coincides, ultimately, with (2) and the convergence is quadratic.

The merit function (74) brings difficulties in connection with nonsymmetric quasi-Newton methods because the direction $-B_k^{-1} F(x_k)$ is not, in general, a descent direction for f. This is one of the reasons why it is important to use good initial Jacobian approximations in this context, whereas diagonal initial Hessian approximations are usually efficient in function minimization. Griewank [44] has proved that Broyden's "good" method, with a suitable line search, also has global convergence properties assuming uniform nonsingularity of the Jacobians. Li and Fukushina [57] introduced a line search for BGM that ensures global and superlinear convergence, if the merit function has bounded level sets and the Jacobians are nonsingular.

Other attempts for globalization of quasi-Newton methods (without nonsingularity assumptions) rely on the exploration of the good descent properties of Newton. Among these we can cite:

1. Hybrid strategies [80,85], in which Broyden's iteration are combined with special iterations which are, essentially, discretizations of Newton iterations.
2. Nonmonotone strategies [34]: here "ordinary" quasi-Newton iterations are accepted, even if the merit function is increased during some iterations, but the algorithm switches to a Newton iteration if a given tolerance is violated.
3. A strategy due to Bonnans and Burdakov [14]: if the sufficient decrease condition is violated the step-length is reduced, but, at the same time, the Jacobian approximation is updated using a secant formula. As a result, the search direction changes during the current iteration and tends to the Newton direction. An antecedent of this idea can be found in [84].

A common drawback of all the globalization strategies based on decreasing a norm is that local-nonglobal minimizers of $f(x)$ are strong attractors of the iterative process. Other norm-minimization related techniques can be found in [53,96,97]. Therefore, globalized algorithms can converge to points in which the Jacobian is singular. Unfortunately, such points are completely useless from the point of view of finding solutions of the nonlinear system. It is easy to see that all the observations related to the Newton direction made in this section, except the ones related to rapid local convergence, are valid for the choice (57) of the Jacobian approximation.

A completely different source of globalization procedures is the homotopic approach, by means of which a sequence of slightly modified problems are solved, in such a way that the first one is trivial and the last one is (1). For example, the "regularizing homotopy", used in [101,102] is

$$H(x,t) = tF(x) + (1-t)(x - x_0). \tag{77}$$

The solution of $H(x,0) = 0$ is, obviously, x_0 and the solution of $H(x,1) = 0$ is the one required in (1). Many methods for tracing the homotopy path are described in the literature. Locally convergent quasi-Newton methods are useful tools in this case since strategies like (77) deal with several nonlinear systems for which good initial estimates are available. See, also, [1,2,90].

8. Results for linear systems

In this section we assume that $F(x) = Ax - b$, $A \in \mathbb{R}^{n \times n}$, $b \in \mathbb{R}^n$. To study the behavior of quasi-Newton methods for linear systems is important under different points of view. On one hand, real-life problems can be linear or nearly linear. On the other hand, the properties of a method in the linear case usually determine the local convergence behavior of the method in the nonlinear case. In a neighborhood of a solution where the Jacobian is nonsingular, the linear approximation of F is dominant and, so, the generated sequence tends to behave as in the linear case. For example, if $F(x) = Ax - b$ and A is nonsingular, Newton's method is well defined and converges in just one iteration. This is the main reason why the local convergence of this method is quadratic.

Until 1979 it was believed that Broyden's methods did not enjoy finite convergence when applied to linear systems. However, in [35] it was proved that Broyden's method and many other methods of the form (66) or (71) also converge in a finite number of steps.

Let us consider the method defined by (7) and (66). Gay's theorem [35] says that, if A and B_0 are nonsingular and $x_0 \in \mathbb{R}^n$ is arbitrary, then $F(x_k) = 0$ for some $k \leqslant 2n$. The convergence of x_k to $x_* \equiv A^{-1}b$ is far from being monotone in any sense.

The local convergence consequences of Gay's theorem for general nonlinear systems are that, under the usual assumptions that guarantee local convergence, methods like BGM, BBM, COLUM and ICUM enjoy $2n$-step quadratic convergence. Therefore, $\|x_{k+2n} - x_*\|/\|x_k - x_*\|^2$ is asymptotically bounded above. This property implies r-superlinear convergence. See [81].

The finite convergence theorem [35] sheds light on theoretical properties of rank-one secant methods but is of little importance for practical large-scale linear problems. The intermediate iterations (x_k with $k < 2n$) are, usually, very poor approximations of the solution so that the full cycle of $2n$ steps is necessary for obtaining a reasonable approximation of x_*. When n is large, a sequence of $2n$ iterations is not affordable for the methods considered in Section 5, since the cost of the kth iteration is proportional to kn, both in terms of time and storage. Therefore, practical implementations of rank-one secant methods for linear systems need modifications of the basic iteration (7). See the discussion in [30].

Some authors [30,77,80] studied variations of BGM for linear systems. Here we survey the results presented in [77], correcting, by the way, some arithmetic typos of that paper. Given $x_0 \in \mathbb{R}^n$ and $B_0 \in \mathbb{R}^{n \times n}$ nonsingular, the linear Broyden method is defined by

$$x_{k+1} = x_k - \lambda_k B_k^{-1} F(x_k), \tag{78}$$

where $\lambda_k \neq 0$ and

$$B_{k+1} = B_k + \eta_k \frac{(y_k - B_k s_k) s_k^{\mathsf{T}}}{s_k^{\mathsf{T}} s_k}. \tag{79}$$

The coefficient $\eta_k \in [0.9, 1.1]$ is such that

$$|\det(B_{k+1})| \geqslant 0.1 |\det(B_k)|. \tag{80}$$

Moré and Trangenstein [80] proved that (80) holds with $\eta_k \in [0.9, 1.1]$ defining $\gamma_k = s_k^{\mathsf{T}} B_k^{-1} y_k / s_k^{\mathsf{T}} s_k$, with $\eta_k = 1$ if $|\gamma_k| \geqslant 0.1$, and $\eta_k = (1 - 0.1 \,\mathrm{sign}(\gamma_k))/(1 - \gamma_k)$ if $|\gamma_k| < 0.1$, where $\mathrm{sign}(0) = 1$. This choice of η_k provides the number closest to unity such that (80) is satisfied. See [80] and Section 6 of this paper.

For the method defined by (78)–(80) it can be proved that

$$\|B_k - A\|_{\mathrm{F}} \leqslant \|B_0 - A\|_{\mathrm{F}}$$

for all $k = 0, 1, 2, \dots$ and

$$\|B_{k+1} - A\|_{\mathrm{F}}^2 \leqslant \|B_k - A\|_{\mathrm{F}}^2 - 0.891 \,\|B_{k+1} - B_k\|_{\mathrm{F}}^2$$

for all $k = 0, 1, 2, \dots$. It follows that the series $\sum \|B_{k+1} - B_k\|_{\mathrm{F}}^2$ is convergent. So, $\|B_{k+1} - B_k\|$ tends to 0.

It can also be proved that the sequence generated by (78)–(80) satisfies

$$\frac{\|x_{k+1} - x_*\|}{\|x_k - x_*\|} \leqslant \frac{\varepsilon_k + |\lambda_k - 1|}{1 - \varepsilon_k} \tag{81}$$

for all $k = 0, 1, 2, \dots$, where $\{\varepsilon_k\}$ is a sequence that tends to zero.

Formula (81) explains the behavior of the error $x_k - x_*$ independently of the convergence of the sequence. In particular, it shows that the sequence is superlinearly convergent if $\lambda_k \to 1$, and that convergence at a linear rate takes place if, eventually, $\lambda_k \in [\sigma, 2 - \sigma]$ for some $\sigma > 0$.

Finally, in [77] it has been proved that $\lambda_k \to 1$ holds when one chooses λ_k as the (nonnull) minimizer of $\|A(x_k + \lambda d_k) - b\|_2^2$ along the direction $d_k \equiv -B_k^{-1} F(x_k)$. If this λ_k is null, we replace it by 1. However, this possible replacement is not necessary for k large enough.

As a result, we have a global and superlinearly convergent BGM-like method for solving linear nonsingular systems. The proposed choice of λ_k has an advantage over the choice $\lambda_k = 1$ in the large-scale case. When λ_k is the one-dimensional minimizer proposed above, the residual norm at the iterate x_{k+1} is smaller than the norm of $Ax_k - b$. Therefore, in terms of the residual norm, the quality of the approximation is improved at every iteration, and an acceptable final approximation can be (perhaps) obtained for $k \ll 2n$. An alternative choice with similar theoretical properties that, in some sense, minimizes a norm of the error, has been considered in [30] and [77].

The effectivity of Broyden-like methods for solving large-scale linear systems is associated to the availability of good preconditioners. If the initial matrix B_0 is defined as the available preconditioner, a small number of iterations can be expected, at least when one uses clever choices of the steplength.

In [30] it has been claimed that these alternatives are competitive with standard Krylov-subspace methods for solving linear systems. However, much research is necessary on this subject both form the theoretical and the practical point of view.

9. Numerical studies

In this section we comment some numerical studies involving the application of quasi-Newton methods for solving large-scale nonlinear systems of equations.

The study [41] involves 7 variably dimensioned nonlinear systems. Six of them are "toy problems" and have been designed with the aim of testing numerical algorithms. The seventh is the discretization of a Poisson equation. The algorithms are Newton's method, the Stationary Newton method, Broyden's "good" method, Broyden's sparse (Schubert) method, the Dennis–Marwil method and three direct-update methods that includes the row-scaling method mentioned in Section 4. Matrix factorizations use the algorithm of George and Ng [37] and a nonmonotone globalization procedure is incorporated.

The study [40] uses 3 discretizations of two-dimensional boundary-value problems with known solutions: Poisson, Bratu and convection–diffusion. The three of them depend on a parameter λ according to which the problem is more or less difficult. If $\lambda = 0$ the problems are linear. If $\lambda \ll 0$, noncoercivity is severe and the discretized problems are very hard. The tested algorithms are Newton, Stationary Newton, BGM and COLUM. All the algorithms have the option of using backtracking to improve global convergence.

The study [34] solves a set of problems given in [63] having similar characteristics to the set of problems of [41]. In addition, a discretization of the driven cavity flow problem is also considered, which has a parameter λ, the Reynolds number, that controls nonlinearity. Finally, the study includes a convection–diffusion problem and a set of artificial problems where Newton's method (without step control) do not converge.

The study [64] includes 30 problems. 16 of them are of the type considered in [41] with some superposition with that set. In addition, the study has countercurrent reactor problems, second-order boundary value problems (including Poisson and convection–diffusion), problems of flow in a channel, swirling flow problems, porous medium problems, a nonlinear biharmonic problem and the driven cavity problem. The objective of this study is to introduce a globalization procedure. The underlying quasi-Newton methods are the discrete Newton method, the Stationary Newton method, the sparse Broyden (Schubert) method, the variation due to Bogle and Perkins [6], Li's method [58], a combination of Li with Schubert, the row-scaling method [41], Broyden's "good" method, COLUM and ICUM.

None of the above cited studies contradicts the common belief that Newton's method is the most robust algorithm for solving nonlinear systems. Concerning globalization procedures, experiments recommend to be cautious, because in many problems the attempts to reduce the sum of squares lead to convergence to local-nonglobal minimizers. As a matter of fact, the simple stabilization procedure that consists in not letting the step-length to be too large (see Section 6) is, frequently, very effective to turn a divergent algorithmic sequence into a convergent one.

When convergence is maintained, quasi-Newton corrections usually improve substantially the performance of Newton's method. The amount of this improvement depends of the Jacobian structure.

In the problems considered in the above cited studies, methods that do not save linear algebra, like Broyden-sparse, must be discarded, since its computational cost per iteration is roughly the same as Newton's. Practically all quasi-Newton corrections are more effective than the Stationary Newton method. According to [64], ICUM ranks first, but there seems to be little difference between this method and BGM or COLUM. Up to our knowledge there are no published numerical studies for large-scale problems where Broyden's "bad" method is included.

10. Conclusions and perspectives

In recent years, quasi-Newton methods for solving square smooth nonlinear systems have been out of the mainstream of numerical analysis research. A popular scientific journal on Numerical Analysis published 4 papers on the subject before 1970, 10 between 1971 and 1980, 11 in the eighties and none from 1991 to 1999. Sometimes, research in a family of numerical techniques becomes out-of-fashion after its incorporation to ordinary practice of problem solvers in Physics, Chemistry, Engineering and Industry. Other times, promising algorithms are completely forgotten, both in research and applications.

The situation of the area surveyed in this paper is perhaps intermediate. The classical paper [25] is cited in most works concerning quasi-Newton methods for nonlinear systems. While this survey was being written it had been cited 361 times in indexed scientific journals. The last 100 citations go from 1992 to the present days. 42 of these citations come from non-mathematical journals. It must be warned that, frequently, the Dennis–Moré paper [25] is cited in connection to quasi-Newton methods for minimization problems, and not for nonlinear systems. Since the everyday practice in Physics, Chemistry and Engineering includes the resolution of nonlinear systems using Newton's method, we are tempted to conclude that the penetration of the quasi-Newton technology in applications, although existing, has not been as intense at the potentiality of the technique deserves.

In the introduction of most quasi-Newton papers, it was stressed that the main motivation was to avoid computation of cumbersome derivatives. However, even before the boom of automatic differentiation, practitioners found that, for many of their problems, computing derivatives was not as difficult or costly as stated in the quasi-Newton literature. They also verified that beginning a quasi-Newton process with $B_0 = I$, or some other arbitrary matrix, very often causes disastrous results and, so, the computation of an initial Jacobian is almost always necessary. Moreover, the programming effort of computing the initial Jacobian is the same as the one necessary for computing all the Jacobians, so the tendency of many practitioners has been to use Newton's method or its stationary variation with refinements.

In practical problems in which the Jacobian can be computed but its structure is too bad for factorization, the modern tendency is to use the inexact-Newton approach [22], in which an iterative linear solver is used for solving the Newtonian linear equation $J(x_k)s = -F(x_k)$ up to some precision which is sufficient to guarantee convergence of the nonlinear solver. Moreover, the inexact-Newton technology fits well with global convergence requirements. Probably, many users felt disappointed when they tried to globalize quasi-Newton methods by the mere introduction of a damping parameter and backtracking procedures.

However, a reasonable scope of problems exists, for which quasi-Newton methods that save linear algebra are quite effective and, probably, outperform inexact-Newton algorithms. This is the case of large-scale problems in which the Jacobian can be computed, its factorization is affordable but it is very costly in comparison to the single updating procedures of rank-one methods. The recipe for those cases is to begin with a Newtonian iteration, and to continue with some cheap rank-one method as far as this is effective. Unfortunately, a code like that must be prepared to return to Newtonian iterations, a disappointing fact for those who hoped that quasi-Newton techniques could always replace Newton.

Quasi-Newton methods for solving large-scale nonlinear systems will be largely used in applications when both numerical analysts and potential users be conscious about their real advantages and limitations. Our point of view is that rank-one algorithms provide, in many problems, efficient and economic ways to refine a basic (first) Newtonian iteration. If we are right, questions often neglected in the quasi-Newton literature, as "when should one restart?" must be answered, in spite of its poor theoretical appeal.

We finish this survey stating 10 open problems, some of which were incidentally mentioned in the text.

1. It is well known that, under the usual nonsingularity and Lipschitz assumptions, the matrices B_k generated by Broyden's "good" method do not necessarily converge to $J(x_*)$. Does this sequence of matrices always have a limit? What happens with the sequences $\{B_k\}$ corresponding to other methods?

2. Convergence theorems for least-change update and other quasi-Newton methods say that there exist $\varepsilon, \delta > 0$ such that $x_k \to x_*$ superlinearly whenever $\|x_0 - x_*\| \leqslant \varepsilon$ and $\|B_0 - J(x_*)\| \leqslant \delta$. Is this superlinear convergence uniform? In other words, for which methods can we prove that "there exist $\varepsilon, \delta > 0$ and a sequence of positive numbers $\varepsilon_k \to 0$ such that whenever $\|x_0 - x_*\| \leqslant \varepsilon$ and $\|B_0 - J(x_*)\| \leqslant \delta$, the sequence x_k converges to x_* and $\|x_{k+1} - x_*\| \leqslant \varepsilon_k \|x_k - x_*\|$ for all k"?

3. Is it possible to prove local convergence without restarts of methods like COLUM and ICUM? What about superlinear convergence?

4. Are there reasonable sufficient conditions under which the convergence of Broyden-like methods for linear systems takes place in less than $2n$ iterations?

5. It is generally accepted that the Dennis–Marwil method (and some other similar direct factorization algorithms) enjoys local convergence only if periodic Jacobian restarts are performed. However, no counterexample showing that local convergence without the restarting condition might not hold is known. Does a counterexample exist in the linear case?

6. Does there exist a cheap and theoretically justified procedure for modifying the LU factorization of B_0 when a null or very small pivot is found?

7. Is it possible to prove that Assumption 4 necessarily holds for the choice (48) of $V(x,z)$?

8. Which are the properties of direct-secant-update and structured quasi-Newton methods when applied to linear systems?

9. The order of convergence of Newton's method with p refinements (the Jacobian is repeated during p consecutive iterations) is $2 + p$. See [81,82,95]. This means that $\|x_{k+p+1} - x_*\|/\|x_k - x_*\|^{2+p}$ is asymptotically bounded. Can something better be expected when, instead of repeating the previous Jacobian, we update it with a secant formula?

10. Many methods in the flourishing interior point field for mathematical programming can be interpreted as clever damped Newton iterations on an homotopic basis. Can they be improved by suitable quasi-Newton updates? (Up to our knowledge, no attempt has been made in this sense, except the one in [28].)

J.M. Martínez / Journal of Computational and Applied Mathematics 124 (2000) 97–121

Acknowledgements

The author is indebted to two anonymous referees for their very careful reading of the first draft of this versions and for many suggestions that led to its improvement.

References

[1] F. Aluffi-Pentini, V. Parisi, F. Zirilli, A differential equations algorithm for nonlinear equations, ACM Trans. Math. Software 10 (1984) 299–316.
[2] F. Aluffi-Pentini, V. Parisi, F. Zirilli, DAFNE: differential-equations algorithm for nonlinear equations, ACM Trans. Math. Software 10 (1984) 317–324.
[3] J.H. Ávila, P. Concus, Update methods for highly structured systems of nonlinear equations, SIAM J. Numer. Anal. 16 (1979) 260–269.
[4] J.G.P. Barnes, An algorithm for solving nonlinear equations based on the secant method, Comput. J. 8 (1965) 66–72.
[5] L. Bittner, Eine Verallgemeinerung des Sekantenverfahrens zur näherungsweisen Berechnung der Nullstellen eines nichtlinearen Gleichngssystems, Will. Z. Tech. Univ. Dresden 9 (1959) 325–329.
[6] I.D.L. Bogle, J.D. Perkins, A new sparsity preserving quasi-Newton update for solving nonlinear equations, SIAM J. Sci. Statist. Comput. 11 (1990) 621–630.
[7] P.N. Brown, Y. Saad, Convergence theory of nonlinear Newton–Krylov algorithms, SIAM J. Optim. 4 (1994) 297–330.
[8] C.G. Broyden, A class of methods for solving nonlinear simultaneous equations, Math. Comput. 19 (1965) 577–593.
[9] C.G. Broyden, Quasi-Newton methods and their applications to function minimization, Math. Comput. 21 (1967) 368–381.
[10] C.G. Broyden, The convergence of an algorithm for solving sparse nonlinear systems, Math. Comp. 19 (1971) 577–593.
[11] C.G. Broyden, J.E. Dennis Jr., J.J. Moré, On the local and superlinear convergence of quasi-Newton methods, J. Inst. Math. Appl. 12 (1973) 223–245.
[12] O. Burdakov, Stable versions of the secant method for solving systems of equations, U.S.S.R. Comput. Math. Math. Phys. 23 (1983) 1–10.
[13] O. Burdakov, On superlinear convergence of some stable variants of the secant method, Z. Angew. Math. Mech. 66 (1986) 615–622.
[14] O. Burdakov, private communication, 1998.
[15] P.H. Calamai, J.J. Moré, Quasi-Newton updates with bounds, SIAM J. Numer. Anal. 24 (1987) 1434–1441.
[16] F.F. Chadee, Sparse quasi-Newton methods and the continuation problem, T.R.S.O.L.85-8, Department of Operations Research, Stanford University, 1985.
[17] T.F. Coleman, B.S. Garbow, J.J. Moré, Software for estimating sparse Jacobian matrices, ACM Trans. Math. Software 11 (1984) 363–378.
[18] T.F. Coleman, J.J. Moré, Estimation of sparse Jacobian matrices and graph coloring problems, SIAM J. Numer. Anal. 20 (1983) 187–209.
[19] A.R. Conn, K. Scheinberg, Ph.L. Toint, Recent progress in unconstrained nonlinear optimization without derivatives, Math. Programming 79 (1997) 397–414.
[20] A.R. Curtis, M.J.D. Powell, J.K. Reid, On the estimation of sparse Jacobian matrices, J. Inst. Math. Appl. 13 (1974) 117–120.
[21] D.W. Decker, C.T. Kelley, Broyden's method for a class of problems having singular Jacobian at the root, SIAM J. Numer. Anal. 22 (1985) 563–574.
[22] R.S. Dembo, S.C. Eisenstat, T. Steihaug, Inexact Newton methods, SIAM J. Numer. Anal. 19 (1982) 400–408.
[23] J.E. Dennis Jr., E.S. Marwil, Direct secant updates of matrix factorizations, Math. Comp. 38 (1982) 459–476.
[24] J.E. Dennis Jr., J.J. Moré, A characterization of superlinear convergence and its application to quasi-Newton methods, Math. Comp. 28 (1974) 549–560.

[25] J.E. Dennis Jr., J.J. Moré, Quasi-Newton methods, motivation and theory, SIAM Rev. 19 (1977) 46–89.

[26] J.E. Dennis Jr., R.B. Schnabel, Least change secant updates for quasi-Newton methods, SIAM Rev. 21 (1979) 443–459.

[27] J.E. Dennis Jr., R.B. Schnabel, Numerical Methods for Unconstrained Optimization and Nonlinear Equations, Prentice-Hall, Englewood Cliffs, NJ, 1983.

[28] J.E. Dennis Jr., M. Morshedi, K. Turner, A variable metric variant of the Karmarkar algorithm for linear programming, Math. Programming 39 (1987) 1–20.

[29] J.E. Dennis Jr., H.F. Walker, Convergence theorems for least-change secant update methods, SIAM J. Numer. Anal. 18 (1981) 949–987.

[30] P. Deuflhard, R. Freund, A. Walter, Fast secant methods for the iterative solution of large nonsymmetric linear systems, Impact Comput. Sci. Eng. 2 (1990) 244–276.

[31] L.C.W. Dixon, Automatic differentiation and parallel processing in optimisation, TR No. 180, The Hatfield Polytechnique, Hatfield, UK, 1987.

[32] I.S. Duff, A.M. Erisman, J.K. Reid, Direct Methods for Sparse Matrices, Oxford Scientific Publications, Oxford, 1989.

[33] S.C. Eisenstat, H.F. Walker, Globally convergent inexact Newton methods, SIAM J. Optim. 4 (1994) 393–422.

[34] A. Friedlander, M.A. Gomes-Ruggiero, D.N. Kozakevich, J.M. Martínez, S.A. Santos, Solving nonlinear systems of equations by means of quasi-Newton methods with a nonmonotone strategy, Optim. Methods Software 8 (1997) 25–51.

[35] D.M. Gay, Some convergence properties of Broyden's method, SIAM J. Numer. Anal. 16 (1979) 623–630.

[36] D.M. Gay, R.B. Schnabel, Solving systems of nonlinear equations by Broyden's method with projected updates, in: O. Mangasarian, R. Meyer, S. Robinson (Eds.), Nonlinear Programming 3, Academic Press, New York, pp. 245 –281.

[37] A. George, E. Ng, Symbolic factorization for sparse Gaussian elimination with partial pivoting, SIAM J. Sci. Statist. Comput. 8 (1987) 877–898.

[38] G.H. Golub, Ch.F. Van Loan, Matrix Computations, The Johns Hopkins University Press, Baltimore, 1989.

[39] M.A. Gomes-Ruggiero, J.M. Martínez, The column-updating method for solving nonlinear equations in Hilbert space, Math. Modelling Numer. Anal. 26 (1992) 309–330.

[40] M.A. Gomes-Ruggiero, D.N. Kozakevich, J.M. Martínez, A numerical study on large-scale nonlinear solvers, Comput. Math. Appl. 32 (1996) 1–13.

[41] M.A. Gomes-Ruggiero, J.M. Martínez, A.C. Moretti, Comparing algorithms for solving sparse nonlinear systems of equations, SIAM J. Sci. Statist. Comput. 13 (1992) 459–483.

[42] W.B. Gragg, G.W. Stewart, A stable variant of the secant method for solving nonlinear equations, SIAM J. Numer. Anal. 13 (1976) 127–140.

[43] A. Griewank, The solution of boundary value problems by Broyden based secant methods, in: J. Noye, R. May (Eds.), Proceedings of the Computational Techniques and Applications Conference, CTAC-85, North-Holland, Amsterdam, 1986.

[44] A. Griewank, The 'global' convergence of Broyden-like methods with a suitable line search, J. Austral. Math. Soc. Ser. B 28 (1986) 75–92.

[45] A. Griewank, Achieving logarithmic growth of temporal and spacial complexity in reverse automatic differentiation, Optim. Methods Software 1 (1992) 35–54.

[46] W.E. Hart, F. Soesianto, On the solution of highly structured nonlinear equations, J. Comput. Appl. Math. 40 (1992) 285–296.

[47] W.E. Hart, S.O.W. Soul, Quasi-Newton methods for discretized nonlinear boundary value problems, J. Inst. Math. Appl. 11 (1973) 351–359.

[48] Z. Huang, E. Spedicato, Numerical testing of quasi-Newton and some other related methods for nonlinear systems, Quaderni del Dipartimento di Matematica, Statistica, Informatica e Applicazioni 4, Universitá degli Studi di Bergamo, Bergamo, Italy, 1994.

[49] C.M. Ip, M.J. Todd, Optimal conditioning and convergence in rank one quasi-Newton updates, SIAM J. Numer. Anal. 25 (1988) 206–221.

[50] M. Iri, Simultaneous computations of functions, partial derivatives and estimates of rounding errors, Complexity and Practicality, Japan J. Appl. Math. 1 (1984) 223–252.

[51] J. Jankowska, Theory of multivariate secant methods, SIAM J. Numer. Anal. 16 (1979) 547–562.

[52] G.W. Johnson, N.H. Austria, A quasi-Newton method employing direct secant updates of matrix factorizations, SIAM J. Numer. Anal. 20 (1983) 315–325.

[53] I.E. Kaporin, O. Axelsson, On a class of nonlinear equation solvers based on the residual norm over a sequence of affine subspaces, SIAM J. Sci. Comput. 16 (1995) 228–249.

[54] G. Kedem, Automatic differentiation of computer programs, ACM Trans. Math. Software 6 (1980) 150–165.

[55] C.T. Kelley, Iterative Methods for Linear and Nonlinear Equations, SIAM, Philadelphia, PA, 1995.

[56] C.T. Kelley, E.W. Sachs, A quasi-Newton method for elliptic boundary value problems, SIAM J. Numer. Anal. 24 (1987) 516–531.

[57] D.H. Li, M. Fukushima, A derivative-free line search and global convergence of Broyden-like method for nonlinear equations, Optim. Methods Software 13 (2000) 181–201.

[58] G. Li, Successive column correction algorithms for solving sparse nonlinear systems of equations, Math. Programming 43 (1989) 187–207.

[59] T.L. Lopes, J.M. Martínez, Combination of the sequential secant method and Broyden's method with projected updates, Computing 25 (1980) 379–386.

[60] V.L.R. Lopes, J.M. Martínez, Convergence properties of the inverse Column-Updating method, Optim. Methods Software 6 (1995) 127–144.

[61] A. Lucia, Partial molar excess properties, null spaces and a new update for the hybrid method of chemical process design, AICHE J. 31 (1995) 558–566.

[62] A. Lucia, D.C. Miller, A. Kumar, Thermodynamically consistent quasi-Newton formulas, A.I.C.H.E. J. 31 (1985) 1381–1388.

[63] L. Lukšan, Inexact trust region method for large sparse systems of nonlinear equations, J. Optim. Theory Appl. 81 (1994) 569–590.

[64] L. Lukšan, J. Vlček, Computational experience with globally convergent descent methods for large sparse systems of nonlinear equations, Optim. Methods Software 8 (1998) 185–199.

[65] Z. Lužanin, N. Krejić, D. Herceg, Parameter selection for Inexact Newton method, Nonlinear Anal. Theory Methods Appl. 30 (1997) 17–24.

[66] J.M. Martínez, Three new algorithms based on the sequential secant method, BIT 19 (1979) 236–243.

[67] J.M. Martínez, On the order of convergence of Broyden–Gay–Schnabel's method, Comment. Math. Univ. Carolin. 19 (1979) 107–118.

[68] J.M. Martínez, A quasi-Newton method with a new updating for the LDU factorization of the approximate Jacobian, Mat. Apl. Comput. 2 (1983) 131–142.

[69] J.M. Martínez, A quasi-Newton method with modification of one column per iteration, Computing 33 (1984) 353–362.

[70] J.M. Martínez, Quasi-Newton methods with factorization scaling for solving sparse nonlinear systems of equations, Computing 38 (1987) 133–141.

[71] J.M. Martínez, A family of quasi-Newton methods for nonlinear equations with direct secant updates of matrix factorizations, SIAM J. Numer. Anal. 27 (1990) 1034–1049.

[72] J.M. Martínez, Local convergence theory of inexact Newton methods based on structured least change updates, Math. Comp. 55 (1990) 143–168.

[73] J.M. Martínez, On the relation between two local convergence theories of least change secant update methods, Math. Comp. 59 (1992) 457–481.

[74] J.M. Martínez, On the convergence of the Column-updating method, Comput. Appl. Math. 12 (1992) 83–94.

[75] J.M. Martínez, Fixed-point quasi-Newton methods, SIAM J. Numer. Anal. 29 (1992) 1413–1434.

[76] J.M. Martínez, L.S. Ochi, Sobre dois métodos de Broyden, Mat. Apl. Comput. 1 (1982) 135–141.

[77] J.M. Martínez, L. Qi, Inexact Newton methods for solving nonsmooth equations, J. Comput. Appl. Math. 60 (1995) 127–145.

[78] J.M. Martínez, M.C. Zambaldi, An inverse Column-updating method for solving large-scale nonlinear systems of equations, Optim. Methods Software 1 (1992) 129–140.

[79] H. Matthies, G. Strang, The solution of nonlinear finite element equations, Internat. J. Numer. Methods Eng. 14 (1979) 1613–1626.

[80] J.J. Moré, J.A. Trangenstein, On the global convergence of Broyden's method, Math. Comp. 30 (1976) 523–540.

[81] J.M. Ortega, W.G. Rheinboldt, Iterative Solution of Nonlinear Equations in Several Variables, Academic Press, New York, 1970.

[82] A.M. Ostrowski, Solution of Equations in Euclidean and Banach Spaces, Academic Press, New York, 1973.

[83] J.R. Paloschi, J.D. Perkins, The updating of LU-factors in quasi-Newton methods, Comput. Chem. Eng. 10 (1986) 241–247.

[84] E. Polak, A globally converging secant method with applications to boundary value problems, SIAM J. Numer. Anal. 11 (1974) 529–537.

[85] M.J.D. Powell, A hybrid method for nonlinear equations, in: P. Rabinowitz (Ed.), Numerical Methods for Nonlinear Algebraic Equations, Gordon and Breach, London, 1970, pp. 87–114.

[86] M.J.D. Powell, Direct search algorithms for optimization calculations, Acta Numer. (1998) 287–336.

[87] L.B. Rall, Automatic Differentiation – Techniques and Applications, Springer Lecture Notes in Computer Science, Vol 120, Springer, Berlin, 1981.

[88] L.B. Rall, Differentiation in PASCAL-SC: type gradient, ACM Trans. Math. Software 10 (1984) 161–184.

[89] L.B. Rall, Optimal implementation of differentiation arithmetic, in: U. Külisch (Ed.), Computer Arithmetic, Scientific Computation and Programming Languages, Teubner, Stuttgart, 1987.

[90] W.C. Rheinboldt, Numerical Analysis of Parametrized Nonlinear Equations, Wiley, New York, 1986.

[91] W.C. Rheinboldt, J.S. Vandergraft, On the local convergence of update methods, SIAM J. Numer. Anal. 11 (1974) 1069–1085.

[92] R.B. Schnabel, Quasi-Newton methods using multiple secant equations, TR CU-CS-247-83, Department of Computer Science, University of Colorado at Boulder, 1983.

[93] L.K. Schubert, Modification of a quasi-Newton method for nonlinear equations with a sparse Jacobian, Math. Comp. 24 (1970) 27–30.

[94] H. Schwetlick, Numerische Lösung Nichtlinearer Gleichungen, Deutscher Verlag der Wissenschaften, Berlin, 1978.

[95] V.E. Shamanskii, A modification of Newton's method, Ukrain Mat. Z. 19 (1967) 133–138.

[96] Y.X. Shi, Solving nonlinear systems using a global-local procedure, Z. Angew. Math. Mech. 76 (1996) 539–540.

[97] Y.X. Shi, A globalization procedure for solving nonlinear systems of equations, Numer. Algorithms 12 (1996) 273–286.

[98] E. Spedicato, J. Greenstadt, On some classes of variationally derived quasi-Newton algorithms for systems of nonlinear algebraic equations, Numer. Math. 29 (1978) 363–380.

[99] R.P. Tewarson, Y. Zhang, Sparse quasi-Newton LDU update, Internat. J. Numer. Methods Eng. 24 (1987) 1093–1100.

[100] S.W. Thomas, Sequential estimation techniques for quasi-Newton algorithms, Technical Report TR 75-227, Cornell University, 1975.

[101] L.T. Watson, A globally convergent algorithm for computing points of C^2 maps, Appl. Math. Comput. 5 (1979) 297–311.

[102] L.T. Watson, S.C. Billups, A.P. Morgan, Algorithm 652: HOMPACK: A suite of codes for globally convergent homotopy algorithms, ACM Trans. Math. Software 13 (1987) 281–310.

[103] P. Wolfe, The secant method for solving nonlinear equations, Comm. ACM 12 (1959) 12–13.

[104] Z. Zlatev, J. Wasniewski, K. Schaumburg, Y12M, Solution of Large and Sparse Systems of Linear Algebraic Equations, Lecture Notes in Computer Science, Vol. 121, Springer, New York, 1981.

ELSEVIER   Journal of Computational and Applied Mathematics 124 (2000) 123–137

JOURNAL OF
COMPUTATIONAL AND
APPLIED MATHEMATICS

www.elsevier.nl/locate/cam

Sequential quadratic programming for large-scale nonlinear optimization ✩

Paul T. Boggs[a],*, Jon W. Tolle[b]

[a] Computational Sciences and Mathematics Research Department, Sandia National Laboratories, Livermore,
CA 94550, USA
[b] Departments of Operations Research and Mathematics, University of North Carolina, Chapel Hill, NC 27514, USA

Received 12 July 1999; received in revised form 17 December 1999

Abstract

The sequential quadratic programming (SQP) algorithm has been one of the most successful general methods for solving nonlinear constrained optimization problems. We provide an introduction to the general method and show its relationship to recent developments in interior-point approaches, emphasizing large-scale aspects. © 2000 Elsevier Science B.V. All rights reserved.

Keywords: Sequential quadratic programming; Nonlinear optimization; Newton methods; Interior-point methods; Local; Trust-region methods convergence; Global convergence

1. Introduction

In this article we consider the general method of Sequential Quadratic Programming (SQP) for solving the nonlinear programming problem

$$\text{minimize}_{x} \quad f(x)$$

$$\text{subject to} \quad h(x) = 0,$$
$$\qquad\qquad\quad g(x) \leqslant 0, \tag{NLP}$$

where $f : \mathscr{R}^n \to \mathscr{R}$, $h : \mathscr{R}^n \to \mathscr{R}^m$, and $g : \mathscr{R}^n \to \mathscr{R}^p$. Broadly defined, the SQP method is a procedure that generates iterates converging to a solution of this problem by solving quadratic programs that are approximations to (NLP). In its many implemented forms, this method has been shown to be a

---

✩ Contribution of Sandia National Laboratories and not subject to copyright in the United States.
* Corresponding author.
E-mail addresses: ptboggs@ca.sandia.gov (P.T. Boggs), tolle@email.unc.edu (J.W. Tolle).

0377-0427/00/$ - see front matter © 2000 Elsevier Science B.V. All rights reserved.
PII: S 0377-0427(00)00429-5

very useful tool for solving nonlinear programs, especially where a significant degree of nonlinearity is present. In this paper our goals are to provide a brief synopsis of the general method, to introduce some of the more recent results, and to provide direction for further investigation. As part of our exposition we relate SQP to the applications of interior-point methods to nonlinear programming. The discussion will be more motivational than rigorous, our emphasis being on the exposition of underlying issues and ideas rather than on detailed theorems and implementation techniques. We will, for the most part, discuss the algorithm and its properties without making any special assumptions about the structure of the problem. While the SQP algorithm is applicable to all sizes of nonlinear programming problems, problems of large scale (i.e., a large number of variables and/or constraints) are the most challenging and therefore the ones where the development of efficient strategies for their solution will have the most impact. Accordingly, our presentation will be slanted towards the procedures which are likely to prove useful for solving large problems. We will provide explicit references for the recent theoretical and computational results, but the literature for SQP is immense and it is beyond the scope of this paper to do it justice. Instead we direct the reader to [5] for a more comprehensive list of sources.

An outline of the paper is as follows: a particular basic formulation of the nonlinear program and its corresponding necessary conditions will be presented in Section 2 followed, in Section 3, by examples of quadratic programming approximations; aspects of the local and global convergence theory will be provided in Sections 4 and 5; in Section 6 various important issues in the solution of the quadratic subproblem will be discussed; and, finally, in Section 7 the important ideas of reduced Hessian SQP methods are presented. A knowledge of basic optimization theory and practice (for example, as developed in [17]) is adequate for following the ideas contained herein.

2. The formulation of the necessary conditions for NLP

We begin by introducing some terminology and notation that is necessary to describe the method; additional terminology and notation will be introduced as it is needed.

Throughout the paper we use bold face letters to represent vectors (both variables and functions) and plain face for scalars and matrices. The subscript or superscript k is used to indicate a kth iterate; similarly an asterisk indicates an optimal solution (or multiplier). We use ∇ to indicate the derivative of a (scalar or vector-valued) function and $\mathcal{H}$ to indicate the Hessian of a scalar function. Sometimes subscripts will be added to indicate the variables with respect to which differentiation is performed; if no subscript is present the differentiation is assumed to be with respect to the vector x only. Unless specified otherwise all norms are assumed to be the Euclidean norm for a vector and the induced operator norm for matrices. We will use the superscript "t" to indicate the transpose of a vector or matrix. Finally, the symbol $x \odot y$ will denote the vector defined by componentwise multiplication of the vectors x and y.

Associated with (NLP) is the *standard Lagrangian function*

$$L(x, u, v) = f(x) + h(x)^{\mathrm{t}}u + g(x)^{\mathrm{t}}v,$$

where u and v are the multiplier (dual) vectors. We denote the active set of inequality constraints at x by $\mathcal{A}(x)$, i.e., $\mathcal{A}(x) = \{i : g_i(x) = 0\}$, and by $D(x)$ the $n \times (m + |\mathcal{A}(x)|)$ matrix whose columns are the gradients (with respect to x) of the equality and active inequality constraints at x. We make the following assumptions concerning (NLP):

A.1 All of the functions in the nonlinear program have Lipschitz continuous second derivatives.

A.2 (NLP) has a feasible (local) solution x^* with optimal multiplier vectors (u^*, v^*) satisfying the first order conditions

$$\nabla_x L(x^*, u^*, v^*) = 0,$$
$$g(x^*) \odot v^* = 0,$$
$$v^* \geqslant 0.$$

A.3 Strict complementary slackness ($i \in \mathscr{A}(x^*)$ implies $v_i^* > 0$) holds.

A.4 The matrix $D(x^*)$ has full column rank.

A.5 The second-order sufficiency condition holds, i.e., for all $y \neq 0$ such that $D(x^*)^t y = 0$ the strict inequality $y^t \mathscr{H} L_* y > 0$ is valid, where $\mathscr{H} L_*$ is the Hessian of the Lagrangian with respect to x evaluated at (x^*, u^*, v^*).

For the analysis that follows it is advantageous to add the vector of slack variables, z, and put the general nonlinear programming problem into slack variable form

$$\underset{(x,z)}{\text{minimize}} \quad f(x)$$

$$\text{subject to} \quad h(x) = 0, \qquad\qquad\qquad\qquad\qquad\qquad\qquad\text{(NLP)}$$
$$\qquad\qquad g(x) + z = 0,$$
$$\qquad\qquad z \geqslant 0.$$

In this form the first-order necessary and feasibility conditions that a solution (x^*, z^*) and its multipliers (u^*, v^*) must satisfy are the following:

$$\nabla f(x^*) + \nabla h(x^*)u^* + \nabla g(x^*)v^* = 0, \tag{1}$$

$$h(x^*) = 0, \tag{2}$$

$$g(x^*) + z^* = 0, \tag{3}$$

$$z^* \odot v^* = 0, \tag{4}$$

$$z^* \geqslant 0, \tag{5}$$

$$v^* \geqslant 0. \tag{6}$$

Defining the *extended Lagrangian function*

$$\mathscr{L}(x, z, u, v) = f(x) + h(x)^t u + (g(x) + z)^t v,$$

we see that solving this version of (NLP) is equivalent to solving the problem

$$\underset{(x,z)}{\text{minimize}} \quad \mathscr{L}(x, z, u, v)$$

$$\text{subject to} \quad h(x) = 0, \qquad\qquad\qquad\qquad\qquad\qquad\qquad\text{(LNLP)}$$
$$\qquad\qquad g(x) + z = 0,$$
$$\qquad\qquad z \geqslant 0$$

for some u and some $v \geqslant 0$ satisfying the complementary slackness conditions $z \odot v = 0$.

The fundamental approach of the SQP method is to solve (NLP) by solving a sequence of quadratic programs that are approximations to (LNLP). In the next section we will provide some of the more common quadratic approximations used in SQP.

3. Examples of quadratic subproblems

The first example, the standard approximation for the basic SQP method, is the Taylor Series approximation to the problem (LNLP). Given an estimate of the variables, (x^k, z^k), and the multiplier vectors, (u^k, v^k), the quadratic program approximation generated by the Taylor Series for a change, (d_x, d_z), in the vectors x^k and z^k is given by

$$\underset{(d_x, d_z)}{\text{minimize}} \quad \tfrac{1}{2} d_x^t \mathscr{H}\mathscr{L}_k d_x + (\nabla_x \mathscr{L}_k)^t d_x + (\nabla_x \mathscr{L}_k)^t d_z$$

$$\text{subject to} \quad \nabla h(x^k)^t d_x = -h(x^k),$$

$$\nabla g(x^k)^t d_x + d_z = -(g(x^k) + z^k),$$

$$d_z \geqslant -z^k,$$

where the subscript on $\mathscr{L}$ indicates that the derivatives are evaluated at (x^k, u^k, v^k). Some simplification of this quadratic program is possible since, when the constraints are satisfied, the gradient of the Lagrangian terms reduce to $\nabla f(x^k)^t d_x$. In addition, $\mathscr{H}\mathscr{L}_k$ is the same as $\mathscr{H}L_k$. The purpose of using the Lagrangian function in the objective function is now clear; even though the constraints are linearized in forming the approximating quadratic program, second order information on the constraint functions is maintained in the objective function via the Hessian of the Lagrangian. In many situations, this Hessian matrix is either unavailable or too costly to evaluate. In these cases a finite difference approximation or a *Quasi-Newton update*, i.e., a matrix that depends on first order information at the preceding iterates, may be used in place of the Hessian. In the latter case, the Hessian approximation is taken to be a positive definite matrix, which makes solving the quadratic subproblem easier. Representing the true Hessian or an approximation thereof by B_k, we can now rewrite the approximating quadratic program as

$$\underset{(d_x, d_z)}{\text{minimize}} \quad \tfrac{1}{2} d_x^t B_k d_x + \nabla f(x^k)^t d_x$$

$$\text{subject to} \quad \nabla h(x^k)^t d_x = -h(x^k), \tag{QP1}$$

$$\nabla g(x^k)^t d_x + d_z = -(g(x^k) + z^k),$$

$$d_z \geqslant -z^k.$$

We note that if the variable d_z is eliminated this quadratic program is a quadratic approximation to the original nonslack form of (NLP).

As a second example of an approximating quadratic program we cast a version of the nonlinear interior-point algorithm into this framework. In this approach, the nonnegative slack variable constraint in (LNLP) is put into the objective function as a log barrier function so that the problem has the form

$$\operatorname*{minimize}_{(\boldsymbol{x},\boldsymbol{z})}\ \mathscr{L}(\boldsymbol{x},\boldsymbol{z},\boldsymbol{u},\boldsymbol{v}) - \rho \sum_{j=1}^{p}(\log(z_j))$$

subject to $\quad \boldsymbol{h}(\boldsymbol{x}) = 0,$
$$\boldsymbol{g}(\boldsymbol{x}) + \boldsymbol{z} = 0,$$

$\qquad\qquad$ (NLPI(ρ))

where ρ is a positive barrier parameter that is chosen to tend to zero in an appropriate manner. The nonnegativity constraint on z is implicitly enforced by the log function. The corresponding quadratic approximation is

$$\operatorname*{minimize}_{(\boldsymbol{d}_x,\boldsymbol{d}_z)}\ \tfrac{1}{2}\boldsymbol{d}_x^{\mathrm{t}}B_k\boldsymbol{d}_x + \tfrac{1}{2}\rho\boldsymbol{d}_z^{\mathrm{t}}Z_k^{-2}\boldsymbol{d}_z + \nabla f(\boldsymbol{x}^k)^{\mathrm{t}}\boldsymbol{d}_x - \rho\mathbf{e}^{\mathrm{t}}Z_k^{-1}\boldsymbol{d}_z$$

subject to $\quad \nabla \boldsymbol{h}(\boldsymbol{x}^k)^{\mathrm{t}}\boldsymbol{d}_x = -\boldsymbol{h}(\boldsymbol{x}^k),$

$$\nabla \boldsymbol{g}(\boldsymbol{x}^k)^{\mathrm{t}}\boldsymbol{d}_x + \boldsymbol{d}_z = -\boldsymbol{g}(\boldsymbol{x}^k + \boldsymbol{z}^k),$$

$\qquad\qquad$ (QP2(ρ))

where $\mathbf{e}$ is the vector of ones, Z_k is the diagonal matrix with components of $\boldsymbol{z}^k$ on the diagonal, and the matrix B_k represents either the Hessian of L or its approximation. The interpretation of the interior-point methods in terms of SQP algorithms is not conventional, but as will be seen it fits naturally into that context in terms of the local convergence analysis. A general interior-point algorithm for solving (NLP) has recently been given in [19].

As a final example we mention a version of an SQP method employing a trust region. In this approach, a constraint limiting the size of the step is included in the constraints. Thus the quadratic subproblem has the form of either (QP1) or (QP2(ρ)) with the added constraint

$$\|(\boldsymbol{d}_x, \boldsymbol{d}_z)\| \leqslant \tau_k. \qquad\qquad (7)$$

Here, as in trust-region algorithms for unconstrained optimization, τ_k is a positive parameter that measures the adequacy of the quadratic approximation to the original problem. A recent work using this approach is [7].

In the implementation of each of these methods, the particular quadratic programs are (approximately) solved to obtain $(\boldsymbol{d}_x, \boldsymbol{d}_z)$. These steps can then be used to compute $\boldsymbol{x}^{k+1}$ and $\boldsymbol{z}^{k+1}$ by

$$\boldsymbol{x}^{k+1} = \boldsymbol{x}^k + \alpha\boldsymbol{d}_x,$$
$$\boldsymbol{z}^{k+1} = \boldsymbol{z}^k + \alpha\boldsymbol{d}_z,$$

where α is a step length parameter that may be used to assure the nonnegativity of $\boldsymbol{z}^{k+1}$ and global convergence (see Section 5). In the case where a trust-region constraint is included, the control of the step length is implicitly included in the solution. Updates, $\boldsymbol{u}^{k+1}$ and $\boldsymbol{v}^{k+1}$, for the multipliers can be determined directly or obtained from the solution of the quadratic programs. We note that there are many methods for solving quadratic programs; for example, active set methods, interior-point methods, and reduced Hessian techniques have all been used in SQP algorithms. Some methods do not lead to estimates of the multiplier vectors directly and so methods for accurately estimating these vectors need to be provided. In Section 6 we discuss some of the methods that are used to solve the quadratic subproblems.

The analysis of the SQP methods can be divided into two distinct, but related, parts: local convergence, which is concerned with the asymptotic rate of the convergence, and global convergence, which concerns the convergence of the iterates when the initial point is not close to the solution. These two issues will be considered in Sections 4 and 5.

4. Local convergence

In this section we discuss the local convergence properties of the SQP method. By local convergence we mean that the algorithm will generate a sequence of iterates that converges to an optimal solution-multiplier vector provided that the initial iterate is sufficiently close to that optimal solution. Associated with this convergence is the asymptotic *rate of convergence* which indicates the rapidity with which the discrepancy between the iterates and the solution goes to zero. Any local convergence results will depend on the details of the implementation of the SQP algorithm and, in particular, on how accurately the quadratic programs are solved. Rather than provide results for any specific implementation, we will use the necessary conditions for the approximating quadratic programs to relate their solutions to the steps taken by Newton's method for solving Eqs. (1)–(6) as described below. Although only locally applicable, Newton's method provides a conventional standard which can be used to measure the asymptotic convergence properties of any particular SQP method. Toward this end, we observe that given a good approximation, (x^k, z^k, u^k, v^k), to the optimal solution and the optimal multiplier vectors of (NLP), Newton's method requires solving the linear system

$$
\begin{bmatrix}
\mathscr{H}L_k & 0 & \nabla h(x^k) & \nabla g(x^k) \\
\nabla h(x^k)^{\mathrm{t}} & 0 & 0 & 0 \\
\nabla g(x^k)^{\mathrm{t}} & I & 0 & 0 \\
0 & V_k & 0 & Z_k
\end{bmatrix}
\begin{bmatrix}
d_x \\
d_z \\
d_u \\
d_v
\end{bmatrix}
=
\begin{bmatrix}
-\nabla_x L_k \\
-h(x^k) \\
-g(x^k) - z^k \\
-Z_k v^k
\end{bmatrix}
\tag{8}
$$

for (d_x, d_z, d_u, d_v) where V_k and Z_k are diagonal matrices whose diagonals are x^k and z^k. The next iterate is then given by

$$
\begin{aligned}
x^{k+1} &= x^k + d_x, \\
z^{k+1} &= z^k + d_z, \\
u^{k+1} &= u^k + d_u, \\
v^{k+1} &= v^k + d_v.
\end{aligned}
\tag{9}
$$

Under our basic assumptions (A.1–A.5) the coefficient matrix for (8) is nonsingular in a neighborhood of the solution and hence Newton's method is well defined and converges quadratically. The nonnegativity restrictions on the slack vector and the inequality multiplier vector are not required to be explicitly enforced because of the local convergence properties of Newton's method.

We first observe that if the B_k are taken as the true Hessians, $\mathscr{H}L_k$, the solutions of the quadratic subproblems (QP1) and (QP2(ρ)) lead to approximations of the Newton step, differing only in the approximation of the complementary slackness conditions. (We note that the local analyses of the trust-region methods generally reduce to that of these two methods since it is assumed that the added constraint will be strictly satisfied as the solution is approached.) As above, we assume that an approximate solution and its corresponding multiplier vectors, (x^k, z^k, u^k, v^k), are known. We denote by $(\boldsymbol{d}_x, \boldsymbol{d}_z, \boldsymbol{u}_{qp}, \boldsymbol{v}_{qp})$ the optimal solutions and multiplier of the quadratic problems (QP1) or (QP2(ρ)). Setting

$$\boldsymbol{d}_u = \boldsymbol{u}_{qp} - \boldsymbol{u}^k,$$
$$\boldsymbol{d}_v = \boldsymbol{v}_{qp} - \boldsymbol{v}_k, \tag{10}$$

the first-order and feasibility conditions for the quadratic programs lead to the first three sets of equations in system (8) together with a fourth set which depends on the particular quadratic approximation. For (QP1) the complementary slackness conditions lead to

$$V_k \boldsymbol{d}_z + Z_k \boldsymbol{d}_v + \boldsymbol{d}_z \odot \boldsymbol{d}_v = -Z_k \boldsymbol{v}^k, \tag{11}$$

while the Lagrangian condition in $\boldsymbol{d}_z$ for (QP2(ρ)) leads to

$$\rho Z_k^{-1} \boldsymbol{d}_z + Z_k \boldsymbol{d}_v = \rho \mathbf{e} - Z_k \boldsymbol{v}^k. \tag{12}$$

Either of these last systems of equations may be taken as a perturbation of the linearized complementary slackness condition in (8) above, i.e., they can be written in the form

$$V_k \boldsymbol{d}_z + Z_k \boldsymbol{d}_v = -Z_k \boldsymbol{v}^k + \mathbf{p}(\boldsymbol{x}^k, \boldsymbol{z}^k, \boldsymbol{v}^k, \boldsymbol{d}_x, \boldsymbol{d}_z, \boldsymbol{d}_v), \tag{13}$$

where $\mathbf{p}$ is some perturbation function. It is worthwhile to point out that a linearization of the complementary slackness condition of the form

$$V_k \boldsymbol{d}_z + Z_k \boldsymbol{d}_v = \rho \mathbf{e} - Z_k \boldsymbol{v}^k \tag{14}$$

is used in place of (12) in most interior-point algorithms. This can also be written in the form (13).

In any case, if the vector $(\boldsymbol{d}_x, \boldsymbol{d}_z, \boldsymbol{u}_{qp}, \boldsymbol{v}_{qp})$ is obtained by solving one of the quadratic subproblems (QP1) or (QP2(ρ)), the vectors $\boldsymbol{d}_u$ and $\boldsymbol{d}_v$ are defined by (10), and new iterates $(\boldsymbol{x}^{k+1}, \boldsymbol{z}^{k+1}, \boldsymbol{u}^{k+1}, \boldsymbol{v}^{k+1})$ are given by (9), then we have a canonical form of an SQP algorithm. The fact that these iterates approximate the Newton iterates provides the underlying motivation for using the SQP method.

If the vector $(\boldsymbol{d}_x, \boldsymbol{d}_z, \boldsymbol{d}_u, \boldsymbol{d}_v)$ is determined by *exactly* solving the particular first-order conditions for the quadratic subproblems (with $B_k = \mathscr{H} L_k$) and the new iterates are obtained using (9), the local convergence theory for SQP methods can be analyzed in terms of a perturbation of the Newton method for solving (1)–(4). Specifically, the results depend on how well the complementary slackness conditions in (8) are approximated by a particular choice of (11), (12), or (14). If there are no inequality constraints the iterates are identical to the Newton iterates for solving (1) and (2). Hence if the initial solution vector $\boldsymbol{x}^0$ and initial multiplier vector $\boldsymbol{u}^0$ are sufficiently close to the optimal solution and multiplier vector then exactly solving the quadratic programs leads to the quadratic convergence of the iterates to the optimal solution-multiplier vector. Note that the initial multiplier vector $\boldsymbol{u}^0$ can always be taken as the least squares solution to the Lagrangian condition when $\boldsymbol{x}^0$ is sufficiently close to the optimal solution.

If inequality constraints are present, a local convergence analysis can be given in the case where the quadratic problem (QP1) is solved exactly by utilizing the fact that when the iterates are sufficiently close to the optimal solution-multiplier vector the active sets at optimality and for (QP1) are the same. Thus the problem essentially reduces to an equality-constrained problem at that point and the Newton theory applies. Quadratic convergence can also be obtained in the interior-point scheme provided that the parameter ρ is chosen appropriately (relevant references can be found in [16]).

From a practical point of view, the local convergence analysis depends on many factors other than the form of the approximation of the complementary slackness conditions; it also depends on the details that determine the specific implementation of an SQP method such as the accuracy of the solution of the quadratic approximation and how the step length parameter is chosen. Here we provide general local convergence results based on the perturbation method described above and then discuss how possible implementations fit (or do not fit) into this scheme. The basic procedure can be described as follows:

- Let the solutions and multipliers to the quadratic subproblem satisfy the system

$$A_k \mathbf{d}_w = -\mathbf{a}^k + \mathbf{p}(\mathbf{w}^k, \mathbf{d}_w).$$ (15)

where

$$\mathbf{w}^k = \begin{bmatrix} x^k \\ z^k \\ u^k \\ v^k \end{bmatrix}, \qquad \mathbf{d}_w = \begin{bmatrix} d_x \\ d_z \\ d_u \\ d_v \end{bmatrix}, \qquad A_k = \begin{bmatrix} B_k & 0 & \nabla h(x^k) & \nabla g(x^k) \\ \nabla h(x^k)^t & 0 & 0 & 0 \\ \nabla g(x^k)^t & I & 0 & 0 \\ 0 & V_k & 0 & Z_k \end{bmatrix},$$

$$\mathbf{a}^k = - \begin{bmatrix} \nabla_x L_k \\ h(x^k) \\ g(x^k) + z^k \\ V_k z^k \end{bmatrix},$$

and $\mathbf{p}$ is a perturbation function that reflects the different approximations of the complementary slackness condition. Differences due to a nonexact solution of the quadratic programs can also be included in this term.
- Update the iterates according to $\mathbf{w}^{k+1} = \mathbf{w}^k + \alpha_k \mathbf{d}_w$ for some $\alpha_k > 0$.

The local convergence of the iterates can then be analyzed by comparing them to the Newton iterates obtained from (8) and (9). The analysis depends on the size of $\mathbf{p}(\mathbf{w}^k, \mathbf{d}_w)$, the values of α_k, and how well B_k approximates $\mathcal{H}L_k$. More complexity is introduced by the fact that the components of $\mathbf{w}^k$ have different local convergence rates. Risking the possibility of oversimplification we will restrict our attention to the convergence rates of the primal variables x^k and z^k. Generally, the multiplier vectors converge at a slower rate than the primal variables.

For the Hessian matrix approximations we make the following assumptions:

B.1 For each k the matrix B_k satisfies the conditions that $y \neq 0$ and $D(x^k)^t y = 0$ imply that $y^t B_k y > 0$.
B.2 There exist constant η_1 and η_2 independent of k such that

$$\|B_{k+1} - \mathcal{H}L_k\| \leq (1 + \eta_1 \sigma_k)\|B_k - \mathcal{H}L_k\| + \eta_2 \sigma_k,$$

where

$$\sigma_k = O(\|\mathbf{w}^{k+1} - \mathbf{w}^*\| + \|\mathbf{w}^k - \mathbf{w}^*\|).$$

Condition B.1 guarantees that the quadratic subproblem with B_k replacing $\mathcal{H}L_k$ has a solution. Condition B.2 is a *bounded deterioration* property on the sequence of matrices that is a common assumption for quasi-Newton methods in solving nonlinear systems and optimization problems. It ensures that the approximations B_k do not wander too far from the true Hessian $\mathcal{H}L_*$ at the solution.

A basic convergence theorem can now be stated. (These are not the weakest conditions under which the results are valid, but they are satisfactory for many, if not most purposes.)

Theorem 4.1. *Suppose that the following conditions are satisfied:*
(i) $\lim_{k\to\infty} \alpha_k = 1$;
(ii) $\mathbf{p}(\mathbf{w}^k, \mathbf{d}_w) = \mathbf{o}(\|\mathbf{w}^k - \mathbf{w}^*\| + \|\mathbf{d}_w\|)$;
(iii) *The sequence* $\{B_k\}$ *satisfies conditions* B.1 *and* B.2.
Then there is an $\varepsilon > 0$ *such that if* $\|\mathbf{w}^0 - \mathbf{w}^*\| < \varepsilon$ *and* $\|B_0 - \mathscr{H}L_*\| < \varepsilon$ *then the sequence* $\{\mathbf{w}^k\}$ *converges to* $\mathbf{w}^*$ *and* $\{(x^k, z^k)\}$ *converges Q-linearly to* (x^*, z^*).

As linear convergence can be so slow as to be unsatisfactory, it is usually desirable to identify conditions under which a faster rate of convergence, namely superlinear convergence, is theoretically possible. Although in practical terms these conditions may not be achievable, they do suggest procedures that can lead to fast linear convergence. The following theorem gives a characterization of superlinear convergence. We use the notation $\mathscr{P}_k$ to denote the projection matrices that project vectors from $\mathscr{R}^n$ onto the null space of $D(x^k)^t$.

Theorem 4.2. *Suppose that the hypotheses of Theorem 4.1 hold. Then the convergence in* $\{(x^k, z^k)\}$ *is Q-superlinear if and only if*

$$\mathscr{P}_k(B_k - \mathscr{H}L_k)(x^{k+1} - x^k) = \mathbf{o}(\|x^{k+1} - x^k\| + \|z^{k+1} - z^k\|). \tag{16}$$

These theorems are a composite of the results of several authors ([6,5,16,9]). As the convergence results depend on conditions that may or may not hold for a particular SQP algorithm we comment on these restrictions separately.

Condition (i) of Theorem 4.1 requires that the step lengths go to unity as the solution is approached. Step lengths less than one may be required in these algorithms for two reasons. First, as is seen in the next section, global convergence considerations generally dictate that the step length be restricted so that some type of merit function be decreased at each step. Typically, the merit function depends only on x and z so that the restriction is on the size of the step (d_x, d_z). Second, the multiplier and slack variables are usually maintained as positive (or at least nonnegative) throughout the iterations so that they will be nonnegative at optimality (which is required for feasibility and the identification of a minimum point). These restrictions put limits on the step length for the steps d_z and d_v. The proof that the step lengths approach unity is very much dependent on the particular algorithm under consideration ([3,16]).

The satisfaction of condition (ii) also depends upon the particular form of the algorithm, specifically on the choice of approximation to the complementary slackness condition, i.e., (11), (12) or (14). It is easy to see that, under our assumptions, (11) satisfies (ii). On the other hand, whether the other approximations do clearly depends on the choice of ρ (as discussed in [16].) In addition, if, as is usually the case in large-scale problems, the quadratic problem is not solved exactly then the convergence analysis can be carried out only if the accuracy to which the solution is computed can be expressed in the form of $\mathbf{p}$.

Finally, the way in which the matrix B_k approximates $\mathscr{H}L_k$ also affects the convergence rate of the iterates. There has been a large amount of research on this issue (see [5] for an earlier review of the literature), but the results are not totally satisfactory. Because it is much easier to solve the quadratic subproblem if the matrix B_k is positive definite, efforts have been made to generalize, for example, the BFGS and DFP updates of unconstrained optimization to the constrained case. These satisfy conditions B.1 and B.2 and work well in the case where the problem (NLP) is convex, but do not satisfy (16) in all cases. A constrained version of the PBS update has been shown to satisfy these two conditions and thus yield superlinear convergence, but has not been considered a satisfactory update owing to the nonconvexity of the quadratic approximation and its poor performance in practice.

5. Global convergence

An algorithm is said to be "globally convergent" if it converges from an arbitrary initial point to a local minimum. The procedures discussed in the preceding section will generally fail if the initial estimate is not near the solution because Newton's method is only locally convergent. To obtain global convergence, it is necessary to have some means of forcing a prospective new iterate, x^{k+1}, to be a better approximation to x^* than is x^k. The standard way of doing this is through the use of a *merit function*. The majority of this section will be devoted to an analysis of merit functions and their properties; at the end we will briefly discuss the idea of a "nonlinear filter", a recent development that provides a radically different approach to obtaining global convergence. It is important to point out that to make an SQP method effective, any procedures implemented to force global convergence should not impede the local convergence rate as the solution is neared.

A merit function is an auxiliary, scalar-valued function, $\phi(x)$, that has the property that if $\phi(x^{k+1}) < \phi(x^k)$, then x^{k+1} is acceptable as the next iterate. To ensure that reduction in ϕ implies progress, one constructs ϕ in such a way that the unconstrained minimizers of ϕ correspond to local solutions of (NLP) and the step d_x generated by the SQP method is a descent direction for ϕ. The natural merit function in unconstrained minimization is the function itself. In constrained optimization, the merit function must blend the need to reduce the objective function with the need to satisfy the constraints. Below we consider the properties of some of the more common examples of merit functions. To simplify the presentation, we first consider equality-constrained problems.

One of the earliest proposed merit functions is the ℓ_1 penalty function given by

$$\phi_1(x; \eta) = f(x) + \eta \|h(x)\|_1,$$

where η is a scalar to be chosen. For a point x^k such that $h(x^k) \neq 0$ and η sufficiently large, reducing ϕ_1 implies that $\|h(x^k)\|_1$ must be reduced. It can be shown that for η sufficiently large an unconstrained minimizer of this function corresponds to a solution of (NLP). This merit function has the disadvantage that it is not differentiable at feasible points.

Smoother merit functions, based on the augmented Lagrangian functions, offer several advantages that have caused them to be extensively studied. We illustrate the class with a simple version given by

$$\phi_F(x; \eta) = f(x) + h(x)^t \bar{u}(x) + \tfrac{1}{2}\eta \|h(x)\|_2^2,$$

where, again, η is a constant to be specified and $\bar{u}(x) = -[\nabla h(x)^t \nabla h(x)]^{-1} \nabla h(x) \nabla f(x)$. Observe that $\bar{u}(x)$ is the least-squares estimate of the multipliers based on the first-order conditions. Thus the first two terms of ϕ_F can be regarded as the Lagrangian and the last term augments the Lagrangian with a penalty term that is zero when the constraints are satisfied.

To show global convergence, we must first make some additional assumptions on the problem. These assumptions allow us to focus on the algorithms and not on the problem structure. They are:

C.1 All iterates x^k lie in a compact set $\mathscr{C}$.
C.2 The columns of $\nabla h(x)$ are linearly independent for all $x \in \mathscr{C}$.

The first assumption simply eliminates the possibility of a sequence of iterates diverging to infinity. This assumption, or something that implies it, is common in almost all global convergence analyses. The second assumption ensures that the linearized constraints are consistent, i.e., that the quadratic programming subproblems can be solved. Further comments on this matter are in Section 6.

Clearly the simple reduction of the merit function is not sufficient to obtain convergence, since then it would be possible for the procedure to stall at a nonoptimal point where α goes to zero. There are a number of conditions (such as the Armijo–Goldstein or Wolfe conditions) that can be imposed to ensure sufficient decrease in the merit function and to keep α bounded away from zero. A relatively simple set of conditions that might be used are given here. Suppose that for a given scalar-valued function $\phi(x)$, the sequence $\{x^k\}$ is generated by $x^{k+1} = x^k + \alpha_k d_x^k$ where d_x^k is a descent direction for ϕ at x^k and is $O\|(\nabla\phi(x^k))\|$. Then, if for fixed $\sigma \in (0,1)$, the α_k are chosen by a backtracking line search to satisfy

$$\phi(x^{k+1}) \leqslant \phi(x^k) + \alpha\sigma_1 \nabla\phi(x^k)^t d_x, \tag{17}$$

it follows that

$$\lim_{k\to\infty} \nabla\phi(x^k) = 0.$$

Thus any limit point of the sequence $\{x^k\}$ is a critical point of ϕ.

It is possible that such a critical point will not be a local minimizer. Since the merit function is being reduced, this situation is rare, but precautions to ensure that a minimum has been achieved can be taken. See [18] for a more complete discussion.

In its simplest form the basic SQP algorithm that uses a merit function can be stated as follows: Solve the quadratic programming approximation for the step d_x; choose α bounded away form 0 to satisfy (17); repeat. Given the above assumptions we can state the following results for the merit function ϕ_F and this prototypical SQP algorithm:

 (i) $x^* \in \mathscr{C}$ is a local minimum of ϕ_F if and only if x^* is a local minimum of (NLP).
 (ii) If x is not a critical point of (NLP), then d_x is a descent direction for ϕ_F.
 (iii) For η sufficiently large condition (17) can be satisfied for α bounded away from zero.
 (iv) For η sufficiently large, the basic SQP algorithm is globally convergent to a critical point of (NLP).

Similar results hold for the merit function ϕ_1, but are slightly more difficult to state due to the nondifferentiability. In fact, such results also hold for the ℓ_p penalty function $\phi_p(x) = f(x) + \eta\|h(x)\|_p$, where $p > 0$. (See [14] for a study of the ℓ_2 merit function.)

As stated earlier, the global convergence procedures should not conflict with the local convergence of Section 4. To achieve superlinear convergence requires, at the least, that the step lengths approach one as a solution is neared and that the Hessian approximations B_k satisfy the condition of Theorem

4.2. This is a difficult issue to resolve since it depends upon the merit function being used, the Hessian approximation, and the conditions for the acceptance of the step (e.g., (17)). For the nondifferentiable merit functions, a step length of one may not be acceptable no matter how close to the solution and no matter how good the Hessian approximation. This is called the Maratos effect [8]. Generally, it is not possible to prove superlinear convergence for a given SQP algorithm; one is forced to prove the weaker result that the merit function will allow a step length of one if superlinear convergence is possible, i.e., if the Hessian approximations satisfy (16). See [3] for an example of such an analysis.

In creating an efficient implementation of an SQP algorithm, it is necessary to make many decisions on the details. Several examples dealing with merit functions illustrate this point. First, a difficulty with the merit functions described above is that they involve a parameter, η, that must be adjusted as the iterations proceed. If the parameter is too large, there is no problem with the theory, but in practice, progress may be substantially slowed. If it is too small, then the merit function may not be adequate. Since the proper size may change from a remote starting guess to the solution, most successful implementations have heuristic adjustment procedures. These procedures usually perform well, but often lack theoretical justification [10,4]. As a second example, it has long been observed in nonlinear optimization that enforcing strict decrease in the merit function can sometimes lead to slow convergence and that allowing some occasional increases could improve the overall performance and even overcome the Maratos effect. One might think that this would destroy the global convergence, but there are ways to implement such a *nonmonotone* strategy that preserve global convergence, for instance, by insisting on sufficient decrease only after every K steps. Finally, some merit functions, including ϕ_F, are expensive to evaluate due to the gradient terms. In such cases, one may consider an approximate merit function at each iteration that is cheaper to evaluate, but sufficient to obtain global convergence [3,4,10].

When there are inequality constraints, constructing a merit function is more complicated. Theoretically, the correct active set will be identified by the quadratic program when the iterates are close enough to the solution. In problems with a large number of inequality constraints, however, it can often take many iterations to determine the correct active set. Thus inequality-constrained problems are, in this sense, harder than those with only equality constraints. Mathematically, inequalities can be partially eliminated by the nonnegative slack-variable techniques used in Section 3. A merit function can be constructed for inequality-constrained problems by using the quantity $\|g(x)_+\|$ where the ith component of $g(x)_+$ is 0 if $g_i(x) \leqslant 0$. This leads to merit functions that are not differentiable. (See [5] for further discussion.)

For SQP algorithms that use the trust-region approach, the step length parameter is not used. In this case the merit function is used in the determination of the trust-region radius, τ, given in (7). In unconstrained optimization, τ represents the radius of a ball about the current iterate in which a quadratic approximation is "trusted" to reflect (NLP) accurately. Similarly, in the constrained case, the trust-region parameter is modified at each step based on the accuracy with which the "predicted" decrease in the merit function fits the actual decrease in the merit function. For details see [7].

As noted above, virtually all merit functions involve a parameter that must be adaptively chosen. Recently the idea of using a "nonlinear filter", in a trust-region method has been suggested as an alternative to a merit function for a problem with inequality constraints. For such a problem the pair (r_k, f_k) is computed at each iterate x^k where

$$r(g(x)) = \max\{0, \max\{g_i(x), \ i = 1, 2, \ldots, m\}\} \tag{18}$$

defines a measure of infeasibility at the vector x. A pair (r_i, f_i) is said to *dominate* the pair (r_j, f_j) if and only if $r_i \leqslant r_j$ and $f_i \leqslant f_j$. This indicates that the pair (r_i, f_i) is at least as good as (r_j, f_j) in that the objective function value is at least as small and the constraint violations are no larger. A *filter* is a list of pairs (r_i, f_i) such that no point in the list dominates any other point. As the algorithm proceeds, a pair (r_k, f_k) is added to the filter if its corresponding filter pair is not dominated by any pair in the filter. If it is added to the filter, all pairs in the filter that it dominates are removed. If the point is not acceptable, the trust-region radius is reduced. The advantage of such a technique is that it does not require the selection and adjustment of any penalty parameter. More details of using this idea can be found in [11].

6. Solvers for quadratic programs

A key aspect of an SQP algorithm is the quadratic program (QP) solver, or equivalently, the solvers for any of the formulations in Section 3. For a QP solver to be effective in the large-scale case, it should have several desirable properties. First, it must be a computationally efficient method and it should be tailored to the specific type of quadratic program arising from (NLP). For large problems, the ability to solve the QPs approximately may lead to substantial improvements in the overall efficiency. Thus there should be criteria that allow the solver to halt early. This, in turn, requires that the SQP method be coordinated with the QP solver in the sense that the approximate solution must still be a descent direction for the merit function or be a useful direction for the SQP algorithm. All QP solvers must detect inconsistent constraints and should take some action to generate a useful step. (Note that trust-region methods in the constrained case can readily cause inconsistencies when the point x_k is not feasible and τ is small.) Remedial action is often accomplished by perturbing the constraints in some way and solving the perturbed problem. Finally, for very large problems, the QP solver should be able to exploit parallelism. The current state of the art suggests that active set or simplex-based methods are not readily parallelizable, whereas interior-point methods lend themselves to parallel environments.

If the final active set for a quadratic program were known, then the solution could be found by solving a single system of linear equations. Thus a standard approach to solving quadratic programs is to use an "active set" method that works from an estimate of the final active set, called the *working set*. The quadratic program is solved assuming that these are equality constraints, ignoring the rest. At this solution, new constraints encountered are added to the working set and some of the current constraints are dropped, depending on the sign of the multipliers. A factorization of the matrix of constraint gradients associated with the working set is usually required [12], but iterative methods can also be used for the linear systems [13]. An advantage of these methods is that the active set from the previous iteration of the SQP algorithm is often a good estimate of the active set at the current iteration. As noted earlier, the active set for (NLP) is identified as the solution is neared, so active set methods tend to be extremely efficient over the final few iterations.

Recently there have been successful implementations of both primal and primal-dual interior-point methods for QPs. (Primal-dual interior-point methods are based on (NLP) (ρ) given in Section 2.) For these methods difficulties can arise in the nonconvex case [19,20]. There is also a purely primal method that works for both convex and nonconvex problems. This method solves a QP by solving a sequence of three-dimensional approximations to the QP [2].

7. Reduced Hessian formulation and applications

In some applications, the optimization problem is an equality-constrained problem characterized by having a large number of equality constraints, m, relative to the number of variables, n, i.e., $n - m$ is relatively small. In this case, the quadratic program (QP1) becomes

$$\underset{\boldsymbol{d}_x}{\text{minimize}}\ \nabla f(\boldsymbol{x}^k)^{\mathrm{t}}\boldsymbol{d}_x + \tfrac{1}{2}\boldsymbol{d}_x^{\mathrm{t}}B_k\boldsymbol{d}_x$$
$$\text{subject to}\ \nabla \boldsymbol{h}(\boldsymbol{x}^k)^{\mathrm{t}}\boldsymbol{d}_x + \boldsymbol{h}(\boldsymbol{x}^k) = 0. \tag{19}$$

Assuming that $\nabla \boldsymbol{h}(\boldsymbol{x}^k)$ has full column rank, let the columns of $Z \in \Re^{n \times (n-m)}$ be a basis for the null space of $\nabla \boldsymbol{h}(\boldsymbol{x}^k)^{\mathrm{t}}$ and let the columns of $Y \in \Re^{n \times m}$ be a basis of the range space of $\nabla \boldsymbol{h}(\boldsymbol{x}^k)$. Decomposing the vector $\boldsymbol{d}_x$ as

$$\boldsymbol{d}_x = Z\boldsymbol{r}_Z + Y\boldsymbol{r}_Y,$$

for vectors $\boldsymbol{r}_Z \in \Re^{n-m}$ and $\boldsymbol{r}_Y \in \Re^m$, the constraint equation in (19) can be written as

$$\nabla \boldsymbol{h}(\boldsymbol{x}^k)^{\mathrm{t}}Y\boldsymbol{r}_Y + \boldsymbol{h}(\boldsymbol{x}^k) = 0,$$

which can be solved to obtain

$$\boldsymbol{r}_Y = -[\nabla \boldsymbol{h}(\boldsymbol{x}^k)^{\mathrm{t}}Y]^{-1}\boldsymbol{h}(\boldsymbol{x}^k).$$

Thus (19) becomes an unconstrained minimization problem in the $(n - m)$ variables $\boldsymbol{r}_Z$ given by

$$\underset{\boldsymbol{r}_Z}{\text{minimize}}\quad \tfrac{1}{2}\boldsymbol{r}_Z^{\mathrm{t}}[Z^{\mathrm{t}}B_kZ]\boldsymbol{r}_Z + (\nabla f(\boldsymbol{x}^k) + B_kY\boldsymbol{r}_Y)^{\mathrm{t}}Z\boldsymbol{r}_Z.$$

The $(n-m)\times(n-m)$ matrix $Z^{\mathrm{t}}B_kZ$ is called the *reduced Hessian* and the $(n-m)$ vector $Z^{\mathrm{t}}(\nabla f(\boldsymbol{x}^k)+ B_kY\boldsymbol{r}_Y)$ is called the *reduced gradient*. The advantage of this formulation is that the reduced Hessian, under our assumptions, is positive definite at the solution. (It is not unique, however, since it depends on the choice of the basis for the null space, Z.) It thus makes sense to approximate the reduced Hessian by updates that maintain positive definiteness. The specific details can vary; see, e.g., [1,12–14].

An application of this idea occurs when solving optimization problems where the objective function and/or the constraint functions require the solution of a partial differential equation (PDE). The function to be optimized depends both on a set of control or design parameters and on a set of state variables. These sets of variables are related through a PDE. To be more specific, let $f(\boldsymbol{x}, \boldsymbol{c})$ be the function to be optimized. Here the vector $\boldsymbol{c} \in \Re^q$ represents the design or control variables and $\boldsymbol{x} \in \Re^n$ represents the state variables. The state variables satisfy a differential equation that is represented as a discretized operator yielding $S(\boldsymbol{x}, \boldsymbol{c}) = 0$, where $S: \Re^{n+q} \to \Re^n$. In this context, it can be assumed that for $\boldsymbol{c}$ restricted to a given set, this equation can be solved for a unique $\boldsymbol{x}(\boldsymbol{c})$. Thus the resulting optimization problem is

$$\underset{(\boldsymbol{x},\boldsymbol{c})}{\text{minimize}}\ f(\boldsymbol{x}, \boldsymbol{c})$$
$$\text{subject to}\ S(\boldsymbol{x}, \boldsymbol{c}) = 0, \tag{20}$$

where there are $(n + q)$ variables and n constraints. Typically q is small compared to n.

The particular structure of the equality constraints allows for a variety of possible versions of the reduced-Hessian SQP algorithm [21]. Similar problem forms arise in related applications including parameter identification and inverse problems. In such applications, there are also some inequality constraints that bound the allowable range of the c components or that restrict some other function of the variables. In some cases, the PDE problems cannot be solved if these constraints are violated and it is necessary to remain feasible with respect to these constraints. A special version of SQP, called FSQP for feasible SQP, is designed to maintain feasibility [15].

References

[1] L.T. Biegler, J. Nocedal, C. Schmid, A reduced Hessian method for large-scale constrained optimization, SIAM J. Optim. 5 (2) (1995) 314–347.

[2] P.T. Boggs, P.D. Domich, J.E. Rogers, An interior-point method for general large scale quadratic programming problems, Ann. Oper. Res. 62 (1996) 419–437.

[3] P.T. Boggs, A.J. Kearsley, J.W. Tolle, A global convergence analysis of an algorithm for large scale nonlinear programming problems, SIAM. J. Optim. 9 (4) (1999) 833–862.

[4] P.T. Boggs. A.J. Kearsley, J.W. Tolle, A practical algorithm for general large scale nonlinear optimization problems, SIAM J. Optim. 9 (3) (1999) 755–778.

[5] P.T. Boggs, J.W. Tolle, Sequential quadratic programming, Acta Numer. 1995 (1995) 1–52.

[6] P.T. Boggs, J.W. Tolle, P. Wang, On the local convergence of quasi-Newton methods for constrained optimizations, SIAM J. Control Optim. 20 (1982) 161–171.

[7] R.H. Byrd, J.C. Gilbert, J. Nocedal, A trust region method based on interior point techniques for nonlinear programming, Technical Report OTC 96-02, Northwestern University, August 1998.

[8] R. Chamberlain, C. Lemarechal, H.C. Pedersen, M.J.D. Powell, The watchdog technique for forcing convergence in algorithms for constrained optimization, Math. Programming Study 16 (1982) 1–17.

[9] R.S. Dembo, S.C. Eisenstat, T. Steihaug, Inexact Newton methods, SIAM J. Numer. Anal. 19 (1982) 400–408.

[10] M. El-Alem, A robust trust-region algorithm with a nonmonotonic penalty parameter scheme for constrained optimization, SIAM J. Optim. 5 (2) (1995) 348–378.

[11] R. Fletcher, S. Leyffer, P. Toint, On the global convergence of an SLP-filter algorithm, Report, University of Dundee, 1998.

[12] P. Gill, W. Murray, M. Saunders, SNOPT: an SQP algorithm for large scale constrained optimization, preprint NA97-2, University of California, San Diego, 1997.

[13] N.I.M. Gould, M.E. Hribar, J. Nocedal, On the solution of equality constrained quadratic programming problems arising in optimization, Technical Report OTC 98/06, Argonne National Laboratory and Northwestern University, 1998.

[14] M. Lalee, J. Nocedal, T. Plantenga, On the implementation of an algorithm for large-scale equality constrained optimization, SIAM J. Optim. 8 (3) (1998) 682–706.

[15] C.T. Lawrence, A.T. Tits, A computationally efficient feasible sequential quadratic programming algorithm, preprint, University of Maryland, 1998.

[16] H.J. Martinez, Z. Parada, R.A. Tapia, On the characterization of q-superlinear convergence of quasi-Newton interior-point methods for nonlinear programming, Bol. Soc. Mat. Mexicana 1 (3) (1995) 137–148.

[17] S.G. Nash, A. Sofer, Linear and Nonlinear Programming, McGraw-Hill, New York, 1995.

[18] J. Nocedal, Theory of algorithms for unconstrained optimization, Acta Numer. 1991 (1992) 199–242.

[19] D.F. Shanno, R.J. Vanderbei, An interior-point algorithm for nonconvex nonlinear programming, preprint, Princeton University, 1998.

[20] R.J. Vanderbei, LOQO: an interior-point code for quadratic programming, Technical Report SOR 94-15, Princeton University, 1994.

[21] D.P. Young, D.E. Keyes, Newton's method and design optimization, preprint, Old Dominion University, 1998.

![NH Elsevier logo]

ELSEVIER

Journal of Computational and Applied Mathematics 124 (2000) 139–154

JOURNAL OF
COMPUTATIONAL AND
APPLIED MATHEMATICS

www.elsevier.nl/locate/cam

Trust region model management in multidisciplinary design optimization

José F. Rodríguez[a], John E. Renaud[b,*], Brett A. Wujek[c], Ravindra V. Tappeta[d]

[a] *Universidad Simón Bolívar, Caracas, Venezuela*
[b] *University of Notre Dame, Notre Dame, IN 46556, USA*
[c] *Engineous Software, Inc., Morrisville, NC, USA*
[d] *General Electric Corporate R & D, Schenectady, NY, USA*

Received 9 August 1999; received in revised form 19 November 1999

Abstract

A common engineering practice is the use of approximation models in place of expensive computer simulations to drive a *multidisciplinary design* process based on nonlinear programming techniques. The use of approximation strategies is designed to reduce the number of detailed, costly computer simulations required during optimization while maintaining the pertinent features of the design problem. *This paper overviews the current state of the art in model management strategies for approximate optimization. Model management strategies coordinate the interaction between the optimization and the fidelity of the approximation models so as to ensure that the process converges to a solution of the original design problem.* Approximations play an important role in multidisciplinary design optimization (MDO) by offering system behavior information at a relatively low cost. Most approximate MDO strategies are sequential, in which an optimization of an approximate problem subject to design variable move limits is iteratively repeated until convergence. The move limits or *trust region* are imposed to restrict the optimization to regions of the design space in which the approximations provide meaningful information. In order to insure convergence of the sequence of approximate optimizations to a Karush–Kuhn–Tucker solution, a trust region model management or move limit strategy is required. In this paper recent developments in approximate MDO strategies and issues of trust region model management in MDO are reviewed. © 2000 Elsevier Science B.V. All rights reserved.

Keywords: Trust region model management; Response surface approximation; Approximate optimization

1. Introduction

The complexity of engineering design has introduced the need to account for interdisciplinary interactions in the design process. This has led to the development of design strategies which provide

---

* Corresponding author.
 E-mail address: john.e.renaud.2@nd.edu (J.E. Renaud).

0377-0427/00/$ - see front matter © 2000 Elsevier Science B.V. All rights reserved.
PII: S 0377-0427(00)00424-6

for multidisciplinary design, in which design variables from different disciplines can be manipulated simultaneously in a coordinated fashion. The increasing demand for improved designs within shorter product development cycle times, requires the incorporation of optimization theory, tools and practices developed in the mathematical community, into the design process. The formal methodologies which incorporate these features are referred to collectively as multidisciplinary design optimization (MDO).

The incorporation of traditional optimization tools into engineering design problems is not an easy task and is still and active area of research. The main challenges are associated with the problem dimensionality and the high computational cost associated with the computation of objective and constraint functions. These two characteristics of engineering systems, along with the organizational issues related to data sharing and inter-discipline communications, prohibit the use of traditional optimization techniques in the optimal design process. Consequently, approximation models must be introduced into the multidisciplinary design methodology and proper model management frameworks must be developed to drive the optimization of these engineering systems.

This paper concentrates on research related to system approximation and the model management strategies used to drive design improvement and the convergence of multidisciplinary design optimization. The paper begins by defining MDO, its characteristics, and its implications in engineering design optimization strategies. With the background of MDO established, the issue of system approximation is visited, making reference to response surface approximation (RSA) methodologies and global design space approximation methodologies based on nonlinear interpolation techniques, which have been vigorously pursued in recent years. Since the focus of this work is approximate model management, Sections 4 and 5 are devoted to this topic. Section 4 focuses on a review of move limit heuristics used to manage approximations in approximate optimization algorithms. Section 5 is dedicated to model management frameworks with strong global convergence properties. Even though Section 5 is primarily focused on trust region based algorithms, alternative rigorous frameworks are also reviewed.

2. Multidisciplinary design optimization

What is MDO? MDO can be described as a methodology for the design of systems where the interaction between several disciplines must be considered, and where the designer is free to significantly affect the system performance in more than one discipline. Comprehensive reviews of MDO are given in a number of publications including [22].

Large-scale engineering design problems, such as an aircraft design or an automobile design, are often characterized by multidisciplinary interactions in which participating disciplines are intrinsically linked to one another. Designers have long recognized the need to decompose such systems into a set of smaller more tractable disciplines. This decomposition is usually based either on the engineering disciplines or on the mathematical models governing the system. As a result, the design of such complex systems often involves the work of many specialists (engineering teams) in various disciplines, each dependent on the work of other groups and knowing little about the analysis and software tools available to the other groups.

Fig. 1 shows the connections between various disciplines (Structures, Aerodynamics, Occupant Dynamics, etc.) in a graph-theoretic format for an automobile system design. Fig. 2 shows the

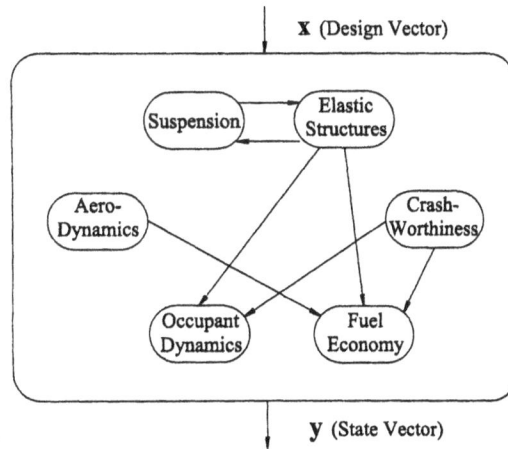

Fig. 1. Interdisciplinary system decomposition with coupling.

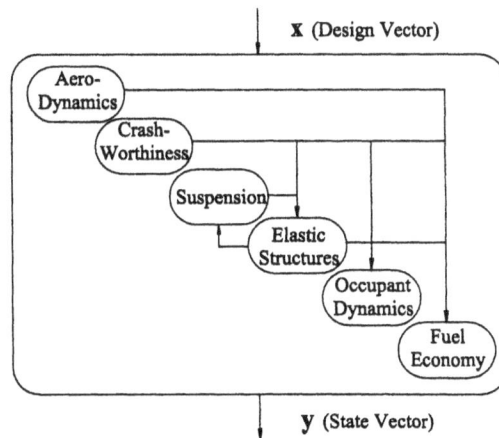

Fig. 2. Dependency diagram of a coupled system.

same automobile system design in a more structured format, called the dependency diagram. The boxes indicate various disciplines (design teams or subspaces) working on a given automobile and the arrows indicate the multidisciplinary interactions between disciplines. Some of the disciplines in an automobile design problem are structures, aerodynamics, occupant dynamics, fuel economy, etc. The arrows in Fig. 2 on the right or upper side are feed-forwards and the arrows on the left lower side are feed-backs. For instance, in the above example, the aerodynamic drag coefficient is fed-forward into fuel efficiency; the crash-worthiness design team feeds crash loads forward into the elastic structures, the occupant dynamics and the fuel economy. Note that there is an iterative loop between suspension and elastic structures since feed-back exists.

Because of this coupling these design teams cannot work in isolation; instead they must work in harmony to arrive at a consistent design. Such systems are known as networked or nonhierarchic systems. A system is a nonhierarchical network if there is no inherent mathematical reason to place one box above another as in hierarchic systems. Note that in Fig. 2, the sequential hierarchic flow is broken by the feed-back between suspension and elastic structures. To obtain a consistent behavior output y for a given initial base line design x it is necessary to invoke an iterative solution to the coupled problem which loops through the feed-backs and feed-forwards until the convergence criteria are met. This iterative solution is referred to as a system analysis (SA). For each SA the disciplines are called in a serial manner and each discipline has to execute its analysis tools a number of times.

If one is merely interested in the proper functioning of the disciplines so as to yield a properly functioning system for the given discipline (as judged by the specifications), the problem is one of sequential design. The design obtained by this approach, however, will not necessarily be a superior solution since the interactions of the subsystem are often overlooked. This leads to less than optimal functioning or, in some cases, nonfunctioning of the system. An iterative approach to design (i.e., MDO) overcomes the problem of a lack of interaction between the various subsystems to a certain extent, leading to a certain degree of optimality, although this requires a considerable amount of time, effort and resources.

Current research in the area of system decomposition in MDO has focused on developing formal measures of accounting for system interactions and couplings. The goal is to improve the iterative design of complex systems by making the process systematic and basing it on a set of consistent mathematical concepts.

The interdisciplinary coupling inherent in MDO tends to present additional challenges beyond those encountered in a single-discipline optimization. It increases the computational burden, and it also increases complexity and creates organizational challenges. The increased computational burden may simply reflect the increased size of the MDO problem, with the number of analysis variables and the number of design variables increasing with each additional discipline. Each disciplinary optimization may have a single-objective function, but the MDO problem may require multiple objectives with an attendant increase in optimization cost. In the MDO of complex systems we also face formidable organizational challenges. The analysis codes for each discipline have to interact with one another for the purpose of system analysis and system optimization. Decisions on the choice of design variables and on whether to use single-level optimization or multi-level optimization have profound effects on the coordination and the transfer of data between analysis codes. These decisions also impact the choice of the optimization tool and the degree of human interaction required.

Traditional single-level optimization of coupled systems requires the optimizer to invoke a SA of the coupled system many times. The application of formal optimization techniques to the design of these systems is often hindered because the number of design variables and constraints is so large that the optimization is both intractable and costly and can easily saturate even the most advanced computers available today. Therefore, the use of approximations to represent the design space is essential to the efficiency of MDO algorithms. Approximations provide information about the system necessary for the optimization process without the cost of executing CPU-intensive analysis tools. Moreover, the use of approximations allows for the temporary decoupling of disciplines which avoids the constant transfer of information among disciplines required during an iterative system analysis. Consequently, most MDO algorithms couple, in an iterative fashion, a traditional optimization code to lower-cost computational models of the objective function and constraints (i.e., system approximation). A solution to the approximate problem is found, a full system analysis is executed at this new design, the approximate model is updated and the process repeated until convergence to a solution of the original problem is achieved. Lower-cost computational models can be categorized as: lower complexity models which are less physically faithful representations of the actual physical problem; and model approximations which are algebraic representations obtained from design sites at which objective and constraints are known (e.g. low-order polynomial response surface approximations and kriging estimates). Most MDO algorithms differ in how the approximate models are built and managed in order to drive convergence to a solution of the original problem.

3. Approximation models

As discussed in the previous section, a common engineering practice is the use of approximation models in place of expensive computer simulations to drive a multidisciplinary design process based on nonlinear programming techniques. Two main alternatives have been investigated in the MDO community to approximate physical systems. The first approach has been the use of a simplified physical representation of the system to obtain less costly simulations as described in [4]. A second alternative for system approximation which has grown in interest in recent years, are RSAs based on polynomial and interpolation models.

Polynomial RSAs employ the statistical techniques of regression analysis and analysis of variance (ANOVA) to determine the approximate function. Consider a function $f(x)$ of n_v design variables, for which its value is known at n_e design sites. A quadratic model, $\tilde{f}(x)$, of the function $f(x)$ at the pth design site is given by

$$f^{(p)} = c_0 + \sum_{i=1}^{n_v} c_i x_i^{(p)} + \sum_{i=1}^{n_v} \sum_{j=1}^{n_v} c_{ij} x_i^{(p)} x_j^{(p)}, \tag{1}$$

where $p = 1, \ldots, n_e$; $f^{(p)}$ is the pth observation; $x_i^{(p)}$ and $x_j^{(p)}$ are the design variables; and c_0, c_i, and c_{ij} are the unknown polynomial coefficients. For the quadratic model, if $c_{ij} = c_{ji}$, there are a total of $n_t = (n_v + 1)(n_v + 2)/2$ unknown coefficients. Therefore, a necessary condition for the proper characterization of model (1) is that $n_s \geqslant n_t$. Under this condition, the estimation problem for the unknown coefficients c_k, $k = 1, \ldots, n_t$, is formulated in matrix form as

$$f = Xc, \tag{2}$$

where f is the vector of n_e observations, and X is a matrix of rank n_t given as

$$X = \begin{bmatrix} 1 & x_1^{(1)} & x_2^{(1)} & \cdots & (x_{n_v}^{(1)})^2 \\ \vdots & \vdots & \vdots & \ddots & \vdots \\ 1 & x_1^{(n_e)} & x_2^{(n_e)} & \cdots & (x_{n_v}^{(n_e)})^2 \end{bmatrix}. \tag{3}$$

Since $n_e \geqslant n_t$, the vector of unknown coefficients c is obtained from a least-squares solution of (2). If the rows of X are linearly independent (i.e., different design sites), the least-squares solution of (2) is unique and is given by

$$\tilde{c} = (X^{\mathrm{T}} X)^{-1} X^{\mathrm{T}} f. \tag{4}$$

When (4) is substituted into (1), values of $f(x)$ can be predicted at any design x. Note that since in general $\tilde{c}$ is obtained from a least-squares solution, the value of the $\tilde{f}(x)$ at the original design sites may be different from the true value $f(x)$ at the same location.

The relatively simple procedure to characterize polynomial RSAs have stimulated their use in approximate optimization algorithms of multidisciplinary systems. However, the simple polynomial representation also limits the accuracy of the RSA to relatively small neighborhoods in nonlinear design spaces.

Global approximations of the design space can be achieved by the use of multipoint approximations which successively improve the model by adding more information to the current approximation [19], or by interpolation and kriging models. In interpolation models, the predicted response $\tilde{f}(x)$ is more strongly influenced by true data close to the current design x than for those points further away. Note that this is not the case in polynomial RSAs where all the n_e observed values of the response are equally weighted. Among the large number of interpolation techniques (i.e., Legendre polynomials, Newton polynomials, splines, etc.), are the class of interpolation techniques based on Bayesian statistics termed kriging models [21]. The conventional kriging model expresses the unknown function as

$$\tilde{f}(x) = \mu + Z(x), \tag{5}$$

where μ is an estimate of the mean of the data, and $Z(x)$ is an Gaussian-random function of zero mean and with variance σ^2 which makes $\tilde{f}(x)$ interpolate $f(x)$ at the observation sites $x^{(p)}$. In order to construct the kriging model, the spatial covariance, $\text{Cov}[Z(x^{(i)}), Z(x^{(j)})]$, has to be specified, which is given as

$$\text{Cov}[Z(x^{(i)}), Z(x^{(j)})] = \sigma^2 R(x^{(i)}, x^{(j)}), \tag{6}$$

where $R(x^{(i)}, x^{(j)})$ is the correlation matrix which is assumed to be the product of one-dimensional exponential correlation functions

$$R(x^{(i)}, x^{(j)}) = \exp\left[-\sum_{k=1}^{n_v} \theta_k (x_k^{(i)} - x_k^{(j)})^2\right], \tag{7}$$

where θ_k is the vector of unknown correlation parameters.

Since the kriging model has to agree with the observed data $f^{(p)}$, the term $Z(x)$ is given by

$$Z(x) = r(x)^{\mathrm{T}} R^{-1}(f - \mu I), \tag{8}$$

where,

$$r(x)^{\mathrm{T}} = [\text{Cov}(x, x^{(1)}), \ldots, \text{Cov}(x, x^{(n_e)})]^{\mathrm{T}},$$
$$R_{ij} = \text{Cov}(Z(x^{(i)}), Z(x^{(j)})),$$
$$f = [f(x^{(1)}), \ldots, f(x^{(n_e)})],$$
$$I = [1, \ldots, 1].$$

The remaining parameters μ, σ^2, and θ_k, in the kriging model are determined via maximum likelihood estimation [12], which is equivalent to maximize over μ, σ^2, and θ_k, the log-likelihood given by

$$L(\mu, \sigma^2, \theta) = -\frac{1}{2}\left[n_e \ln(2\pi) + n_e \ln(\sigma^2) + \ln(|R|) + \frac{1}{\sigma^2}(f - \mu I)^{\mathrm{T}} R^{-1}(f - \mu I)\right]. \tag{9}$$

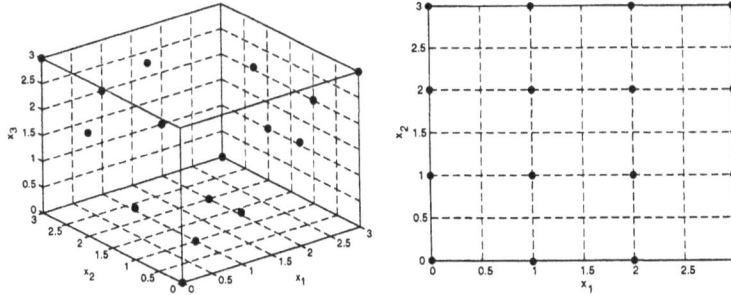

Fig. 3. Orthogonal array.

Maximization over μ and σ^2 yields

$$\mu(\boldsymbol{\theta}) = \frac{\boldsymbol{I}^{\mathrm{T}}\boldsymbol{R}^{-1}\boldsymbol{f}}{\boldsymbol{I}^{\mathrm{T}}\boldsymbol{R}\boldsymbol{I}} \tag{10}$$

and

$$\sigma^2(\boldsymbol{\theta}) = \frac{1}{n_e}(\boldsymbol{f} - \mu\boldsymbol{I})^{\mathrm{T}}\boldsymbol{R}^{-1}(\boldsymbol{f} - \mu\boldsymbol{I}). \tag{11}$$

Substitution of (10) and (11) into (9), reduces the estimation of $\boldsymbol{\theta}$ to the following optimization problem:

$$\max_{\boldsymbol{\theta}\in\mathbb{R}^{n_c}} \quad (-1/2)[n_e\ln(\sigma^2(\boldsymbol{\theta})) + \ln|\boldsymbol{R}|], \tag{12}$$

$$\boldsymbol{\theta} \geqslant 0.$$

Thus, by solving this maximization problem, the kriging model (5) is completely defined. Note that for any design x, the predicted value $\tilde{f}(x)$ given by (5) is a linear combination of the data $f^{(p)}$, where the coefficients depend on the value of x. Moreover, these coefficients weigh the contribution of sampling points nearby x more than those points farther away.

In both RSA methodologies discussed above, one of the important issues for the construction of accurate models is choosing a proper set of initial design sites. For the case of polynomial RSAs a variety of techniques have been used to generate the data required to obtain the unknown coefficients c. Sampling patterns based on design of experiments methodology (i.e., factorial and fractional factorial experiments), as well as optimal sampling techniques based on D-optimal criteria [7] have been extensively used. Also, variable fidelity data generated during concurrent subspace optimizations (CSSOs) have been employed to build local quadratic RSAs of multidisciplinary systems [26]. In the case of interpolation models, it is important that the experimental designs fill the design space in order for the kriging to be accurate. For this reason, optimal sampling strategies [21] based on design and analysis of computer experiments (DACE) are usually implemented. However, fractional factorial experiments based on orthogonal arrays (OAs) [17] also offer a valid alternative for providing the initial design sites, due to their appealing space-filling property called strength. An OA of strength t, for a k level experiment, represents, for every subset of t independent variables, a k^t grid. This is shown in Fig. 3, where the design sites for a four-level factorial OA of strength 2 for three independent variables are depicted.

Note that even though the design sites represent a "cloud" in the whole of the design space, the grid pattern is observed in the projections. This property gives the designer the confidence that the design are infiltrating the design space.

Through the use of design space approximations, optimization of large complex systems is made more practicable. It is important that these approximations accurately portray the design space so that the infeasible region is avoided and the design objective is continuously improving. In the case of local approximations, they will tend to stray from the actual system response surface as the design moves away from the data point(s) about which the approximation was formed. Therefore, design variable move limits are imposed to restrict the approximate optimization to regions of the design space in which the approximations are accurate. After each sequence of approximate optimization, the approximations of system behavior are updated with new information about the current design. Thus, many iterations of such algorithms may be required before convergence of the optimization process is achieved, and every additional iteration adds to the cost of the process. In light of this, a primary concern in developing an approximate optimization strategy is the proper choice of a move limit management strategy.

4. Model management

The standard form of a nonlinear optimization process is shown in Eqs. (13)–(16). The lower and upper bounds ($x_{(L)}$ and $x_{(U)}$) in Eq. (16) are global variable bounds imposed by the designer:

$$\min \quad f(x), \quad x = [x_1, \ldots, x_n] \tag{13}$$

$$\text{s.t.} \quad g_i(x) \geqslant 0, \quad i = 1, \ldots, p, \tag{14}$$

$$h_j(x) = 0, \quad j = 1, \ldots, m, \tag{15}$$

$$x_{(L)} \leqslant x \leqslant x_{(U)}. \tag{16}$$

Approximate optimization algorithms typically build approximations of the objective $\tilde{f}(x)$ and the constraints $\tilde{g}(x)$ and $\tilde{h}(x)$ and then solve a sequence of approximate optimizations as given in Eqs. (17)–(20). In Eq. (20) additional move limits ($x_{(L)}^{(s)}$ and $x_{(U)}^{(s)}$) about the current design iterate $x^{(s)}$ are placed on the design variables in an attempt to ensure approximation accuracy.

$$\min \quad \tilde{f}(x), \quad x = [x_1, \ldots, x_n] \tag{17}$$

$$\text{s.t.} \quad \tilde{g}_i(x) \geqslant 0, \quad i = 1, \ldots, p, \tag{18}$$

$$\tilde{h}_j(x) = 0, \quad j = 1, \ldots, m, \tag{19}$$

$$\max[x_{(L)}, x_{(L)}^{(s)}] \leqslant x \leqslant \min[x_{(U)}^{(s)}, x_{(U)}]. \tag{20}$$

These limits are temporary bounds applied at each design iterate and may change as the optimization proceeds, but they are always restricted by the original global bounds of the problem ($x_{(L)}$ and $x_{(U)}$). If the allowed changes in the design variables are too liberal, the discrepancy between the approximations and the actual response surface may eventually become unacceptable and adversely affect the optimization process. Common maladies are cycling about a minimum or about a constraint boundary, inability to recover from an infeasible region, and visiting physically unrealistic designs. If the allowed changes in the design variables are too stringent, the overall progress of the optimization will be unnecessarily slowed and more algorithm iterations will be required, increasing the cost of the process.

It can be reasoned that the amount that design variables should be allowed to change at any given time is related to the nature of the design space at the current location, the accuracy of the current approximation, and/or possibly even the history of previous movements. Thus, the development of a strategy to account for the aforementioned concerns in setting design variable move limits is a worthy task.

Many move limit strategies or model management schemes have been employed in an effort to insure that design decisions made based on lower fidelity information (i.e., response surface approximations) will yield improvements in the actual system. For example, in [26] fixed percentage move limits are used to manage the approximate design optimization. A strategy which accounts for the history of the design is offered in [23]. The method combines global move limit adjustment with individual move limit adjustment strategies. Global move limits are set based on the accuracy of the approximation and maximum constraint violation. Individual move limits, introduced to overcome the problem of premature or slow convergence caused by the global move limit strategy, are adjusted if the given design variable hits its move limit bound. Even though the strategy accounts for individual move limits for each design variable, the move limit settings are driven by bounds rather than overall design improvement. Bloebaum et al. [5] introduce a move limit methodology based on design space sensitivity. The strategy works by giving less restrictive move limits to those variables that have larger impact on design improvement. In this methodology, the impact of a particular design variable on design improvement is measured by an effectiveness coefficient based on the sensitivity of the objective and constraints to the design variables. The resulting effectiveness coefficients define an effectiveness space in which upper and lower bounds are used to assign move limits to each design variable. Chen [8] proposes a number of different methods for calculating design variable move limits for use in the sequential linear programming (SLP) algorithm. The Chen method utilizes linear approximations of the constraints to determine when the bounds will be reached. The strategy is applicable to both equality and inequality constraints. One of the disadvantages of the strategy is that it does not account for the nonlinearities of the constraints and objective functions which might lead to difficulties in convergence. Fadel et al. [14] also developed a strategy which relies on gradient information to adjust individual move limits in the design variables. The method is based on the two-point exponential approximation (TPEA) which is an extension of the Taylor series which accounts for the matching of the derivatives at consecutive design points through an exponential correction factor. The exponents computed from the TPEA are used as a measure of nonlinearity of the objective and constraints with respect to each design variable. These exponents are then used to compute individual move limits. In most cases these model management or move limit strategies are heuristic and in general lead to improved designs but not necessarily converged Karush–Kuhn–Tucker (KKT) designs.

In each of the aforementioned studies the primary consideration has been that the application of the approximate optimization strategies should lead to improved designs. Improved designs are obtained in each of the approximate optimization studies as measured by the design objective employed for each respective study. The attribute of achieving improved designs is laudable and obviously relevant for practicing designers. However, from a mathematics stand point, it is equally important that approximate optimization methods be developed that insure convergence to a Karush–Kuhn–Tucker solution.

A model management strategy which falls into this category is the trust region approach. This is a classical method prevalent in nonlinear programming [13] which provides a framework for adaptively managing the amount of movement allowed in the design space using approximate models. Trust region methods were originally introduced as a way of ensuring global convergence for Newton-like methods. The name is derived from the fact that the trust region defines the region in which one may trust the approximate model to accurately portray the actual design space. Quite simply, the trust region is merely another name for the region of the design space defined by the design variable move limits. Recent work has focused on the development of trust region models management strategies to insure convergence to KKT designs when employing approximate optimization strategies.

5. Trust region model management

Trust region methods were originally introduced to apply to modern nonlinear unconstrained optimization algorithms with a robust global behavior. Robust global behavior infers the mathematical assurance that the optimization algorithm will converge to a stationary point or local optimum of the problem, regardless of the initial iterate.

In trust region methods, a second-order approximation, $\tilde{f}(x)$, of the objective, $f(x)$, is successively minimized with the trust region regulating the length of the steps in each iteration. The global convergence properties of these methods relies primarily, on the assumption that the approximation $\tilde{f}(x)$ and the actual function $f(x)$ match up to the first order, i.e.,

$$\tilde{f}(x^{(s)}) = f(x^{(s)}), \tag{21}$$

$$\nabla \tilde{f}(x^{(s)}) = \nabla f(x^{(s)}). \tag{22}$$

On the other hand, the length of the steps or size of the trust region is controlled based on how well the quadratic model predicts the decrease in f. A reliability index, $\rho^{(s)}$, which monitors how well the current approximation represents the actual design space is defined as

$$\rho^{(s)} = \frac{f(x^{(s)}) - f(x^{(s+1)})}{\tilde{f}(x^{(s)}) - \tilde{f}(x^{(s+1)})}. \tag{23}$$

This is simply the ratio of the actual change in the function to the change predicted by the approximation. After each optimization iteration (s), the trust region radius is updated according to the following principles:

1. If the ratio is negative or small, the iteration is considered unsuccessful since either the actual objective increased (it is known that $\tilde{f}$ will not increase) or it did decrease, but not nearly as much as predicted by the approximation. In either case, the approximation is certainly poor and the trust region must be reduced. For the case of a negative ratio, most algorithms actually reset the design to the previous iterate and repeat that optimization iteration. This is done in order to guarantee convergence of the algorithm.

2. Conversely, if the ratio is large, a reasonable decrease in the objective function has been observed relative to the approximate decrease, and the iteration is considered successful. It should be noted that if the ratio is significantly larger than one, the objective function actually decreased more than had been predicted, and the approximation is actually a poor representation of the design space. However, since this scenario is actually favorable as more reduction is gained than expected, an increase in the trust region radius is justified.
3. Finally, if the ratio is an intermediate value, the wisest choice of action may be to leave the size of the trust region as it is.

Mathematically, the above rules for updating the trust region radius may be described by choosing constants to define the ranges of the ratio value for which reduction or enlargement are necessary. The positive constants $R_1 < R_2 < 1$ and $c_1 < 1$, $c_2 > 1$ are chosen so that the trust region radius is updated as

$$\Delta^{(s+1)} = \begin{cases} c_1 \Delta^{(s)} & \text{if } \rho^{(s)} < R_1, \\ c_2 \Delta^{(s)} & \text{if } \rho^{(s)} > R_2, \\ \Delta^{(s)} & \text{otherwise.} \end{cases} \tag{24}$$

Typical values used for the limiting range values are $R_1 = 0.25$ and $R_2 = 0.75$. The trust region multiplication factors c_1 and c_2 are usually chosen to be in the range between 0.25 and 0.5 for c_1, and 2 for c_2 suggests 0.5 and 2. These limits are usually adjusted for the cases when the steps $\|x^{(s+1)} - x^{(s)}\|$ are smaller than the current trust region radius $\rho^{(s)}$.

In [13], an additional mechanism for regulating the trust region is given to adjust the size of the radius to be consistent with the magnitude of the steps taken. When the step taken to solve the approximate optimization problem is a Newton step (a step to the minimum of the quadratic approximation) which is shorter than the current trust region radius, the radius is immediately reduced to the length of the Newton step.

A choice for the initial trust region size is left for the user to determine. It may be based on knowledge of the problem or on some other criteria such as the length of the Cauchy step, a step to the minimum of the approximation in the steepest descent direction. Another option is to choose the initial radius to be proportional to the norm of the gradient of the objective, $\Delta^{(0)} = \alpha \| \mathrm{d}f/\mathrm{d}x |_{x^0} \|$, although proper choice of the proportionality constant α must still be dealt with. Although an algorithm may recover from a bad initial trust region radius, this value has an effect on the efficiency of the algorithm since extra iterations may be required.

Besides the arbitrary setting of the limiting range constants (R_1, R_2) and the adjustment factors (c_1, c_2), one obvious drawback to this method is the fact that it is defined for *unconstrained* optimization. It necessarily compares the change in a single quantity to the predicted change in that quantity. For use in *constrained* optimization, this is often overcome by the use of a penalty function, such as the augmented Lagrangian function, Φ, in which the constraints of the problem are included with the original objective function to form a new objective function of the form

$$\Phi(x, \lambda, w) = f(x) + \lambda^{\mathrm{T}} g(x) + \tfrac{1}{2} g(x)^{\mathrm{T}} W g(x), \tag{25}$$

where λ is a vector of Lagrange multipliers and W is a diagonal matrix of penalty weighting terms. This type of problem modification is acceptable as long as the optimization algorithm may be adapted to approximate and use the desired penalty function. In many cases, however, the problem is formulated such that approximations are available only for the objective function and constraints individually. These approximations can be combined to form an approximate penalty function, with the only additional burden being the calculation and updating of penalty parameters (Lagrange multipliers, etc.).

The most promising aspect of the trust region approach is its excellent global convergence properties. Global convergence is defined as the ability to converge to a local optimum from any starting point. In [9], a proof of global convergence is offered for constrained approximate optimization algorithms in which design variable movement from iteration to iteration is governed by the trust region. This characteristic is important to avoid cycling and to provide steady convergence in the design process.

One can see that the trust region method encompasses many of the ideas of the move limit strategies previously discussed. It uses a constant reduction/enlargement scheme based on the history of the previous iteration and on how well the convergence of the approximation conformed to the actual convergence. That is, the model used to approximate the design space might be found to perform exceptionally well, so that use of a model with less accuracy but lower computational cost could be considered. In CSSO for example, it might be found that the quadratic formed from the subspace data is extremely accurate in the coordination procedure. As a result it may be decided that either less data is required in the next iteration or the subspace optimizations may be bypassed altogether and SLA performed until it is determined that a quadratic model is again required. Thus, it can be seen that a trade-off exists between reducing the required level of approximation or expanding the trust region for use with a highly accurate model.

The strong convergence properties of trust region methods make them an ideal tool for model management. Several model management frameworks for approximate constrained minimization based on trust region methods have been developed in the last few years. In these algorithms the high-fidelity analysis tools interact with lower fidelity system approximations during the optimization process through a trust region model management strategy which controls the accuracy of the approximation. Fig. 4 shows a flow chart of the provably convergent trust region model management algorithm for constrained approximate minimization using RSAs developed in [20].

In this particular model management framework, the problem (13)–(16) is solved by successively minimizing quadratic approximations of the augmented Lagrangian. The quadratic approximation, satisfying (21)–(22) is built using variable-fidelity data generated using the concurrent subspace approach of Wujek et al. [26]. The performance of this algorithm has been improved by introducing a continuous relationship between ρ and the trust region adjustment factor [25]. The Wujek and Renaud approach provides a more consistent and flexible mechanism for controlling the error in the variable fidelity approximations used in [20]. The strategy developed provides a logical quantitative measure for enlarging/reducing the size of the trust region based on gradient information. Alexandrov et al. [3] develop a framework for managing approximation models in unconstrained minimization. The Alexandrov et al. trust region based optimization algorithm is not restricted to quadratic models of the objective function as in the case of traditional trust region methods, but the algorithm requires the models to satisfy conditions (21)–(22) which are enforced using a β-correlation approach. The framework allows the approximation model to change during the optimization, which adds flexibility over traditional trust region algorithms in the sense that it makes improvement of the model by either changing the model itself or adjusting the trust region radius, a model management decision. Extension of the work in [3], to the case of constrained optimization is performed in [1]. In this

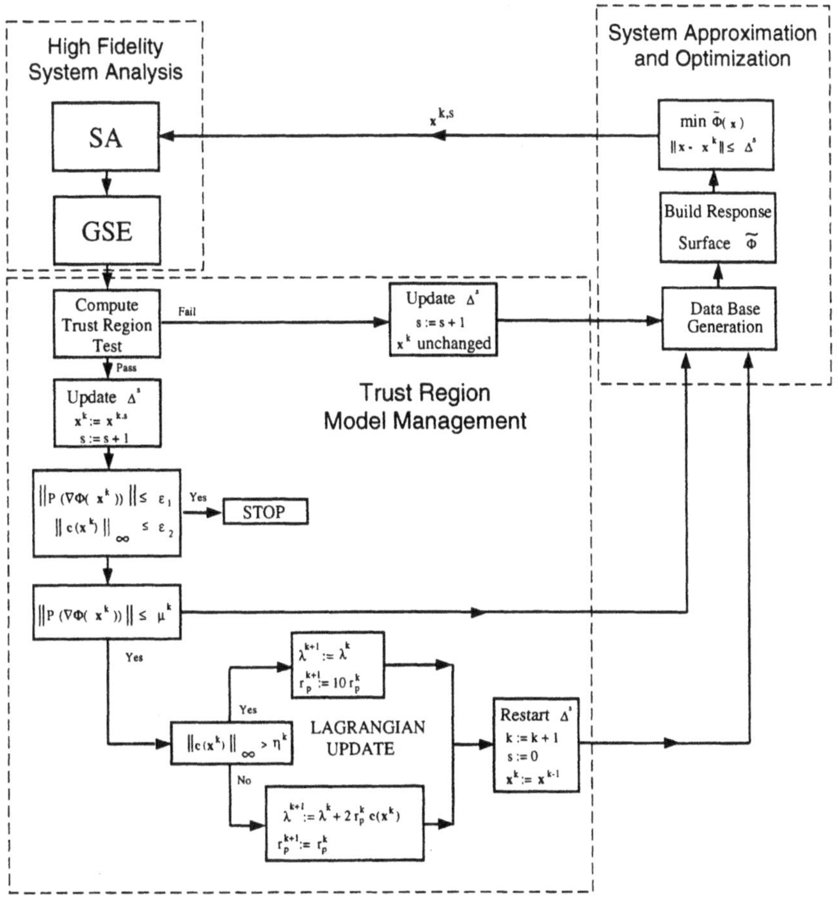

Fig. 4. Trust region augmented Lagrangian model management algorithm.

framework, the problem (13)–(16) is solved by successively optimizing linear system approximations using a methodology based on a class of multilevel methods for constrained optimization described in [2]. The length of the trial steps is also controlled by a trust region which provides this algorithm with the global convergence characteristics. The framework has been tested on a number of analytical problems and is currently being implemented on a subset of the MDO Test Suite [18].

All of the algorithms and frameworks for model management described above assume that first-order information for building the approximations is available. However, this is not always the case. Evaluation of objective functions and constraints may imply a high computational cost, which for a typical multidisciplinary engineering problem might require hours even in highly parallelized platforms. Besides this practical limitation, accuracy is another issue. Values obtained for the objective and constraints may have few correct significant digits and be noisy which makes gradient calculations difficult using the finite difference method. Consequently, the development of derivative-free model management frameworks has captured great interest among the MDO community. Conn et al. [10] propose a derivative-free algorithm for minimizing a function without constraints. The algorithm uses an interpolation set to build a quadratic model which is minimized in a trust-region framework.

The key point of their algorithm is the definition of the interpolation set in the neighborhood of the current iterate, since it controls the accuracy with which the true objective function is approximated. The "goodness" of the interpolation set is measured by the linear independence of the interpolating basis as defined in [10]. The algorithm has been proven to be globally convergent when Newton fundamental polynomials are used [10]. Computational results for a number of analytical problems and for the Boeing helicopter problem [11], are encouraging. In a recent paper, Booker et al. [6] have proposed a framework for optimization of expensive functions by surrogates. The methodology, which is able to handle nonlinear optimization problems with variable bounds, is based on global convergence results for pattern search methods presented by Torczon [24] and Lewis and Torczon [15,16]. The framework has been developed to accommodate a variety of surrogates: (i) lower complexity physical models of the high-fidelity simulation; (ii) approximation of the high-fidelity simulation by interpolating or smoothing known values of the objective; or (iii) model-approximation hybrids. The framework has been successfully applied to the Boeing helicopter problem [6] where surrogates constructed by kriging were used. Initial interpolating sites for the kriging model were selected using design and analysis of computer experiments (DACE).

6. Closure

As computers advance in speed, more efficient data sharing and exchange algorithms are developed. One observes that an increasing number of discipline sets are being encompassed in actual engineering optimization process. Problem complexity is observed to grow at a pace which taxes the limits of the advances in processing powers. Therefore, the dimensionality and complexity of MDO problems may always necessitate the use of approximations and decomposition strategies to make the optimization a practical task. In this regard, RSAs have emerged as useful tools for reducing computational cost while handling designs of variable complexity in an efficient manner. Both local and global RSA strategies based on local and accumulated design data of variable fidelity have been proposed and investigated. Moreover, in recent years, there has been a growing interest in the use of global approximations based on kriging and interpolating models built with high-fidelity information at design sites usually given by DACE methodologies.

The introduction of approximations in MDO has led to the need to develop efficient model management strategies. These model management strategies have focused on controlling the model accuracy within the context of optimization. In practice, many of these strategies were based on heuristics and design experience which only guarantee improved designs; however a vigorous movement toward the generation of converged designs rather than improved designs in MDO is evidenced in the significant number of references devoted to the development of MDO frameworks with strong global convergence properties. These MDO frameworks, which use a variety of system approximations, are based on the common trust region methodology for managing the approximate models.

One observes an increasing collaboration between the mathematical and engineering communities in the development of new MDO frameworks. This collaboration has lead to algorithms which provide for the needs of designers while incorporating the rigor and robustness of mathematical programming. Consequently, many new algorithms and frameworks being developed account for the limitations and data communication difficulties usually encountered in multidisciplinary engineering applications, without sacrificing convergence properties.

References

[1] N. Alexandrov, On managing the use of surrogates in general nonlinear optimization and MDO, Proceedings of the Seventh AIAA/USAF/NASA/ISSMO Multidisciplinary Analysis & Optimization Symposium, Saint Louis, MO, 1998, Paper 98-4798, pp. 720–729.

[2] N. Alexandrov, J.E. Dennis, A class of general trust-region multilevel algorithms for nonlinear constrained optimization, submitted for publication.

[3] N. Alexandrov, J.E. Dennis, R.M. Lewis, V. Torczon, A trust region framework for managing the use of approximation models in optimization, J. Structural Optim. 15 (1998) 16–23.

[4] J.F. Barthelemy, R.T. Haftka, Approximation concepts for optimum structural design – a review, Structural Optim. 5 (1993) 129–144.

[5] C.L. Bloebaum, W. Hong, A. Peck, Improved moved limit strategy for approximate optimization, Proceedings of the Fifth AIAA/USAF/NASA/ISSMO Symposium, Panama City, FL, Paper 94-4337-CP, 1994, pp. 843–850.

[6] A.J. Booker, J.E. Dennis, P.D. Frank, D.B. Serafini, V. Torcson, M.W. Trosset, A rigorous framework for optimization of expensive functions by surrogates, Struct. Optim. 17 (1999) 1–13.

[7] G.E. Box, N.R. Draper, Empirical Model-Building and Response Surfaces, Wiley, New York, 1987.

[8] T.Y. Chen, Calculation of the move limits for the sequential linear programming method, Internat. J. Numer. Methods Engrg. 36 (1993) 2661–2679.

[9] A.R. Conn, N.I.M. Gould, Ph.L. Toint, A globally convergent augmented Lagrangian algorithm for optimization with general constraints and simple bounds, SIAM J. Numer. Anal. 28 (2) (1991) 545–572.

[10] A.R. Conn, K. Scheinberg, Ph.L. Toint, On the convergence of derivative-free methods for unconstraint optimization, in: A. Iserles, A. Buhmann (Eds.), Approximation Theory and Optimization: Tributes to M.J.D. Powell, Cambridge University Press, Cambridge, 1997, pp. 83–108.

[11] A.R. Conn, K. Scheinberg, Ph.L. Toint, A derivative-free optimization algorithm in practice, Proceedings of the Seventh AIAA/USAF/NASA/ISSMO Multidisciplinary Analysis & Optimization Symposium, Saint Louis, MO, 1998, Paper 98-4718, pp. 129–139.

[12] C. Currin, T. Mitchell, M. Morris, D. Ylvisaker, Bayesian prediction of deterministic functions, with applications to the design and analysis of experiments, J. Amer. Statist. Assoc. 86 (416) (1991) 953–963.

[13] J.E. Dennis, R.B. Schnabel, Numerical Methods for Unconstrained Optimization and Nonlinear Equations, Prentice-Hall, Engelwood Cliffs, NJ, 1983.

[14] G.M. Fadel, M.F. Riley, J.F.M. Barthelemy, Two point exponential approximation method for structural optimizations, Struct. Optim. 2 (1990) 117–124.

[15] R.M. Lewis, V. Torczon, A globally convergent augmented Lagrangian pattern search algorithm for optimization with general constraints and simple bounds, Technical Report 98-31, ICASE, NASA Langley Research Center, Hampton, VA 23681-2199, 1998.

[16] R.M. Lewis, V. Torczon, Pattern search algorithms for bound constrained minimization, SIAM J. Optim. 9 (1999) 1082–1099.

[17] A.B. Owen, Orthogonal arrays for computer experiments, integration and visualization, Statist. Sinica 2 (1992) 439–452.

[18] S. Padula, N. Alexandrov, L. Green, MDO test suite at NASA Langley Research Center, Proceedings of the Sixth AIAA/NASA/ISSMO Symposium on Multidisciplinary Analysis & Optimization, Bellevue, WA, 1996, Paper 96-4028.

[19] J. Rasmussen, Nonlinear programming by cumulative approximation refinement, Struct. Optim. 15 (1998) 1–7.

[20] J.F. Rodríguez, J.E. Renaud, L.T. Watson, Convergence of trust region augmented Lagrangian methods using variable fidelity approximation data, Struct. Optim. 15 (1998) 141–156.

[21] J. Sacks, W.J. Welch, T.J. Mitchell, H.P. Wynn, Design and analysis of computer experiments, Statist. Sci. 4 (1989) 409–435.

[22] J. Sobieszczanski-Sobieski, R.T. Haftka, Multidisciplinary aerospace optimization: survey of recent developments, Struct. Optim. 14 (1997) 1–23, Paper No. AIAA 96-0711.

[23] H.L. Thomas, G.N. Vanderplaats, Y.K. Shyy, A study of move limit adjustment strategies in the approximation concepts approach to structural synthesis, Proceedings of the Fourth AIAA/USAF/NASA/ISSMO Symposium on Multidisciplinary Analysis & Optimization, Cleveland, OH, 1992, pp. 507–512.

[24] V. Torczon, On the convergence of pattern search algorithms, SIAM J. Optim. 7 (1997) 1–25.
[25] B.A. Wujek, J.E. Renaud, A new adaptive move limit management strategy for approximate optimization, Part 1, AIAA J. 36 (10) (1998) 1911–1921.
[26] B.A. Wujek, J.E. Renaud, S.M. Batill, A concurrent engineering approach for multidisciplinary design in a distributed computing environment, in: N. Alexandrov, M.Y. Hussaini (Eds.), Multidisciplinary Design Optimization: State-of-the-Art, Proceedings in Applied Mathematics, Vol. 80, SIAM, Philadelphia, 1997, pp. 189–208.

JOURNAL OF
COMPUTATIONAL AND
APPLIED MATHEMATICS

Journal of Computational and Applied Mathematics 124 (2000) 155–170

www.elsevier.nl/locate/cam

ELSEVIER

ABS algorithms for linear equations and optimization

Emilio Spedicato[a, *], Zunquan Xia[b], Liwei Zhang[b]

[a]*Department of Mathematics, University of Bergamo, 24129 Bergamo, Italy*
[b]*Department of Applied Mathematics, Dalian University of Technology, Dalian, China*

Received 20 May 1999; received in revised form 18 November 1999

Abstract

In this paper we review basic properties and the main achievements obtained by the class of ABS methods, developed since 1981 to solve linear and nonlinear algebraic equations and optimization problems. © 2000 Elsevier Science B.V. All rights reserved.

Keywords: Linear equations; Optimization; ABS methods; Quasi-Newton methods; Linear programming; Feasible direction methods; KT equations; Interior point methods

1. The scaled ABS class: general properties

ABS methods were introduced by Abaffy et al. [2], in a paper considering the solution of linear equations via what is now called the *basic or unscaled ABS class*. The basic ABS class was later generalized to the so-called *scaled ABS class* and subsequently applied to linear least-squares, nonlinear equations and optimization problems, as described by Abaffy and Spedicato [4] and Zhang et al. [43]. Recent work has also extended ABS methods to the solution of diophantine equations, with possible applications to combinatorial optimization. There are presently over 350 papers in the ABS field, see the bibliography by Spedicato and Nicolai [24]. In this paper we review the basic properties of ABS methods for solving linear determined or underdetermined systems and some of their applications to optimization problems.

Let us consider the general linear (determined or underdetermined) system, where $rank(A)$ is arbitrary,

$$Ax = b, \quad x \in \mathbb{R}^n, \ b \in \mathbb{R}^m, \ m \leqslant n \tag{1}$$

---

* Corresponding author.
E-mail address: emilio@unibg.it (E. Spedicato).

0377-0427/00/$ - see front matter © 2000 Elsevier Science B.V. All rights reserved.
PII: S 0377-0427(00)00419-2

or

$$a_i^{\mathrm{T}} x - b_i = 0, \quad i = 1, \ldots, m, \tag{2}$$

where

$$A = \begin{bmatrix} a_1^{\mathrm{T}} \\ \cdots \\ a_m^{\mathrm{T}} \end{bmatrix}. \tag{3}$$

The steps of the scaled ABS class algorithms are the following:

(A) Let $x_1 \in \mathbb{R}^n$ be arbitrary, $H_1 \in \mathbb{R}^{n,n}$ be nonsingular arbitrary, v_1 be an arbitrary nonzero vector in $\mathbb{R}^m$. Set $i = 1$.

(B) Compute the residual $r_i = Ax_i - b$. If $r_i = 0$ stop (x_i solves the problem.) Otherwise compute $s_i = H_i A^{\mathrm{T}} v_i$. If $s_i \neq 0$, then go to (C). If $s_i = 0$ and $\tau = v_i^{\mathrm{T}} r_i = 0$, then set $x_{i+1} = x_i$, $H_{i+1} = H_i$ and go to (F). Otherwise stop (the system has no solution).

(C) Compute the search vector p_i by

$$p_i = H_i^{\mathrm{T}} z_i, \tag{4}$$

where $z_i \in \mathbb{R}^n$ is arbitrary save for the condition

$$v_i^{\mathrm{T}} A H_i^{\mathrm{T}} z_i \neq 0. \tag{5}$$

(D) Update the estimate of the solution by

$$x_{i+1} = x_i - \alpha_i p_i, \tag{6}$$

where the stepsize α_i is given by

$$\alpha_i = v_i^{\mathrm{T}} r_i / r_i^{\mathrm{T}} A p_i. \tag{7}$$

(E) Update the matrix H_i by

$$H_{i+1} = H_i - H_i A^{\mathrm{T}} v_i w_i^{\mathrm{T}} H_i / w_i^{\mathrm{T}} H_i A^{\mathrm{T}} v_i, \tag{8}$$

where $w_i \in \mathbb{R}^n$ is arbitrary save for the condition

$$w_i^{\mathrm{T}} H_i A^{\mathrm{T}} v_i \neq 0. \tag{9}$$

(F) If $i = m$, then stop (x_{m+1} solves the system). Otherwise define v_{i+1} as an arbitrary vector in $\mathbb{R}^m$ but linearly independent of $v_1, \ldots, v_i$. Increment i by one and go to (B).

The matrices H_i appearing in step (E) are generalizations of (oblique) projection matrices. They probably first appeared in a book by Wedderburn [34]. They have been named *Abaffians* since the First International Conference on ABS methods (Luoyang, China, 1991) and this name will be used here. It should be noted that there are several alternative formulations of the linear algebra of the above process, using, e.g., vectors instead of the square matrix H_i (see [4]). Such alternative formulations differ in storage requirement and arithmetic cost and the choice for the most convenient formulation may depend on the relative values of m and n.

The above recursion defines a class of algorithms, each particular method being determined by the choice of the parameters H_1, v_i, z_i, w_i. The *basic ABS class* is obtained by taking $v_i = e_i$, e_i being the ith unitary vector in $\mathbb{R}^m$. Parameters w_i, z_i, H_1 have been introduced, respectively, by Abaffy, Broyden and Spedicato, whose initials are referred to in the name of the class. It is possible to show that the scaled ABS class is a complete realization of the so-called Petrov–Galerkin iteration for solving a linear system, where the iteration has the form $x_{i+1} = x_i - \alpha_i p_i$ with α_i, p_i chosen so that $r_{i+1}^T v_j = 0$, $j = 1, \dots, i$, holds, the vectors v_j being arbitrary and linearly independent. It appears that all deterministic algorithms in the literature having finite termination on a linear system are members of the scaled ABS class. Moreover, the quasi-Newton methods of the Broyden class, which (under some conditions) are known to have termination in at most $2n$ steps, can be embedded in the ABS class. The iterate of index $2i - 1$ generated by Broyden's iteration corresponds to the ith iterate of a certain algorithm in the scaled ABS class.

Referring to the monograph of Abaffy and Spedicato [4] for proofs, we give some properties of methods of the scaled ABS class, assuming, for simplicity, that A has full rank:

- Define $V_i = (v_1, \dots, v_i)$, $W_i = (w_1, \dots, w_i)$. Then $H_{i+1} A^T V_i = 0$, $H_{i+1}^T W_i = 0$, meaning that vectors $A^T v_j, w_j$, $j = 1, \dots, i$, span the null spaces of H_{i+1} and its transpose, respectively.
- The vectors $H_i A^T v_i$, $H_i^T w_i$ are nonzero if and only if a_i, w_i are linearly independent of $a_1, \dots, a_{i-1}$, $w_1, \dots, w_{i-1}$, respectively.
- Define $P_i = (p_1, \dots, p_i)$. Then the implicit factorization $V_i^T A_i^T P_i = L_i$ holds, where L_i is nonsingular lower triangular. From this relation, if $m = n$, one obtains the following semiexplicit factorization of the inverse, with $P = P_n$, $V = V_n$, $L = L_n$:

$$A^{-1} = PL^{-1}V^T. \tag{10}$$

For several choices of the matrix V the matrix L is diagonal, hence formula (10) gives a fully explicit factorization of the inverse as a byproduct of the ABS solution of a linear system, a property that does not hold for the classical solvers. It can also be shown that all possible factorizations of form (10) can be obtained by proper parameter choices in the scaled ABS class, another completeness result.

- Define S_i and R_i by $S_i = (s_1, \dots, s_i)$, $R_i = (r_1, \dots, r_i)$, where $s_i = H_i A^T v_i$, $r_i = H_i^T w_i$. Then the Abaffian can be written in the form $H_{i+1} = H_1 - S_i R_i^T$ and the vectors s_i, r_i can be built via a Gram–Schmidt-type iteration involving the previous vectors (the search vector p_i can be built in a similar way). This representation of the Abaffian in terms of $2i$ vectors is computationally convenient when the number of equations is much less than the number of variables. Notice that there is also a representation in terms of $n - i$ vectors.
- A compact formula for the Abaffian in terms of the parameter matrices is the following:

$$H_{i+1} = H_1 - H_1 A^T V_i (W_i^T H_1 A^T V_i)^{-1} W_i^T H_1. \tag{11}$$

Letting $V = V_m$, $W = W_m$, one can show that the parameter matrices H_1, V, W are admissible (i.e. are such that condition (9) is satisfied) iff the matrix $Q = V^T A H_1^T W$ is strongly nonsingular (i.e., is LU factorizable). Notice that this condition can always be satisfied by suitable exchanges of the columns of V or W, equivalent to a row or a column pivoting on the matrix Q. If Q is strongly nonsingular and we take, as is done in all algorithms so far considered, $z_i = w_i$, then condition (5) is also satisfied.

It can be shown that the *scaled ABS class* corresponds to applying (implicitly) the unscaled ABS algorithm to the scaled (or preconditioned) system $V^T A x = V^T b$, where V is an arbitrary nonsingular matrix of order m. Therefore, the scaled ABS class is also complete with respect to all possible left preconditioning matrices, which in the ABS context are defined implicitly and dynamically (only the ith column of V is needed at the ith iteration, and it can also be a function of the previous column choices).

2. Subclasses of the ABS class

In the monograph of Abaffy and Spedicato [4] nine subclasses are considered of the scaled ABS class. Here we mention only two of them.

(a) The *conjugate direction subclass*. This class is obtained by setting $v_i = p_i$. It is well defined under the condition (sufficient but not necessary) that A is symmetric and positive definite. It contains the implicit Choleski algorithm and the ABS versions of the Hestenes–Stiefel and the Lanczos algorithms (i.e., algorithms that generate the same sequence x_i as these classical methods in exact arithmetic). This class generates all possible algorithms whose search directions are A-conjugate. The vector x_{i+1} minimizes the energy or A-weighted Euclidean norm of the error over $x_1 + \mathrm{Span}(p_1, \ldots, p_i)$. If $x_1 = 0$ then the solution is approached monotonically from below in the energy norm.

(b) The *orthogonally scaled subclass*. This class is obtained by setting $v_i = A p_i$. It is well defined if A has full column rank and remains well defined even if m is greater than n. It contains the ABS formulation of the QR algorithm (the so-called *implicit QR algorithm*), the GMRES and the conjugate residual algorithms. The scaling vectors are orthogonal and the search vectors are AA^{T}-conjugate. The vector x_{i+1} minimizes the Euclidean norm of the residual over $x_1 + \mathrm{Span}(p_1, \ldots, p_i)$. It can be shown that the methods in this class can be applied to overdetermined systems of $m > n$ equations, where in n steps they obtain the solution in the least-squares sense.

3. The Huang algorithm, the implicit LU and the implicit LX algorithms

Specific algorithms of the scaled ABS class are obtained by choosing the parameters. Here we consider three important particular algorithms in the basic ABS class.

The *Huang algorithm* is obtained by the parameter choices $H_1 = I$, $z_i = w_i = a_i$, $v_i = e_i$. This method was initially proposed by Huang [21], who claimed that it was very accurate on ill conditioned problems. It is this method whose generalization has led to the ABS class. A mathematically equivalent, but numerically more stable, see Broyden [8], formulation of this algorithm is the so-called *modified Huang algorithm* where the search vectors and the Abaffians are given by formulas $p_i = H_i(H_i a_i)$ and $H_{i+1} = H_i - p_i p_i^{\mathrm{T}} / p_i^{\mathrm{T}} p_i$. Some properties of this algorithm follow.

(a) The search vectors are orthogonal and are the same as would be obtained by applying the classical Gram–Schmidt orthogonalization procedure to the rows of A. The modified Huang algorithm is related to, but is not numerically identical with, the reorthogonalized Gram–Schmidt algorithm of Daniel et al. [9].

(b) If x_1 is the zero vector, then the vector x_{i+1} is the solution with least Euclidean norm of the first i equations. The solution x^+ with least Euclidean norm of the whole system is approached monotonically and from below by the sequence x_i. Zhang [38] has shown that the Huang algorithm can be applied, via the active set strategy of Goldfarb and Idnani [19], to systems of linear inequalities. The process in a finite number of steps either finds the solution with least Euclidean norm or determines that the system has no solution.

(c) While the error growth in the Huang algorithm is governed by the square of the number $\eta_i = \|a_i\|/\|H_i a_i\|$, which is certainly large for some i if A is ill conditioned, the error growth depends only on η_i if p_i or H_i are defined as in the modified Huang algorithm and, to first order, there is no error growth for the modified Huang algorithm.

(d) Numerical experiments (see [30]), have shown that the modified Huang algorithm is very stable, usually giving better accuracy in the computed solution than both the implicit LU algorithm and the classical LU factorization method. The modified Huang algorithm has a higher operation count, varying between $1.5n^3$ and $2.5n^3$, depending on which formulation of the ABS algorithm is used (the count for the Huang algorithm varies between n^3 and $1.5n^3$).

The *implicit LU algorithm* is given by the choices $H_1 = I$, $z_i = w_i = v_i = e_i$. Some of its properties are

(a) The algorithm is well defined iff A is regular (i.e., all principal submatrices are nonsingular). Otherwise column pivoting has to be performed (or, if $m = n$, equation pivoting).

(b) The Abaffian H_{i+1} has the following structure, with $K_i \in \mathbb{R}^{n-i,i}$:

$$H_{i+1} = \begin{bmatrix} 0 & 0 \\ \cdots & \cdots \\ 0 & 0 \\ K_i & I_{n-i} \end{bmatrix}. \tag{12}$$

(c) Only the first i components of p_i can be nonzero and the ith component is unity. Hence the matrix P_i is unit upper triangular, so that the implicit factorization $A = LP^{-1}$ is of the LU type, with units on the diagonal.

(d) Only K_i has to be updated. The algorithm requires $nm^2 - 2m^3/3$ multiplications plus lower-order terms. Hence, for $m = n$ there are $n^3/3$ multiplications plus lower-order terms, which is the same cost as for the classical LU factorization or Gaussian elimination (which are two essentially equivalent processes).

(e) The main storage requirement is the storage of K_i, which has at most $n^2/4$. This is half the storage needed by Gaussian elimination and a quarter the storage needed by the LU factorization algorithm (assuming that A is not overwritten). Hence, the implicit LU algorithm is computationally better than the classical Gaussian elimination or LU algorithm, having the same arithmetic cost but using less memory.

(f) The implicit LU algorithm, implemented in the case $m = n$ with row pivoting, has been shown in experiments of Bertocchi and Spedicato [6] to be numerically stable and in experiments of Bodon [7] on the vector processor Alliant FX 80 (with eight processors) to be about twice as fast as than the LAPACK implementation of the classical LU algorithm.

The *implicit LX algorithm* is defined by the choices $H_1 = I$, $v_i = e_i$, $z_i = w_i = e_{k_i}$, where k_i is an integer, $1 \leqslant k_i \leqslant n$, such that

$$e_{k_i}^T H_i a_i \neq 0. \tag{13}$$

Notice that by a general property of the ABS class for A with full rank there is at least one index k_i such that (13) is satisfied. For stability reasons it may be recommended to select k_i such that $\eta_i = |e_{k_i}^T H_i a_i|$ is maximized. The following properties are valid for the implicit LX algorithm. Let N be the set of integers from 1 to n, $N = (1, 2, \ldots, n)$. Let B_i be the set of indices $k_1, \ldots, k_i$ chosen for the parameters of the implicit LX algorithm up to the step i. Let N_i be the set $N \backslash B_i$. Then

(a) The index k_i is selected from N_{i-1}.

(b) The rows of H_{i+1} of index $k \in B_i$ are null rows.

(c) The vector p_i has $n - i$ zero components; its k_ith component is equal to one.

(d) If $x_1 = 0$, then x_{i+1} is a basic type solution of the first i equations, whose nonzero components may lie only in the positions corresponding to the indices $k \in B_i$.

(e) The columns of H_{i+1} of index $k \in N_i$ are the unit vectors e_k, while the columns of H_{i+1} of index $k \in B_i$ have zero components in the jth position, with $j \in B_i$, implying that only $i(n - i)$ elements of such columns have to be computed.

(f) At the ith step $i(n - i)$ multiplications are needed to compute $H_i a_i$ and $i(n - i)$ to update the nontrivial part of H_i. Hence the total number of multiplications is the same as for the implicit LU algorithm (i.e., $n^3/3$), but no pivoting is necessary, reflecting the fact that no condition is required on the matrix A.

(g) The storage requirement is the same as for the implicit LU algorithm, i.e., at most $n^2/4$. Hence the implicit LX algorithm shares the same storage advantage of the implicit LU algorithm over the classical LU algorithm, with the additional advantage of not requiring pivoting.

(h) Numerical experiments by Mirnia [23] have shown that the implicit LX method gives usually better accuracy, in terms of error in the computed solution, than the implicit LU algorithm and often even than the modified Huang algorithm. In terms of size of the final residual, its accuracy is comparable to that of the LU algorithm as implemented (with row pivoting) in the MATLAB or LAPACK libraries, but it is better again in terms of error in the solution.

4. Other ABS linear solvers and implementational details

ABS reformulations have been obtained for most algorithms proposed in the literature. The availability of several formulations of the linear algebra of the ABS process allows alternative formulations of each method, with possibly different values of overhead, storage and different properties of numerical stability, vectorization and parallelization. The reprojection technique, already seen in the case of the modified Huang algorithm and based upon the identities $H_i q = H_i(H_i q)$, $H_i^T = H_i^T(H_i^T q)$, valid for any vector q if $H_1 = I$, remarkably improves the stability of the algorithm, the increase in the number of operations depending on the algorithm and the number of reprojections made. The ABS versions of the Hestenes–Stiefel and the Craig algorithms for instance are very stable under the above reprojection. The *implicit QR algorithm*, defined by the choices $H_1 = I$, $v_i = A p_i$, $z_i = w_i = e_i$ can be implemented in a very stable way using the reprojection in both the definition of the search vector and the scaling vector. It should also be noticed that the classical iterative refinement procedure, which amounts to a Newton iteration on the system $Ax - b = 0$ using the approximate factors of A, can be reformulated in the ABS context using the previously defined search vectors p_i. Experiments of Mirnia [23] have shown that ABS refinement works excellently.

For problems with special structure ABS methods can often be implemented taking into account the effect of the structure on the Abaffian matrix, which often tends to reflect the structure of the matrix A. Several cases of structured problems are discussed by Abaffy and Spedicato [4] and in later papers. As an example, one can show that if A has a banded structure, the same is true for the Abaffian matrix generated by the implicit LU, the implicit QR and the Huang algorithm, albeit the band size is increased. If A is symmetric positive definite (SPD) and has a so-called nested dissection (ND) structure, the same is true for the Abaffian matrix. In this case the implementation of the implicit LU algorithm uses much less storage, for large n, than the Choleski algorithm. For matrices having the Kuhn–Tucker (KT) structure large classes of ABS methods have been developed, many of them better either in storage or in arithmetic cost than classical methods. For matrices with general sparsity patterns little is presently known about minimizing the fill-in in the Abaffian matrix. Careful use of BLAS4 routines can however substantially reduce the number of operations and make the ABS implementation competitive with a sparse implementation of say the LU factorization (e.g., by the code MA28) for values of n up to about 1000.

It is possible to implement the ABS process also in block form, where several equations, instead of just one, are dealt with at each step. The block formulation does not damage the numerical accuracy and can lead to reduction of overhead on special problems or to faster implementations on vector or parallel computers.

Finally, infinite iterative methods can be obtained by the finite ABS methods via two ways. The first one involves restarting the iteration after $k < m$ steps, so that the storage will be of order $2kn$ if the representation of the Abaffian in terms of $2i$ vectors is used. The second approach uses only a limited number of terms in the Gram–Schmidt-type processes that are alternative formulations of the ABS procedure. For both cases convergence at a linear rate has been established using the technique developed by Dennis and Turner [10]. The infinite iteration methods obtained by these approaches define a very large class of methods, that contains not only all Krylov-space-type methods of the literature, but also non-Krylov-type methods such as the Gauss–Seidel, the De La Garza and the Kackmartz methods, with their generalizations.

5. Applications to optimization; the unconstrained optimization case

We will now present some applications of ABS methods to optimization problems. In this section we describe a class of ABS related methods for unconstrained optimization. In Section 6 we show how ABS methods provide the general solution of the quasi-Newton equation, with sparsity and symmetry conditions and we discuss how SPD solutions can be obtained. In Section 7 we present several special ABS methods for solving the Kuhn–Tucker equations. In Section 8 we consider the application of the implicit LX algorithm to the linear programming (LP) problem. In Section 9 we present ABS approaches to the general linearly constrained optimization problem, which unify linear and nonlinear problems.

ABS methods can be applied directly to the unconstrained minimization of a function $f(x)$. They use the iteration $x_{i+1} = x_i - \alpha_i H_i^{\mathrm{T}} z_i$, where H_i is reset after n or less steps and z_i is chosen so that the descent condition holds, i.e., $g_i^{\mathrm{T}} H_i^{\mathrm{T}} z_i > 0$, with $g_i = \nabla f(x_i)$. If $f(x)$ is quadratic, one can identify the matrix A in the Abaffian update formula with the Hessian of $f(x)$. Defining a perturbed point x' by $x' = x_i - \beta v_i$ one has on quadratic functions $g' = g - \beta A v_i$, and hence the update of the Abaffian takes the form $H_{i+1} = H_i - H_i y_i w_i^{\mathrm{T}} H_i / w_i^{\mathrm{T}} H_i y_i$, where $y_i = g' - g_i$, so that no information is needed about second derivatives. The above-defined class has termination on quadratic functions and local superlinear (n-step Q-quadratic) rate of convergence on general functions. It is a special case of a class of projection methods developed by Psenichny and Danilin [25]. Almost no numerical results are available about the performance of the methods in this class.

6. Applications to quasi-Newton updates

ABS methods have been used to provide the general solution of the quasi-Newton equation, with the additional conditions of symmetry, sparsity and positive definiteness. While the general solution of only the quasi-Newton equation has already been given by Adachi [5], the explicit formulas obtained for the sparse symmetric case are new, and so is the way of constructing sparse SPD updates.

Let us consider the transpose form of the quasi-Newton equation defining the new approximation to a Jacobian or a Hessian

$$d^{\mathrm{T}} B' = y^{\mathrm{T}}, \tag{14}$$

where $d = x' - x$, $y = g' - g$. We observe that (14) can be seen as a set of n linear underdetermined systems, each one having just one equation and differing only on the right-hand side. Hence, the general solution can be obtained by one step of the ABS method. It can be written in the following way:

$$B' = B - s(B^{\mathrm{T}} d - y)^{\mathrm{T}} / d^{\mathrm{T}} s + (I - s d^{\mathrm{T}} / d^{\mathrm{T}} s) Q, \tag{15}$$

where $Q \in \mathbb{R}^{n,n}$ is arbitrary, and $s \in \mathbb{R}^n$ is arbitrary subject to $s^{\mathrm{T}} d \neq 0$. Formula (15), derived by Spedicato and Xia [31], is equivalent to a formula of Adachi [5].

Now the conditions that some elements of B' should be zero, or have constant value or that B' should be symmetric can be written as additional linear constraints. This if b_i' is the ith column of B', we require

$$(b_i')^{\mathrm{T}} e_k = \eta_{ij}, \tag{16}$$

where $\eta_{ij} = 0$ implies sparsity, $\eta_{ij} = constant$ implies that some elements do not change their value and $\eta_{ij} = \eta_{ji}$ implies symmetry. The ABS algorithm can deal with these extra conditions. Spedicato and Zhao [33] give the solution in explicit form, columnwise in presence of symmetry. By adding the condition that the diagonal elements be sufficiently large, it is possible to obtain formulas where B' is quasi-positive definite or quasi-diagonally dominant, in the sense that the principal submatrix of order $n - 1$ is positive definite or diagonally dominant. It is not possible in general to force B' to be SPD, since SPD solutions may not exist, which is reflected in the fact that no additional conditions can be put on the last diagonal element, since the last column is fully determined by the $n - 1$ symmetry conditions and the quasi-Newton equation. This result can however be exploited to provide SPD approximations by imbedding the original minimization problem of n variables in a problem of $n + 1$ variables, whose solution with respect to the first n variables is the original solution. This imbedding modifies the quasi-Newton equation so that SPD solutions exist. Numerical results on the performance of the proposed sparse quasi-Newton methods are not yet available.

7. ABS methods for KT equations

The Kuhn–Tucker (KT) equations, which should more appropriately be named Kantorovich–Karush–Kuhn–Tucker (KKKT) equations, are a special linear system, based on the optimality conditions of the problem of minimizing a quadratic function with Hessian G subject to the linear equality constraint $Cp=c$. The system has the following form, with $G \in \mathbb{R}^{n,n}$, $C \in \mathbb{R}^{m,n}$, $p, g \in \mathbb{R}^n$, $z, c \in \mathbb{R}^m$:

$$\begin{bmatrix} G & C^{\mathrm{T}} \\ C & 0 \end{bmatrix} \begin{pmatrix} p \\ z \end{pmatrix} = \begin{pmatrix} g \\ c \end{pmatrix}. \tag{17}$$

If G is nonsingular, the coefficient matrix is nonsingular iff $CG^{-1}C^{\mathrm{T}}$ is nonsingular. Usually G is nonsingular, symmetric and positive definite, but this assumption, required by several classical solvers, is not necessary for the ABS solvers.

ABS classes for solving the KT problem can be derived in several ways. Observe that system (17) is equivalent to the two subsystems

$$Gp + C^{\mathrm{T}}z = g, \tag{18}$$

$$Cp = c. \tag{19}$$

The general solution of subsystem (19) can be written in the following ABS form:

$$p = p_{m+1} + H_{m+1}^{\mathrm{T}}q \tag{20}$$

with q arbitrary. The parameter choices made to construct p_{m+1} and H_{m+1} are arbitrary and define therefore a class of algorithms.

Since the KT equations have a unique solution, there must be a choice of q in (20) which makes p the unique n-dimensional subvector defined by the first n components of the solution x. Notice that since H_{m+1} is singular, q is not uniquely defined (but would be uniquely defined if one takes the representation of the Abaffian in terms of $n - m$ vectors).

By multiplying Eq. (18) on the left by H_{m+1} and using the ABS property $H_{m+1}C^{\mathrm{T}} = 0$, we obtain the equation

$$H_{m+1}Gp = H_{m+1}g \tag{21}$$

which does not contain z. Now there are two possibilities for determining p.

(A1) Consider the system formed by Eqs. (19) and (21). Such a system is solvable but overdetermined. Since $\mathrm{rank}(H_{m+1}) = n - m$, m equations are recognized as dependent and are eliminated in step (B) of any ABS algorithm applied to this system.

(A2) In Eq. (21) replace p by the general solution (20) to give

$$H_{m+1}GH_{m+1}^{\mathrm{T}}q = H_{m+1}g - H_{m+1}Gp_{m+1}. \tag{22}$$

The above system can be solved by any ABS method for a particular solution q, m equations being again removed at step (B) of the ABS algorithm as linearly dependent.

Once p is determined, one can determine z in two ways, namely,

(B1) Solve by any ABS method the overdetermined compatible system

$$C^{\mathrm{T}}z = g - Gp \tag{23}$$

by removing at step (B) of the ABS algorithm the $(n - m)$-dependent equations.
(B2) Let $P = (p_1,\ldots,p_m)$ be the matrix whose columns are the search vectors generated on the system $Cp = c$. Now $CP = L$, with L nonsingular lower diagonal. Multiplying Eq. (23) on the left by P^{T} we obtain a triangular system, defining z uniquely as

$$L^{\mathrm{T}}z = P^{\mathrm{T}}g - P^{\mathrm{T}}Gp. \tag{24}$$

Extensive numerical testing has evaluated the accuracy of the above ABS algorithms for KT equations for certain choices of the ABS parameters (corresponding to the implicit LU algorithm with row pivoting and the modified Huang algorithm). The methods have been tested against classical methods, in particular the method of Aasen and methods using the QR factorization. The experiments have shown that some ABS methods are the most accurate, in both residual and solution error; moreover some ABS algorithms are cheaper in storage and in overhead, up to one order, especially for the case when m is close to n. In particular, two methods based upon the implicit LU algorithm not only have turned out to be more accurate, especially in residual error, than the method of Aasen and the method using QR factorization via Houselder matrices, but they are also cheaper in number of operations (the method of Aasen has a lower storage for small m but a higher storage for large m).

In many interior point methods the main computational cost is to compute the solution for a sequence of KT problems where only G, which is diagonal, changes. In such a case the ABS methods, which initially work on the matrix C, which is unchanged, have an advantage, particularly when m is large, where the dominant cubic term decreases with m and disappears for $m = n$, so that the overhead is dominated by second-order terms. Again numerical experiments show that some ABS methods are more accurate than the classical ones (for details see [28]).

8. Reformulation of the simplex method via the implicit LX algorithm

The implicit LX algorithm has a natural application to a reformulation of the simplex method for the LP problem in standard form, i.e., the problem of minimizing $c^{\mathrm{T}}x$, subject to $Ax = b$, $x \geqslant 0$.

The applicability of the implicit LX method is a consequence of the fact that the iterate x_{i+1} generated by the method, started from the zero vector, is a basic type vector, with a unit component in the position k_i. Nonidentically zero components correspond to indices $j \in B_i$, where B_i is the set of indices of the unit vectors chosen as the z_i, w_i parameters, i.e., the set $B_i = (k_1,\ldots,k_i)$, while the components of x_{i+1} with indices in the set $N_i = N/B_i$ are identically zero, where $N = (1,\ldots,n)$. Therefore if the nonzero components are nonnegative, the point defines a vertex of the polytope containing the feasible points defined by the constraints of the LP problem.

In the simplex method one moves from one vertex to another, according to some rules and usually reducing at each step the value of the function $c^T x$. The direction along which one moves from one vertex to another is an edge direction of the polytope and is determined by solving a linear system, whose coefficient matrix A_B, the *basic matrix*, is defined by m linearly independent columns of the matrix A, called the *basic columns*. Usually such a system is solved by LU factorization or occasionally by the QR method (see [14]). The new vertex is associated with a new basic matrix A_B', which is obtained by replacing one of the columns in A_B by a column of the matrix A_N, which comprises the columns of A that do not belong to A_B. The most efficient algorithm for solving the modified system, after the column interchange, is the method of Forrest and Goldfarb [15], requiring m^2 multiplications. Notice that the classical simplex method requires m^2 storage for the matrix A_B plus mn storage for the matrix A, which must be kept in general to provide the columns for the exchange.

The application of the implicit LX method to the simplex method, developed by Xia [35], Zhang and Xia [42], Spedicato et al. [32], Feng et al. [13], exploits the fact that in the implicit LX algorithm the interchange of a jth column in A_B with a kth column in A_N corresponds to the interchange of a previously chosen parameter vector $z_j = w_j = e_j$ with a new parameter $z_k = w_k = e_k$. This operation is a special case of the perturbation of the Abaffian after a change in the parameters and can be done using a general formula of Zhang [39], without explicit use of the kth column in A_N. Moreover, all quantities needed for the construction of the search direction (the edge direction) and for the interchange criteria can as well be obtained without explicit use of the columns of A. Hence it follows that the ABS approach needs only the storage of the matrix H_{m+1}, which, in the case of the implicit LX algorithm, is of size at most $n^2/4$. Therefore, for values of m close to n the storage required by the ABS formulation is about 8 times less than for the classical simplex method.

Here we give the basic formulae of the simplex method in both the classical and the ABS formulation. The column in A_N replacing an old column in A_B is often taken as the column with minimal relative cost. In terms of the ABS formulation this is equivalent to minimizing the scalar $\eta_i = c^T H^T e_i$ with respect to $i \in N_m$. Let N^* be the index chosen in this way. The column in A_B to be exchanged is usually chosen with the criterion of the maximum displacement along an edge which keeps the basic variables nonnegative. Define $\omega_i = x^T e_i / e_i^T H^T e_{N^*}$, where x is the current basic feasible solution. Then the above criterion is equivalent to minimizing ω_i with respect the set of indices $i \in B_m$ such that

$$e_i^T H^T e_{N^*} > 0. \tag{25}$$

Notice that $H^T e_{N^*} \neq 0$ and that an index i such that (25) is satisfied always exists, unless x is a solution of the LP problem.

The update of the Abaffian after the interchange of the unit vectors, which corresponds to the update of the LU or QR factors after the interchange of the basic with the nonbasic column, is given by the following formula:

$$H' = H - (H e_{B^*} - e_{B^*}) e_{N^*}^T H / e_{N^*}^T H e_{B^*}. \tag{26}$$

The search direction d, which in the classical formulation is obtained by solving the system $A_B d = -A e_{N^*}$, is given by $d = H_{m+1}^T e_{N^*}$, hence at no cost. Finally, the relative cost vector r, classically given by $r = c - A^T A_B^{-1} c_B$, where c_B consists of the components of c with indices corresponding to those of the basic columns, is simply given by $r = H_{m+1} c$.

Let us now consider the computational cost of update (26). Since He_{B^*} has at most $n - m$ nonzero components, while $H^{\mathrm{T}}e_{N^*}$ has at most m, no more than $m(n - m)$ multiplications are required. The update is most expensive for $m = n/2$ and gets cheaper the smaller m is or the closer it is to n. In the dual steepest edge method of Forrest and Goldfarb [15] the overhead for replacing a column is m^2, hence formula (26) is faster for $m > n/2$ and is recommended on overhead considerations for m sufficiently large. However we notice that ABS updates having a $O(m^2)$ cost can also be obtained by using the representation of the Abaffian in terms of $2m$ vectors. No computational experience has been obtained till now on the new ABS formulation of the simplex method.

Finally, a generalization of the simplex method, based upon the use of the Huang algorithm started with a suitable singular matrix, has been developed by Zhang [40]. In this formulation the solution is approached by points lying on a face of the polytope. Whenever the point hits a vertex the remaining iterates move among vertices and the method is reduced to the simplex method.

9. ABS unification of feasible direction methods for minimization with linear constraints

ABS algorithms can be used to provide a unification of feasible point methods for nonlinear minimization with linear constraints, including as a special case the LP problem. Let us first consider the problem $\min f(x)$, $x \in \mathbb{R}^n$, subject to $Ax = b$, $A \in \mathbb{R}^{m,n}$, $m \leqslant n$, rank$(A) = m$.

Let x_1 be a feasible starting point. An iteration of the form $x_{i+1} = x_i - \alpha_i d_i$, the search direction will generate feasible points iff

$$Ad_i = 0. \tag{27}$$

Solving the underdetermined system (27) for d_i by the ABS algorithm, the solution can be written in the following form, taking, without loss of generality, the zero vector as a special solution

$$d_i = H_{m+1}^{\mathrm{T}}q. \tag{28}$$

Here the matrix H_{m+1} depends on the arbitrary choice of the parameters H_1, w_i and v_i used in solving (27) and $q \in R^n$ is arbitrary. Hence the general feasible direction iteration has the form

$$x_{i+1} = x_i - \alpha_i H_{m+1}^{\mathrm{T}}q. \tag{29}$$

The search direction is a descent direction iff $d^{\mathrm{T}}\nabla f(x) = q^{\mathrm{T}}H_{m+1}\nabla f(x) > 0$. Such a condition can always be satisfied by choice of q unless $H_{m+1}\nabla f(x) = 0$, which implies, from the null space structure of H_{m+1}, that $\nabla f(x) = A^{\mathrm{T}}\lambda$ for some λ. In this case x_{i+1} is a KT point and λ is the vector of the Lagrange multipliers. When x_{i+1} is not a KT point, it is easy to see that the search direction is a descent directions if we select q by formula

$$q = WH_{m+1}\nabla f(x)m, \tag{30}$$

where W is a symmetric and positive-definite matrix.

Particular well-known algorithms from the literature are obtained by the following choices of q, with $W = I$:

(a) The *reduced gradient method* of Wolfe. Here H_{m+1} is constructed by the implicit LU (or the implicit LX) algorithm.
(b) The *gradient projection method* of Rosen. Here H_{m+1} is built using the Huang algorithm.

(c) The *method of Goldfarb and Idnani*. Here H_{m+1} is built via the modification of the Huang algorithm where H_1 is a symmetric positive definite matrix approximating the inverse Hessian of $f(x)$.

Using the above ABS representations several formulations can be obtained of these classical algorithms, each one having in general different storage requirements or arithmetic costs. No numerical experience with such ABS methods is yet available but is expected that, similar to the case of methods for KKT equations, some of these formulations are better than the classical ones.

If there are inequalities two approaches are possible.

(A) The *active set* approach. In this approach the set of linear equality constraints is modified at every iteration by adding and/or dropping some of the linear inequality constraints. Adding or deleting a single constraint can be done, for every ABS algorithm, in order $O(nm)$ operations (see [39]). In the ABS reformulation of the method of Goldfarb and Idnani [19] the initial matrix is related to a quasi-Newton approximation of the Hessian and an efficient update of the Abaffian after a change in the initial matrix is discussed by Xia et al. [37].

(B) The *standard form* approach. In this approach, by introducing slack variables, the problem with both types of linear constraints is written in the equivalent form $\min f(x)$, subject to $Ax = b$, $x \geqslant 0$.

The following general iteration, started with a feasible point x_1, generates a sequence of feasible points for the problem in standard form

$$x_{i+1} = x_i - \alpha_i \beta_i H_{m+1} \nabla f(x). \tag{31}$$

In (30) the parameter α_i can be chosen by a line search along the vector $H_{m+1} \nabla f(x)$, while the relaxation parameter $\beta_i > 0$ is selected to prevent the new point having negative components.

If $f(x)$ is nonlinear, then H_{m+1} can usually be determined once and for all at the first step, since $\nabla f(x)$ generally changes from iteration to iteration and this will modify the search direction in (31) (unless the change in the gradient happens to be in the null space of H_{m+1}). If however $f(x) = c^T x$ is linear we must change H_{m+1} in order to modify the search direction used in (31). As observed before, the simplex method is obtained by constructing H_{m+1} with the implicit LX algorithm, every step of the method corresponding to a change of the parameters e_{k_i}. It can be shown (see [36]), that the method of Karmarkar (equivalent to an earlier method of Evtushenko [12]) corresponds to using the generalized Huang algorithm, with initial matrix $H_1 = \text{Diag}(x_i)$ changing from iteration to iteration. Another method, faster than Karmarkar's having superlinear against linear rate of convergence and $O(\sqrt{n})$ against $O(n)$ complexity, again first proposed by Evtushenko, is obtained by the generalized Huang algorithm with initial matrix $H_1 = \text{Diag}(x_i^2)$.

10. Final remarks and conclusions

In the above review we have given a partial presentation of the many results in the field of ABS methods, documented in more than 350 papers. We have completely skipped the following topics where ABS methods have provided important results.

(1) For linear least squares large classes of algorithms have been obtained; extensive numerical experiments have shown that several ABS methods are more accurate than the QR-based codes available in standard libraries (see [26,27]).

(2) For nonlinear algebraic equations ABS methods provide a generalization and include as special cases both Newton and Brent methods (see [3]); they can be implemented using only, about half of the Jacobian matrix, as is the case for Brent method (see [22,29]). Some of these methods have better global convergence properties than Newton method (see [16]), and they can keep a quadratic rate of convergence even if the Jacobian is singular at the solution (see [17]).

(3) ABS methods have potential applications to the computation of eigenvalues (see [1]). Zhang [41] particular deals with the computation of the inertia in KKT matrices, an important problem for algorithms for the general quadratic programming problem using the active set strategy proposed by Gould [20] and Gill et al. [18].

(4) ABS methods can find a special integer solution and provide the set of all integer solutions of a general Diophantine linear system (see [11]). The ABS approach to this problem provides inter alia an elegant characterization of the integer solvability of such a system that is a natural generalization of the classical result well known for a single Diophantine equation in n variables. Work is on progress on using the ABS approach for 0–1 problems and integer solution of systems of integer inequalities and integer LP.

(5) Finally, work is in progress to develop and document ABSPACK, a package of ABS algorithms for linear systems, linear least squares, nonlinear systems and several optimization problems. A first release of the package is expected in 2002.

We think that the ABS approach, for reasons not to be discussed here still little known by mathematicians, deserves to be more widely known and to be further developed in view of the following undisputable achievements.

(6) It provides a unifying framework for a large field of numerical methods, by itself an important philosophical attainement.

(7) Due to alternative formulations of the linear algebra of the ABS process, several implementations are possible of a given algorithm (here we identify an algorithm with the sequence of approximations to the solution that it generates in exact arithmetic), each one with possible advantages in certain circumstances.

(8) Considerable numerical testing has shown many ABS methods to be preferable to more classical algorithms, justifying the interest in ABSPACK. We have solved real industrial problems using ABS methods where all other methods from commercial codes had failed.

(9) It has led to the solution of some open problems in the literature, e.g., the explicit formulation of sparse symmetric quasi-Newton updates. The implicit LX algorithm provides a general solver of linear equations that is better than Gaussian elimination (same arithmetic cost, less intrinsic storage, regularity not needed) and that provides a much better implementation of the simplex method than via LU factorization for problems where the number of constraints is greater than half the number of variables.

The most interesting challenge to the power of ABS methods is whether by suitable choice of the available parameters an elementary proof could be provided of Fermat's last theorem in the context of linear Diophantine systems.

Acknowledgements

Work performed in the framework of a research supported partially by MURST Programma Co-finanziato 1997.

References

[1] J. Abaffy, An ABS algorithm for general computation of eigenvalues, Communication, Second International Conference of ABS Methods, Beijing, June 1995.

[2] J. Abaffy, C.G. Broyden, E. Spedicato, A class of direct methods for linear systems, Numer. Math. 45 (1984) 361–376.

[3] J. Abaffy, A. Galantai, E. Spedicato, The local convergence of ABS methods for non linear algebraic equations, Numer. Math. 51 (1987) 429–439.

[4] J. Abaffy, E. Spedicato, ABS Projection Algorithms: Mathematical Techniques for Linear and Nonlinear Equations, Ellis Horwood, Chichester, 1989.

[5] N. Adachi, On variable metric algorithms, J. Optim. Theory Appl. 7 (1971) 391–409.

[6] M. Bertocchi, E. Spedicato, Performance of the implicit Gauss–Choleski algorithm of the ABS class on the IBM 3090 VF, Proceedings of the 10th Symposium on Algorithms, Strbske Pleso, 1989, pp. 30–40.

[7] E. Bodon, Numerical experiments on the ABS algorithms for linear systems of equations, Report DMSIA 93/17, University of Bergamo, 1993.

[8] C.G. Broyden, On the numerical stability of Huang's and related methods, J. Optim. Theory Appl. 47 (1985) 401–412.

[9] J. Daniel, W.B. Gragg, L. Kaufman, G.W. Stewart, Reorthogonalized and stable algorithms for updating the Gram–Schmidt QR factorization, Math. Comput. 30 (1976) 772–795.

[10] J. Dennis, K. Turner, Generalized conjugate directions, Linear Algebra Appl. 88/89 (1987) 187–209.

[11] H. Esmaeili, N. Mahdavi-Amiri, E. Spedicvato, Solution of Diophantine linear systems via the ABS methods, Report DMSIA 99/29, University of Bergamo, 1999.

[12] Y. Evtushenko, Two numerical methods of solving nonlinear programming problems, Sov. Dok. Akad. Nauk 251 (1974) 420–423.

[13] E. Feng, X.M. Wang, X.L. Wang, On the application of the ABS algorithm to linear programming and linear complementarity, Optim. Meth. Software 8 (1997) 133–142.

[14] R. Fletcher, Dense factors of sparse matrices, in: A. Iserles, M.D. Buhmann (Eds.), Approximation Theory and Optimization, Cambridge University Press, Cambridge, 1997, pp. 145–166.

[15] J.J.H. Forrest, D. Goldfarb, Steepest edge simplex algorithms for linear programming, Math. Programming 57 (1992) 341–374.

[16] A. Galantai, The global convergence of ABS methods for a class of nonlinear problems, Optim. Methods Software 4 (1995) 283–295.

[17] R. Ge, Z. Xia, An ABS algorithm for solving singular nonlinear systems, Report DMSIA 36/96, University of Bergamo, 1996.

[18] P.E. Gill, W. Murray, M.A. Saunders, M.H. Wright, Inertia controlling methods for general quadratic programming, SIAM Rev. 33 (1991) 1–36.

[19] D. Goldfarb, A. Idnani, A numerically stable dual method for solving strictly convex quadratic programming, Math. Program. 27 (1983) 1–33.

[20] N.I.M. Gould, On practical conditions for the existence and uniqueness of solutions to the general equality quadratic programming problem, Math. Programming 32 (1985) 90–99.

[21] H.Y. Huang, A direct method for the general solution of a system of linear equations, J. Optim. Theory Appl. 16 (1975) 429–445.

[22] A. Jeney, Discretization of a subclass of ABS methods for solving nonlinear systems of equations, Alkalmazott Mat. Lapok 15 (1990) 353–364 (in Hungarian).

[23] K. Mirnia, Numerical experiments with iterative refinement of solutions of linear equations by ABS methods, Report DMSIA 32/96, University of Bergamo, 1996.

[24] S. Nicolai, E. Spedicato, A bibliography of the ABS methods, Optim. Methods Software 8 (1997) 171–183.

[25] B.N. Psenichny, Y.M. Danilin, Numerical Methods in Extremal Problems, MIR, Moscow, 1978.

[26] E. Spedicato, E. Bodon, Numerical behaviour of the implicit QR algorithm in the ABS class for linear least squares, Ric. Oper. 22 (1992) 43–55.

[27] E. Spedicato, E. Bodon, Solution of linear least squares via the ABS algorithm, Math. Programming 58 (1993) 111–136.

[28] E. Spedicato, Z. Chen, E. Bodon, ABS methods for KT equations, in: G. Di Pillo, F. Giannessi (Eds.), Nonlinear Optimization and Applications, Plenum Press, New York, 1996, pp. 345–359.

[29] E. Spedicato, Z. Chen, N. Deng, A class of difference ABS-type algorithms for a nonlinear system of equations, Numer. Linear Algebra Appl. 13 (1994) 313–329.

[30] E. Spedicato, M.T. Vespucci, Variations on the Gram–Schmidt and the Huang algorithms for linear systems: a numerical study, Appl. Math. 2 (1993) 81–100.

[31] E. Spedicato, Z. Xia, Finding general solutions of the quasi-Newton equation in the ABS approach, Optim. Methods Software 1 (1992) 273–281.

[32] E. Spedicato, Z. Xia, L. Zhang, Reformulation of the simplex algorithm via the ABS algorithm, preprint, University of Bergamo, 1995.

[33] E. Spedicato, J. Zhao, Explicit general solution of the quasi-Newton equation with sparsity and symmetry, Optim. Methods Software 2 (1993) 311–319.

[34] J.H.M. Wedderburn, Lectures on Matrices, Colloquium Publications, American Mathematical Society, New York, 1934.

[35] Z. Xia, ABS reformulation of some versions of the simplex method for linear programming, Report DMSIA 10/95, University of Bergamo, 1995.

[36] Z. Xia, ABS generalization and formulation of the interior point method, preprint, University of Bergamo, 1995.

[37] Z. Xia, Y. Liu, L. Zhang, Application of a representation of ABS updating matrices to linearly constrained optimization, Northeast Oper. Res. 7 (1992) 1–9.

[38] L. Zhang, An algorithm for the least Euclidean norm solution of a linear system of inequalities via the Huang ABS algorithm and the Goldfarb–Idnani strategy, Report DMSIA 95/2, University of Bergamo, 1995.

[39] L. Zhang, Updating of Abaffian matrices under perturbation in W and A, Report DMSIA 95/16, University of Bergamo, 1995.

[40] L. Zhang, On the ABS algorithm with singular initial matrix and its application to linear programming, Optim. Meth. Software 8 (1997) 143–156.

[41] L. Zhang, Computing inertias of KKT matrix and reduced Hessian via the ABS algorithm for applications to quadratic programming, Report DMSIA 99/7, University of Bergamo, 1999.

[42] L. Zhang, Z.H. Xia, Application of the implicit LX algorithm to the simplex method, Report DMSIA 9/95, University of Bergamo, 1995.

[43] L. Zhang, Z. Xia, E. Feng, Introduction to ABS Methods in Optimization, Dalian University of Technology Press, Dalian, 1999 (in Chinese).

![N·H logo] ELSEVIER

Journal of Computational and Applied Mathematics 124 (2000) 171–190

JOURNAL OF
COMPUTATIONAL AND
APPLIED MATHEMATICS

www.elsevier.nl/locate/cam

Automatic differentiation of algorithms

Michael Bartholomew-Biggs, Steven Brown, Bruce Christianson*, Laurence Dixon

Numerical Optimisation Centre, University of Hertfordshire, Hatfield, UK

Received 30 November 1999; received in revised form 28 February 2000

Abstract

We introduce the basic notions of automatic differentiation, describe some extensions which are of interest in the context of nonlinear optimization and give some illustrative examples. © 2000 Elsevier Science B.V. All rights reserved.

Keywords: Adjoint programming; Algorithm; Automatic differentiation; Checkpoints; Error analysis; Function approximation; Implicit equations; Interval analysis; Nonlinear optimization; Optimal control; Parallelism; Penalty functions; Program transformation; Variable momentum

1. Introduction

Automatic differentiation (AD) is a set of techniques for transforming a program that calculates numerical values of a function, into a program which calculates numerical values for derivatives of that function with about the same accuracy and efficiency as the function values themselves. The derivatives sought may be first order (the gradient of a target function, or the Jacobian of a set of constraints), higher order (Hessian times direction vector or a truncated Taylor series), or nested (calculating $\nabla_x F(x, f(x), f'(x))$ for given f and F).

Many nonlinear optimization techniques exploit gradient and curvature information about the target and constraint functions being calculated. Derivatives also play a key role in sensitivity analysis (model validation), inverse problems (data assimilation) and simulation (design parameter choice).

These derivatives can be estimated using divided differences, but such estimates are prone to truncation error when the differencing intervals are numerically large, and to round-off error when they are small. In addition, the run-time requirements of a divided difference approach are often unacceptably high, particularly for problems with a large number (thousands) of independent variables.

---

* Corresponding author.
E-mail address: b.christianson@herts.ac.uk (B. Christianson).

0377-0427/00/$ - see front matter © 2000 Elsevier Science B.V. All rights reserved.
PII: S 0377-0427(00)00422-2

The manual development of code for evaluating analytic derivatives of a function is a tedious and error-prone activity. Of course, symbol manipulation programs can differentiate individual equations, but the code for evaluating a function of interest typically has a nontrivial control flow, involving conditional statements, loops, and subroutine calls, as well as data structures which may be updated many times during the evaluation process. Particularly if the underlying program is subject to continual structural change, it is generally desirable to automate, at least in part, the process of transforming it into a program that calculates derivative values, and this was the initial motivation for the development of AD.

The basic process of AD is to take the text of a program (called the underlying program) which calculates a numerical value, and to transform it into the text of program (called the transformed program) which calculates the desired derivative values. The transformed program carries out these derivative calculations by repeated use of the chain rule from elementary calculus, but applied to floating point numerical values rather than to symbolic expressions.

The transformation process may be carried out by a compiler-like tool, or by operator overloading. Tools using the latter approach are simpler to build, but produce code which is less efficient to run.

The compiler-like transformation of a pre-existing program is not the whole story of AD however. The efficient transformation of programs which include the solution of complicated sub-problems often benefits from user insight into the problem structure, and conversely the conceptual framework imposed by AD often gives users insight into more efficient ways of coding the underlying program. Consequently, the term AD has stretched to cover the user-driven transformation of abstract algorithms, as well as the automatic transformations of concrete programs.

In this paper we give a rapid review of the basic techniques of AD, followed by a quick tour of a few extensions and examples with which we have been personally involved and which we consider interesting from the standpoint of nonlinear optimization. This paper does not attempt to give a history of AD (see [20]), nor does it give a complete account of the foundations of AD (a full account from a mathematical point of view is given in the excellent book [17]). Neither do we attempt to make a systematic survey of prior or current work in the field (such as that in the blue and green books [19,3]), nor of the many tools which are available.[1] A great deal of work which we regard as central to the discipline is not mentioned at all in this paper, for reasons of space. Nevertheless, we hope to impart a flavour of AD, to give the reader some idea of what goes on inside an AD tool, and to develop an initial insight into the effect which certain lines of research in AD may eventually have upon what such tools can accomplish for optimization.

The rest of this paper is organized as follows. In the first part of the paper we develop the two basic building blocks of AD, the forward and reverse accumulation modes. The forward mode is set out in the next section, and the reverse mode in Section 4, following the introduction of the ancillary notion of a Wengert list in Section 3. The reverse mode can be implemented directly by overloading, but the more efficient program transformation approach requires the adjoint program construction techniques set out in Section 5. Section 5 also introduces the important concepts of checkpointing and pre-accumulation.

---

[1] See for example www.mcs.anl.gov/Projects/autodiff/AD_Tools

The second part of the paper outlines some extensions of the basic techniques of AD. Section 6 introduces some of the issues raised for AD by function approximation techniques such as discretization and iterative solution of subproblems. The case of implicit equation solution is considered in more detail in Section 7. Section 8 deals with the use of the reverse mode to obtain automatic error estimates, and Section 9 considers the extension of AD to second and higher derivatives.

The final part of the paper begins in Section 10 with a discussion of the differences between the overloading and code translation approaches to AD implementation. This is followed by two examples, which are used to illustrate the earlier theory and to provide a concrete setting for some of the discussion: a discrete-time optimal control problem in Section 11 and a constrained nonlinear optimization with exact penalty function in Section 12. Some reflections upon the future impact of AD-related research are set out in the final section.

2. Forward accumulation

Suppose that we have an underlying program (or a subroutine) f, which takes n independent variables x_i as inputs, and produces m dependent variables y_i as outputs, and that we wish to obtain numerical values for the Jacobian $J = f' = [\partial y_i / \partial x_j]$ given particular values for the x_i.

The forward accumulation technique associates with each floating point program variable v a vector $\dot{v}$ of floating point derivative values. Conceptually the simplest, *Cartesian*, case is when each dot vector $\dot{v}$ contains one component for each independent variable x_i and component i contains the corresponding derivative $\partial v / \partial x_i$ so that

$$\dot{v} = \nabla_x v.$$

More generally, the number r of vector components may differ from the number of independent variables, and component i may contain an arbitrary directional derivative, or tangent vector, of the form $p_i \cdot \nabla_x v$ corresponding to the tangent direction given by the n-vector p_i.

In the cartesian case, we initialize the dot vector $\dot{x}_i$ corresponding to the independent variable x_i by setting $\dot{x}_i = e_i$, the ith cartesian unit vector. We write this loosely as $[\dot{x}] := I_n$. More generally, we initialize $\dot{x}_i$ to the ith row of the tangent direction bundle $P = [p]_{n \times r}$.

Each operation which assigns a value to a floating point variable must be augmented by an operation to assign correct floating point values to the corresponding dot vector, for example, the operation

$$v_3 := v_1 * \sin(v_2)$$

must be augmented by the assignment

$$\dot{v}_3 := v_1 * \cos(v_2) * \dot{v}_2 + \dot{v}_1 * \sin(v_2).$$

It is straightforward to see how to modify the underlying program so that it calculates the dot-vector values directly itself. We can use an operator-overloading approach, or we can systematically transform the source code. The source translation approach requires a greater initial investment in development, but has certain advantages from the viewpoint of efficiency, which we discuss further in Section 10 below.

In an overloading approach, the pair $(v, \dot{v})$ can be combined into a new user-defined data type called a *doublet*. Appropriate overloaded operations corresponding to the usual floating point operations can be defined to manipulate the dot values in accordance with the chain rule. All active floating point program variables[2] can be re-declared to be of this doublet type. The derivative operations and storage management will automatically occur even though the text of the evaluation program is unchanged.

In a source-translation approach the code which declares and manipulates storage space and values for active program variables v can be augmented by code to declare and manipulate storage space and values for $\dot{v}$ in tandem.[3]

If suitable processors are available, the components of $\dot{v}$ can be calculated in parallel. If the structure of the problem is such that the $\dot{v}$ are sparse, then they can be implemented as sparse vectors. If the number of nonzero components is large, then it will usually be more efficient to evaluate them in batches,[4] with the underlying function evaluation repeated for each batch.

3. Wengert lists

In order to describe the reverse accumulation technique, we need to untangle the relationship between a mathematical variable and a program variable. In this section we describe for this purpose an abstraction called a *Wengert list* [24]. We can think of a Wengert list as a trace of a particular run of a program, with specific values for the inputs. The only statements which occur in the Wengert list are assignment statements to nonoverwritable variables called Wengert variables. The Wengert list abstracts away from all control-flow considerations: all loops are unrolled, all procedure calls are inlined, and all conditional statements are replaced by the taken branch. Consequently, the Wengert list may be many times longer than the text of the program to which it corresponds.

The Wengert list also abstracts away from all considerations of storage management. Each assignment statement in the Wengert list has a different variable on the left-hand side. Thus, a single program variable may correspond to many different Wengert variables, one Wengert variable for each occasion upon which a value is assigned to the program variable. The Wengert list can be considered as a straight-line program for evaluating y from x without overwriting any variable after it has been initialized. Alternatively, a Wengert list can be viewed as an unordered set of mathematical equations expressing functional dependencies between Wengert variables and which could be differentiated symbolically.[5] The length of the Wengert list, and hence the number of Wengert variables for which storage is required, is proportional to the run time of the underlying program.

---

[2] A program variable is *active* if it both depends upon an independent variable, and influences the value of a dependent variable, for some possible control flow of the program.

[3] It is prudent to place the assignment to $\dot{v}$ before that for v in the transformed code, since the variable v on the left-hand side of an assignment statement may also appear on the right, and the old value of v rather than the new is required to evaluate $\dot{v}$. In most modern computer languages parameter passing mechanisms, array index calculations, and pointer manipulation make it difficult to determine at compile time whether two variables are the same.

[4] The batch size is chosen so that the overhead of repeating the function evaluation, amortized over the size of the batch, just balances the thrashing caused by the growth of the working set with the batch size.

[5] The Wengert list can also be viewed as a linearization of the computational graph.

In general, a Wengert list has the following form:

for i **from** 1 **upto** n **do**
    $v_i := x_i$
enddo
for i **from** $n+1$ **upto** N **do**
    $v_i := f_i(v_{\tau_i 1}, \ldots, v_{\tau_i n_i})$
enddo
for i **from** $N+1$ **upto** $N+m$ **do**
    $y_{i-N} := v_{i-m}$
enddo
$\{(y_1, \ldots, y_m) = f(x_1, \ldots, x_n)\}$

where for each $i > n$, n_i is the arity of f_i and τ_i is a map from $\{1, \ldots, n_i\}$ into $\{1, \ldots, i-1\}$.

In this formulation, we allow the functions f_i to be arbitrary differentiable scalar-valued functions. However we could, by introducing additional Wengert variables, ensure that the functions f_i were all of a certain simple form: for example, we could allow only unary operations (operations on single arguments) together with binary addition.[6] Alternatively, we could allow more general vector- or matrix-valued functions for f_i.

In what follows, we frequently write down derivative expressions such as $\partial y_j / \partial v_i$. This is a slight abuse of notation, since each intermediate variable v_i depends functionally upon the input variables x_i. Purists who wish to avoid any ambiguity about whether a variable is dependent or independent can replace the assignment $v_i := f_i(v_{\tau_i 1}, \ldots, v_{\tau_i n_i})$ by the identity $v_i = f_i(v_{\tau_i 1}, \ldots, v_{\tau_i n_i}) + u_i$ where the u_i are additional independent variables with value zero, and consider $\partial y_j / \partial u_i$ when we write $\partial y_j / \partial v_i$.

4. Reverse accumulation

The reverse accumulation technique associates with each floating point program variable v a vector $\bar{v}$ of floating point derivative values.[7]

Conceptually, the simplest case is when each of these bar vectors contains one component for each dependent variable, and component i contains the corresponding derivative $\partial y_i / \partial v$, so that

$$\bar{v} = D_v y.$$

More generally, the number s of vector components may differ from the number of dependent variables, and component i may contain an arbitrary adjoint derivative, or co-tangent vector, of the form $q_i \cdot D_v y$. corresponding to the co-tangent direction given by the m-vector q_i.

Each operation which assigns a value to a floating point variable must be augmented by an operation to assign correct floating point values to the corresponding bar vectors, according to the

---

[6] Multiplication by a constant and squaring are unary operations, and binary multiplication can be defined by $a * b = 2^{-2} * [(a+b)^2 - (a-b)^2]$. To avoid cancellation error, the operands can be dynamically scaled by opposite powers of two, which cancel in the derivative formulae.

[7] Formally, $\dot{v}$ is a column vector and $\bar{v}$ is a row vector.

chain rule: for example the operation

$$v_3 := v_1 * \sin(v_2)$$

corresponds to the assignments

$$\bar{v}_1 := \bar{v}_3 * \sin(v_2), \quad \bar{v}_2 := \bar{v}_3 * v_1 * \cos(v_2).$$

In contrast with the forward case, the bar vectors $\bar{v}_i$ cannot be calculated in the same sequence as the variable values v_i, but must be evaluated in the opposite (or reverse) order.

In the simplest case, we initialize the bar vector $\bar{y}_i$ corresponding to the dependent variable y_i by setting $\bar{y}_i := e_i^T$, the ith cartesian unit vector. We write this loosely as $[\bar{y}] := I_m$. More generally, we initialize $\bar{y}_i$ to the ith column of the co-tangent direction bundle $Q = [q]_{s \times m}$.

In this section we explain how to reverse accumulate the adjoint variables $\bar{v}$ for programs expressed in the form of a Wengert list. In Section 5 which follows, we extend these techniques to more general programs with variable assignment and control flow. Examination of the Wengert list yields the following algorithm for computing the adjoint variables:

> **for** i **from** 1 **upto** n **do**
> $v_i := x_i$
> $\bar{v}_i := 0.0$
> **enddo**
> **for** i **from** $n+1$ **upto** N **do**
> $v_i := f_i(v_{\tau_i 1}, \ldots, v_{\tau_i n_i})$
> $\bar{v}_i := 0.0$
> **enddo**
> **for** i **from** $N+1$ **upto** $N+m$ **do**
> $y_{i-N} := v_{i-m}$
> $\bar{v}_{i-m} := \bar{y}_{i-N}$
> **enddo**
> **for** i **from** N **downto** $n+1$ **do**
>   **for** j **from** 1 **to** n_i **do**
> $\bar{v}_{\tau_i j} := \bar{v}_{\tau_i j} + \bar{v}_i * (D_j f_i)(v_{\tau_i 1}, \ldots, v_{\tau_i n_i})$
>   **enddo**
> **enddo**
> **for** i **from** n **downto** 1 **do**
> $\bar{x}_i := \bar{v}_i$
> **enddo**
> $\{(\bar{x}_1^T, \ldots, \bar{x}_n^T) = Q f'(x_1, \ldots, x_n)\}$

The adjoint variables are incremented rather than simply assigned because although a Wengert variable can be written only once, it can be read several times. At each point at which it enters the subsequent calculation it can affect the dependent variables, and the relevant adjoint value is the sum of all such effects.

Of particular interest is the case $m=1$ where there is only one-dependent variable. Programs which calculate a single scalar-valued objective or *target* function arise in unconstrained problems or when constraints are incorporated using penalty or barrier functions. In this case the $\bar{v}_i$ are scalars, and it is clear that reverse mode AD allows the entire gradient vector to be extracted, to the same level of precision as the function, for about the cost of three function evaluations, *regardless of the number of independent variables*. This fact, which deserves to be more widely known than it appears to be, follows from the consideration that the computational cost of evaluating Df_i for elementary f_i is generally no greater than that of evaluating f_i itself.

In the case where overloading is used, it is a relatively simple matter to modify the underlying program so that it builds its own Wengert list of elementary floating point operations, with each overloaded operation appending the next list item to a data structure as a side effect. The reverse pass over this list can then be invoked by calling a separate routine. The Jacobian values of $D_j f_i$ can be saved on a stack on the way forward, and used in reverse order on the way back. Alternatively, the values of the program variables can be saved whenever they are overwritten, and the restored values used to calculate the $D_j f_i$ on the way back.

The high storage requirement of such a naive approach to reverse mode AD is prohibitive for large problems. However, for many small-to-medium size problems the relatively cheap cost of secondary storage, the efficiency of virtual memory, and the fact that access to the Wengert list can be made essentially serial, means that the naive implementation approach is viable.

However, it is also possible (and more efficient in both run time and storage space, see Section 10 below) to implement the reverse method by transforming the underlying program into an adjoint program with the "opposite" control flow, and we consider how to do this in the next section. This transformation enables the more subtle analysis of the trade-offs between *storing* results that will be needed later and *recomputing* them, which is required by larger problems. The judicious use of recomputation usually allows reverse mode AD to be done with a storage requirement that is only a small factor larger than that required by the underlying program. Furthermore, the recomputations can generally be done in parallel in such a way that the overall run time is not increased. We consider this issue further in the next section, and give an example in Section 11.

5. Adjoint program construction

In this section we sketch how to transform code so as to enable the calculation of adjoint values. We have no space here to describe the informatics involved, so we simply set out the transformation process as if it were being done by hand. The initial task of AD is to automate this process of program transformation, by the development of compiler-like tools and appropriate operating system interfaces. We assume that the underlying program has been augmented to save partial derivative values or overwritten variable values on the way forward, and consider the structure of the program, called the *adjoint program*, required to carry out the reverse pass.

5.1. Variables and assignment statements

The adjoint program declares and manipulates adjoint program variables, which may be vectors or scalars. Exactly one adjoint program variable $\bar{v}$ is required for each program variable v in the underlying program. This follows from the observation that, if two Wengert variables correspond to successive values of the same program variable on the way forward, then their adjoints can share the same storage on the way back: a program variable value which has been overwritten can no longer influence the dependent variable values and so has adjoint value zero, while a program variable value which has not yet been assigned corresponds to an adjoint value which will not be used again and so can be discarded.

Hence, to the program assignment statement: "$v_i := f_i(v_{\tau_i 1}, \ldots, v_{\tau_i n_i})$" corresponds the adjoint code:

$$\bar{t} := \bar{v}_i$$
$$\bar{v}_i := 0.0$$
for j **from** 1 **to** n_i **do**
$$\bar{v}_{\tau_i j} := \bar{v}_{\tau_i j} + \bar{t} * (D_j f_i)(v_{\tau_i 1}, \ldots, v_{\tau_i n_i})$$
enddo

Here $\bar{t}$ is a "temporary" adjoint variable, introduced to allow for the fact that in the underlying program, in contrast to the Wengert list, the variable v_i on the left of an assignment statement may also appear on the right.[8]

An alternative to saving partial derivative values on the way forward is to calculate them on the reverse pass.[9] This requires some of the overwritten values of program variables to be saved on the way forward so that they can be restored at the corresponding point on the reverse pass. Specifically, if the overwritten value appears as an argument to a nonaffine function f_i then it must be saved. Sometimes we can avoid the need to store and restore floating point program variable values by inverting the calculation which produced them [22] but roundoff makes this difficult in general.[10] Alternatively, we can re-calculate the overwritten program variable values from checkpoints as described in Section 5.5 below.

5.2. Sequence of statements

The statements in the adjoint program consist of the adjoints of the statements in the underlying program, but in reverse order, so that the adjoint of "$S_1; S_2; S_3$" is "$\bar{S}_3; \bar{S}_2; \bar{S}_1$" We have already seen how to adjoin assignment statements. We indicate below how to adjoin statements affecting control flow.

5.3. Procedure and function calls

The adjoint of a procedure call is a call to the adjoint procedure. The adjoint procedure $\bar{P}$ contains the adjoints of the statements in the underlying procedure P, in the reverse order. Out parameters become in parameters and in parameters become in–out. Functions can be treated as procedures with an additional out-parameter.

---

[8] It is prudent to do this in all cases since, as mentioned before, parameter passing mechanisms, array index calculations, and pointer manipulation make it difficult to determine at compile time whether two variables are the same.

[9] Whichever alternative is adopted, access to the archived values is serial and predictable, so high latency secondary storage can be used provided the burst bandwidth is sufficiently high.

[10] Although it is worth noting that if $w_{i+1} = f_i(w_i)$ for all i where w is the state vector, and if g_i is an approximate inverse of f_{i-1}, then the transformation $\tilde{f}_i$ defined by $w_{i+1} = f_i(w_i) + w_{i-1} - g_i(w_i)$ approximates f_i to the same degree and f_{i-1} has exact inverse $\tilde{g}_i$ given by $w_{i-1} = g_i(w_i) + w_{i+1} - f_i(w_i)$.

When there is a need to trade storage space against recomputation, procedure boundaries provide a natural point at which to do so. According to the orthodox view, in a well-designed program the number of times variables are updated across procedure-call boundaries, either as global variables or as parameters, is low relative to the number of program variable updates which occur within the procedure. This allows space to be saved using *pre-accumulation* or *checkpointing*.

5.4. Preaccumulation

Pre-accumulation involves treating the entire procedure as a (possibly vector-valued) elementary operation f_i and storing the partial derivative f_i' on the stack instead of storing partial derivative values for the complete set of internal operations. This Jacobian f_i' can be evaluated at the time when f_i is called, by a recursive application of forward or reverse mode AD to the procedure.[11] This leads to a substantial space saving when the space occupied by f_i' is small relative to the number of internal operations of the procedure f_i.

Although extra multiplications are required to incorporate f_i' on the outer reverse pass, the total operation count may be actually reduced, depending upon the number of procedure inputs and outputs n_i and m_i relative to n and m [9]. Similar considerations apply to the exploitation of structural sparsity. In a parallel processing environment, preaccumulation can shorten the elapsed time of a calculation even when $m = 1$, because the pre-accumulation of f_i' can be done in parallel and so moved off the critical path of the calculation [5].

5.5. Checkpointing

A checkpoint is a complete record of the program state at a particular point of execution.[12] *Incremental* checkpointing across a procedure boundary is a matter of noting what changes are made to the environment by the procedure via parameters and global variables, in such a way that these changes can be quickly undone and reapplied (toggled) to a previously recorded checkpoint.[13] When a checkpoint has been taken at the entry point of a procedure call, then the complete internal record of variable values overwritten by the procedure can be discarded and the storage saved, since these values can now be recomputed from the checkpoint. According to orthodoxy, in a well-designed program an incremental checkpoint across a procedure call boundary should require only a small proportion of the space occupied by the entire program state. On a reverse pass, the adjoint procedure $\bar{P}$ begins by toggling the program state from the exit state to the entry state using the incremental checkpoint, then calls the augmented version of the underlying procedure P to re-create the internal record before proceeding with reverse accumulation. In the parallel processing case this re-creation

---

[11] From the linear algebra viewpoint, the Wengert list expresses the Jacobian as a product of large, sparse matrices, one for each f_i. Forward and reverse accumulation correspond to multiplying these sparse matrices from left to right, or from right to left. There is a huge body of recent interesting work on the optimal order in which to interlock forward and reverse accumulation steps to optimize the operation count, which we do not have space to touch on here. A good conceptual overview of the issues is given in [18]. See also [11].

[12] A checkpoint includes the program counter and a snapshot of the procedure calling history (run-time stack), as well as the program variable values.

[13] For example a "fork" can be used to take an incremental checkpoint if the operating system uses a lazy copy-on-write scheme for the virtual memory pages in the process run-time slack.

process for P can be moved off the critical path by allowing it to be started sufficiently early to be ready when required. Finally, in the case of nested procedure calls, subroutines P_1, P_2, etc., called by P need not be re-evaluated when P is re-evaluated: provided the incremental checkpoint for P_i is available, evaluation of P_i can be replaced by a state toggle from the entry to the exit state.

These two techniques of preaccumulation and checkpointing can be combined. For further details see [23].

5.6. Conditional statements and loops

The adjoint of the conditional statement "if c then S_1 else S_2 endif" is the statement "if c then $\bar{S}_1$ else $\bar{S}_2$ endif". If either of the statements $S1$ or $S2$ could affect the value of the condition c, then the value of c can be pushed on a stack on the way forward, and popped on the way back, just like the partial derivative values or overwritten program variable values.

The adjoint of a loop is also a loop. In case of a **for** loop, the adjoint is a **for** loop in reverse order. In case of a **while** loop, the adjoint loop performs the adjoint of the loop body the same number of times as on the way forward. We can either determine a precondition to identify the first iteration of the forward loop, which is the last iteration of the backward loop [15, Chapter 21], or we can store the number of iterations that was actually performed, analogous to the **if** statement. [14]

Where loop iterations are independent, they can be done in the same order as on the way forward and array subscripts can be calculated in the same way as on the way forward. Otherwise the array index calculations must be reversed: sometimes this is possible, since roundoff is not an issue, but in the worst case the index values have to be stored in sequence on the way forward and restored on the reverse pass, just like overwritten floating point variables.

Loop iterations also form good boundaries at which to consider checkpointing and pre-accumulation. Loops which perform temporal evolution or some other form of in-place state-space update (such as ODE evolution or optimal control) are particularly good candidates for checkpointing (for example see Section 11 below).

Loops to perform array operations can be regarded as single steps and replaced by the corresponding adjoint step. For example, the matrix operation $X := Y * Z$ corresponds to the adjoint operations $\bar{Y} := \bar{Y} + Z * \bar{X}$; $\bar{Z} := \bar{Z} + \bar{X} * Y$, where we adopt the convention that adjoint matrix components are of transpose shape relative to the underlying matrix.

Loops which perform equation solving are of particular interest, since in general we do not need to record the process by which the solution was found (see Section 7 below).

5.7. Input and output

For reads and writes to a sequential file, called say "foo", the adjoint operations are straight-forward, and similar to those for variable assignment. The adjoint to "read (v,foo)" is "write $(\bar{v}, \text{foobar})$; $\bar{v} := 0.0$" and the adjoint to "write (v,foo)" is "read $(\bar{t}, \text{foobar})$; $\bar{v} := \bar{v} + \bar{t}$". For random access files the situation is a little more fraught, see [13] for a good account of what is involved.

---

[14] Often the sequence of values to be stored exhibits a regular pattern, in which case standard data compression techniques such as Huffman encoding can be applied to reduce the space required.

In the parallel processing case similar considerations apply to inter-processor communication: sends can be regarded as writes and receives as reads.[15]

6. Approximating differentiable functions

A question which we often need to consider explicitly is "if we calculate an approximation f_n to a function f_*, when do we want the derivative of f_n and when do we need an approximation to the derivative of f_*?" This is an important question, because the fact that f_n approximates f_* does not imply that f_n' approximates f_*' to the same order, or even in the limit. This is particularly apparent when piecewise-defined functions are glued together using if-statements.[16] For example the code

if $x = 0.0$ then $y := 0$ else $y := (1 - \cos(x))/x$ endif

will give the derivative value $\partial y/\partial x = 0.0$ for $x = 0.0$ instead of the presumably intended value of 0.5. For forward or reverse mode AD to work correctly in this case the programmer could have written:

if $x = 0.0$ then $y := x/2$ else $y := (1 - \cos(x))/x$ endif.

A similar problem occurs when using a while loop: different numbers of iterations give different branches of a piecewise function. Differentiate an iterative approximation $v := \phi(x, v)$ and the derivatives $\dot{v}$ may not converge, or may lag behind the convergence of v. For example, suppose the starting value for v is exactly right: then the while loop is skipped and we have $\dot{v} = 0$. The situation with reverse accumulation is even more problematic if we take the naive approach of differentiating the approximation function which we coded without having considered at the time when we coded it the requirement that it also approximate the derivative [14].

Clearly the derivatives must be incorporated into the stopping criterion in some way. A lot is now known about how to do this, but in many cases it is better to construct an iterative approximation to the derivative of the function to which the underlying iterative approximation is converging, rather than to differentiate the underlying approximation function directly. We consider this issue further in the next section, but point out that methods suitable for an interval-valued approach appear to have some potential to reconcile these two agendas.[17]

Another source of inaccuracy is introduced by discretizing a continuous problem. In this case, it is usually best to differentiate the discretization used, since verification of descent criteria (e.g. Wolfe conditions) and the introduction of devices to enforce global convergence of Newton-like methods should be applied to the numerical values actually being calculated.[18] However this policy requires

---

[15] Actually, there is an interesting dualism between trying to find the optimal decomposition of a program into parallel parts to minimize run time and IPC, and the optimal checkpointing schedule to minimize the overwrite stack and incremental checkpoint sizes.

[16] In contrast with differencing, an AD tool can produce a warning when an intermediate variable v is too close to a cut value, by looking at $\dot{v}$ in the light of the given tolerance for x or at $\bar{v}$ in the light of the required tolerance for y, cf. Section 8.

[17] Consider the properties of an algorithm which produces a joint enclosure for the true and approximation function values.

[18] Convergence under dynamic refinement of the discretization typically relies upon an unstated compactness result. Again, consideration of enclosure properties suggests that interval methods have some potential here in the context of AD.

the discretization to be suitable for derivatives as well as for function values, which is a nontrivial additional constraint upon the modelling process.

7. Iteration and equation solving

Many computations $y := f(x)$ include as a subproblem the solution of implicit equations, of the form $\psi(u,v) = 0$ where u, v are p- and q-vectors of knowns and unknowns, respectively, and ψ is a well-behaved q-vector valued map. In the linear case, these subproblems take the form of solving $Av = b$ for v where A and b are functions of x.

The underlying program contains code for solving these implicit equations, and it would be possible to treat this solver code as a black box, and to apply AD to it mechanistically. In some cases, as we saw in the previous section, this will not produce the derivative values which are required: in other cases it will produce correct, but very inefficient, derivative code. It is usually advantageous for the AD translation process to identify explicitly the equations being solved, and to provide or invoke a solution code for the corresponding derivative equations which exploits shared values between the two equation solutions.

For example, if $Av = b$ then $A\dot{v} = \dot{b} - \dot{A}v$ for each tangent direction. Similarly, the adjoint operations corresponding to solving $Av = b$ for v are $\bar{b} := \bar{b} + z$; $\bar{A} := \bar{A} - vz$ where z is the solution to $zA = \bar{v}$ for the corresponding co-tangent direction. [19] If the underlying program forms an LU-decomposition of A in order to solve the original equations, then this can be exploited to obtain $\dot{v}$ from v, or $\bar{A}, \bar{b}$ from $\bar{v}$, at a much lower cost than simply applying AD to the equation solver: typically the operation count becomes of order q^2 rather than q^3, which in many cases means that the derivatives effectively become free [10].

For the nonlinear case of solving $\psi(u,v) = 0$ for v, an iterative scheme $v := \phi(u,v)$ will generally be used. Now $\dot{v}$ must satisfy [1] the linear equations $\psi'_v \dot{v} = -\psi'_u \dot{u}$. Similarly [7] the adjoint operation corresponding to solving $\psi(u,v) = 0$ is $\bar{u} := \bar{u} - z\psi'_u$ where z is the solution to the linear equations $z\psi'_v = \bar{v}$. We could use AD to form the matrices ψ'_u, ψ'_v explicitly but if, for example, Newton's method is used as the iterative scheme ϕ for solving the underlying nonlinear equations, then the relevant matrices will already have been formed and factorized. Conversely, explicit formation of the derivatives produces information that can be used to improve the solution of the underlying equations, possibly at the next trial point of the function under evaluation. Similar remarks apply to preconditioning.

8. Automatic error analysis

It is useful to know when a function value has converged as accurately as rounding error will allow. Consider again the Wengert list of Section 3, with the items in the form $v_i := f_i(v_{\tau_i 1}, \dots, v_{\tau_i n_i}) + u_i$. Suppose that the f_i instead of being floating point operations are actually smooth operations on

---

[19] We follow the convention that the elements of $\bar{A}$ have the transpose form to A. If pairs of floating point real variables are being interpreted as floating point complex numbers, then the adjoint values are conjugated as well as transposed: $y = f(v)$ and $\bar{y} = 1.0 + i0.0$ implies $\bar{v} = f'(v)^*$ since $\bar{v}_{re} + i\bar{v}_{im} = \partial y_{re}/\partial v_{re} + i\partial y_{re}/\partial v_{im} = \partial y_{re}/\partial v_{re} - i\partial y_{im}/\partial v_{re} = \partial y^*/\partial v$ by the Cauchy–Riemann equations.

infinite precision real numbers, and that the u_i rather than being zero are the errors introduced by round-off and normalization. Then the difference $y - \hat{y}$ between the calculated value of y and the true value $\hat{y}$ is to first order equal to $\sum_{i=n+1}^{N} u_i \bar{v}_i$. If the errors u_i are statistically independent and from symmetric distributions, and we have a priori or a posteriori error bounds $|u_i| < \Delta_i$ then the Euclidean norm $\|y - \hat{y}\|_2$ is almost certainly bounded by $4\sqrt{\sum_{i=n+1}^{N} \Delta_i^2 \|\bar{v}_i\|_2^2}$, see [20, Section 12]. Similarly, the use of interval analysis and the L_1-norm gives a validated and asymptotically tight error bound [20].

Optimization algorithms almost always evaluate target functions more than once in regions where the exact target value is critical. Where an iterative solution is being used for a subproblem, therefore, it is natural to ask: when is the solution accurate enough to enable the routine evaluating the outer function to make a correct decision, and conversely how should the solution from the previous outer evaluation be used to initialize the subproblem solution, and how accurate will the resulting derivatives be?

Reverse accumulation provides some assistance with questions of this type [7]. Suppose that v is an approximate solution to $\psi(u, v) = 0$ and the exact solution is $\hat{v}$, and let the corresponding values for the dependent variables be $y, \hat{y}$. Set $w := \psi(u, v)$, then $\hat{y} = y - zw + O(\|w\|^2)$ provided z is chosen to satisfy $\|\bar{v} - z\psi_v\| < \|w\|$, and in this case $\bar{u}$ is accurate to order $\|w\|$. In the linear case $Av = b$, $w = Av - b$, giving $\hat{y} = y - zw$ to order $\|w\|^2$ provided z satisfies $\|zA - \bar{v}\| \leqslant \|w\|$.

9. Higher derivatives

We can apply first-order forward or reverse mode AD repeatedly, to obtain higher-order derivative values.[20] For example, applying the forward mode twice gives matrices $\ddot{v}$ with $[\ddot{y}] = P^{\mathrm{T}} f'' P$. In the case of a single independent variable, we can generalize this to calculate truncated Taylor series in a particular direction. These are potentially very useful when performing line searches. When $n > 1$ we can interpolate Taylor series to obtain derivatives of arbitrary order [4], for example,

$$\frac{\partial^2}{\partial x_1 \partial x_2} = \frac{1}{4}\left[\left(\frac{\partial}{\partial x_1} + \frac{\partial}{\partial x_2}\right)^2 - \left(\frac{\partial}{\partial x_1} - \frac{\partial}{\partial x_2}\right)^2\right].$$

We can also obtain second derivative information by combining the forward and reverse modes. In outline, we take the program $y := f(x)$, transform it using reverse mode to give the adjoint program $\bar{x} := \bar{y} f'(x)$, and then transform this using the forward mode to give the program $\dot{\bar{x}} := \dot{\bar{y}} f'(x) + \bar{y} f''(x) \dot{x}$. If we set $\bar{y} = I_m$, $\dot{x} = I_n$, $\dot{\bar{y}} = 0_{nm}$ then this gives $\dot{\bar{x}} = f''(x)$. However, sometimes it is useful to set other initial values for quantities such as $\dot{\bar{y}}$, e.g., if a projected Hessian is required, or as in the example of Section 11 below.

This approach of applying forward to reverse is particularly efficient in the case $m = 1$ of a single target variable, in which case we obtain a complete Hessian $H = f''$ at a cost of about $6n$ evaluations of f, or a projected Hessian at even lower cost. If we are using a truncated Newton or conjugate gradient algorithm, or some form of gradient descent algorithm with a variable momentum term, then

---

[20] We can regard initialized tangent or co-tangent components in differentiated code as being additional independent variables in their own right. Subsequent code differentiation is simplified by use of identities such as $\partial v_j / \partial v_i = \partial \dot{v}_j / \partial \dot{v}_i = \partial \bar{v}_i / \partial \bar{v}_j$; $\partial \bar{v}_j / \partial v_i = \partial \bar{v}_i / \partial v_j$; etc.

it is very useful to be able to evaluate terms like Hp at a computational cost which is independent of n.

Applying reverse mode to forward differentiated code produces the same calculation, and hence the same result, as applying forward to reverse. All that happens is that the dots change places on the barred variables, [21] so that $\bar{\dot{v}}$ corresponds to $\dot{\bar{v}}$ and $\dot{v}$ corresponds to $\dot{v}$.

Reverse differentiation of reverse differentiated code can always be replaced by forward differentiation of the original forward code. There is therefore never any need to adjoin adjoint code. For example, suppose we want to differentiate the scalar function $y := F(g(v))$ where $g = f'$ and v is a function of x. Evaluate $y := f(v)$, set $\bar{y} := 1.0$ and reverse gives $\bar{v} = f'(v)$. Set $w := \bar{v}$ and evaluation of $y := F(w)$ is straightforward. But how do we obtain $\partial y / \partial x$?

Setting $\bar{y} := 1.0$ and reversing F gives $\bar{w} := F'(w)$. Instead of adjoining $w := \bar{v}$ by setting $\bar{\bar{v}} := \bar{w}$, which would require us to adjoin the adjoint code for g to get the value for $\bar{x}$, we set $\dot{v} := \bar{w}$ and then forward and reverse through f gives $\bar{v} := \dot{\bar{v}}$ from which we can obtain $\bar{x}$ as usual. This is the numerically correct assignment, since $\dot{\bar{v}} = f''(v)\dot{v} = \bar{w}f''(v) = F'(f'(v))f''(v)$.

We can also fix tangent or co-tangent directions to be derivatives of other functions: for example if $y := f(x)$ then setting $\dot{x} := \bar{x}$ and repeating the evaluation of y and $\bar{x}$ gives the quantity $\dot{\bar{x}} = Hg$ where $H = f''(x)$, $g = f'(x)$. Accurate quantities of this type are useful in many gradient descent algorithms, including Truncated Newton.

10. Overloading and program transformation

The overloading approach is quick to implement, but suffers from a number of disadvantages. Most compilers implement expressions containing overloaded operators exactly as they are written, without performing any compile-time optimization on the expression. For example, the assignment

$$y := a * \sin(a * x * *2 + b * x + c) + b * \cos(a * x * *2 + b * x + c)$$

contains the shared subexpression $a * x * *2 + b * x + c$, which need only be evaluated once, and which would be more efficiently evaluated as $(a * x + b) * x + c$. Consequently, an overloaded doublet implementation will be considerably less efficient than the optimized underlying floating point implementation, even before the costs of the extra floating point operations are taken into account. [22]

Nevertheless, there is no better way to understand AD than to implement a baby AD tool using operator overloading and for many small-to-medium size problems such a tool is adequate.

Transforming the underlying program to a new source program, rather than augmenting it using overloaded operators, allows the compiler to perform optimization on the derivative calculations as well as upon the underlying calculations. For example, when adjoining the assignment to y, the

---

[21] Conceptually, different sets of dots and bars are used, corresponding to different tangent and co-tangent variables. Strictly, we should use a tensor derivative notation for repeated differentiation.

[22] There are good reasons for this literal-minded compilation. Overloaded operators may have complex side-effects involving global state, and in any case cannot generally be assumed to have the same semantics as their built-in counterparts. For example, matrix multiplication is not commutative, octonian multiplication is not associative, intervals do not satisfy the distributive law, and common subexpressions involving random oracles must be recomputed for each occurrence. Most overloaded operator languages give the user no way to tell the compiler which optimizing transformations are safe.

derivatives of sin and cos are already available, and the derivative of the argument can be obtained by adding the two available quantities $a*x$ and $a*x+b$.

With a language translation approach, a great deal more can also be done to automate the dependency analysis required to determine which variables are active, although when array indices or pointers are manipulated in a complex way at run time, the translator must make a conservative assumption, or rely upon user-inserted directives. Deferring choices until run time almost inevitably produces code which runs more slowly than when the decision can be made at compile time.

The output from the translator is input to an optimizing compiler, so there is generally no need for the code to be particularly efficient; rather, the translator must produce code which it is easy for the compiler to analyse and optimize. This requirement is certainly compatible with making the transformed code intelligible to humans, and users have become accustomed to being able to write source code in a form that is intelligible to them, and to rely upon the compiler to re-arrange it into a form which is efficient before producing object code.

11. Pantoja's algorithm and checkpointing

In this section, we show how automatic differentiation can be combined with Pantoja's algorithm and a checkpointing technique in such a way as to allow accurate evaluation of the Newton direction for a discrete-time optimal control problem at an extremely low computational cost [8]. The purpose of this example is to show the combined use of forward and reverse mode AD to produce Hessian information, and to illustrate how checkpointing can be combined with parallel processing to reduce the run-time storage requirement to something feasible.

Consider the following discrete-time optimal control problem: choose independent control variables $x_i \in \mathbb{R}^p$ to minimize the scalar target function

$$y = F(v_N) \quad \text{where } v_{i+1} = f_i(x_i, v_i) \quad \text{for } 0 \leqslant i < N$$

and v_0 is some fixed constant. Each f_i is a smooth map from $\mathbb{R}^p \times \mathbb{R}^q \to \mathbb{R}^q$ and F is a smooth map from the state space $\mathbb{R}^q$ to $\mathbb{R}$: the states v_i may include running totals of cost functions which are composed into y by F.

Starting with stored value for x_i: $0 \leqslant i < N$, we seek the Newton direction, i.e., vectors $t_i \in \mathbb{R}^p$ such that

$$\sum_{j=0}^{N-1} \left[\frac{\partial^2 y}{\partial x_i \, \partial x_j} \right] t_j + \frac{\partial y}{\partial x_i} = 0 \quad \text{for } 0 \leqslant i < N.$$

Pantoja [21] gives an algorithm for calculating the Newton direction exactly. However, his algorithm involves the solution of linear equations with coefficients given by recursive identities such as

$$A_i = [f'_{v,i}]^{\mathrm{T}} D_{i+1} [f'_{v,i}] + \bar{v}_{i+1} [f''_{vv,i}],$$

$$B_i = [f'_{x,i}]^{\mathrm{T}} D_{i+1} [f'_{v,i}] + \bar{v}_{i+1} [f''_{xv,i}],$$

$$C_i = [f'_{x,i}]^{\mathrm{T}} D_{i+1} [f'_{x,i}] + \bar{v}_{i+1} [f''_{xx,i}],$$

$$D_i = A_i - B_i^{\mathrm{T}} C_i^{-1} B_i, \quad \bar{v}_i = \bar{v}_{i+1} [f'_{x,i}]$$

which in turn requires the accurate evaluation of terms containing second derivatives of f_i. Fortunately AD can be applied to the original code for evaluating F in such a way that the values $\dot{\bar{x}} = \dot{\bar{y}}f'(x) + \bar{y}f''(x)\dot{x}$ are exactly the quantities required [8]. A primary benefit of AD here is the elimination of the labour of forming and differentiating adjoint equations by hand, however the total flop-cost of the AD-form of the algorithm is of the same order as $6(p+q)$ evaluations of the target function y, regardless of the number of timesteps N.

Algorithm (Pantoja with AD)
(1) For i from 1 upto N, calculate and store v_i.
(2) Evaluate $a_N = \bar{v}_N = F'(v_N)$, $D_N = [F''(v_N)]$ as described in Section 9 above.
(3) For each i from $N-1$ down to 0 calculate q-vectors $\bar{v}_i$, a_i and a $q \times q$ matrix D_i as follows:
    (3.1) Define dot vectors of length $p+q$ by
$$\begin{bmatrix} \dot{x}_i \\ \dot{v}_i \end{bmatrix} = \begin{bmatrix} I_p & O \\ O & I_q \end{bmatrix}.$$
    (3.2) Evaluate $v_{i+1} = f_i(x_i, v_i)$ using forward mode AD, so that
$$[\dot{v}_{i+1}] = [f'_{x,i} \quad f'_{v,i}].$$
    (3.3) Set $\bar{v}_{i+1}$ to the value supplied by the previous iteration and set
$$[\dot{\bar{v}}_{i+1}] := [D_{i+1}f'_{x,i} \quad D_{i+1}f'_{v,i}].$$
    (3.4) Apply the forward mode of AD to the forward calculation $v_{i+1} := f_i(x_i, v_i)$ and then to the adjoint calculation $[\bar{x}_i \; \bar{v}_i] := \bar{v}_{i+1}f_i(x_i, v_i)$, giving the matrix
$$\begin{bmatrix} \dot{\bar{x}}_i \\ \dot{\bar{v}}_i \end{bmatrix} = \begin{bmatrix} C_i & B_i \\ B_i^{\mathrm{T}} & A_i \end{bmatrix}.$$
    (3.5) Row reduce this to obtain
$$\begin{bmatrix} I & C_i^{-1}B_i \\ O & A_i - B_i^{\mathrm{T}}C_i^{-1}B_i \end{bmatrix} = \begin{bmatrix} I & E_i \\ O & D_i \end{bmatrix}$$
    and at the same time calculate the vectors
$$a_i = a_{i+1}([f'_{v,i}] - [f'_{x,i}]E_i), \quad c_i^{\mathrm{T}} = -a_{i+1}[f'_{x,i}]C_i^{-1}.$$
    Now $\bar{v}_i$, a_i, D_i are available for the next iteration.
    (3.6) Store the values $\dot{v}_{i+1}$, $\bar{x}_i$, E_i, c_i.
(4) For each i from 0 up to $N-1$ calculate $t_i \in \mathbb{R}^p$, $s_{i+1} \in \mathbb{R}^q$ by
$$s_0 = 0, \quad t_i = c_i - E_i s_i, \quad s_{i+1} = [f'_{x,i}]t_i + [f'_{v,i}]s_i$$
Now t_i is the Newton direction.
STOP

Many other solution techniques which use state-control feedback can be implemented as simple modifications of this algorithm. For example, differential dynamic programming (DDP) replaces the vector a_i by $\bar{v}_i$ in the calculation for c_i. AD in principle allows algorithms of this form, combined with the techniques for differentiating implicit equation solutions, to be applied to differential equations.

11.1. Reducing the storage requirement

By using the state values v_i as checkpoints, we can reduce the storage requirement of the reverse mode to that required for a single timestep f_i together with one checkpoint per timestep. Each checkpoint requires storage for the state vector v_i together with the values $\dot{v}_{i+1}, \bar{x}_i, E_i, c_i$.

However, a much more efficient use of checkpoint storage than this is possible. For example, suppose that N is a million. If we store values for $x_i, \bar{x}_i, D_i$ whenever i is a multiple of a thousand, then we can re-compute the values of E_i, c_i etc. when we need them, in groups of a thousand at a time. This doubles the total computational effort required but reduces the storage requirement from a million full checkpoints to a thousand primary plus a thousand additional checkpoints.

This line of argument can be developed further: with a third level of checkpoint we require three times the computational cost, but storage for only 300 checkpoints. With six levels these numbers are 6 and 60, and with $20 = \log_2 N$ levels of checkpoint we require just $\log_2 N$ times the computational effort together with storage of $\log_2 N$ checkpoints. For this example the storage requirement for reverse accumulation is therefore less than the storage already required to hold the values of the control variables.

In fact, by spacing the checkpoints irregularly we can halve these requirements [16]. If we have several processors available, we can use them to re-calculate the various levels of checkpoint in parallel with the main algorithm so that the required values are ready just in time. It is instructive to work out in detail the schedule for doing this in such a way that the overall runtime does not increase as the storage requirement reduces [2].

12. Fletcher's ideal penalty function

In this section we show how AD can be used to evaluate and differentiate a parameter-free form of a penalty function introduced in [12]. The purpose of this example is to illustrate the differentiation of functions which combine nested subproblem solution with the calculation of gradients of other functions.

Consider the constrained optimization problem: optimize $f(x)$ subject to $k(x) = 0$ where f, k are smooth maps $\mathbb{R}^n \to \mathbb{R}$ and $\mathbb{R}^n \to \mathbb{R}^q$, respectively. Set $g = f'$, $N = k'$ to be the function gradient and constraint normals, and define $\lambda(x), \mu(x) \in \mathbb{R}^q$ by the equations

$$NN^{\mathrm{T}} \lambda = Ng, \quad NN^{\mathrm{T}} \mu = k.$$

Now define $v(x) \in \mathbb{R}^n$ by $v = N^{\mathrm{T}} \mu$ and $F : \mathbb{R}^n \to \mathbb{R}$ by

$$F(x) = f(x - v(x)) + \sum_{i=1}^{q} \left[\lambda_i(x) k_i(x - v(x)) + \tfrac{1}{2} v_i^2(x) \right].$$

Under mild conditions we have [6] that (i) if x^* is a constrained local minimum of f subject to $k = 0$ then x^* is an unconstrained local minimum of F and conversely (ii) if x^* is an unconstrained local minimum of F satisfying $k = 0$ then x^* is a constrained local minimum of f. It follows that if x^* is a constrained minimum of f subject to $k = 0$ then there is a neighbourhood of x^* in which x^* is the only unconstrained local minimum of F, and minimizing F in this neighbourhood will find x^*.

The penalty function F also has the desirable property that near a minimum point the penalty function has the same curvature as the Lagrangian of the target function in directions tangent to the constraint manifold, and unit positive curvature in directions normal to the constraint manifold. Thus, F has numerical conditioning similar to that of the target function f and constraints k from which F is constructed.

We can evaluate F as follows. Solve the equation $NN^T\mu = k$ for μ using AD to evaluate NN^T. For example, we could set $y := k(x)$; $\bar{y} := [I_q]$ and reverse to get $\bar{x} = N^T$. Then set $v := N^T\mu$. Similarly, λ is the solution of $NN^T\lambda = Ng$, where reverse accumulation gives g. Now it is a simple matter to compute the value of F.

We can use AD to obtain the gradient and directional Hessians of F, and these can be used by optimization software to find a local minimum point x^* of F which corresponds to the solution of the original constrained problem. For example, the adjoint of the step "solve $NN^T\mu = k$ for μ" is "solve $\xi NN^T = \bar{\mu}$ for ξ then set $\bar{k} := \bar{k} + \xi$, $\bar{N} := \bar{N} - v\xi - (\mu\xi N)^T$", and the adjoint of the step $N := \bar{x}^T$ is to set $\dot{x} := \bar{N}$ then go forward and reverse through the calculation of k and set $\bar{x} := \bar{x} + \dot{\bar{x}}$.

If q is large we may prefer an iterative method of solving the linear equations for n and λ such as conjugate gradient, which in turn requires evaluations of vectors such as $NN^T p$. Reverse accumulation also allows automatic error estimates to be made for the effect of truncating a subproblem solution upon the calculated function value as described in Section 8 above. This allows us to solve the equations for λ and μ with just sufficient accuracy to ensure that the calculated value of $F(x)$ is correct to the required accuracy (specified in advance) at each iteration step of the optimization algorithm. We can even apply AD to the implicit equations defining x^* so as to perform an automatic error analysis or to determine sensitivities of the solution.

13. Future directions

Several themes for future developments emerge from this. AD has largely achieved its initial agenda of producing fast, accurate derivative code without the costly and error-prone intervention of well-intentioned humans. An analogy can be drawn with the experiences gained by automating the process of translating computer programs from high-level language descriptions into machine code, and from this perspective the future of AD is increasing bound up with the process of compiler-writing and language translation generally. More and more scientific compilers will contain AD algorithms, or at least hooks to allow AD algorithms to be invoked during the compilation process. A great deal of research still remains to be done in this area, particularly in the case of parallelizing compilers, but increasingly the task of AD in this context is to formulate the program transformation problem in terms which enable it to be solved by existing and emerging compiler-generator tools.

Although a great number of AD users are content simply to apply AD to their existing code, this is not the end of the story. At the opposite extreme from the legacy-code user are those doing research into nonlinear optimization algorithms. Taking (for brevity of exposition) a somewhat combative stance, we could assert that many optimization algorithms were initially designed upon the implicit assumption that gradient information was, by its nature, expensive and inaccurate relative to the function evaluation. Second derivative information was likely to be even worse, and any algorithm which required third- or higher-order derivative information was not viable. The current state of AD implies that even quite mild forms of this position are no longer tenable.

While many traditional algorithms work extremely well even in very large dimensions when given accurate derivatives,[23] the contribution of AD to algorithm design remains open. Certainly the ability of reverse accumulation to give complete, accurate gradient and directional Hessian vectors at a cost of a few function evaluations, regardless of the problem dimension, influences the choice of algorithm and the globalization strategy for problems in very large dimension, and we identify this as one context in which AD is likely to develop further from a theoretical point of view. The interaction between AD and interval analysis is another interesting arena for future development.

Many by-products produced during reverse accumulation are of a type which could naturally be exploited during the optimization process by an algorithm with knowledge of the target function's structure, and conversely explicit representation of such structure would in many cases allow an AD tool to operate more effectively. In particular, when equation solution is a sub-problem, there is a benefit to coding the equations being solved as well as the code to solve them, even if the solution code never evaluates the equations, in order to allow the residuals and their derivatives to be used by the AD tool. Likewise there is a benefit to signaling explicitly to an AD tool the accuracy to which derivatives are required, and the use of which they will subsequently be put.

Perhaps the most ambitious way forward for the next few years is the development of AD as a conceptual tool to allow users to capture and express their insights into the nature and structure of the algorithms which their programs instantiate, and to develop new ways of representing these algorithms beyond those offered by current programming languages, in such a way that these insights can be automatically exploited by the environment in which their programs run.

References

[1] M. Bartholomew-Biggs, Using forward accumulation for automatic differentiation of implicitly defined functions, Comput. Optim. Appl. 9 (1998) 65–84.

[2] J. Benary, Parallelism in the reverse mode, in: M. Berz et al. (Eds.), Computational Differentiation: Techniques, Applications and Tools, Society for Industrial and Applied Mathematics, Philadelphia, PA, 1996, pp. 137–147.

[3] M. Berz et al., Computational Differentiation: Techniques, Applications and Tools, Society for Industrial and Applied Mathematics, Philadelphia, PA, 1996.

[4] C. Bischof et al., Structured second- and higher-order derivatives through univariate Taylor series, Optim. Methods Software 2 (1993) 211–232.

[5] S. Brown, B. Christianson, Automatic differentiation of computer programs in a parallel computing environment, in: H. Power, J.C. Long (Eds.), Applications of High Performance Computing in Engineering, Vol. V, Computational Mechanics Publications, Southampton, UK, 1997, pp. 169–178.

[6] B. Christianson, A geometric approach to Fletcher's ideal penalty function, J. Optim. Theory Appl. 84 (2) (1993) 433–441.

[7] B. Christianson, Reverse accumulation and implicit functions, Optim. Methods Software 9 (4) (1998) 307–322.

[8] B. Christianson, Cheap Newton steps for optimal control problems: automatic differentiation and Pantoja's algorithm, Optim. Methods Software 10 (5) (1999) 729–743.

[9] B. Christianson, L. Dixon, S. Brown, Sharing storage using dirty vectors, in: M. Berz et al. (Eds.), Computational Differentiation: Techniques, Applications and Tools, Society for Industrial and Applied Mathematics, Philadelphia, PA, 1996, pp. 107–115.

---

[23] In fairness, it should be mentioned that some traditional algorithms rely upon the inaccuracy of supplied derivatives in order to avoid saddle points, which tend to proliferate in large dimensions.

[10] B. Christianson et al., Giving reverse differentiation a helping hand, Optim. Methods Software 8 (1) (1997) 53–67.

[11] L. Dixon, Use of automatic differentiation for calculating Hessians and Newton steps, in: A. Griewank, G. Corliss (Eds.), Automatic Differentiation of Algorithms, Society for Industrial and Applied Mathematics, Philadelphia, PA, 1991, pp. 114–125.

[12] R. Fletcher, A class of methods for nonlinear programming with termination and convergence properties, in: J. Abadie (Ed.), Integer and Nonlinear Programming, North-Holland, Amsterdam, 1970, pp. 157–175.

[13] R. Geiring, T. Kaminski, Recipes for adjoint code construction, ACM Trans. Math. Software 24 (4) (1998) 437–474.

[14] J.C. Gilbert, Automatic differentiation and iterative processes, Optim. Methods Software 1 (1992) 13–21.

[15] D. Gries, The Science of Programming, Springer, Berlin, 1981.

[16] A. Griewank, Achieving logarithmic growth of temporal and spatial complexity in reverse automatic differentiation, Optim. Methods Software 1 (1) (1992) 35–54.

[17] A. Griewank, Evaluating Derivatives: Principles and Techniques of Algorithmic Differentiation, Society for Industrial and Applied Mathematics, Philadlphia, PA, 2000.

[18] A. Griewank, S. Reese, On the calculation of Jacobian matrices by the Markowitz rule, in: A. Griewank, G. Corliss (Eds.), Automatic Differentiation of Algorithms, Society for Industrial and Applied Mathematics, Philadelphia, PA, 1991, pp. 126–135.

[19] A. Griewank, G. Corliss (Eds.), Automatic Differentiation of Algorithms, Society for Industrial and Applied Mathematics, Philadelphia, PA, 1991.

[20] M. Iri, History of automatic differentiation and rounding error estimation, in: A. Griewank, G. Corliss (Eds.), Automatic Differentiation of Algorithms, Society for Industrial and Applied Mathematics, Philadelphia, PA, 1991, pp. 3–16.

[21] J.F.A. deO. Pantoja, Differential dynamic programming and Newton's method, Internat. J. Control 47 (5) (1988) 1539–1553.

[22] D. Shiraev, Fast automatic differentiation for vector processors and reduction of the spatial complexity in a source translation environment, Ph.D. Dissertation, Karlsruhe, 1993.

[23] Yu.M. Volin, G.M. Ostrovskii, Automatic computation of derivatives with the use of the multilevel differentiation technique, Comput. Math. Appl. 11 (11) (1985) 1099–1114.

[24] R.E. Wengert, A simple automatic derivative evaluation program, Comm. ACM 7 (1964) 463–464.

N·H

ELSEVIER

Journal of Computational and Applied Mathematics 124 (2000) 191–207

JOURNAL OF
COMPUTATIONAL AND
APPLIED MATHEMATICS

www.elsevier.nl/locate/cam

Direct search methods: then and now

Robert Michael Lewis[a,1], Virginia Torczon[a,b,*,2] , Michael W. Trosset[c,a]

[a]ICASE, Mail Stop 132C, NASA Langley Research Center, Hampton, VA 23681-2199, USA
[b]Department of Computer Science, College of William & Mary, P.O. Box 8795, Williamsburg, VA 23187-8795, USA
[c]Department of Mathematics, College of William & Mary, P.O. Box 8795, Williamsburg, VA 23187-8795, USA

Received 1 July 1999; received in revised form 23 February 2000

Abstract

We discuss direct search methods for unconstrained optimization. We give a modern perspective on this classical family of derivative-free algorithms, focusing on the development of direct search methods during their golden age from 1960 to 1971. We discuss how direct search methods are characterized by the absence of the construction of a model of the objective. We then consider a number of the classical direct search methods and discuss what research in the intervening years has uncovered about these algorithms. In particular, while the original direct search methods were consciously based on straightforward heuristics, more recent analysis has shown that in most — but not all — cases these heuristics actually suffice to ensure global convergence of at least one subsequence of the sequence of iterates to a first-order stationary point of the objective function. © 2000 Elsevier Science B.V. All rights reserved.

Keywords: Derivative-free optimization; Direct search methods; Pattern search methods

1. Introduction

Robert Hooke and T.A. Jeeves coined the phrase "direct search" in a paper that appeared in 1961 in the Journal of the Association of Computing Machinery [12]. They provided the following description of direct search in the introduction to their paper:

---

* Corresponding author.
E-mail addresses: buckaroo@icase.edu (R.M. Lewis), va@cs.wm.edu (V. Torczon), trosset@math.wm.edu (M.W. Trosset).

[1] This research was supported by the National Aeronautics and Space Administration under NASA Contract No. NAS1-97046.
[2] This research was supported by the National Science Foundation under Grant CCR-9734044 and by the National Aeronautics and Space Administration under NASA Contract No. NAS1-97046, while the author was in residence at the Institute for Computer Applications in Science and Engineering (ICASE).

0377-0427/00/$ - see front matter © 2000 Elsevier Science B.V. All rights reserved.
PII: S 0377-0427(00)00423-4

We use the phrase "direct search" to describe sequential examination of trial solutions involving comparison of each trial solution with the "best" obtained up to that time together with a strategy for determining (as a function of earlier results) what the next trial solution will be. The phrase implies our preference, based on experience, for straightforward search strategies which employ no techniques of classical analysis except where there is a demonstrable advantage in doing so.

To a modern reader, this preference for avoiding techniques of classical analysis "except where there is a demonstrable advantage in doing so" quite likely sounds odd. After all, the success of quasi-Newton methods, when applicable, is now undisputed. But consider the historical context of the remark by Hooke and Jeeves. Hooke and Jeeves' paper appeared five years before what are now referred to as the Armijo–Goldstein–Wolfe conditions were introduced and used to show how the method of steepest descent could be modified to ensure global convergence [1,11,29]. Their paper appeared only two years after Davidon's unpublished report on using secant updates to derive quasi-Newton methods [8], and two years before Fletcher and Powell published a similar idea in The Computer Journal [10]. So in 1961, this preference on the part of Hooke and Jeeves was not without justification.

Forty years later, the question we now ask is: why are direct search methods still in use? Surely, this seemingly hodge-podge collection of methods based on heuristics, which generally appeared without any attempt at a theoretical justification, should have been superseded by more "modern" approaches to numerical optimization.

To a large extent direct search methods *have* been replaced by more sophisticated techniques. As the field of numerical optimization has matured, and software has appeared which eases the ability of consumers to make use of these more sophisticated numerical techniques, many users now routinely rely on some variant of a globalized quasi-Newton method.

Yet direct search methods persist for several good reasons. First and foremost, direct search methods have remained popular because they work well in practice. In fact, many of the direct search methods are based on surprisingly sound heuristics that fairly recent analysis demonstrates guarantee global convergence behavior analogous to the results known for globalized quasi-Newton techniques. Direct search methods succeed because many of them — including the direct search method of Hooke and Jeeves — can be shown to rely on techniques of classical analysis in ways that are not readily apparent from their original specifications.

Second, quasi-Newton methods are not applicable to all nonlinear optimization problems. Direct search methods have succeeded when more elaborate approaches failed. Features unique to direct search methods often avoid the pitfalls that can plague more sophisticated approaches.

Third, direct search methods can be the method of first recourse, even among well-informed users. The reason is simple enough: direct search methods are reasonably straightforward to implement and can be applied almost immediately to many nonlinear optimization problems. The requirements from a user are minimal and the algorithms themselves require the setting of few parameters. It is not unusual for complex optimization problems to require further software development before quasi-Newton methods can be applied (e.g., the development of procedures to compute derivatives or the proper choice of perturbation for finite-difference approximations to gradients). For such problems, it can make sense to begin the search for a minimizer using a direct search method with known global convergence properties, while undertaking the preparations for the quasi-Newton method. When the preparations for the quasi-Newton method have been completed, the best known result from the direct search calculation can be used as a "hot start" for one of the quasi-Newton approaches, which enjoy superior local convergence properties. Such hybrid optimization strategies are as old as the direct search methods themselves [21].

We have three goals in this review. First, we want to outline the features of direct search that distinguish these methods from other approaches to nonlinear optimization. Understanding these features will go a long way toward explaining their continued success. Second, as part of our categorization of direct search, we suggest three basic approaches to devising direct search methods and explain how the better known classical techniques fit into one of these three camps. Finally, we review what is now known about the convergence properties of direct search methods. The heuristics that first motivated the development of these techniques have proven, with time, to embody enough structure to allow — in most instances — analysis based on now standard techniques. We are never quite sure if the original authors appreciated just how reliable their techniques would prove to be; we would like to believe they did. Nevertheless, we are always impressed by new insights to be gleaned from the discussions to be found in the original papers. We enjoy the perspective of forty intervening years of optimization research. Our intent is to use this hindsight to place direct search methods on a firm standing as one of many useful classes of techniques available for solving nonlinear optimization problems.

Our discussion of direct search algorithms is by no means exhaustive, focusing on those developed during the dozen years from 1960 to 1971. Space also does not permit an exhaustive bibliography. Consequently, we apologize in advance for omitting reference to a great deal of interesting work.

2. What is "direct search"?

For simplicity, we restrict our attention in the paper to unconstrained minimization:

$$\text{minimize } f(x), \tag{2.1}$$

where $f : \mathbb{R}^n \to \mathbb{R}$. We assume that f is continuously differentiable, but that information about the gradient of f is either unavailable or unreliable.

Because direct search methods neither compute nor approximate derivatives, they are often described as "derivative-free". However, as argued in [27], this description does not fully characterize what constitutes "direct search".

Historically, most approaches to optimization have appealed to a familiar "technique of classical analysis", the Taylor's series expansion of the objective function. In fact, one can classify most methods for numerical optimization according to how many terms of the expansion are exploited. Newton's method, which assumes the availability of first and second derivatives and uses the second-order Taylor polynomial to construct local quadratic approximations of f, is a second-order method. Steepest descent, which assumes the availability of first derivatives and uses the first-order Taylor polynomial to construct local linear approximations of f, is a first-order method. In this taxonomy, "zero-order methods" do not require derivative information and do not construct approximations of f. They are direct search methods, which indeed are often called zero-order methods in the engineering optimization community.

Direct search methods rely exclusively on values of the objective function, but even this property is not enough to distinguish them from other optimization methods. For example, suppose that one would like to use steepest descent, but that gradients are not available. In this case, it is customary to replace the actual gradient with an estimated gradient. If it is possible to observe exact values of the objective function, then the gradient is usually estimated by finite differencing. This is the case of numerical optimization, with which we are concerned herein. If function evaluation is uncertain, then the gradient is usually estimated by designing an appropriate experiment and performing a regression analysis. This occurs, for instance, in *response surface methodology* in stochastic optimization. Response surface methodology played a crucial role in the pre-history of direct search methods, a point to which we return shortly. Both approaches rely exclusively on values of the objective function, yet each is properly classified as a first-order method. What, then, is a direct search method? What exactly does it mean to say that direct search methods neither compute nor approximate derivatives?

Although instructive, we believe that a taxonomy based on Taylor expansions diverts attention from the basic issue. As in [27], we prefer here to emphasize the construction of approximations, not the mechanism by which they are constructed. The optimization literature contains numerous examples of methods that do not require derivative information and approximate the objective function without recourse to Taylor expansions. Such methods are "derivative-free", but they are not direct searches. What is the distinction?

Hooke and Jeeves considered that direct search involves the comparison of each trial solution with the best previous solution. Thus, a distinguishing characterization of direct search methods (at least in the case of unconstrained optimization) is that they do not require numerical function values: the relative rank of objective values is sufficient. That is, direct search methods for unconstrained optimization depend on the objective function only through the relative ranks of a countable set of function values. This means that direct search methods can accept new iterates that produce simple decrease in the objective. This is in contrast to the Armijo–Goldstein–Wolfe conditions for quasi-Newton line search algorithms, which require that a sufficient decrease condition be satisfied. Another consequence of this characterization of direct search is that it precludes the usual ways of approximating f, since access to numerical function values is not presumed.

There are other reasons to distinguish direct search methods within the larger class of derivative-free methods. We have already remarked that response surface methodology constructs local approximations of f by regression. Response surface methodology was proposed in 1951, in the seminal paper [4], as a variant of steepest descent (actually steepest ascent, since the authors were maximizing). In 1957, concerned with the problem of improving industrial processes and the shortage of technical personnel, Box [3] outlined a less sophisticated procedure called *evolutionary operation*. Response surface methodology relied on esoteric experimental designs, regression, and steepest ascent; evolutionary operation relied on simple designs and the direct comparison of observed function values. Spendley et al. [21] subsequently observed that the designs in [3] could be replaced with simplex designs and suggested that evolutionary operation could be automated and used for numerical optimization. As discussed in Section 3.2, their algorithm is still in use and is the progenitor of the simplex algorithm of Nelder and Mead [17], the most famous of all direct search methods. Thus, the distinction that G.E.P. Box drew in the 1950s, between response surface methodology and evolutionary operation, between approximating f and comparing values of f, played a crucial role in the development of direct search methods.

3. Classical direct search methods

We organize the popular direct search methods for unconstrained minimization into three basic categories. For a variety of reasons, we focus on the classical direct search methods, those developed during the period 1960–1971. The restriction is part practical, part historical.

On the practical side, we will make the distinction between *pattern search methods, simplex methods* (and here we do *not* mean the simplex method for linear programming), and *methods with adaptive sets of search directions*. The direct search methods that one finds described most often in texts can be partitioned relatively neatly into these three categories. Furthermore, the early developments in direct search methods more or less set the stage for subsequent algorithmic developments. While a wealth of variants on these three basic approaches to designing direct search methods have appeared in subsequent years — largely in the applications literature — these newer methods are modifications of the basic themes that had already been established by 1971. Once we understand the motivating principles behind each of the three approaches, it is a relatively straightforward matter to devise variations on these three themes.

There are also historical reasons for restricting our attention to the algorithmic developments in the 1960s. Throughout those years, direct search methods enjoyed attention in the numerical optimization community. The algorithms proposed were then (and are now) of considerable practical importance. As their discipline matured, however, numerical optimizers became less interested in heuristics and more interested in formal theories of convergence. At a joint IMA/NPL conference that took place at the National Physics Laboratory in England in January 1971, Swann [23] surveyed the status of direct search methods and concluded with this apologia:

Although the methods described above have been developed heuristically and no proofs of convergence have been derived for them, in practice they have generally proved to be robust and reliable in that only rarely do they fail to locate at least a local minimum of a given function, although sometimes the rate of convergence can be very slow.

Swann's remarks address an unfortunate perception that would dominate the research community for years to come: that whatever successes they enjoy in practice, direct search methods are theoretically suspect. Ironically, in the same year as Swann's survey, convergence results for direct search methods began to appear, though they seem not to have been widely known, as we discuss shortly. Only recently, in the late 1990s, as computational experience has evolved and further analysis has been developed, has this perception changed [30].

3.1. Pattern search

In his belated preface for ANL 5990 [8], Davidon described one of the most basic of pattern search algorithms, one so simple that it goes without attribution:

Enrico Fermi and Nicholas Metropolis used one of the first digital computers, the Los Alamos Maniac, to determine which values of certain theoretical parameters (phase shifts) best fit experimental data (scattering cross sections). They varied one theoretical parameter at a time by steps of the same magnitude, and when no such increase or decrease in any one parameter further improved the fit to the experimental data, they halved the step size and repeated the process until the steps were deemed sufficiently small. Their simple procedure was slow but sure, ...

Pattern search methods are characterized by a series of *exploratory moves* that consider the behavior of the objective function at a pattern of points, all of which lie on a rational lattice. In the example described above, the unit coordinate vectors form a basis for the lattice and the current magnitude of the steps (it is convenient to refer to this quantity as Δ_k) dictates the resolution of the lattice. The exploratory moves consist of a systematic strategy for visiting the points in the lattice in the immediate vicinity of the current iterate.

It is instructive to note several features of the procedure used by Fermi and Metropolis. First, it does not model the underlying objective function. Each time that a parameter was varied, the scientists asked: was there improvement in the fit to the experimental data. A simple "yes" or "no" answer determined which move would be made. Thus, the procedure is a direct search. Second, the parameters were varied by steps of predetermined magnitude. When the step size was reduced, it was multiplied by one-half, thereby ensuring that all iterates remained on a rational lattice. This is the key feature that makes the direct search a pattern search. Third, the step size was reduced *only* when no increase or decrease in any one parameter further improved the fit, thus ensuring that the step sizes were not decreased prematurely. This feature is another part of the formal definition of pattern search in [26] and is crucial to the convergence analysis presented therein.

3.1.1. Early analysis

By 1971, a proof of global convergence for this simple algorithm existed in the optimization text [18], where the technique goes by the name *method of local variations*. Specifically, Polak proved the following result:

Theorem 3.1. *If $\{x_k\}$ is a sequence constructed by the method of local variations, then any accumulation point x' of $\{x_k\}$ satisfies $\nabla f(x') = 0$. (By assumption, $f(x)$ is at least once continuously differentiable.)*

Polak's result is as strong as any of the contemporaneous global convergence results for either steepest descent or a globalized quasi-Newton method. However, to establish global convergence for these latter methods, one must enforce either sufficient decrease conditions (the Armijo–Goldstein–Wolfe conditions) or a fraction of Cauchy decrease condition — all of which rely on explicit numerical function values, as well as explicit approximations to the directional derivative at the current iterate. What is remarkable is that we have neither for direct search methods, yet can prove convergence.

What Polak clearly realized, though his proof does not make explicit use of this fact, is that all of the iterates for the method of local variations lie on a rational lattice (one glance at the figure on p. 43 of his text confirms his insight). The effect, as he notes, is that the method can construct only a *finite* number of intermediate points before reducing the step size by one-half. Thus the algorithm "cannot jam up at a point" — precisely, the pathology of premature convergence that the Armijo–Goldstein–Wolfe conditions are designed to preclude.

Polak was not alone in recognizing that pattern search methods contain sufficient structure to support a global convergence result. In the same year, Céa also published an optimization text [7] in which he provided a proof of global convergence for the pattern search algorithm of Hooke and Jeeves [12]. The assumptions used to establish convergence were stronger (in addition to the assumption that $f \in C^1$, it is assumed that f is strictly convex and that $f(x) \to +\infty$ as $\|x\| \to +\infty$). Nevertheless, it is established that the sequence of iterates produced by the method of Hooke and Jeeves converges to the unique minimizer of f — again with an algorithm that has no explicit recourse to the directional derivative and for which ranking information is sufficient.

Both Polak's and Céa's results rely on the fact that when either of these two algorithms reach the stage where the decision is made to reduce Δ_k, which controls the length of the steps, sufficient information about the local behavior of the objective has been acquired to ensure that the reduction is not premature. Specifically, neither the method of local variations nor the pattern search algorithm of Hooke and Jeeves allow Δ_k to be reduced until it has been verified that

$$f(x_k) \leqslant f(x_k \pm \Delta_k e_i), \quad i = \{1, \ldots, n\},$$

where e_i denotes the ith unit coordinate vector. This plays a critical role in both analyses. As long as x_k is not a stationary point of f, then at least one of the $2n$ directions defined by $\pm e_i$, $i \in \{1, \ldots, n\}$ must be a direction of descent. Thus, once Δ_k is sufficiently small, we are guaranteed that either $f(x_k + \Delta_k e_i) < f(x_k)$ or $f(x_k - \Delta_k e_i) < f(x_k)$ for at least one $i \in \{1, \ldots, n\}$.

The other early analysis worth noting is that of Berman [2]. In light of later developments, Berman's work is interesting precisely because he realized that if he made explicit use of a rational lattice structure, he could construct algorithms that produce minimizers to continuous nonlinear functions that might not be differentiable. For example, if f is continuous and strongly unimodal, he argues that convergence to a minimizer is guaranteed.

In the algorithms formulated and analyzed by Berman, the rational lattice plays an explicit role. The lattice L determined by x_0 (the initial iterate) and Δ_0 (the initial resolution of the lattice) is defined by $L(x_0, \Delta_0) = \{x \mid x = x_0 + \Delta_0 \lambda, \; \lambda \in \Lambda\}$, where Λ is the lattice of integral points of $\mathbb{R}^n$. Particularly important is the fact that the lattices used successively to approximate the minimizer have the following property: if $L_k = L(x_k, \Delta_k)$, where $\Delta_k = \Delta_0 / \tau^k$ and $\tau > 1$ denotes a positive integer, then $L_k \subset L_{k+1}$. The important ramification of this fact is that $\{x_0, x_1, x_2, \ldots, x_k\} \subset L_{k+1}$, for any choice of k, thus ensuring the finiteness property to which Polak alludes, and which also plays an important role in the more recent analysis for pattern search.

Before moving on to the more recent results, however, we close with some observations about this early work. First, it is with no small degree of irony that we note that all three results [2,7,18] are contemporaneous with Swann's remark that no proofs of convergence had been derived for direct search methods. However, each of these results was developed in isolation. None of the three authors appears to have been aware of the work of the others; none of the works contains citations of the other two and there is nothing in the discussion surrounding each result to suggest that any one of the authors was aware of the more-or-less simultaneous developments by the other two. Furthermore, these results have passed largely unknown and unreferenced in the nonlinear optimization literature. They have not been part of the "common wisdom" and so it was not unusual, until quite recently, to still hear claims that direct search methods had "been developed heuristically and no proofs of convergence have been derived for them".

Yet all the critical pieces needed for a more general convergence theory of pattern search had been identified by 1971. The work of Polak and Céa was more modest in scope in that each was proving convergence for a single, extant algorithm, already widely in use. Berman's work was more ambitious in that he was defining a general principle with the intent of deriving any number of new algorithms tailored to particular assumptions about the problem to be solved. What remained to be realized was that all this work could be unified under one analysis — and generalized even further to allow more algorithmic perturbations.

3.1.2. Recent analysis

Recently, a general theory for pattern search [26] extended a global convergence analysis [25] of the multidirectional search algorithm [24]. Like the simplex algorithms of Section 3.2, multidirectional search proceeds by reflecting a simplex ($n+1$ points in $\mathbb{R}^n$) through the centroid of one of the faces. However, unlike the simplex methods discussed in Section 3.2, multidirectional search is also a pattern search.

In fact, the essential ingredients of the general theory has already been identified in [2,7,18]. First, the pattern of points from which one selects trial points at which to evaluate the objective function must be sufficiently rich to ensure at least one direction of descent if x_k is not a stationary point of f. For Céa and Polak, this meant a pattern that included points of the form $x_k' = x_k \pm \Delta_k e_i$, $i \in \{1,\ldots,n\}$, where the e_i are the unit coordinate vectors. For Berman, it meant requiring Λ to be the lattice integral points of $\mathbb{R}^n$, i.e., requiring that the basis for the lattice be the identity matrix $I \in \mathbb{R}^{n \times n}$.

In [26], these conditions were relaxed to allow any nonsingular matrix $B \in \mathbb{R}^{n \times n}$ to be the basis for the lattice. In fact, we can allow patterns of the form $x_k' = x_k + \Delta_k B \gamma_k'$, where γ_k' is an integral vector, so that the direction of the step is determined by forming an integral combination of the columns of B. The special cases studied by Céa and Polak are easily recovered by choosing $B \equiv I$ and $\gamma_k' = \pm e_i$, $i \in \{1,\ldots,n\}$.

Second, an essential ingredient of each of the analyses is the requirement that Δ_k not be reduced if the objective function can be decreased by moving to one of the x_k'. Generalizations of this requirement were considered in [26,15]. This restriction acts to prevent premature convergence to a nonstationary point.

Finally, we restrict the manner by which Δ_k is rescaled. The conventional choice, used by both Céa and Polak, is to divide Δ_k by two, so that $\Delta_k = \Delta_0/2^k$. Somewhat more generally, Berman allowed dividing by any integer $\tau > 1$, so that (for example) one could have $\Delta_k = \Delta_0/3^k$. In fact, even greater generality is possible. For $\tau > 1$, we allow $\Delta_{k+1} = \tau^w \Delta_k$, where w is any integer in a designated finite set. Then there are three possibilities:

1. $w < 0$. This decreases Δ_k, which is only permitted under certain conditions (see above). When it is permitted, then $L_k \subset L_{k+1}$, the relation considered by Berman.
2. $w = 0$. This leaves Δ_k unchanged, so that $L_k = L_{k+1}$.
3. $w > 0$. This increases Δ_k, so that $L_{k+1} \subset L_k$.

It turns out that what matters is not the relation of L_k to L_{k+1}, but the assurance that there exists a single lattice $L_i \in \{L_0, L_1, \ldots, L_k, L_{k+1}\}$, for which $L_j \subseteq L_i$ for all $j = 0, \ldots, k+1$. This implies that $\{x_0, \ldots, x_k\} \subset L_i$, which in turn plays a crucial role in the convergence analysis.

Exploiting the essential ingredients that we have identified, one can derive a general theory of global convergence. The following result says that at least one subsequence of iterates converges to a stationary point of the objective function.

Theorem 3.2. *Assume that* $L(x_0) = \{x \mid f(x) \leqslant f(x_0)\}$ *is compact and that* f *is continuously differentiable on a neighborhood of* $L(x_0)$. *Then for the sequence of iterates* $\{x_k\}$ *produced by a generalized pattern search algorithm,*

$$\liminf_{k \to +\infty} \| \nabla f(x_k) \| = 0.$$

Under only slightly stronger hypotheses, one can show that every limit point of $\{x_k\}$ is a stationary point of f, generalizing Polak's convergence result. Details of the analysis can be found in [26,15]; [14] provides an expository discussion of the basic argument.

3.2. Simplex search

Simplex search methods are characterized by the simple device that they use to guide the search.

The first of the simplex methods is due to Spendley et al. [21] in a paper that appeared in 1962. They were motivated by the fact that earlier direct search methods required anywhere from $2n$ to 2^n objective evaluations to complete the search for improvement on the iterate. Their observation was that it should take no more than $n+1$ values of the objective to identify a downhill (or uphill) direction. This makes sense, since $n+1$ points in the graph of $f(x)$ determine a plane, and $n+1$ values of $f(x)$ would be needed to estimate $\nabla f(x)$ via finite differences. At the same time, $n+1$ points determine a simplex. This leads to the basic idea of simplex search: construct a nondegenerate simplex in $\mathbb{R}^n$ and use the simplex to drive the search.

A simplex is a set of $n+1$ points in $\mathbb{R}^n$. Thus one has a triangle in $\mathbb{R}^2$, a tetrahedron in $\mathbb{R}^3$, etc. A nondegenerate simplex is one for which the set of edges adjacent to any vertex in the simplex forms a basis for the space. In other words, we want to be sure that any point in the domain of the search can be constructed by taking linear combinations of the edges adjacent to any given vertex.

Not only does the simplex provide a frugal design for sampling the space, it has the added feature that if one replaces a vertex by reflecting it through the centroid of the opposite face, then the result is also a simplex, as shown in Fig. 1. This, too, is a frugal feature because it means that one can proceed parsimoniously, reflecting one vertex at a time, in the search for an optimizer.

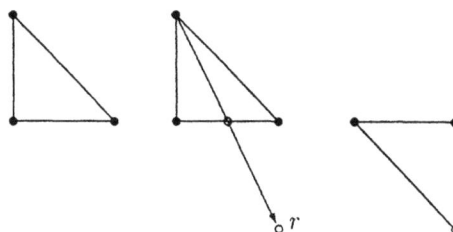

Fig. 1. The original simplex, the reflection of one vertex through the centroid of the opposite face, and the resulting reflection simplex.

Once an initial simplex is constructed, the single move specified in the original Spendley et al. simplex algorithm is that of reflection. This move first identifies the "worst" vertex in the simplex (i.e., the one with the least desirable objective value) and then reflects the worst simplex through the centroid of the opposite face. If the reflected vertex is still the worst vertex, then next choose the "next worst" vertex and repeat the process. (A quick review of Fig. 1 should confirm that if the reflected vertex is not better than the next worst vertex, then if the "worst" vertex is once again chosen for reflection, it will simply be reflected back to where it started, thus creating an infinite cycle.)

The ultimate goals are either to replace the "best" vertex (i.e., the one with the most desirable objective value) or to ascertain that the best vertex is a candidate for a minimizer. Until then, the algorithm keeps moving the simplex by flipping some vertex (other than the best vertex) through the centroid of the opposite face.

The basic heuristic is straightforward in the extreme: we move a "worse" vertex in the general direction of the remaining vertices (as represented by the centroid of the remaining vertices), with the expectation of eventual improvement in the value of the objective at the best vertex. The questions then become: do we have a new candidate for a minimizer and are we at or near a minimizer?

The first question is easy to answer. When a reflected vertex produces strict decrease on the value of the objective at the best vertex, we have a new candidate for a minimizer; once again the simple decrease rule is in effect.

The answer to the second question is decidedly more ambiguous. In the original paper, Spendley, Hext, and Himsworth illustrate — in two dimensions — a circling sequence of simplices that could be interpreted as indicating that the neighborhood of a minimizer has been identified. We see a similar example in Fig. 2, where a sequence of five reflections brings the search back to where it started, without replacing x_k, thus suggesting that x_k may be in the neighborhood of a stationary point.

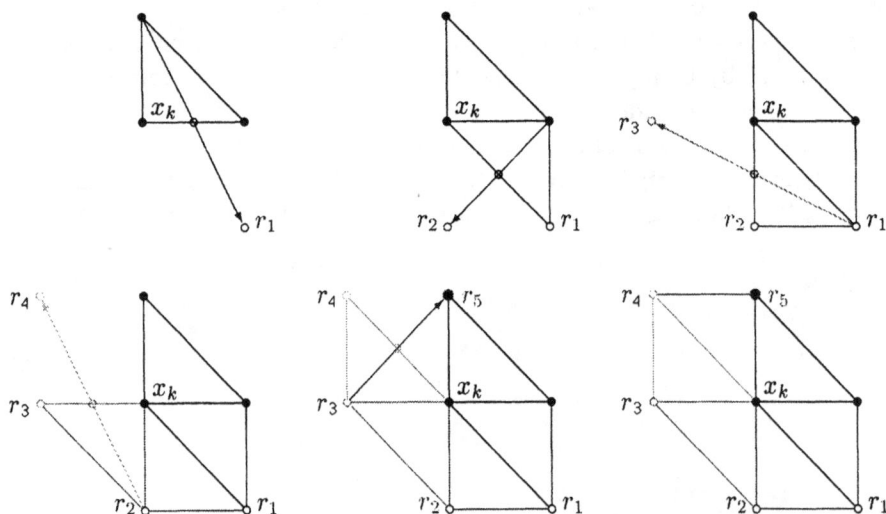

Fig. 2. A sequence of reflections $\{r_1, r_2, r_3, r_4, r_5\}$, each of which fails to replace the best vertex x_k, which brings the search back to the simplex from which this sequence started.

The picture in two dimensions is somewhat misleading since the fifth reflection maps back onto the worst vertex in the original simplex — a situation that only occurs in either one or two dimensions. So Spendley, Hext, and Himsworth give a heuristic formula for when the simplex has flipped around the current best vertex long enough to conclude that the neighborhood of a minimizer has been identified. When this situation has been detected, they suggest two alternatives: either reduce the lengths of the edges adjacent to the "best" vertex and resume the search or resort to a higher-order method to obtain faster local convergence.

The contribution of Nelder and Mead [17] was to turn simplex search into an optimization algorithm with additional moves designed to accelerate the search. In particular, it was already well-understood that the reflection move preserved the original shape of the simplex — regardless of the dimension. What Nelder and Mead proposed was to supplement the basic reflection move with additional options designed to accelerate the search by deforming the simplex in a way that they suggested would better adapt to the features of the objective function. To this end, they added what are known as expansion and contraction moves, as shown in Fig. 3.

We leave the full details of the logic of the algorithm to others; a particularly clear and careful description, using modern algorithmic notation, can be found in [13]. For our purposes, what is important to note is that the expansion step allows for a more aggressive move by doubling the length of the step from the centroid to the reflection point, whereas the contraction steps allow for more conservative moves by halving the length of the step from the centroid to either the reflection point or the worst vertex. Furthermore, in addition to allowing these adaptations within a single iteration, these new possibilities have repercussions for future iterations as they deform (or, as the rationale goes, adapt) the shape of the original simplex.

Nelder and Mead also resolved the question of what to do if none of the steps tried bring acceptable improvement by adding a shrink step: when all else fails, reduce the lengths of the edges adjacent to the current best vertex by half, as is also illustrated in Fig. 3.

The Nelder–Mead simplex algorithm has enjoyed enduring popularity. Of all the direct search methods, the Nelder–Mead simplex algorithm is the one most often found in numerical software packages. The original paper by Nelder and Mead is a Science Citation Index classic, with several thousand references across the scientific literature in journals ranging from *Acta Anaesthesiologica*

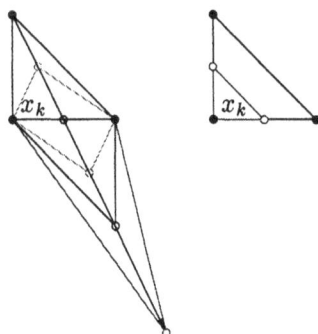

Fig. 3. The original simplex, with the reflection, expansion, and two possible contraction simplices, along with the shrink step toward the best vertex x_k, when all else fails.

Scandinavica to *Zhurnal Fizicheskio Khimii*. In fact, there is an entire book from the chemical engineering community devoted to simplex search for optimization [28].

So why bother with looking any further? Why not rely exclusively on the Nelder–Mead simplex method if one is going to employ a direct search method? The answer: there is the outstanding question regarding the robustness of the Nelder–Mead simplex method that has long troubled numerical optimizers. When the method works, it can work very well indeed, often finding a solution in far fewer evaluations of the objective function than other direct search methods. But it can also fail. One can see this in the applications literature, fairly early on, frequently reported as no more than "slow" convergence. A systematic study of Nelder–Mead, when applied to a suite of standard optimization test problems, also reported occasional convergence to a nonstationary point of the function [24]; the one consistent observation to be made was that in these instances the deformation of the simplex meant that the search direction (i.e., the direction defined along the worst vertex toward the centroid of the remaining vertices) became numerically orthogonal to the gradient.

These observations about the behavior of Nelder–Mead in practice led to two, relatively recent, investigations. The first [13], strives to investigate what *can* be proven about the asymptotic behavior of Nelder–Mead. The results show that in $\mathbb{R}^1$, the algorithm is robust; under standard assumptions, convergence to a stationary point is guaranteed. Some general properties in higher dimensions can also be proven, but none that guarantee global convergence for problems in higher dimensions.

This is not surprising in light of a second recent result by McKinnon [16]. He shows with several examples that limits exist on proving global convergence for Nelder–Mead: to wit, the algorithm can fail on smooth (C^2) convex objectives in two dimensions.

This leaves us in the unsatisfactory situation of reporting that no general convergence results exist for the simplex methods of either Spendley et al. or Nelder and Mead — despite the fact that they are two of the most popular and widely used of the direct search methods. Further, McKinnon's examples indicate that it will not be possible to prove global convergence for the Nelder–Mead simplex algorithm in higher dimensions. On the other hand, the mechanism that leads to failure in McKinnon's counterexample does not seem to be the mechanism by which Nelder–Mead typically fails in practice. This leaves the question of why Nelder–Mead fails in practice unresolved.

3.3. Methods with adaptive sets of search directions

The last family of classical methods we consider includes Rosenbrock's and Powell's methods. These algorithms attempt to accelerate the search by constructing directions designed to use information about the curvature of the objective obtained during the course of the search.

3.3.1. Rosenbrock's method

Of these methods, the first was due to Rosenbrock [20]. Rosenbrock's method was quite consciously derived to cope with the peculiar features of Rosenbrock's famous "banana function", the minimizer of which lies inside a narrow, curved valley. Rosenbrock's method proceeds by a series of stages, each of which consists of a number of exploratory searches along a set of directions that are fixed for the given stage, but which are updated from stage to stage to make use of information acquired about the objective.

The initial stage of Rosenbrock's method begins with the coordinate directions as the search directions. It then conducts searches along these directions, cycling over each in turn, moving to new

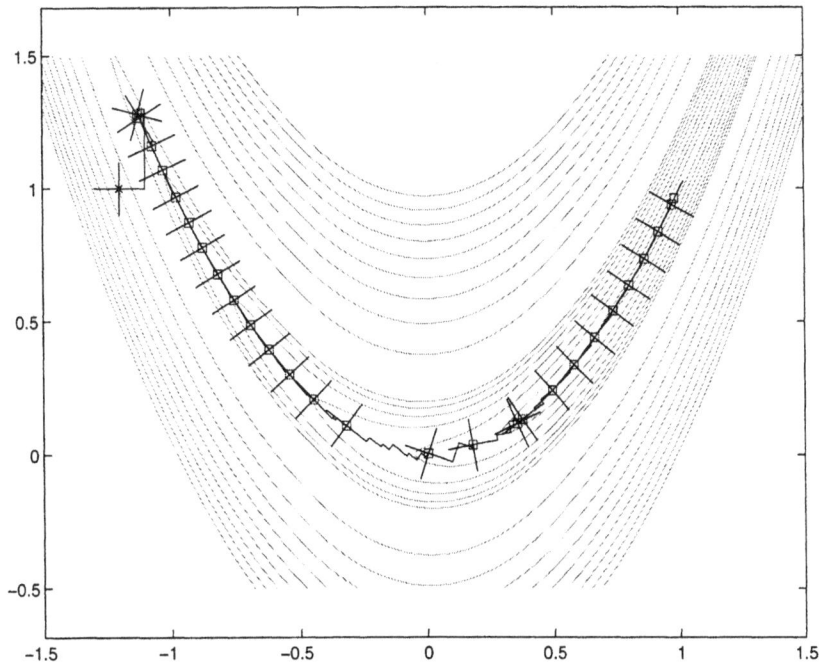

Fig. 4. Rosenbrock's algorithm in action.

iterates that yield successful steps (an unsuccessful step being one that leads to a less desirable value of the objective). This continues until there has been at least one successful and one unsuccessful step in each search direction. Once this occurs, the current stage terminates. As is the case for direct search methods, numerical values of the objective are not necessary in this process. If the objective at any of these steps is perceived as being an improvement over the objective at the current best point, we move to the new point.

At the next stage, rather than repeating the search process with the *same* set of orthogonal vectors, as is done for the method of local variations, Rosenbrock *rotates* the set of directions to capture information about the objective ascertained during the course of the earlier moves. Specifically, he takes advantage of the fact that a nonzero step from the iterate at the beginning of the previous stage to the iterate at the start of the new stage suggests a good direction of descent — or, at the very least, a promising direction — so in the new stage, he makes sure that this particular direction is included in the set of directions along which the search will be conducted. (This heuristic is particularly apt for following the bottom of the valley that leads to the minimizer of the banana function.) Rosenbrock imposes the condition that the set of search directions always be an orthogonal set of *n* vectors so that the set of vectors remains nicely linearly independent. The new set of orthonormal vectors is generated using the Gram-Schmidt orthonormalization procedure, with the "promising" direction from the just-completed stage used as the first vector in the orthonormalization process.

Rosenbrock's method as applied to his banana function is depicted in Fig. 4. The iterate at the beginning of each stage is indicated with a square. Superimposed on these iterates are the search directions for the new stage. Note how quickly the search adapts to the narrow valley; within three stages the search directions reflect this feature. Also notice how the search directions change to allow the algorithm to turn the corner in the valley and continue to the solution.

Updating the set of search directions for Rosenbrock's method entails slightly more complexity than that which appears in any of the other two families of direct search methods we have surveyed. On the other hand, the example of the banana function makes the motivation for this additional work clear: adapting the entire set of search directions takes advantage of what has been learned about the objective during the course of the search.

3.3.2. The variant of Davies, Swann, and Campey

A refinement to Rosenbrock's algorithm was proposed in [22].[3] Davies et al. noted that there was merit to carrying out a sequence of more sophisticated one-dimensional searches along each of the search directions than those performed in Rosenbrock's original algorithm.

As described in [23], the more elaborate line search of Davies et al. first takes steps of increasing multiples of some fixed value Δ along a direction from the prescribed set until a bracket for the (one-dimensional) minimizer is obtained. This still corresponds to our definition of a direct search method.

However, once a bracket for the one-dimensional minimizer has been found, a "single quadratic interpolation is made to predict the position of the minimum more closely" [23]. This is the construction of a model of the objective, and to do this, numerical values for the objective must be in hand. Thus, this final move within an iteration disqualifies the method of Davies, Swann, and Campey as a direct search method by our characterization. Nonetheless, this strategy is undeniably appealing, and its authors aver that this variant of Rosenbrock's method is more generally efficient than the original [6].

3.3.3. Powell's method

In a paper appearing the same year as the report in [22], Powell [19] outlined a method for finding minimizers without calculating derivatives. By our definition, it is a derivative-free, rather than a direct search method, for modeling is at the heart of the approach. The explicit goal is to ensure that if the method is applied to a convex quadratic function, conjugate directions are chosen with the goal of accelerating convergence. In this sense, Powell's algorithm may be viewed as a derivative-free version of nonlinear conjugate gradients.

Like Rosenbrock's method, Powell's method proceeds in stages. Each stage consists of a sequence of $n+1$ one-dimensional searches. The one-dimensional searches are conducted by finding the exact minimizer of a quadratic interpolant computed for each direction (hence our classification of the method as a derivative-free, but not direct search, method). The first n searches are along each of a set of linearly independent directions. The last search is along the direction connecting the point obtained at the end of the first n searches with the starting point of the stage. At the end of the stage, one of the first n search directions is replaced by the last search direction. The process then repeats at the next stage.

Powell showed that if the objective is a convex quadratic, then the set of directions added at the last step of each stage forms a set of conjugate directions (provided they remain linearly

---

[3] A paper the authors have been unable to locate. The authors would be very much obliged to any reader who has a copy of the original report and would forward a photocopy to us.

independent). Powell used this, in turn, to show that his method possessed what was known then as the "Q-property". An algorithm has the Q-property if it will find the minimizer of a convex quadratic in a finite number of iterations. That is, the Q-property is the finite termination property for convex quadratics such as that exhibited by the conjugate gradient algorithm. In the case of Powell's method, one obtains finite termination in n stages for convex quadratics.

Zangwill [31] gave a modification of Powell's method that avoids the possibility of linearly dependent search directions. Zangwill further proved convergence to minimizers of strictly convex functions (though not in a finite number of steps).

To the best of our knowledge, Powell's method marks the first time that either a direct search or a derivative-free method appeared with any attendant convergence analysis. The appeal of the *explicit* modeling of the objective such as that used in the line-searches in Powell's method is clear: it makes possible strong statements about the behavior of the optimization method. We can expect the algorithm to quickly converge to a minimizer once in a neighborhood of a solution on which the objective is essentially quadratic.

Finite termination on quadratic objectives was a frequently expressed concern within the optimization community during the 1960s and 1970s. The contemporary numerical results produced by the optimization community (for analytical, closed-form objective functions, it should be noted) evidence this concern. Most reports of the time [5,9] confirm the supposed superiority of the modeling-based approach, with guaranteed finite termination as embodied in Powell's derivative-free conjugate directions algorithm.

Yet forty years later, direct search methods, "which employ no techniques of analysis except where there is a demonstrable advantage in doing so", remain popular, as indicated by any number of measures: satisfied users, literature citations, and available software. What explains this apparently contradictory historical development?

4. Conclusion

Direct search methods remain popular because of their simplicity, flexibility, and reliability. Looking back at the initial development of direct search methods from a remove of forty years, we can firmly place what is now known and understood about these algorithms in a broader context.

With the exception of the simplex-based methods specifically discussed in Section 3.2, direct search methods are robust. Analytical results now exist to demonstrate that under assumptions comparable to those commonly used to analyze the global behavior of algorithms for solving unconstrained nonlinear optimization problems, direct search methods can be shown to satisfy the first-order necessary conditions for a minimizer (i.e., convergence to a stationary point). This seems remarkable given that direct search methods neither require nor explicitly estimate derivative information; in fact, one obtains these guarantees even when using only ranking information. The fact that most of the direct search methods require a set of directions that span the search space is enough to guarantee that sufficient information about the local behavior of the function exists to safely reduce the step length after the full set of directions has been queried.

Following the lead of Spendley et al. [21], we like to think of direct search methods as "methods of steep descent". These authors made it quite clear that their algorithm was designed to be related to the method of steepest descent (actually steepest ascent, since the authors were maximizing). Although no explicit representation of the gradient is formed, enough local information is obtained by sampling to ensure that a downhill direction (though not necessarily the steepest downhill direction) can be identified. Spendley et al. also intuited that steep descent would be needed to ensure what we now call global convergence; furthermore, they recognized the need to switch to higher-order methods to obtain fast local convergence.

This brings us to the second point to be made about the classical direct search methods. They do not enjoy finite termination on quadratic objectives or rapid local convergence. For this, one needs to capture the local curvature of the objective, and this necessarily requires some manner of modeling — hence, the undeniable appeal of modeling-based approaches. However, modeling introduces additional restrictions that may not always be appropriate in the settings in which direct search methods are used: specifically, the need to have explicit numerical function values of sufficient reliability to allow interpolation or some other form of approximation. In truth, the jury is still out on the effectiveness of adding this additional layer of information to devise derivative-free methods that also approximate curvature (second-order) information. Several groups of researchers are currently looking for a derivative-free analog of the elegant trust region globalization techniques for quasi-Newton methods that switch seamlessly between favoring the Cauchy (steepest-descent) direction to ensure global convergence and the Newton direction to ensure fast local convergence.

We close with the observation that, since nonlinear optimization problems come in all forms, there is no "one-size-fits-all" algorithm that can successfully solve all problems. Direct search methods are sometimes used — inappropriately — as the method of first recourse when other optimization techniques would be more suitable. But direct search methods are also used — appropriately — as the methods of last recourse, when other approaches have been tried and failed. Any practical optimizer would be well-advised to include direct search methods among their many tools of the trade. Analysis now confirms what practitioners in many different fields have long recognized: a carefully chosen, carefully implemented direct search method can be an effective tool for solving many nonlinear optimization problems.

References

[1] L. Armijo, Minimization of functions having Lipschitz continuous first partial derivatives, Pacific J. Math. 16 (1966) 1–3.

[2] G. Berman, Lattice approximations to the minima of functions of several variables, J. Assoc. Comput. Mach. 16 (1969) 286–294.

[3] G.E.P. Box, Evolutionary operation: a method for increasing industrial productivity, Appl. Statist. 6 (1957) 81–101.

[4] G.E.P. Box, K.B. Wilson, On the experimental attainment of optimum conditions, J. Roy. Statist. Soc. Ser. B XIII (1951) 1–45.

[5] M.J. Box, A comparison of several current optimization methods, and the use of transformations in constrained problems, Comput. J. 9 (1966) 67–77.

[6] M.J. Box, D. Davies, W.H. Swann, Non-Linear Optimization Techniques, ICI Monograph, No. 5, Oliver & Boyd, Edinburgh, 1969.

[7] J. Céa, Optimisation: Théorie et Algorithmes, Dunod, Paris, 1971.

[8] W.C. Davidon, Variable metric method for minimization, SIAM J. Optim. 1 (1991) 1–17 (the article was originally published as Argonne National Laboratory Research and Development Report 5990, May 1959 (revised November 1959)).

[9] R. Fletcher, Function minimization without evaluating derivatives — a review, Comput. J. 8 (1965) 33–41.

[10] R. Fletcher, M.J.D. Powell, A rapidly convergent descent method for minimization, Comput. J. 6 (1963) 163–168.

[11] A.A. Goldstein, Constructive Real Analysis, Harper & Row, New York, 1967.

[12] R. Hooke, T.A. Jeeves, Direct search solution of numerical and statistical problems, J. Assoc. Comput. Mach. 8 (1961) 212–229.

[13] J.C. Lagarias, J.A. Reeds, M.H. Wright, P.E. Wright, Convergence properties of the Nelder–Mead simplex method in low dimensions, SIAM J. Optim. 9 (1998) 112–147.

[14] R.M. Lewis, V. Torczon, M.W. Trosset, Why pattern search works, Optima (1998) 1–7.

[15] R.M. Lewis, V.J. Torczon, Rank ordering and positive bases in pattern search algorithms, Technical Report 96-71, Institute for Computer Applications in Science and Engineering, Mail Stop 132C, NASA Langley Research Center, Hampton, VA 23681–2199, 1996.

[16] K.I.M. McKinnon, Convergence of the Nelder–Mead simplex method to a nonstationary point, SIAM J. Optim. 9 (1998) 148–158.

[17] J.A. Nelder, R. Mead, A simplex method for function minimization, Comput. J. 7 (1965) 308–313.

[18] E. Polak, Computational Methods in Optimization: A Unified Approach, Academic Press, New York, 1971.

[19] M.J.D. Powell, An efficient method for finding the minimum of a function of several variables without calculating derivatives, Comput. J. 7 (1964) 155–162.

[20] H.H. Rosenbrock, An automatic method for finding the greatest or least value of a function, Comput. J. 3 (1960) 175–184.

[21] W. Spendley, G.R. Hext, F.R. Himsworth, Sequential application of simplex designs in optimisation and evolutionary operation, Technometrics 4 (1962) 441–461.

[22] W.H. Swann, Report on the development of a new direct search method of optimization, Research Note 64/3, I.C.I. Central Instrument Lab, 1964.

[23] W.H. Swann, Direct search methods, in: W. Murray (Ed.), Numerical Methods for Unconstrained Optimization, Academic Press, New York, 1972, pp. 13–28.

[24] V. Torczon, Multi-directional search: a direct search algorithm for parallel machines, Ph.D. Thesis, Department of Mathematical Sciences, Rice University, Houston, Texas, 1989.

[25] V. Torczon, On the convergence of the multidirectional search algorithm, SIAM J. Optim. 1 (1991) 123–145.

[26] V. Torczon, On the convergence of pattern search algorithms, SIAM J. Optim. 7 (1997) 1–25.

[27] M.W. Trosset, I know it when I see it: toward a definition of direct search methods, SIAG/OPT Views-and-News 9 (1997) 7–10.

[28] F.H. Walters, L.R. Parker Jr., S.L. Morgan, S.N. Deming, Sequential Simplex Optimization, Chemometrics Series, CRC Press, Boca Raton, FL, 1991.

[29] P. Wolfe, Convergence conditions for ascent methods, SIAM Rev. 11 (1969) 226–235.

[30] M.H. Wright, Direct search methods: once scorned, now respectable, in: D.F. Griffiths, G.A. Watson (Eds.), Numerical Analysis 1995, Addison-Wesley Longman, Reading, MA, 1996, pp. 191–208.

[31] W.I. Zangwill, Minimizing a function without calculating derivatives, Comput. J. 10 (1967) 293–296.

Journal of Computational and Applied Mathematics 124 (2000) 209–228

JOURNAL OF
COMPUTATIONAL AND
APPLIED MATHEMATICS

www.elsevier.nl/locate/cam

ELSEVIER

Recent developments and trends in global optimization

Panos M. Pardalos[a],[*], H. Edwin Romeijn[a], Hoang Tuy[b]

[a]*Center for Applied Optimization, Department of Industrial and Systems Engineering, University of Florida,
303 Weil Hall, P.O. Box 116595, Gainesville, FL 32611-6595, USA*
[b]*Institute of Mathematics, P.O. Box 631, Bo Ho, Hanoi, Viet Nam*

Received 29 December 1999; received in revised form 3 February 2000

Abstract

Many optimization problems in engineering and science require solutions that are globally optimal. These optimization problems are characterized by the nonconvexity of the feasible domain or the objective function and may involve continuous and/or discrete variables. In this paper we highlight some recent results and discuss current research trends on deterministic and stochastic global optimization and global continuous approaches to discrete optimization. © 2000 Elsevier Science B.V. All rights reserved.

Keywords: Global optimization; Deterministic methods; Stochastic methods; Monotonicity; Minimax; Integer programming

1. Introduction

Our time is witnessing the rapid growth of a new field, global optimization. Many new theoretical, algorithmic, and computational contributions of global optimization have been used to solve many problems in science and engineering. Global optimization problems abound in the mathematical modeling of real-world systems for a very broad spectrum of applications. Such applications include finance, allocation and location problems, operations research, statistics, structural optimization, engineering design, network and transportation problems, chip design and database problems, nuclear and mechanical design, chemical engineering design and control, and molecular biology. Discrete optimization problems form a special class of global optimization problems. For apparently historical reasons there is an artificial separation of continuous and discrete optimization problems. From our point of view, the major difference between optimization problems is based on the presence

---

[*] Corresponding author. Tel.: +1-352392-9011; fax: +1-352392-3537.
E-mail addresses: pardalos@cao.ise.ufl.edu (P.M. Pardalos), romeijn@ise.ufl.edu (H.E. Romeijn), htuy@hn.vnn.vn (H. Tuy).

0377-0427/00/$ - see front matter © 2000 Elsevier Science B.V. All rights reserved.
PII: S 0377-0427(00)00425-8

or absence of convexity. Since in most optimization problems convexity of the objective function or the feasible domain is not easily recognizable, we may assume that the problem is nonconvex. Traditional approaches of nonlinear programming have been very successful in computing stationary points and locally optimal solutions. Since multi-extremal problems may have an exponential number of local minima, traditional nonlinear programming approaches are inadequate.

In this paper we focus on some recent developments and research trends in global optimization. It is inevitable that much of the material is related to the work of the authors. The first part of the paper covers material regarding deterministic approaches to global optimization with focus on d.c. and monotonic optimization, as well as continuous approaches to discrete optimization problems. The second part discusses stochastic approaches and metaheuristics. Some specific examples are discussed to illustrate the richness of the new techniques. For a more extensive set of references, we refer the reader to Pardalos et al. [18].

2. Deterministic approaches

Deterministic approaches are those which exploit analytical properties of the problem to generate a deterministic sequence of points (finitely or infinitely) converging to a global optimal solution. Two analytical properties: *convexity* and *monotonicity*, have been most successfully exploited, giving rise to two important research trends: *d.c. optimization* (dealing with problems described by means of differences of convex functions or sets) and *monotonic optimization* (dealing with problems described by means of functions monotonically increasing or decreasing along rays). Among these problems a subclass constituted by *quadratic and polynomial programs* has in the last few years attracted a growing level of attention due to many practical applications. We devote a section on the topic of continuous approaches to *discrete problems*, showing a number of such approaches that show a lot of promise. Finally, we discuss *general continuous optimization* problems, i.e., problems with very little information available on their mathematical structure. These have always formed the biggest challenges to global optimizers.

2.1. D.C. optimization

It is common knowledge that when both the objective function and the constraint set are convex, the problem can be solved by efficient algorithms. Difficulties arise only when the objective function or the constraint set fails to be convex. Fortunately, however, in most nonconvex problems of interest, convexity is present in some limited or "opposite" sense. Specifically, a wide variety of optimization problems encountered in practice can be cast in the form

$$\min \ f_0(x) \equiv f_{0,1}(x) - f_{0,2}(x)$$
$$\text{s.t} \quad f_i(x) \equiv f_{i,1}(x) - f_{i,2}(x) \leqslant 0 \quad (i = 1, \ldots, m), \tag{1}$$
$$x \in X \subset \mathbb{R}^n,$$

where X is a compact convex set and $f_{i,1}(x), f_{i,2}(x)$, $(i=0,1,\ldots,m)$, are convex functions. A function representable as a difference of two convex functions is called a *d.c. function*, so a problem of form

(1) is referred to as *d.c. optimization problem*, or a *d.c. program* for short. The following properties explain why most optimization problems can be described as d.c. programs:

1. any twice continuously differentiable function (in particular any polynomial) is d.c. on any compact convex set in $\mathbb{R}^n$;
2. any closed set $S \subset \mathbb{R}^n$ can be represented as the solution set of a d.c. inequality: $S = \{x \in \mathbb{R}^n \mid g_S(x) - \|x\|^2 \leqslant 0\}$ where $g_S(x)$ is a continuous convex function on $\mathbb{R}^n$;
3. if $f_1(x), \ldots, f_m(x)$ are d.c. then the functions $\sum_i \alpha_i f_i(x)$ ($\alpha_i \in \mathbb{R}$), $\max_{i,\ldots,m} f_i(x)$ and $\min_{i=1,\ldots,m} f_i(x)$ are also d.c.

Using these properties, it can be proven that, in principle, every continuous optimization problem can be reduced to a d.c. program with a linear objective function and no more than *one* convex and *one* reverse convex constraint (for details, see [10]).

A typical, and in fact one of the most intensely studied problems of global optimization, is the linearly constrained concave minimization problem (sometimes referred to as the *concave programming* problem under linear constraints), which seeks to globally minimize a concave function $c(x)$ over a polyhedron $D \subset \mathbb{R}^n$:

$$\min\{c(x) \mid x \in D\}, \quad D = \{x \in \mathbb{R}^n \mid Ax \leqslant b\}. \tag{2}$$

Despite the relative simplicity of its formulation, this problem has a surprisingly diverse range of direct and indirect applications. Over more than three decades since it was first studied, many ideas and methods proposed for solving it have been refined and extended to more general d.c. optimization problems. Furthermore, many d.c. optimization methods use concave minimization algorithms as subroutines. For a review of concave minimization methods and d.c. optimization methods up to 1994 we refer the interested reader to Horst and Pardalos [9].

The most important property of a concave function is that *its minimum over a polytope is achieved at a vertex (extreme point)*. Based on this property, one can find the minimum of a concave function $c(x)$ over a compact convex set D by inductively constructing a nested sequence of polytopes $P_1 \supset P_2 \supset \cdots \supset D$ such that P_{k+1} is obtain from P_k by imposing just an additional linear constraint, chosen so as to ensure that

$$\min\{c(x) \mid x \in P_k\} \nearrow \min\{c(x) \mid x \in D\}.$$

Starting from P_1 with a readily available vertex set V_1, one can then derive V_k for all $k = 2, 3, \ldots$ by an efficient procedure (see, e.g., [11]), hence compute $x^k \in \arg\min\{c(x) \mid x \in V_k\} = \arg\min\{c(x) \mid x \in P_k\}$ and obtain the global optimal solution $\bar{x}$ as any accumulation point of the sequence $x^1, x^2, \ldots$. Although conceptually very simple, this *outer approximation* method has been used successfully in a number of applications including problems of design centering and continuous location. Originally devised for convex programs, it was extended to concave programs, then to reverse convex and d.c. programs (see [9]) as well as monotonic optimization problems.

A major difficulty of nonlinear programming methods is that using these methods one may get trapped at a local minimizer, or even a stationary point. Therefore, a fundamental issue in global optimization is to *transcend local optimality*, or more generally to *transcend the incumbent*, i.e., given a feasible solution $\bar{x}$ (the best feasible solution available), to check whether $\bar{x}$ is globally optimal, and if it is not, to compute a better feasible solution. It turned out that for any d.c.

optimization problem, transcending the incumbent reduces to solving a subproblem of the form

$$(\text{DC})_{\bar{x}} \quad \text{Find any } x \in D \setminus C \text{ or else prove that } D \subset C, \tag{3}$$

where C, D are convex sets and C, or D, depends on $\bar{x}$. Therefore, if a method is available for solving $(\text{DC})_{\bar{x}}$, then the global optimization problem can be solved according to a two phase scheme as follows.

Phase 0: Let z^0 be an initial feasible solution. Set $k = 0$.

Phase 1: (Local phase) Starting from z^k and using any suitable, relatively inexpensive, local method search for a feasible point x^k at least as good as z^k.

Phase 2: (Global phase) Solve $(\text{DC})_{x^k}$. If the global optimality of x^k (with the given tolerance) is established, stop. If a feasible point z is obtained such that $f(z) < f(x^k)$, then set $z^{k+1} \leftarrow z$, increment k and return to Phase 1.

This approach provides a unified view on global optimization algorithms. Furthermore, since all kinds of local, heuristics or stochastic searches can be used in Phase 1, this approach allows a practical combination of deterministic global methods with other search methods to enhance efficiency.

It should be noted that in several cases, the conversion of a problem to the d.c. form allows a substantial simplification of its computational analysis. For example, in its standard formulation the classical Weber's problem with p facilities and N users in location theory involves up to $p(N+2)$ variables together with a complicated objective function, while the d.c. formulation uses only $2p$ variables and a much simpler objective function, which allows a very efficient solution method. In other circumstances, however, direct methods may be more convenient for exploiting a particular mathematical structure, such as the network or the multilevel ones (see [6,9]).

As in combinatorial optimization, the most popular methods in global optimization use *branch and bound* principles. Specifically, to solve problem (1), the space is partitioned into polyhedral subsets (called *partition sets*) which may be simplices (simplicial partition), (hyper)rectangles (rectangular partition), or cones (conical partition). For every partition set M a *lower bound* $\beta(M)$ is estimated for the minimum of the objective function $f(x)$ over the feasible points in M, i.e., for

$$f_M^* \equiv \min\{f_0(x) \mid f_i(x) \leqslant 0 \ (i = 1, \ldots, m), \ x \in M \cap X\}. \tag{4}$$

Then, on the basis of the information currently available, some partition sets M are discarded from further consideration as nonpromising, while the most promising partition set (usually the one with smallest $\beta(M)$) is selected and further partitioned. This gives rise to a more refined partition of the space, and the process is repeated. To compute a lower bound $\beta(M)$ for (4) a common method is to relax (4) to a convex problem

$$\min\{\varphi_0^M(x) \mid \varphi_i^M(x) \leqslant 0 \ (i = 1, \ldots, m), \ x \in M \cap X\}, \tag{5}$$

where $\varphi_i^M(x)$ is a suitably chosen convex minorant (underestimator) of $f_i(x)$ on M (for $i = 0, 1, \ldots, m$), i.e., $\varphi_i^M(x)$ is convex on M and $\varphi_i^M(x) \leqslant f_i(x)$ for all $x \in M$. For instance, if M is an n-simplex then a convex minorant of $f_i(x) = f_{i,1}(x) - f_{i,2}(x)$ on M is given by the function $\varphi_i^M(x) = f_{i,1}(x) - \ell_i^M(x)$ where $\ell_i^M(x)$ denotes the affine function that agrees with $f_{i,2}(x)$ at every vertex v of M.

In many cases, conical subdivision is more convenient than simplicial or rectangular subdivisions. In conical algorithms (which date back to the middle 1960s) the space is partitioned into cones with a common vertex x^0 and having each exactly n edges. If the objective function $f_0(x)$ is concave and

$\bar{x}$ is the current incumbent, while $\pi(x - x^0) \leqslant 1$ is the halfspace containing x^0 and bounded by the hyperplane passing through the n intersection points of the edges of M with the surface $f_0(x) = f_0(\bar{x})$, then, obviously, $f_0(x) \geqslant f_0(\bar{x})$ for every $x \in M$ satisfying $\pi(x - x^0) \leqslant 1$, so the halfspace $\pi(x - x^0) \leqslant 1$ cuts off a region of M no longer of interest for us. (This halfspace is often referred to as a *concavity cut* at x^0.) Therefore, a lower bound of $f_0(x)$ over the feasible points in M can be computed by considering only the feasible portion in $M \cap \{x \mid \pi(x - x^0) \geqslant 1\}$ (see [11,20]).

Various methods of estimating lower bounds have been proposed in the literature, each trying to exploit the specific structure of the problem under study. Aside form convex relaxation, most bounding methods use Lagrange relaxation, dualization, cutting and range reduction techniques.

While for discrete optimization branch and bound algorithms are always finite, for global optimization they converge only under certain consistency conditions between branching and bounding. Furthermore, the convergence speed may depend upon the branching rule, more precisely upon the way a partition set is further partitioned.

It should also be noted that the computational burden of a branch and bound process usually increases exponentially with the dimension of the space in which branching is performed. Therefore, for the efficiency of a branch and bound procedure, it is important to have branching performed in a space of lowest possible dimension. For example, if a problem becomes convex when certain variables x_i $(i = 1, \ldots, p)$, with $p < n$, are fixed (these are called *complicating variables*), one should try to branch upon x_i $(i = 1, \ldots, p)$ and not upon all $x_1, \ldots, x_n$.

2.2. Monotonic optimization

Monotonicity with respect to some variables (partial monotonicity) or to all variables (total monotonicity) is a natural property exhibited by many problems encountered in mathematical modeling of real-world systems in a broad range of economic, engineering and other activities. To provide a tool for the numerical study of these problems a number of "monotonicity principles" have been formulated whose usefulness has been demonstrated in quite a few papers on optimal design (see [17]). Of particular interest are the cases when monotonicity is coupled with convexity or reverse convexity, as it happens in *multiplicative programming* [9], *C-programming*, and, more generally, in so-called *low-rank nonconvex problems* [13], i.e., roughly speaking, problems with relatively few "complicating variables". During the last decade, parametric methods and other duality-based decomposition approaches have been developed that can now solve these problems rather fast, provided the number of complicating variables is reasonably small.

The most difficult monotonic optimization problems are those in which the monotonic structure does not involve any partial convexity or reverse convexity. These are problems of the form

$$\min\{f(x) \mid g(x) \leqslant 1 \leqslant h(x), \ x \in \mathbb{R}_+^n\}, \tag{6}$$

where every function involved is only supposed to be *increasing*, i.e., monotonely nondecreasing on every ray in the nonnegative orthant (so $f(x)$, say, is *increasing* if $f(x') \geqslant f(x)$ whenever $0 \leqslant x \leqslant x'$). Under some additional assumptions, by considering abstract convexity, certain special cases of this problem can be tackled by a so-called generalized outer approximation strategy. However, the most important advantage offered by the "pure" monotonic structure is that it provides global information which can be used to simplify the problem by limiting the global search to a much restricted region of the feasible domain. In fact, as the objective function in (6) is increasing, once a feasible point z

is known one can ignore the whole orthant $z + \mathbb{R}^n_+$ because no better feasible solution can be found in this set. Analogously, as the function $g(x)$ ($h(x)$, respectively) is increasing, once a point z is known to be infeasible to the constraint $g(x) \leqslant 1$ ($h(x) \geqslant 1$, respectively), the whole orthant $z + \mathbb{R}^n_+$ (the whole rectangle $0 \leqslant x \leqslant z$, respectively) can be discarded from further consideration. Based on these observations, efficient methods of outer approximation or branch and bound type can be devised for handling monotonicity.

A set G is said to be *normal* if it is of the form $G = \bigcup_{z \in Z} [0, z]$ (union of a collection of boxes $[0, z]$, $z \in Z$), which is the case if there exists an increasing function $g(x)$ such that $G = \{x \in \mathbb{R}^n_+ \mid g(x) \leqslant 1\}$. If Z is finite, the normal set is called a *polyblock*. Just as a compact convex set is the intersection of a nested sequence of polytopes, a compact normal set is the intersection of a nested sequence of polyblocks. Using this fact, a characterization of the structure of the solution set of a monotonic system can be established which allows efficient numerical analysis of monotonic inequalities and monotonic optimization problems. More importantly, this polyblock approximation method can be extended to solve optimization problems involving *differences* of *increasing* functions (*d.i. functions*), i.e., problems of form (1), where all the functions $f_{i,1}, f_{i,2}$ are increasing. Since any polynomial of n variables can be written as a difference of two polynomials with positive coefficients, i.e., a difference of two increasing functions on $\mathbb{R}^n_+$, it follows from Weierstrass Theorem that the set of d.i. functions on $[0, b] = \{x \in \mathbb{R}^n \mid 0 \leqslant x \leqslant b\}$ is dense in $C_{[0, b]}$. Therefore, the range of applicability of d.i. optimization includes polynomial programming (in particular nonconvex quadratic programming) as well as many other classes of global and combinatorial optimization problems.

2.3. Quadratic and polynomial programming

A quadratic program is a problem (1) in which all the functions $f_i(x)$ ($i = 0, 1, \ldots, m$) are quadratic, i.e.,

$$f_i(x) = \tfrac{1}{2} x^{\mathrm{T}} Q^i x + x^{\mathrm{T}} c^i + d_i. \tag{7}$$

The importance of quadratic programs stems from several facts.

1. Quadratic functions are the simplest smooth functions whose derivatives are readily available and easy to manipulate.
2. Any twice differentiable function can be approximated by a quadratic function in the neighborhood of a given point, so in a sense quadratic models are the most natural.
3. Numerous applications in economics, engineering, and other fields lead to quadratic nonconvex optimization problems. Furthermore, many combinatorial problems can be reformulated as quadratic programs because any set of 0–1 constraints like $x_i \in \{0, 1\}$ ($i = 1, \ldots, p$) is equivalent to the set of quadratic constraints $\sum_{i=1}^{p} x_i(x_i - 1) \geqslant 0$, $0 \leqslant x_i \leqslant 1$ ($i = 1, \ldots, p$) (see Section 2.4.1).

A few nonconvex quadratic programs can be solved by quite efficient algorithms. Among these, the most noticeable are problems with a low nonconvexity rank, including multiplicative programs [9], and also problems with at most one local nonglobal optimal solution, such as the problem of minimizing an indefinite quadratic function over an ellipsoid. Aside from these few exceptions, nonconvex quadratic programs are, as a rule, very hard problems for which the most suitable approach seems to be branch and bound. As we argued, a basic issue in branch and bound methods is to compute a lower bound of f_M^* (see (4)) for any given partition set M. This is usually achieved through a relaxation of the subproblem (4). Since a quadratic function $f_i(x)$ can be written as an explicit d.c. function $f_i(x) = g_i(x) - r_i\|x\|^2$ where r_i is the spectral radius of its defining matrix Q_i, a convex minorant of $f_i(x)$ over a rectangle $M = [p, q]$ is

$$\varphi_i^M(x) = f_i(x) + r_i\|x\|^2 - r_i \sum_{j=1}^{n} (x_j - p_j)(x_j - q_j). \tag{8}$$

A convex relaxation of (4) is thus obtained by substituting $\varphi_i^M(x)$ for $f_i(x)$ $(i = 0, 1, \ldots, m)$. Alternatively, if r_i is not readily available, one can observe that $x^T Q^i x = \sum_{j,k} Q_{j,k}^i x_j x_k$, so a convex minorant of $f_i(x)$ on $[p, q]$ can also be obtained by replacing each nonlinear term $x_j x_k$ (or $-x_j x_k$, respectively) by its convex envelope on $[p_j, q_j] \times [p_k, q_k]$, i.e., by $\max\{p_k x_j + p_j x_k - p_j p_k, q_k x_j + q_j x_k - q_j q_k\}$ $(-\min\{q_k x_j + p_j x_k - p_j q_k, p_k x_j + q_j x_k - q_j p_k\}$, respectively).

Aside from convex relaxation which is easy to obtain, but not always efficient, several other relaxations have been proposed: Lagrange relaxation, reformulation-convexification (RC), and semidefinite programming (SDP) relaxation. In these relaxations, one assumes that all the constraints are quadratic, which is innocuous because if $M = [p, q]$ and $X \cap M = \{x \mid x^T a^i \leqslant \alpha_i, \ i = 1, \ldots, r\}$ then the constraint $x \in X \cap M$, is equivalent to

$$\left. \begin{array}{l} (x^T a^i - \alpha_i)(x_j - p_j) \leqslant 0 \\ (x^T a^i - \alpha_i)(q_j - x_j) \leqslant 0 \end{array} \right\} \quad (i = 1, \ldots, r; \ j = 1, \ldots, n). \tag{9}$$

By writing system (9) as $f_i(x) \leqslant 0$ $(i = m + 1, \ldots, N)$, the Lagrangian of problem (4) is $L(x, u) = f_0(x) + \sum_{i=1}^{N} u_i f_i(x)$. It is well known that $\psi(u) \equiv \inf_{x \in \mathbb{R}^n} L(x, u) \leqslant f_M^*$ for every $u \geqslant 0$, hence $\sup_{u \geqslant 0} \psi(u) \leqslant f_M^*$ and since it can easily be seen that $\psi(u) = -\infty$ if $L(x, u)$ is nonconvex in x, one has

$$\beta_{LR}(M) = \sup_{u \geqslant 0:\, L(x,u) \text{ is convex in } x} \ \inf_{x \in \mathbb{R}^n} L(x, u) \leqslant f_M^*.$$

Taking account of (7) it can be proved that this bound equals

$$\max\left\{ t \left| \begin{pmatrix} Q_0 & c_0 \\ c_0^T & d_0 - t \end{pmatrix} + u_1 \begin{pmatrix} Q_1 & c_1 \\ c_1^T & d_1 \end{pmatrix} + \cdots + u_N \begin{pmatrix} Q_N & c_N \\ c_N^T & d_N \end{pmatrix} \succcurlyeq 0, \ u \geqslant 0 \right. \right\}, \tag{10}$$

where for any symmetric $n \times n$ matrix Q the notation $Q \succcurlyeq 0$ means that Q is positive semidefinite. Note that if $Q_0, Q_1, \ldots, Q_N$ are symmetric $n \times n$ matrices, then an inequality of the form $Q(x) \equiv Q_0 + \sum_{j=1}^{N} x_j Q_j \succcurlyeq 0$, referred to as a *linear matrix inequality*, is actually a convex inequality, since $\{x \mid Q(x) \succcurlyeq 0\} = \bigcap_{y \in \mathbb{R}^n} \{x \mid y^T Q(x) y \geqslant 0\}$. Therefore, (10) is a convex program, called a *semidefinite program* (SDP).

Introducing new variables $w_{jk} = x_j x_k$ and substituting w_{jk} for $x_j x_k$ in the expanded form of $f_i(x)$ we can write (4) as

$$\min\{L_0(x, w) \mid L_i(x, w) \leqslant 0 \ (i = 1, \ldots, N), w_{jk} = x_j x_k \ (j, k = 1, \ldots, n)\}, \tag{11}$$

where $L_i(x, w)$ are affine functions of (x, w). Therefore, a lower bound of (4) is also given by

$$\beta_{RL}(M) = \min\{L_0(x, w) \mid L_i(x, w) \leqslant 0, \ i = 1, \ldots, N\}$$

(reformulation-linearization relaxation). If we denote by W the $n \times n$ matrix with elements w_{jk} then the condition $w_{jk} = x_j x_k$, i.e., $W = xx^T$ is equivalent to

$$W \succcurlyeq xx^T, \quad \mathrm{tr}(W - xx^T) \leqslant 0,$$

where the first inequality is convex and the second reverse convex. Using this observation more refined SDP relaxations have been proposed in the literature. The increasing interest in SDP relaxations is motivated by the fact that more and more efficient interior point methods can be developed for solving large-scale SDPs.

The above relaxation methods can be extended to polynomial programs

$$\min\{P_0(x) \,|\, P_i(x) \leqslant 0 \ (i = 1, \ldots, m), \ x \in X\},$$

where all P_i $(i = 0, 1, \ldots, m)$ are polynomials. It should be noted, however, that a polynomial program is actually a d.i. optimization problem since each polynomial is the difference of two polynomials with positive coefficients. Therefore, polynomial programs can also be solved by the above-mentioned (polyblock approximation) method of monotonic optimization.

2.4. Global optimization approaches to discrete problems

Discrete (or combinatorial) optimization problems, that is, problems with a discrete feasible domain and/or a discrete domain objective function, model a large spectrum of applications in computer science, operations research and engineering.

Solution methods for discrete optimization problems can be classified into combinatorial and continuous approaches. A typical combinatorial approach generates a sequence of states, which represent a partial solution, drawn from a discrete finite set. Continuous approaches for solving discrete optimization problems are based on different equivalent characterizations in a continuous space. These characterizations include equivalent continuous formulations, or continuous relaxations (including semidefinite programming), that is, embeddings of the discrete domain in a larger continuous space.

There are many ways to formulate discrete problems as equivalent continuous problems or to embed the discrete feasible domain in a larger continuous space (relaxation). The surprising variety of continuous approaches reveal interesting theoretical properties which can be explored to develop new algorithms for computing (sub)optimal solutions to discrete optimization problems.

2.4.1. Equivalence of mixed integer programming and LCP

The simplest nonconvex constraints are the 0–1 integer constraints. Integer constraints are equivalent to continuous nonconvex constraints. For example, $z \in \{0, 1\} \Leftrightarrow z + w = 1, z \geqslant 0, w \geqslant 0, zw = 0$ or in another approach $z \in \{0, 1\} \Leftrightarrow z - z^2 = z(1 - z) = 0$. Therefore, it seems that there is no significant difference between discrete and continuous optimization. However, there is a considerable difference (in terms of problem complexity) between convex and nonconvex optimization problems. Next, we show that the mixed integer feasibility problem is equivalent to the complementarity problem. The complementarity conditions which are present in optimality conditions reveal deep connections with discrete optimization.

We consider the general linear complementarity problem (LCP) of finding a vector $x \in \mathbb{R}^n$ such that

$$Mx + q \geqslant 0, \quad x \geqslant 0, \quad x^T Mx + q^T x = 0,$$

(or proving that such an x does not exist) where M is an $n \times n$ rational matrix and $q \in \mathbb{R}^n$ is a rational vector. For given data M and q, the problem is generally denoted by LCP(M,q). The LCP unifies a number of important problems in operations research. In particular, it generalizes the primal–dual linear programming problem, convex quadratic programming, and bimatrix games.

For the general matrix M, where $S = \{x \mid Mx + q \geq 0,\ x \geq 0\}$ can be bounded or unbounded, the LCP can always be solved by solving a specific 0–1, linear, mixed-integer problem with n zero-one variables. Consider the following mixed 0–1 integer problem (MIP):

$$\max_{\alpha, y, z} \alpha$$

$$\text{s.t.} \quad 0 \leq My + \alpha q \leq e - z,$$

$$\alpha \geq 0,\ 0 \leq y \leq z,$$

$$z \in \{0,1\}^n,$$

where $e \in \mathbb{R}^l$ is the vector of all 1's. Let (α^*, y^*, z^*) be any optimal solution of (MIP). If $\alpha^* > 0$, then $x^* = y^*/\alpha^*$ solves the LCP. If in the optimal solution $\alpha^* = 0$, then the LCP has no solution. In fact, every feasible point (α, y, z) of (MIP), with $\alpha > 0$, corresponds to a solution of LCP (see [10]).

On the other hand, the mixed integer feasibility problem can be formulated as an LCP. Given matrices $A_{n \times n}$, $B_{n \times l}$ and a vector $b \in \mathbb{R}^n$ with rational entries, the mixed integer feasibility problem is to find (x, z), such that $x \in \mathbb{R}^n$, $x \geq 0$, $z \in \{0,1\}^l$ that satisfy $Ax + Bz = b$.

The condition $z_i \in \{0,1\}$ is equivalent to

$$z_i + w_i = 1, \quad z_i \geq 0, \quad w_i \geq 0, \quad z_i w_i = 0.$$

With this transformation z_i is a continuous variable and for each z_i a new continuous variable w_i is introduced. In addition, let $s, t \in \mathbb{R}^n$ be such that

$$s = Ax + Bz - b \geq 0, \qquad t = -Ax - Bz + b \geq 0.$$

The only way for these two inequalities to be satisfied is to have $s = t = 0$, which implies that $Ax + Bz = b$. Then, the mixed integer feasibility problem can be reduced to the problem of finding a solution of the LCP: Find v, y such that

$$v \geq 0, \quad y \geq 0, \quad v^T y = 0, \quad v = My + q,$$

where

$$y = \begin{pmatrix} z \\ x \\ \theta \end{pmatrix}, \quad v = \begin{pmatrix} w \\ s \\ t \end{pmatrix}, \quad M = \begin{pmatrix} -I & 0 & 0 \\ B & A & 0 \\ -B & -A & 0 \end{pmatrix}, \quad q = \begin{pmatrix} e \\ b \\ -b \end{pmatrix},$$

where $\theta \in \mathbb{R}^n$.

2.4.2. Satisfiability problems

The satisfiability problem (SAT) is central in mathematical logic, computing theory, and many industrial application problems (see [3]). Problems in computer vision, VLSI design, databases, automated reasoning, computer-aided design and manufacturing, involve the solution of instances of the satisfiability problem. Furthermore, SAT is the basic problem in computational complexity. Developing efficient exact algorithms and heuristics for satisfiability problems can lead to general approaches for solving combinatorial optimization problems.

Let $\mathscr{C}_1, \mathscr{C}_2, \ldots, \mathscr{C}_n$ be n clauses, involving m Boolean variables $x_1, x_2, \ldots, x_m$, which can take on only the values true or false (1 or 0). Define clause i to be

$$\mathscr{C}_i = \bigvee_{j=1}^{m_i} l_{ij},$$

where the literals $l_{ij} \in \{x_i, \bar{x}_i \mid i = 1, \ldots, m\}$, and $\bar{x}_i$ is the negation of x_i.

In the Satisfiability Problem in *Conjuctive Normal Form (CNF)*

$$F(x) \equiv \bigwedge_{i=1}^{n} \mathscr{C}_i = \bigwedge_{i=1}^{n} \left(\bigvee_{j=1}^{m_i} l_{ij} \right),$$

one is to determine the assignment of truth values to the m variables that satisfy all n clauses.

Given a *CNF* formula $F(x)$ from $\{0,1\}^m$ to $\{0,1\}$ with n clauses $C_1, \ldots, C_n$, we define a real function $f(y)$ from $\mathbb{R}^m$ to $\mathbb{R}$ that transforms the SAT problem into an unconstrained global optimization problem. Next we describe two different global optimization approaches.

Nondifferentiable unconstrained global optimization:

$$\min_{y \in \mathbb{R}^m} f(y),$$

where

$$f(y) = \sum_{i=1}^{n} c_i(y).$$

A clause function $c_i(y)$ is a product of m *literal functions* $q_{ij}(y_j)$ $(j = 1, \ldots, m)$:

$$c_i = \prod_{j=1}^{m} q_{ij}(y_j),$$

where

$$q_{ij}(y_j) = \begin{cases} |y_j - 1| & \text{if literal } x_j \text{ is in clause } C_i, \\ |y_j + 1| & \text{if literal } \bar{x}_j \text{ is in clause } C_i, \\ 1 & \text{if neither } x_j \text{ nor } \bar{x}_j \text{ is in } C_i. \end{cases}$$

The correspondence between x and y is defined as follows (for $i = 1, \ldots, m$):

$$x_i = \begin{cases} 1 & \text{if } y_i = 1, \\ 0 & \text{if } y_i = -1, \\ \text{undefined} & \text{otherwise.} \end{cases}$$

Clearly, $F(x)$ is true if and only if $f(y) = 0$ on the corresponding $y \in \{-1, 1\}^m$.

Polynomial unconstrained global optimization:

$$\min_{y \in \mathbb{R}^m} f(y)$$

where

$$f(y) = \sum_{i=1}^{n} c_i(y).$$

A clause function $c_i(y)$ is a product of m *literal functions* $q_{ij}(y_j)$ $(j = 1, \ldots, m)$:

$$c_i = \prod_{j=1}^{m} q_{ij}(y_j),$$

where

$$q_{ij}(y_j) = \begin{cases} (y_j - 1)^{2p} & \text{if } x_j \text{ is in clause } C_i, \\ (y_j + 1)^{2p} & \text{if } \bar{x}_j \text{ is in clause } C_i, \\ 1 & \text{if neither } x_j \text{ nor } \bar{x}_j \text{ is in } C_i, \end{cases}$$

where p is a positive integer.

The correspondence between x and y is defined as follows (for $i = 1, \ldots, m$):

$$x_i = \begin{cases} 1 & \text{if } y_i = 1, \\ 0 & \text{if } y_i = -1, \\ \text{undefined} & \text{otherwise.} \end{cases}$$

Clearly, $F(x)$ is true if and only if $f(y) = 0$ on the corresponding $y \in \{-1, 1\}^m$.

These models transform the SAT problem from a discrete, constrained decision problem into an unconstrained global optimization problem. A good property of the transformation is that these models establish a correspondence between the global minimum points of the objective function and the solutions of the original SAT problem. A *CNF* $F(x)$ is true *if and only if* $f(y)$ takes the global minimum value 0.

2.4.3. Minimax optimization

In recent years, new powerful techniques for minimax global optimization problems gave birth to new approaches for studying difficult combinatorial optimization problems [5]. Classical minimax theory initiated by Von Neumann, together with duality and saddle point analysis, has played a critical role in optimization, game theory and best approximation. However, minimax appears in a very wide area of disciplines. Recently, continuous minimax theory has been applied in many diverse problems such as Steiner trees, network flow, combinatorial group testing, and other combinatorial problems. The famous Gilbert and Pollak conjecture about Steiner trees was resolved using a new continuous minimax approach.

This new approach, based on a nontrivial new minimax result, was introduced in Du and Hwang's proof of the Steiner tree conjecture [4]. The center part of this approach is a new theorem about the following minimax problem:

$$\text{global} \min_{x \in X} \left(\max_{i \in I} f_i(x) \right),$$

where X is a convex region X in n-dimensional Euclidean space $\mathbb{R}^n$, I is a finite index set, and $f_i(x)$'s are continuous functions over X.

A subset Z of X is called an *extreme subset* of X if

$$\left. \begin{array}{l} x, y \in X, \\ \lambda x + (1 - \lambda)y \in Z \text{ for some } 0 < \lambda < 1, \end{array} \right\} \Rightarrow x, y \in Z.$$

With this definition, the Du–Hwang result can be stated as follows: Let $g(x) = \max_{i \in I} f_i(x)$. If every $f_i(x)$ is a concave function, then the minimum value of $g(x)$ over the polytope X is achieved at some point x^* satisfying the following condition:

There exists an extreme subset Z of X such that $x^* \in Z$ and the set $I(x^*)$ $(\equiv \{i \mid g(x^*) = f_i(x^*)\})$ is maximal over Z.

In addition, the following continuous version has been recently proved: Let $f(x, y)$ be a continuous function on $X \times Y$ where X is a polytope in $\mathbb{R}^m$ and Y is a compact set in $\mathbb{R}^n$. Let $g(x) = \max_{y \in Y} f(x, y)$. If $f(x, y)$ is concave with respect to x, then the minimum value of $g(x)$ over X is achieved at some point $\hat{x}$ satisfying the following condition:

There exists an extreme subset Z of X such that $\hat{x} \in Z$ and the set $I(\hat{x})$ $(\equiv \{y \mid g(\hat{x}) = f(\hat{x}, y)\})$ is maximal over Z.

As an example of how these results can be applied to combinatorial optimization problems, we mention the problem of packing circles in a square. What is the maximum radius of n equal circles that can be packed into a unit square? This problem is equivalent to the following: How should n points be arranged into a unit square such that the minimum distance between them is greatest? Let $x_1, x_2, \ldots, x_n$ be the n points. We can write the second problem in the following form:

$$\min_{x_i \in [0,1] \times [0,1]} \max_{1 \leqslant i < j \leqslant n} -\|x_i - x_j\|.$$

For fixed i and j, $\|x_i - x_j\|$ is clearly a convex function. Thus, the above result of Du and Hwang can be applied to it. New results have been obtained by using the above minimax formulation (see [16]).

2.5. General continuous optimization

The most challenging optimization problem is that of finding the global minimum (or maximum) of a continuous function $f(x)$ over a compact convex set $S \subset \mathbb{R}^n$. Despite its difficulty, this problem started to be investigated by a number of authors from the early 1970s. However, most methods in this period dealt with unconstrained minimization of smooth functions and were able to handle only problems of just one or two dimensions. Attempts to solve constrained problems of higher dimensions by deterministic methods have begun only in recent years.

To make the problem tractable, some further assumption, aside from continuity, is necessary. A quite common assumption is that the objective function $f(x)$ as well as the constraint functions $g_i(x) \leqslant 0$ $(i = 1, \ldots, m)$, defining the feasible set S satisfy a Lipschitz condition. A comprehensive review of *Lipschitz optimization* can be found in [9], where the best known methods for univariate as well as multivariate problems are discussed and compared experimentally. Note that if K is the Lipschitz constant of $f(x)$ then for any n-simplex M and any arbitrarily chosen $x_M \in M$ one has $f(x) \geqslant f(x_M) - K \|x - x_M\|$. Hence the affine function $\varphi_M(x)$ that agrees with $\psi(x) \equiv f(x_M) - K \|x - x_M\|$ at every vertex of M is an affine minorant of $f(x)$ on M satisfying $\sup_{x \in M} [f(x) - \varphi_M(x)] \to 0$ as diam $M \to 0$. Assuming for simplicity that S is a box, one can then solve the Lipschitz optimization problem by a simplicial branch and bound method in which a lower bound of $f_M^* \equiv \min\{f(x) \mid x \in M\}$ is computed by solving the relaxed linear problem $\min\{\varphi_M(x) \mid x \in M \cap S\}$. A similar rectangular branch and bound method is applicable when the function $f(x)$ is separable.

As mentioned earlier, the core of a global optimization problem is the subproblem of transcending local optimality, namely: given a local minimum $\bar{x}$, find a better feasible solution (i.e., escape from this local minimum), or show that $\bar{x}$ is a global minimum. In the so-called *modified function approaches*, this subproblem is solved by replacing the original function with a properly modified function such that a local search procedure, started from $\bar{x}$ and applied to the modified function will lead to a lower minimum, if there is one. However, a modified function satisfying the required conditions is very difficult to construct. In its two typical versions, this modified function (a *tunneling function* or a *filled function*) depends on parameters whose correct values, in many cases, can be determined only by trial and error. In a more recent method (TRUST, terminal repeller unconstrained subenergy tunneling [6]), the modified function combines two concepts, subenergy tunneling and non-Lipschitz terminal repeller, so as to transform the current local minimum of $f(x)$ into a global maximum while preserving all lower local minima. Thus, when gradient descent is applied to this modified function, the new system escapes the current local minimum to a lower valley of $f(x)$ with a lower local minimum. Benchmark results reported in [6] show that this method is faster and more accurate than previously reported techniques, although its successful implementation still heavily depends on the appropriate setting of parameters.

Another approach to continuous global optimization consists in generating a set of paths such that at least one global minimum is known a priori to lie on one of these paths. In most cases these paths are solution trajectories to ordinary differential equations of first or second order. For a review of *trajectory methods* that implement this multistart path following strategy, see [9]. A numerical implementation of an extended continuous Newton method, together with some experimental results, are described in [9] as well.

In contrast to modified function methods and trajectory methods which consider mostly unconstrained problems, the so-called *relief indicator method* [20] deals with continuous constrained global optimization. Using the fact already mentioned that any closed set in $\mathbb{R}^n$ is the solution set of a d.c. inequality, it is shown in [20] that the subproblem of transcending an incumbent $\bar{x}$ with $f(\bar{x}) = \alpha$ can be reduced to a concave minimization problem of the form $\min\{t - \|x\|^2 \mid h(\alpha, x) \leqslant t\}$, where $h(\alpha, x)$ is some convex function in x whose subdifferential can be easily computed in most cases of interest. The last property, together with the fact $h(\alpha, x) \leqslant h(\alpha' x)$ for $\alpha' \leqslant \alpha$, allow the problem to be solved either by outer approximation, or branch and bound or by a combination of both.

Experience has shown that for solving continuous optimization problems with little information available on the mathematical structure, branch and bound is usually the best approach. In the *interval methods* [9] based on branch and bound, bounds are obtained from an interval arithmetical evaluation of the functions involved, so that the solution data (minimizers, optimal value) is included in boxes at any stage of the algorithm. These techniques, first developed for unconstrained optimization, have been extended to constrained optimization, with a beneficial use of recent progress on subdivision strategies. One of the obvious advantages of the interval approach is that interval arithmetic provides a tool for estimating and automatically controlling all kinds of errors, especially rounding errors, truncation errors, etc.

3. Stochastic approaches

As mentioned in Section 2.5, the most challenging global optimization problems are problems without any known structure that can be used, so-called *black-box* optimization problems. In other words, the problem is to globally minimize (say) a continuous function f over a compact set $S \subseteq \mathbb{R}^n$. Stochastic methods, i.e., methods for which the outcome is random, are particularly suited for problems that possess no known structure that can be exploited. These methods generally require little or no additional assumptions on the optimization problem, at the expense of at most being able to provide a probabilistic convergence guarantee.

The three main classes of stochastic methods are: two-phase methods, random search methods, and random function methods. We will briefly review each of those classes, and discuss recent developments for algorithms in these three classes. We conclude with a brief description of metaheuristics for global optimization.

3.1. Two-phase methods

Two-phase methods consist of a *global phase*, in which the function is evaluated in a number of randomly sampled points in the feasible region, and a *local phase*, in which these sample points are manipulated, e.g., by means of local searches, to yield a candidate global optimum.

3.1.1. Multistart and its traditional variants

Most two-phase methods can be viewed as variants of the so-called *Multistart* algorithm. The global phase of this algorithm consists of generating a sample of points from a uniform distribution over the feasible region S. In the local phase a local search procedure is applied to each of these points, yielding various local optima. The best local optimum found is the resulting estimate of the global optimum. These methods are most successful for problems with relatively few local optima, and enough structure that efficient local search algorithms exist.

The global phase, without adding any local searches, is called the *Pure Random Search* (PRS) algorithm. The sequence of record values (i.e., the sequence of best function values) generated by this algorithm converges with probability one to the global optimum value. This fundamental results lies at the basis of asymptotic convergence results for many stochastic methods for global optimization. For example, it is easy to see that this result implies asymptotic convergence with probability one to the global optimum for the Multistart algorithm.

Many variants of the Multistart algorithm have been proposed to increase its efficiency — in particular by attempting to find each local optimum only once. Examples are *clustering methods*, which try to identify the different regions of attraction of the local optima, and start a local search only from a single point in each (estimated) region of attraction. Unfortunately, these methods often cannot be shown to be convergent. The *Multi-Level Single-Linkage* (MLSL) algorithm [9] is a method which combines the computational efficiency of clustering methods with the theoretical virtues of Mutlistart. The local search procedure is applied to every sample point, *except* if there is another sample point within some judiciously chosen critical distance with a better function value. The key is to choose the critical distance in such a way that the method converges with minimal effort. For the large class of global optimization problems possessing only a finite number of local optima, a critical distance (as a function of the iteration number) was derived in such a way that (i) the total number of local searches started by the algorithm is finite with probability one, and (ii) any local optimum will be found within a finite number of iterations with probability one. Recently, it has been shown that the assumptions underlying the MLSL algorithm can be relaxed while retaining the same theoretical properties.

3.1.2. Random Linkage

A major limitation of the MLSL method is that the theoretical results only hold if starting a local search from a point that is near the boundary of the feasible region is disallowed. This can be significant in higher dimensions, where almost all sample points are near the boundary of the feasible region. Another disadvantage is the fact that the complete sample of candidate points needs to be stored for the duration of the algorithm, since the algorithm may revise an initial decision not to start a local search from a given sample point in a later stage.

The class of *Random Linkage* algorithms, introduced in [15], overcomes both these disadvantages. The algorithms proceed by, in each iteration, sampling a single point from the uniform distribution over S. Then, a local search is started from that point with a probability depending on the distance to the closet point with better objective function value. For the case of global minimization, the algorithm then reads:

Random Linkage

Step 0: Set $k = 0$.
Step 1: Sample a single point X_{k+1} from the uniform distribution over S.
Step 2: Start a local search from X_{k+1} with probability

$$p_k(\delta_k(X_{k+1}))$$

with

$$\delta_k(x) \equiv \max\{\|x - X_j\| : j = 1, \ldots, k, \ f(X_j) < f(x)\}.$$

Step 3: Increment k and return to Step 1.

Here $\{p_k\}$ is a family of nondecreasing acceptance probability functions with $p_k(0) = 0$ and $p_k(x) \leqslant 1$ for all $x > 0$. This class of algorithms includes PRS and Multistart, and can be chosen to approximate MLSL. Conditions on the class of acceptance probability functions so that the total number of local searches applied is finite, while retaining asymptotic convergence to the global optimum, can be derived. Note that the property of MLSL that any local optimum is found within a finite number of iterations with probability one is not shared by Random Linkage. However, this can be viewed as an advantage rather than a disadvantage since the only local optimum that we are really interested in finding is the global one!

3.2. Random search methods

The class of random search methods consists of algorithms which generate a sequence of points in the feasible region following some prespecified probability distribution, or sequence of probability distributions. These algorithms are very flexible, in that they can even be applied to ill-structured problems for which no efficient local search procedures exist. In addition, they can be very successful in the early stages of studying a class of (practical) problems, before investing time in studying and exploiting the structural properties of the class of problems under consideration.

The most basic algorithm from this class is the PRS algorithm mentioned in the previous section. More sophisticated methods adaptively update the distribution from which a sample point is generated, based on the observed sample points.

3.2.1. Adaptive search methods

On a conceptual level, various Adaptive Search algorithms have been introduced. These algorithms are conceptual in the sense that no efficient implementation exists as yet. However, the theoretical results that can be obtained for these algorithms are interesting in their own right, and they have inspired new, or provided theoretical support for existing, practical approaches to global optimization.

The first algorithm from this class is *Pure Adaptive Search* (PAS). It differs from PRS only by forcing improvement in each iteration. In particular, each iteration point is generated from the level set corresponding to the previous iteration point. This conceptually simple modification of PRS has the property that the number of iteration points needed to approximate the global optimum increases only *linearly* in the dimension of the problem (for the class of Lipschitz global optimization problems over a convex domain, see [21]), as opposed to the *exponentially* increasing number of iteration points needed by PRS. Recently, it was shown that it even suffices to *approximate* the uniform distribution over the improving level set in each iteration, thereby showing that the complexity of PAS is only marginally worse (by a factor equal to the dimension of the problem) than the complexity of generating an approximately uniformly distributed point in a given set.

A related class of algorithms is simply called *Adaptive Search*. In these algorithms points are sampled from the entire feasible region in each iteration, but the distribution from which they are sampled changes adaptively. This class of algorithms has been shown to inspire efficient simulated annealing algorithms (which in fact is a random search algorithm itself, but will be discussed in Section 3.4 on metaheuristics). Another variant of PAS is called *Hesitant Adaptive Search*. This conceptual algorithm has been introduced as a theoretical analyzable approximation of *Pure Localization Search*. This algorithm is, in spirit, a randomized analogue of the Piyavskii–Shubert algorithm and its higher dimensional extensions, and in itself is an attempt at finding an efficiently implementable approximation of PAS.

The basic idea behind localization search algorithms is to avoid the difficult problem in PAS of generating an iteration point from the improving level set. As an approximation, a point is generated uniformly from a superset of the improving level set, where this superset has the desirable property that a uniformly distributed point in that set can be efficiently generated. The challenge is of course to find a close enough approximation of the level set while retaining computational efficiency of the sampling step.

3.3. Random function methods

An interesting alternative for the above methods is the random function approach. In the random function approach the objective function f is assumed to be a sample path of an a priori defined stochastic process, which is defined as a probability distribution on a class of functions. Then any question which can be asked about f can just as readily be asked about its random counterpart. For example, the stochastic process implicitly defines probability distributions of various interesting quantities, such as the number of local optima, and the joint distribution of the location and function value of the global optimum, or the size of its region of attraction. The, often irreconcilable, dilemma in this approach is that, on the one hand, an a priori stochastic process should be specified which is consistent with known properties of f, such as, for example, continuity or differentiability. On the other hand, the process should be mathematically tractable.

The random function approach has proven quite unsuccessful in solving traditional global optimization problems, where it is unable to compete with deterministic methods, or conceptually simpler stochastic methods. One of the reasons is that the determination of the next candidate point to evaluate is itself a global optimization problem, and as such difficult to solve. However, it has recently been shown (see [12]) that the method can quite successfully be applied to global optimization problems where the evaluation of the objective function is very expensive. This is frequently the case in industrial design problems, where the objective function value is often the result of an extensive and expensive simulation. The main problem with the stochastic method mentioned above does not apply here, since the global optimization problem that needs to be solved to find the next iteration point, although difficult, is much easier than a single objective function evaluation.

As mentioned above, an important choice to be made in the random function approach is the stochastic process used. Such a stochastic process is mainly characterized by a correlation function between the function values at each pair of points in the feasible region. Often, this covariance function is of the form

$$R(x^1, x^2) = \exp(-d(x^1, x^2)),$$

where d is a distance function. Common choices are

$$d(x^1, x^2) = -\tfrac{1}{2}\|x^1 - x^2\|^2,$$

$$d(x^1, x^2) = -\|x^1 - x^2\|.$$

Flexibility can be added to this by choosing

$$d(x^1, x^2) = -\sum_{i=1}^{n} \theta_i |x_i^1 - x_i^2|^{p_i},$$

where $\theta_i \geqslant 0$ are scaling parameters, and $p_i \in [1, 2]$ are smoothness parameters (see [12]). Suitable values of these parameters are estimated during the course of the algorithm. The next iteration point is chosen by globally maximizing the expected improvement over the current record value that will be made in the next iteration. Finally, a validation scheme is suggested to test the suitability of the chosen stochastic process. A failure of this test indicates that the optimization problem (in particular, the objective function) should be transformed to yield better performance of the algorithm.

3.4. Metaheuristics

Metaheuristics are methods that are often based on processes observed in physics or biology. Such heuristics have proven very successful in the last one or two decades in solving hard combinatorial optimization problems. With the exception of simulated annealing, their application to global optimization problems is fairly limited. In the next sections we will discuss three main categories of metaheuristics: simulated annealing, tabu search, and genetic algorithms.

3.4.1. Simulated annealing

Simulated annealing is a random search technique that avoids getting trapped in local minima by accepting, in addition to transitions corresponding to an improvement in objective function value, also transitions corresponding to a worse objective function value. The latter is done in a limited way by means of a probabilistic acceptance criterion, such that the probability of accepting a deterioration decreasing as the algorithm progresses. The deteriorations make it possible to move away from local optima. This method originated from an analogy with the physical annealing process of finding low-energy states of a solid in a heat bath. The advantage of simulated annealing is that it is very easily implementable, robust, and applicable to a very general class of global optimization problems. The simulated annealing algorithm, applied to a minimization problem, reads:

Simulated Annealing

Step 0: Choose $X_0 \in S$ and $T_0 \in R_+$, and set $k = 0$ and $y_0 = f(X_0)$.
Step 1: Generate a point x according to some distribution $R(X_k, \cdot)$ on S.
Step 2: With probability

$$\min(1, e^{(f(X_k) - f(x))/T_k})$$

set $X_{k+1} = x$. Otherwise, set $X_{k+1} = X_k$.
Step 3: Choose T_{k+1}, increment k and return to Step 1.

The most challenging aspect of simulated annealing has been to theoretically support the proposed algorithms by providing an asymptotic convergence guarantee. A necessary condition for convergence of a simulated annealing algorithm is that the cooling schedule in Step 3 converges to zero. We can distinguish between *deterministic* cooling schedules, where the value of T_{k+1} depends on k only, and *adaptive* cooling schedules, where the value of T_{k+1} is random and depends on the iteration points generated so far by the algorithm. The choice for a particular cooling schedule is difficult issue, and should be problem dependent.

The Hide-and-Seek algorithm (see [19]) was the first simulated annealing algorithm for global optimization for which asymptotic convergence was proven formally. The first general result can be found in [1], where convergence with probability one to the global optimum is proved under very mild conditions on the cooling schedule, for simulated annealing algorithms where the candidate point generator has *global reach*, i.e., any subset of the feasible region with positive Lebesgue measure can be reached with positive probability from any iteration point. More recently, significant contributions have been reported in [14], where conditions under which simulated annealing algorithms that do not exhibit global reach converge to the global optimum are derived.

Although much progress has been made, the main problem remains that it is often very difficult to choose the parameter values of a particular simulated annealing algorithm in such a way that convergence can be guaranteed, even if sufficient conditions for convergence exists. A main challenge is thus to make the sufficient conditions more explicit for given algorithms and problems instances.

3.4.2. Tabu search

The tabu search technique has been very successfully applied to combinatorial optimization problems (see [7]). It was basically designed as a deterministic algorithm, but can easily be randomized, and thereby be viewed as an algorithm related to simulated annealing. The idea is that, the set of all candidate solutions that can be generated in a given iteration, should not only depend on the current iteration point, but should be modified by excluding a subset of candidate solutions that are *tabu* (taboo). The definition of which candidate solutions are tabu depends on moves that have been made between recent iteration points.

There are relatively few examples of applications of this method to continuous optimization problem. A notable exception is [2], where successful experimental results are reported on a set of widely used global optimization test problems.

Although it has computationally proven successful, the major disadvantage of the tabu search technique (both for combinatorial and continuous optimization) remains the lack of convergence results. The main challenge in this area is therefore to address this issue, thereby lifting the status of this technique from a purely heuristic one to an algorithm that will eventually reach the global optimum solution.

3.4.3. Genetic algorithm

Genetic algorithms are based on the idea of *survival of the fittest* that is observed in nature. In an optimization setting, a *population* of candidate points is manipulated by means of selection, crossover, and mutation operators. In the selection phase, certain members of the population are identified that will generate offspring. The crossover operator is applied to a pair of selected population members to create offspring, and the mutation operator is used as a slight modification of this offspring, or of remaining members of the population (see [8]).

Like the class of simulated annealing algorithms, this class of algorithms originated as a technique applied to combinatorial optimization algorithms. Application of this technique to continuous optimization problem is relatively limited. A major unresolved issue is how to encode feasible solutions. A straightforward generalization of the genetic algorithms for discrete optimization would call for a binary encoding of solutions. However, recent results suggest that the traditional real encoding may be superior to a binary encoding.

As for tabu search methods, the main challenge in this area is to provide some kind of convergence guarantee. Until that time, genetic algorithms will remain heuristics for solving global optimization problems.

4. Concluding remarks

In this paper we have described the state of the art and recent trends and development in certain areas of global optimization. We have shown examples of classes of global optimization problems for which the structure can successfully be exploited to yield efficient solution techniques. We have also shown that the idea of solving binary or mixed-integer programming problems as continuous global optimization problems is quite promising. Finally, we have reviewed stochastic methods for global optimization, which should mainly be applied to ill-structured problems or problems that do not exhibit any known structure, indicated some recent developments, and identified their major shortcomings and issues for future research.

References

[1] C.J.P. Bélisle, Convergence theorems for a class of simulated annealing algorithms on $\mathbb{R}^d$, J. Appl. Probab. 29 (4) (1992) 885–895.

[2] D. Cvijović, J. Klinowski, Taboo search: an approach to the multiple minima problem, Science 267 (1995) 664–666.

[3] D.-Z. Du, J. Gu, P.M. Pardalos (Eds.), Satisfiable Problem: Theory and Applications, DIMACS Series, Vol. 35, American Mathematical Society, Providence, RI, 1997.

[4] D.-Z. Du, F.K. Hwang, An approach for proving lower bounds: solution of Gilbert–Pollak's conjecture on Steiner ratio, Proceedings of the 31st FOCS, 1990, pp. 76–85.

[5] D.-Z. Du, P.M. Pardalos (Eds.), Minimax and Applications, Kluwer Academic Publishers, Dordrecht, The Netherlands, 1994.

[6] C.A. Floudas, P.M. Pardalos (Eds.), State of the Art in Global Optimization, Kluwer Academic Publishers, Dordrecht, The Netherlands, 1996.

[7] F. Glover, M. Laguna, Tabu Search, Kluwer Academic Publishers, Boston, MA, 1997.

[8] D. Goldberg, Genetic Algorithms in Search, Optimization, and Machine Learning, Addison-Wesley, San Mateo, CA, 1989.

[9] R. Horst, P.M. Pardalos (Eds.), Handbook of Global Optimization, Kluwer Academic Publishers, Dordrecht, The Netherlands, 1995.

[10] R. Horst, P.M. Pardalos, N.V. Thoai, Introduction to Global Optimization, Kluwer Academic Publishers, Dordrecht, The Netherlands, 1995.

[11] R. Horst, H. Tuy, Global Optimization (Deterministic Approaches), 3rd ed., Springer, Berlin, Germany, 1994.

[12] D.R. Jones, M. Schonlau, W.J. Welch, Efficient global optimization of expensive black-box functions, J. Global Optim. 13 (4) (1998) 455–492.

[13] H. Konno, P.T. Thach, H. Tuy, Optimization on Low Rank Nonconvex Structures, Kluwer Academic Publishers, Dordrecht, The Netherlands, 1997.

[14] M. Locatelli, convergence properties of simulated annealing for continuous global optimization, J. Appl. Probab. 33 (1996) 1127–1140.

[15] M. Locatelli, F. Schoen, Random Linkage: a family of acceptance/rejection algorithms for global optimisation, Math. Programming 85 (2) (1999) 379–396.

[16] C. Maranas, C.A. Floudas, P.M. Pardalos, New results in the packing of equal circles in a square, Discrete Math. 142 (1995) 287–293.

[17] P. Papalambros, D.J. Wilde, Principles of Optimal Design — Modeling and Computation, Cambridge University Press, New York, 1986.

[18] P.M. Pardalos, H.E. Romeijn, H. Tuy, Recent developments and trends in global optimization, Research Report 99-15, Department of Industrial and Systems Engineering, University of Florida, Gainesville, FL, 1999.

[19] H.E. Romeijn, R.L. Smith, Simulated annealing for constrained global optimization, J. Global Optim. 5 (1994) 101–126.

[20] H. Tuy, Convex Analysis and Global Optimization, Kluwer Academic Publishers, Dordrecht, The Netherlands, 1998.

[21] Z.B. Zabinsky, R.L. Smith, Pure adaptive search in global optimization, Math. Programming 53 (3) (1992) 323–338.

![NH logo]

ELSEVIER

Journal of Computational and Applied Mathematics 124 (2000) 229–244

JOURNAL OF
COMPUTATIONAL AND
APPLIED MATHEMATICS

www.elsevier.nl/locate/cam

Numerical continuation methods: a perspective

Werner C. Rheinboldt

Department of Mathematics, University of Pittsburgh, Pittsburgh, PA 15260, USA

Received 1 June 1999; received in revised form 16 June 1999

Abstract

In this historical perspective the principal numerical approaches to continuation methods are outlined in the framework of the mathematical sources that contributed to their development, notably homotopy and degree theory, simplicial complexes and mappings, submanifolds defined by submersions, and singularity and foldpoint theory. © 2000 Elsevier Science B.V. All rights reserved.

1. Introduction

The term *numerical continuation methods*, as it is typically used, covers a variety of topics which — while related — exhibit also considerable differences. This is already reflected in some of the alternate terminology that has been used, such as *imbedding methods, homotopy methods, parameter variation methods*, or *incremental methods*, just to name a few.

In order to provide an overview from a historical viewpoint, it appears that the general structure of the area is illuminated best by focusing first on the principal underlying mathematical sources that have contributed to its development. Accordingly, in the first two sections we concentrate on (i) homotopy and degree theory, (ii) simplicial complexes and mappings, (iii) submanifolds defined by submersions, and (iv) singularity and foldpoint theory. Then the subsequent sections address some of the numerical approaches growing out of this theoretical basis. Since methods based on (ii) above; that is, notably, the piecewise linear methods, are covered in another article [5] in this volume, this area will not be addressed here any further. Clearly, in a brief article as this one, only the bare outlines of the theoretical and computational topics can be sketched and many aspects had to be left out. An effort was made to give references to sources that provide not only further details but also relevant bibliographic data to the large literature in the area.

E-mail address: wcrheint@pitt.edu (W.C. Rheinboldt).

0377-0427/00/$ - see front matter © 2000 Elsevier Science B.V. All rights reserved.
PII: S 0377-0427(00)00428-3

2. Theoretical sources: homotopies

An important task in many applications is the solution of nonlinear equations defined on finite- or infinite-dimensional spaces. In order to avoid technical details we restrict here the discussion to the finite-dimensional case

$$F(y) = 0, \quad F : \mathbb{R}^m \to \mathbb{R}^m. \tag{1}$$

The computational approximation of a solution $y^* \in \mathbb{R}^m$ of (1) typically requires the application of some iterative process. However, except in rare circumstances, such a process will converge to y^* only when started from a point (or points) in a neighborhood of the desired — but, of course, unknown — point y^*. In other words, for an effective overall solution process we need tools for localizing the area of the expected solution and for constructing acceptable starting data for the iteration.

Evidently, localization requires the determination of a suitably small domain which is guaranteed to contain a solution — hopefully the one we are interested in. This represents a problem about the existence of solutions of (1). Among the many approaches for addressing it, an important one — dating back to the second half of the 19th century — is the use of homotopies.

2.1. Homotopies and Brouwer degree

Let $\Omega \in \mathbb{R}^m$ be a given open set and $C(\bar{\Omega})$ the set of all continuous mappings from the closure $\bar{\Omega}$ into $\mathbb{R}^m$. Two members $F_0, F_1 \in C(\bar{\Omega})$ are *homotopic* if there exists a continuous mapping

$$H : \bar{\Omega} \times [0,1] \to \mathbb{R}^m \tag{2}$$

such that $H(y,0) = F_0(y)$, $H(y,1) = F_1(y)$ for all $y \in \bar{\Omega}$. This introduces an equivalence relation on $C(\bar{\Omega})$. The topic of homotopy theory is the study of properties of the functions in $C(\bar{\Omega})$ that are preserved under this equivalence relation.

Among the properties of interest is, of course, the solvability of Eq. (1) defined by homotopic members of $C(\bar{\Omega})$. An important tool is here the concept of the degree of a mapping. Without entering into historical details we mention only that the concept of a local degree; that is, a degree with respect to a neighborhood of an isolated solution, was introduced by L. Kronecker in 1869. The extension of this local concept to a degree in the large was given by L. Brouwer in 1912. Then, in 1934, the seminal work of J. Leray and J. Schauder opened up the generalization to mappings on infinite-dimensional spaces. We refer, e.g., to [2,33,26] for some details and references.

The Brouwer degree is by nature a topological concept but it can also be defined analytically. We sketch only the general idea. For any C^1-map F from some open set of $\mathbb{R}^n$ into $\mathbb{R}^m$ a vector $z \in \mathbb{R}^m$ is called a *regular value* of F if $DF(y)$ has maximal rank $\min(n,m)$ for all $y \in F^{-1}(z)$. Let Ω be a bounded set and consider a mapping $F \in C(\bar{\Omega}) \cap C^1(\Omega)$ and some regular value $z \in \mathbb{R}^m$ of F. Then the cardinality of $F^{-1}(z)$ must be finite and the *degree* of F with respect to Ω and y can be defined as

$$\deg(F, \Omega, z) := \sum_v \operatorname{sign} \det(DF(y^v)), \tag{3}$$

where the sum is taken over all $y^v \in F^{-1}(z)$. Now, appropriately defined approximations can be used to obtain an extended definition of the degree for any $F \in C(\bar{\Omega})$ and $b \notin F(\partial\Omega)$. For the details we refer, e.g., to [33] or [26].

The following result lists some relevant properties of the Brouwer degree.

Theorem 1 (Homotopy invariance). *Let* $\Omega \subset \mathbb{R}^m$ *be bounded and open.*

(i) *If* $deg(F, \Omega, z) \neq 0$ *for* $F \in C(\bar{\Omega})$ *and* $z \notin F(\delta\Omega)$ *then* $F(y) = z$ *has a solution in* Ω.

(ii) *If for some mapping* (2) *the restricted mappings* $H_t := H(\cdot, t)$ *are in* $C(\bar{\Omega})$ *for each* $t \in [0, 1]$ *and*

$$z \notin H_t(\partial\Omega), \quad \forall t \in [0, 1], \tag{4}$$

then $deg(H_t, \Omega, z)$ *is constant for* $t \in [0, 1]$.

These results show that we can deduce solvability properties for a map F_1 from corresponding known facts about another homotopic map F_0. This represents a powerful tool for the establishment of existence results and for the development of computational methods. In that connection, we note that condition (4) is indeed essential as the following example shows:

$$H : [-1, 1] \times [0, 1] \to \mathbb{R}^1, \quad H(y, t) := y^2 - \tfrac{1}{2} + t, \; z = 0. \tag{5}$$

Here $H_0(y) = 0$ has two distinct roots in $[-1, 1]$ while $H_1(y) = 0$ has none. Theorem 1(ii) does not hold because $0 \in H_1(\partial\Omega)$.

2.2. Simplicial approximations

The above homotopy results constitute a theoretical source of an important subclass of continuation methods, the so-called *piecewise linear methods*. In order to see this we begin with a summary of some basic definitions.

An k-dimensional simplex (or simply k-simplex), σ^k in $\mathbb{R}^n$, $n \geq k \geq 0$, is the closed, convex hull, $\sigma^k = co(u^0, \ldots, u^k)$, of $k + 1$ points $u^0, \ldots, u^k \in \mathbb{R}^n$ that are not contained in any affine subspace of dimension less than k. These points form the *vertex set* $vert(\sigma^k) = \{u^0, \ldots, u^k\}$ of σ^k. The *barycenter* of σ^k is the point $x = [1/(k+1)](u_0 + \cdots + u_k)$ and the *diameter* of σ^k is defined as $diam(\sigma^k) = \max\{\|u^j - u^i\|_2 : i, j = 0, \ldots, k\}$. An ℓ-simplex $\sigma^\ell \in \mathbb{R}^n$ is an ℓ-*face* of σ^k if $vert(\sigma^\ell) \subset vert(\sigma^k)$. The unique k-face is σ^k itself and the 0-faces are the vertices.

A (finite) *simplicial complex* of dimension k is a finite set $\mathscr{S}$ of k-simplices[1] in $\mathbb{R}^n$ with the two properties

(a) If $\sigma \in \mathscr{S}$ then all its faces belong to $\mathscr{S}$ as well,

(b) for $\sigma^1, \sigma^2 \in \mathscr{S}$, $\sigma_1 \cap \sigma_2$ is either empty or a common face.

For a simplicial complex $\mathscr{S}$, the *carrier* is the set $|\mathscr{S}| = \{x \in \mathbb{R}^n : x \in \sigma \text{ for some } \sigma \in \mathscr{S}\}$, and $vert(\mathscr{S}) = \{x \in \mathbb{R}^n : x \in vert(\sigma) \text{ for some } \sigma \in \mathscr{S}\}$ is the *vertex set*. Since $\mathscr{S}$ is assumed to be finite, the carrier $|\mathscr{S}|$ must be a compact subset of $\mathbb{R}^n$ and the diameter $diam(\mathscr{S})$ is well defined as the largest diameter of the simplices of $\mathscr{S}$.

---

[1] We exclude here complexes of simplices with different dimensions, usually permitted in topology.

Let $\mathscr{S}$ and $\mathscr{T}$ be two simplicial complexes of $\mathbb{R}^n$ (not necessarily of the same dimension). A mapping $K : \mathscr{S} \to \mathscr{T}$ is a *simplicial map* if it maps every simplex of $\mathscr{S}$ affinely onto a simplex of $\mathscr{T}$. Since an affine map of a simplex is fully defined by the images of its vertices, it follows that a simplicial map $K : \mathscr{S} \to \mathscr{T}$ is fully defined by specifying the image $K(x) \in \mathrm{vert}(\mathscr{T})$ of every $x \in \mathrm{vert}(\mathscr{S})$. Note that the images of different vertices of $\mathscr{S}$ need not be distinct. Clearly a simplicial map $K : \mathscr{S} \to \mathscr{T}$ induces a continuous mapping from $|\mathscr{S}|$ to $|\mathscr{T}|$.

A simplicial complex $\mathscr{T}$ is a *subdivision* of the simplicial complex $\mathscr{S}$ if $|\mathscr{S}| = |\mathscr{T}|$ and each simplex of $\mathscr{T}$ is contained in a simplex of $\mathscr{S}$. A subdivision of $\mathscr{S}$ is fully specified once subdivisions of each of its simplices are provided. For example, let $\mathrm{vert}(\sigma^k) = \{u^0, \dots, u^k\}$ be the vertex set of a k-simplex σ^k and u^* its barycenter. Then, for each j, $0 \leqslant j \leqslant k$, the k-simplex $\sigma^{(k,j)}$ with the vertex set $(\mathrm{vert}(\sigma^k) \setminus u^j) \cup u^*$ is contained in σ^k and the collection of the $k+1$ simplices $\sigma^{(k,j)}$, $j = 0, \dots, k$, forms a subdivision of σ^k, the so-called *barycentric subdivision*. Evidently, by repeated barycentric subdivision, complexes with arbitrarily small diameter can be generated. This holds also for various other types of subdivisions (see, e.g., [37]).

The following basic result about approximations of continuous mappings by simplicial mappings is proved, e.g., in [2]. It provides the intended connection with the results of the previous subsection.

Theorem 2. *Let $\mathscr{S}$ and $\mathscr{T}$ be simplicial complexes of $\mathbb{R}^n$ and $\mathbb{R}^m$, respectively, and suppose that $\{\mathscr{S}^r\}_{r=1}^{\infty}$ is a sequence of successive subdivisions of $\mathscr{S}$ for which the diameter tends to zero when $r \to \infty$. If $F : |\mathscr{S}| \to |\mathscr{T}|$ is a continuous mapping, then, for any $\varepsilon > 0$, there exists a sufficiently large r and a simplicial map $K_r : \mathscr{S}^r \to \mathscr{T}$ such that*

$$\max_{x \in |\mathscr{S}|} \|F(x) - K_r(x)\|_2 \leqslant \varepsilon.$$

Moreover, there is a continuous homotopy $H : |\mathscr{S}| \times [0,1] \to \mathbb{R}^m$ such that $H(x,0) = K_r(x)$, $H(x,1) = F(x)$.

In other words, a continuous mapping F between the carries of the complexes $\mathscr{S}$ and $\mathscr{T}$ can be approximated arbitrarily closely by a simplicial mapping between these complexes which, at the same time, is homotopic to F. Hence, in particular, Theorem 1 can be applied here. This represents, in essence, the theoretical basis of the mentioned piecewise linear continuation methods.

3. Theoretical sources: manifolds

In applications nonlinear equations (1) typically arise as models of physical systems which almost always involve various parameters. While some of these parameters can be fixed, for others we often may know only a possible range. Then interest centers in detecting any significant changes in the behavior of the solutions when these parameters are varied, as for instance, when a mechanical structure buckles.

Problems of this type require the changeable parameters to be incorporated in the specification of the equations. In other words, in place of (1), we have to consider now equations of the form

$$F(y, \lambda) = 0, \quad F : \mathbb{R}^m \times \mathbb{R}^d \to \mathbb{R}^m, \quad d > 0, \tag{6}$$

where $y \in \mathbb{R}^m$ typically represents a state vector and $\lambda \in \mathbb{R}^d$ is the parameter vector. In working with such systems, it is often desirable to combine the vectors y and λ into a single vector $x \in \mathbb{R}^n$ of dimension $n = m + d$. This means that (6) is written in the form

$$F(x) = 0, \quad F : \mathbb{R}^n \to \mathbb{R}^m \tag{7}$$

and that the *parameter splitting*

$$\mathbb{R}^n = \mathbb{R}^m \times \mathbb{R}^d, \quad x = (y, \lambda), \quad y \in \mathbb{R}^m, \quad \lambda \in \mathbb{R}^d \tag{8}$$

is disregarded.

For equations of form (7) (as well as (6)) it rarely makes sense to focus on the determination of a specific solution $x \in F^{-1}(0)$. Instead, as noted, interest centers on analyzing the properties of relevant parts of the solution set $\mathcal{M} = F^{-1}(0)$. In most cases, this set has the structure of a differentiable submanifold of $\mathbb{R}^n$. In the study of equilibrium problems in engineering this is often reflected by the use of the term 'equilibrium surface', although, rarely, any mathematical characterization of the manifold structure of $\mathcal{M}$ is provided.

3.1. Submanifolds of $\mathbb{R}^n$

In this section we summarize some relevant definitions and results about manifold and refer for details, e.g., to [1]. Here the dimensions n, m are assumed to be given such that $n = m + d$, $d > 0$, and ρ denotes a *positive* integer or ∞.

When $F : \mathbb{R}^n \to \mathbb{R}^m$ is of class C^ρ on a open set $\Omega \subset \mathbb{R}^n$, then F is an *immersion* or *submersion* at a point $x^0 \in \Omega$ if its first derivative $DF(x^0) \in \mathscr{L}(\mathbb{R}^n, \mathbb{R}^m)$ is a one-to-one mapping or a mapping onto $\mathbb{R}^m$, respectively. More generally, F is an immersion or submersion on a subset $\Omega_0 \subset \Omega$ if it has that property at each point of Ω_0.

We use the following characterization of submanifolds of $\mathbb{R}^n$.

Definition 3. A subset $\mathcal{M} \subset \mathbb{R}^n$ is a d-dimensional C^ρ-submanifold of $\mathbb{R}^n$ if $\mathcal{M}$ is nonempty and for every $x^0 \in \mathcal{M}$ there exists an open neighborhood $\mathscr{U}$ of x^0 in $\mathbb{R}^n$ and a submersion $F : \mathscr{U} \mapsto \mathbb{R}^m$ of class C^ρ such that $\mathcal{M} \cap \mathscr{U} = F^{-1}(0)$.

An equivalent definition utilizes the concept of a *local parametrization*:

Definition 4. Let $\mathcal{M}$ be a nonempty subset of $\mathbb{R}^n$. A local d-dimensional C^ρ parametrization of $\mathcal{M}$ is a pair $(\mathscr{U}, \phi)$ where $\mathscr{U} \subset \mathbb{R}^d$ is a nonempty open set and $\phi : \mathscr{U} \mapsto \mathbb{R}^m$ a mapping of class C^ρ such that

  (i) $\phi(\mathscr{U})$ is an open subset of $\mathcal{M}$ (under the induced topology of $\mathbb{R}^n$) and ϕ is a homeomorphism of $\mathscr{U}$ onto $\phi(\mathscr{U})$,
  (ii) ϕ is an immersion on $\mathscr{U}$.

If $x^0 \in \mathcal{M}$ and $(\mathscr{U}, \phi)$ is a local d-dimensional C^ρ parametrization of $\mathcal{M}$ such that $x^0 \in \phi(\mathscr{U})$, then $(\mathscr{U}, \phi)$ is called a local d-dimensional C^ρ parametrization of $\mathcal{M}$ near x^0.

Theorem 5. *A nonempty subset $\mathcal{M} \subset \mathbb{R}^n$ is a C^p-submanifold of $\mathbb{R}^n$ of dimension d if and only if for every $x^0 \in \mathcal{M}$ there exists a local d-dimensional C^p parametrization of $\mathcal{M}$ near x^0. When $\mathcal{M}$ is a d-dimensional C^p-submanifold of $\mathbb{R}^n$ then any local C^p parametrization of $\mathcal{M}$ is necessarily d-dimensional.*

The following result is central to our discussion.

Theorem 6 (Submersion Theorem). *Suppose that, for the C^p mapping $F : \mathbb{R}^n \to \mathbb{R}^m$ on the open set Ω, the set $\mathcal{M} = F^{-1}(0)$ is not empty and F is a submersion on $\mathcal{M}$. Then $\mathcal{M}$ is a d-dimensional C^p-submanifold of $\mathbb{R}^n$.*

In our setting we can define tangent spaces of submanifolds of $\mathbb{R}^n$ as follows:

Definition 7. Let $\mathcal{M}$ be a d-dimensional C^p-submanifold of $\mathbb{R}^n$. For any $x^0 \in \mathcal{M}$ the tangent space $T_{x^0}\mathcal{M}$ of $\mathcal{M}$ at x^0 is the d-dimensional linear subspace of $\mathbb{R}^n$ defined by

$$T_{x^0}\mathcal{M} := \mathrm{rge}\ D\phi(\phi^{-1}(x^0)), \tag{9}$$

where $(\mathcal{U}, \phi)$ is any local C^p parametrization of $\mathcal{M}$ near x^0. The subset $T\mathcal{M} = \bigcup_{x \in \mathcal{M}} [\{x\} \times T_x \mathcal{M}]$ of $\mathbb{R}^n \times \mathbb{R}^n$ is the tangent bundle of $\mathcal{M}$.

This definition is independent of the choice of the local parametrization. In the setting of the Submersion Theorem 6 this follows directly from the fact that then $T_{x^0}\mathcal{M} = \ker DF(x^0)$. The following result provides a basis for the computational evaluation of local parametrizations.

Theorem 8. *Under the conditions of the submersion theorem on the mapping F let $U \in \mathcal{L}(\mathbb{R}^d, \mathbb{R}^n)$ be an isomorphism from $\mathbb{R}^d$ onto a d-dimensional linear subspace $T \subset \mathbb{R}^n$. Then the mapping*

$$K : \mathbb{R}^n \to \mathbb{R}^k \times \mathbb{R}^d, \quad K(x) := (F(x), U^T x), \quad \forall x \in \mathbb{R}^n, \tag{10}$$

is a local diffeomorphism on an open neighborhood of $x^c \in \mathcal{M}$ in $\mathbb{R}^n$ if and only if

$$T_{x^c}\mathcal{M} \cap T^\perp = \{0\}. \tag{11}$$

Let $j : \mathbb{R}^d \to \mathbb{R}^n \times \mathbb{R}^d$ denote the canonical injection that maps $\mathbb{R}^d$ isomorphically to $\{0\} \times \mathbb{R}^d$. If (11) holds at x^c then there exists an open set $\mathcal{U}^d$ of $\mathbb{R}^d$ such that the pair $(\mathcal{U}^d, \phi)$, defined with the mapping $\phi = K^{-1} \circ j : \mathcal{U}^d \to \mathbb{R}^n$, is a local parametrization of $\mathcal{M}$ near x^c.

We call a d-dimensional linear subspace $T \subset \mathbb{R}^n$ a *coordinate subspace* of $\mathcal{M}$ at $x^c \in \mathcal{M}$ if (11) holds. At any point $x^c \in \mathcal{M}$ an obvious choice for a coordinate subspace is $T = T_{x^c}\mathcal{M}$, the *tangential coordinate space* of $\mathcal{M}$ at that point. When we work with equations of the form (6); that, is when the parameter splitting (8) is available, then the parameter space $T = \{0\} \times \mathbb{R}^d$ is another possible choice of coordinate subspace. In that case, the point x^c where (11) fails to hold often have special significance.

A frequent approach in the study of the solution manifold $\mathcal{M} = F^{-1}(0)$ of Eq. (7) is to work with suitable paths on $\mathcal{M}$. Such a path may be specified by means of an augmented system

$$G(x) = 0, \quad G(x) := \begin{pmatrix} F(x) \\ \Gamma(x) \end{pmatrix}, \quad \Gamma : \mathbb{R}^n \to \mathbb{R}^{d-1} \tag{12}$$

of $n - 1$ equations and n variables. If G is of class C^p on the open set Ω and a submersion on the solution set $G^{-1}(0)$, then this set is indeed a one-dimensional submanifold of $\mathcal{M}$ provided, of course, it is not empty. Thus, (12) is a problem of form (7) with $d = 1$.

Here there is certainly some similarity with the homotopy mappings (2). But the geometric meaning is very different. In fact, (2) relates the two mappings $H_0 := H(\cdot, 0)$ and $H_1 := H(\cdot, 1)$, each with their own solution set, and the t-variable does have a very specific meaning. On the other hand, (12) defines a path that connects certain points of $\mathcal{M}$; that is, solutions of (7). Moreover, unless the augmenting mapping Γ in (12) is suitably chosen, no component of x can be singled out in any way. Of course, in many cases Γ does have a special form. For instance, when the parameter splitting (8) is available then we may use, $\Gamma(y, \lambda) := (\lambda_1 - \lambda_1^0, \ldots, \lambda_{k-1} - \lambda_{k-1}^0, \lambda_{k+1} - \lambda_{k+1}^0, \ldots, \lambda_d - \lambda_d^0)^{\mathrm{T}}$, whence only the parameter component $\mu = \lambda_k$ remains variable and we may reduce (12) to an equation

$$\tilde{G}(y, \mu) = 0, \quad \tilde{G} : \mathbb{R}^m \times \mathbb{R}^1 \to \mathbb{R}^m. \tag{13}$$

3.2. Singularities

When in example (5) the subspace of the parameter t is used to define a local parametrization, then condition (11) requires that $y = 0$. In other words, the t-parametrization fails at the point $(0, \frac{1}{2})$ where the two t-parametrized solution paths $(\pm\sqrt{\frac{1}{2} - t}, t)$ meet; that is, where the number of solutions of the form $y = y(t)$ of (5) reduces to one.

This is the simplest example of a bifurcation phenomenon. Loosely speaking, in this setting, bifurcation theory concerns the study of parametrized equations with multiple solutions and, in particular, the study of changes in the number of solutions when a parameter varies. Typically, such equations arise in applications modeling the equilibrium behavior of physical systems and then bifurcations signify a critical change in the system such as the, already mentioned, collapse of a mechanical structure. Accordingly, during the past several decades, the literature on bifurcation theory and the computation of bifurcation points has grown rapidly. It is beyond the framework of this presentation to enter into any details of the wide range of results.

As shown in [17,18] an important approach to bifurcation studies is via the more general study of singularities of stable mappings. In essence, for equations of form (13) involving a scalar parameter, the theory addresses problems of the following type: (i) The identification of (usually simpler) equations that are in a certain sense equivalent to the original one with the aim of recognizing equations of a particular qualitative type, (ii) the enumeration of all qualitatively different perturbations of a given equation, in particular, in terms of a so-called *universal unfolding*, (iii) the classification of qualitatively different equations that may occur, e.g., by considering the *codimension*; that is, the number of parameters needed in the universal unfolding.

In problems involving submanifolds that are defined as the solution set $\mathcal{M} := F^{-1}(0)$ of a (smooth) submersion $F : \mathbb{R}^n \to \mathbb{R}^m$, $n = m + d$, bifurcation phenomena are closely related to the occurrence of certain foldpoints. Generally, a point $x^c \in \mathcal{M}$ is a *foldpoint* with respect to a given d-dimensional coordinate subspace $T \subset \mathbb{R}^n$ of $\mathcal{M}$ if condition (11) fails at that point. When the parameter splitting (8) is available, the parameter space may be used as the coordinate subspace and, typically in applications, it turns out that the corresponding foldpoints are exactly the points where the solution behaviour shows drastic changes. Moreover, these points can be shown to be bifurcation points in the sense of the above-mentioned theory.

A general study and classification of foldpoints was given in [16]. This led to applications in various settings. In particular, in [27] a connection with the second fundamental tensor of the manifold was established and used for the computations of certain types of foldpoints. In [13] a different approach led to a new method for a particular subclass of foldpoints and in [19] this method was extended to problems with symmetries.

There is also a close connection between foldpoints and the general sensitivity problem for parameterized equations. If for Eq. (13) with a scalar parameter μ the solution can be written in the form $(y(\mu), \mu)$ then, traditionally, the derivative $Dy(\mu)$ is defined as a measure of the sensitivity of the solution with respect to the parameter. In [30] it was shown that this concept can be generalized to submanifolds $\mathcal{M} := F^{-1}(0)$ defined by a (smooth) submersion $F : \mathbb{R}^n \to \mathbb{R}^m$, $n = m + d$, $d \geqslant 1$. For a given local parametrization near a point $x^0 \in \mathcal{M}$ specified by the coordinate space $T \subset \mathbb{R}^n$ the sensitivity is a linear mapping Σ_T from $\mathbb{R}^d$ into $\mathbb{R}^m$. We will not give the details but note that in the above special case Σ_T reduces to $Dy(\mu)$. Moreover, as shown in [30] the Euclidean norm of Σ_T satisfies

$$||\Sigma_T||_2 = \frac{\mathrm{dist}(T, T_{x^0}\mathcal{M})}{[1 - \mathrm{dist}(T, T_{x^0}\mathcal{M})]^{1/2}}, \tag{14}$$

where, as usual, the distance $\mathrm{dist}(S_1, S_2)$ between any two, equi-dimensional linear subspaces S_1 and S_2 of $\mathbb{R}^n$ is the norm-difference $||P_1 - P_2||_2$ of the orthogonal projections P_j of $\mathbb{R}^n$ onto S_j, $j = 1, 2$. Eq. (14) shows that, in essence, the sensitivity Σ_T at $x^0 \in \mathcal{M}$ represents a measure of the distance between the local coordinate space T and the tangent space $T_{x^0}\mathcal{M}$. In other words, $||\Sigma_T||_2$ specifies how close X^0 is to the nearest foldpoint of $\mathcal{M}$ with respect to the local basis defined by T. This has been shown to be an valuable tool for identifying computationally the location of foldpoints.

4. Parametrized methods

The title of this section is intended to refer broadly to numerical methods for equations involving a *scalar* parameters. This includes the equations $H(y, t) = 0$ defined by any homotopy mapping (2) as well as those of the form (6) with a one-dimensional parameter $\lambda \in \mathbb{R}^1$. The emphasis will be on methods that do not utilize explicitly any manifold structure of the solution set even if that structure exists.

4.1. Incremental methods

For the equation $H(y, t) = 0$ defined by a continuous homotopy mapping (2), the Homotopy Invariance Theorem 1 provides information about the existence of solutions (y, t) for any given $t \in [0, 1]$ but not about, say, the continuous dependence of y upon t. For that we require additional conditions on H. Let

$$H : \Omega \times \mathcal{J} \to \mathbb{R}^m, \quad \Omega \subset \mathbb{R}^m, \quad \mathcal{J} \subset \mathbb{R}^1, \tag{15}$$

be a C^ρ, $\rho \geqslant 2$, mapping on the product of an open set $\mathcal{D}$ and an open interval $\mathcal{J}$. Then, for any solution $(y^0, t_0) \in \Omega \times \mathcal{J}$ where $D_y H(y^0, t_0)$ is nonsingular, the implicit function theorem ensures the existence of a $C^{\rho-1}$ mapping $\eta : \mathcal{J} \to \Omega$ on some interval $\mathcal{J}_0 \subset \mathcal{J}$ containing t_0 such that $H(\eta(t), t) = 0$ for $t \in \mathcal{J}_0$. Moreover, by shrinking $\mathcal{J}_0$ if needed, $D_y H(\eta(t), t)$ will be nonsingular for t in that interval.

By repeating the process with different points in the t-intervals, we obtain ultimately a solution curve $\eta : \mathcal{J}^* \to \Omega$ on an open interval $\mathcal{J}^* \subset \mathcal{J}$ that is maximal under set-inclusion. At the endpoints of this maximal interval the process stops either because the derivative $D_y H$ becomes singular (as in the example (5)) or the curve leaves the set $\mathcal{D}$.

Various iterative processes can be used to compute any point $\eta(t)$ for given $t \in \mathcal{J}^*$ on the solution curve. For instance, Newton's method

$$u^{j+1} = u^j - [D_y H(u^j, t)]^{-1} H(u^j, t), \quad j = 0, 1, \ldots, \tag{16}$$

converges to $\eta(t)$ if only $\|u^0 - \eta(t)\|$ is sufficiently small. This can be guaranteed by proceeding in small t-steps along the curve. In other words, we start say from a known point (y^0, t_0) and, for $i = 1, 2, \ldots$, compute a sequence (y^i, t_i) of approximations of $(\eta(t_i), t_i)$, $t_i \in \mathcal{J}^*$, with sufficiently small t-steps $h_i = t_i - t_{i-1}$. Once (y^i, t_i) is available, (16) can be started, for instance, with $u^0 = y^i$ and $t = t_{i+1} := t_i + h_{i+1}$. Alternately, some extrapolation of the prior computed points can be generated as a starting point.

This is the basic concept of the so-called incremental methods which date back at least to the work of Lahaye [22]. Over the years numerous variations of these processes have been proposed. This includes the use of a wide variety of iterative methods besides (16) and of numerous improved algorithms for the starting points. We refer here only to the extensive literature cited, e.g., in [26,20,3,4,32].

4.2. Continuation by differentiation

Consider again a mapping (15) under the same conditions as stated there. Suppose that there exists a continuous mapping $\eta : \mathcal{J}^* \to \Omega$ which is at least C^1 on some interval $\mathcal{J}^* \subset \mathcal{J}$ and satisfies

$$H(\eta(t), t) = 0, \quad \forall t \in \mathcal{J}^*. \tag{17}$$

Then, with $y^0 = \eta(t_0)$ for given $t_0 \in \mathcal{J}^*$, it follows that $y = \eta(t)$ is a solution of the initial value problem

$$D_y H(y, t) \dot{y} + D_t H(y, t) = 0, \quad y(t_0) = y^0. \tag{18}$$

Conversely, for any solution $y = \eta(t)$ of (18) on an interval $\mathcal{J}^* \subset \mathcal{J}$ such that $t_0 \in \mathcal{J}^*$ and $H(y^0, t_0) = 0$, the integral mean value theorem implies that (17) holds.

In a lengthy list of papers during a decade starting about 1952, D. Davidenko utilized the ODE (18) for the solution of a wide variety of problems including, not only nonlinear equations, but also integral equations, matrix inversion problems, determinant evaluations, and matrix eigenvalue problems (see [26] for some references). This has led occasionally to the use of the term 'Davidenko equation' for (18).

Evidently, if $D_y H(y,t)$ is nonsingular for all $(y,t) \in \Omega \times \mathcal{J}$, then the classical theory of explicit ODEs ensures the existence of solutions of (18) through any point (y^0, t_0) in this domain for which $H(y^0, t_0) = 0$. Moreover, these solutions are known to terminate only at boundary points of the set $\Omega \times \mathcal{J}$. But, if the derivative $D_y H(y,t)$ is allowed to become singular at certain points of the domain, then (18) becomes an implicit ODE and the standard ODE theory no longer applies in general. However, if we assume now that rank $DH(y,t) = m$ (whence $H^{-1}(0)$ has a manifold structure) then (18) turns out to be equivalent with an explicit ODE. Of course, its solutions can no longer be written globally as functions of t. For ease of notation, in the following result from [28] we drop the explicit t-representation and hence introduce again the combined vector $x = (y,t) \in \mathbb{R}^n$, $n = m+1$.

Theorem 9. *Suppose that $F : \mathcal{D} \to \mathbb{R}^{n-1}$ is C^1 on some open set $\mathcal{D} \subset \mathbb{R}^n$ and that* rank $DF(x) = n - 1$, $\forall x \in \mathcal{D}$. *Then, for each $x \in \mathcal{D}$ there exists a unique $u_x \in \mathbb{R}^n$ such that*

$$DF(x)u_x = 0, \quad \|u_x\|_2 = 1, \quad \det \begin{pmatrix} DF(x) \\ u_x^{\mathrm{T}} \end{pmatrix} > 0 \tag{19}$$

and the mapping

$$\Psi : \mathcal{D} \mapsto \mathbb{R}^n, \quad \Psi(x) = u_x, \quad \forall x \in \mathcal{D} \tag{20}$$

is locally Lipschitz on $\mathcal{D}$.

The mapping G of (20) defines the autonomous initial value problem

$$\frac{\mathrm{d}}{\mathrm{d}\tau} x = \Psi(x), \quad x(0) = x^0 \in \mathcal{D}. \tag{21}$$

By the local Lipschitz continuity of Ψ, standard ODE theory guarantees that (21) has for any $x^0 \in \mathcal{D}$ a unique C^1-solution $x : \mathcal{J} \to \mathcal{D}$ which is defined on an open interval $\mathcal{J}$ with $0 \in J$ that is maximal with respect to set inclusion. Moreover, if $s \in \partial \mathcal{J}$ is finite then $x(\tau) \to \partial \mathcal{D}$ or $\|x(\tau)\|_2 \to \infty$ as $\tau \to s, \tau \in \mathcal{J}$. Any solution $x = x(\tau)$ of (21) satisfies $DF(x(\tau))\dot{x}(\tau) = DF(x(\tau))\Psi(x(\tau)) = 0$ which implies, as before, that $F(x(\tau)) = F(x^0)$ for $\tau \in \mathcal{J}$.

Clearly, this is a more general result than the earlier one for (18). It can be combined with condition (4) of the homotopy invariance theorem 1 to avoid 'degeneracies' in the solution path of a given homotopy. The basis for this is the so-called Sard theorem and its generalizations covered, e.g., in some detail in [1]. The following result represents a very specified case:

Theorem 10. *Let $D_n \subset \mathbb{R}^n$ and $\mathcal{D}_k \subset \mathbb{R}^k$ be open sets and $F : \mathcal{D}_n \times \mathcal{D}_k \to \mathbb{R}^m$, $n \geq m$, a C^∞-map which has $z \in \mathbb{R}^m$ as regular value. Then for almost all $u \in \mathcal{D}_k$ (in the sense of Lebesgue measure) the restricted map $F_u := F(\cdot, u) : \mathcal{D}_n \to \mathbb{R}^m$ has z as regular value.*

As an application we sketch an example that follows results of Chow et al., [12]. For a bounded map $G \in C^\infty(\Omega) \cap C(\bar{\Omega})$, where $\Omega \subset \mathbb{R}^m$ is a bounded, open set, assume that $0 \in \mathbb{R}^m$ is a regular value. With $u \in \Omega$ consider the homotopy mapping

$$\hat{H} : \Omega \times \Omega \times \mathbb{R}^1 \to \mathbb{R}^m, \quad \hat{H}(y,u,t) := y - u - t(G(y) - u).$$

Then, clearly, rank $D\hat{H}(y, u, t) = m$ on the domain of $\hat{H}$; that is, $0 \in \mathbb{R}^m$ is a regular value of $\hat{H}$. Hence, for almost all $u \in \Omega$, the restricted map

$$H : \Omega \times \mathbb{R} \to \mathbb{R}^m, \quad H(y, t) := \hat{H}(y, u, t)$$

has 0 as regular value. By Theorem 9 this implies that there exists a unique solution $\tau \in \mathscr{J} \mapsto (y(\tau), t(\tau)) \in \Omega \times \mathbb{R}$ of $H(y, t) = 0$ which satisfies $(y(0), t(0)) = (u, 0)$ and is defined on some maximal open interval $\mathscr{J}$ containing the origin. We consider the solution path in the cylindrical domain $\mathscr{D} := \bar{\Omega} \times [0, 1]$. From (19) it follows that at the starting point $(u, 0)$ the path is not tangential to $\mathbb{R}^m \times \{0\}$ and enters $\mathscr{D}$. Hence, it can terminate only on $\partial \mathscr{D}$. Since $H(y, 0) = 0$ has only the solution $(u, 0)$, we see that the path cannot return to $\mathbb{R}^m \times \{0\}$ and hence must reach the set $(\bar{\Omega} \times \{1\}) \cup (\partial \Omega \times (0, 1))$. Now suppose that condition (4) of the homotopy invariance theorem 1 holds. Then we obtain that the solution path must reach the set $\bar{\Omega} \times \{1\}$ at a fixed point of G in Ω.

In line with the title of [12] the concept underlying this approach has been generally called the *probability-one homotopy paradigm*. It constitutes the principal theoretical foundation of the extensive and widely used HOMPACK package which incorporates many of the solution procedures indicated here. The curve tracing algorithms of this package were described in [39] and for the latest version we refer to [40] where also other relevant references are included.

In a sense the topological degree arguments of Theorem 1 are here replaced by analytic arguments involving inverse images of points in the range of the mapping under consideration. There exists an extensive literature centered on this idea. It includes in particular, various results on numerically implementable fixed point theorems. For an overview and comprehensive bibliography we refer to [3].

5. Manifold methods

As in Section 3 let $F : \mathbb{R}^n \to \mathbb{R}^m$, $n = m + d$, be a C^ρ mapping, $\rho \geq 1$, on the open set $\Omega \subset \mathbb{R}^n$ and suppose that the set $\mathscr{M} = F^{-1}(0)$ is not empty and F is a submersion of $\mathscr{M}$. Hence $\mathscr{M}$ is a d-dimensional C^ρ-submanifold of $\mathbb{R}^n$.

The computational tasks involved with such an implicitly defined manifold differ considerably from those arising in connection with manifolds defined in explicit, parametric form as they occur, e.g., in computational graphics. In fact, unlike in the latter case, for implicitly defined manifolds the algorithms for determining local parametrizations and their derivatives still need to be made available. A collection of algorithms was given in [31] for performing a range of essential tasks on general, implicitly specified submanifolds of a finite-dimensional space. This includes algorithms for determining local parametrizations and their derivatives, and for evaluating quantities related to the curvature and to sensitivity measures. The methods were implemented as a FORTRAN 77 package, called MANPAK. We discuss here only briefly one of these algorithms, namely, for the computation of local parametrizations.

Theorem 8 readily becomes a computational procedure for local parametrizations by the introduction of bases. Suppose that on $\mathbb{R}^n$ and $\mathbb{R}^d$ the canonical bases are used and that the vectors $u^1, \ldots, u^d \in \mathbb{R}^n$ form an orthonormal basis of the given coordinate subspace T of $\mathscr{M}$ at x^c. Then the matrix representation of the mapping U is the $n \times d$ matrix, denoted by U_c, with the vectors $u^1, \ldots, u^d$

Table 1
Algorithm GPHI

Input: $\{y, x^c, U_c, DK(x^c), \text{tolerances}\}$
$x := x^c + U_c y$;
while: 'iterates do not meet tolerances'
  evaluate $F(x)$;

  solve $DK(x^c)w = \begin{pmatrix} F(x) \\ 0 \end{pmatrix}$ for $w \in \mathbb{R}^n$;

  $x := x - w$;
  if 'divergence detected' **then return** *fail*;
endwhile
Output: $\{\phi(y) := x\}$.

---

as columns. It is advantageous to shift the open set $\mathscr{U}^d$ such that $\phi(0) = x^c$. Now, componentwise, the nonlinear mapping K of (10) assumes the form

$$K : \mathbb{R}^n \to \mathbb{R}^n, \quad K(x) = \begin{pmatrix} F(x) \\ U_c^T(x - x^c) \end{pmatrix}, \quad \forall x \in \Omega \subset \mathbb{R}^n. \tag{22}$$

By definition of ϕ we have $K(\phi(y)) = jy$ for all $y \in \mathscr{U}^d$, thus, the evaluation of $x = \phi(y)$ requires finding zeroes of the nonlinear mapping $K_y(x) := K(x) - jy$. For this a chord Newton method works well in practice. With the special choice $x^0 = x^c + U_c y$, the iterates satisfy $0 = U_c^T(x^k - x^c) - y = U_c^T(x^k - x_0)$ which implies that the process can be applied in the form

$$x^{k+1} := x^k - DK(x^c)^{-1} \begin{pmatrix} F(x^k) \\ 0 \end{pmatrix}, \quad x^0 = x^c + U_c y, \tag{23}$$

with a y-dependence only at the starting point. By standard results, it follows that, for any y near the origin of $\mathbb{R}^d$, the algorithm of Table 1 produces the point $x = \phi(y)$ in the local parametrization $(\mathscr{U}^d, \phi)$ near x^c defined by Theorem 8. For further methods related to local parametrizations and for determining bases of suitable coordinate spaces we refer to [31].

As a special case suppose now that $d = 1$ and hence that $\mathscr{M}$ is a one-dimensional C^ρ-submanifold of $\mathbb{R}^n$. Note that $\mathscr{M}$ may well have several connected component. The continuation methods for the computation of $\mathscr{M}$ begin from a given point x^0 on $\mathscr{M}$ and then produce a sequence of points x^k, $k = 0, 1, 2, \ldots$, on or near $\mathscr{M}$. In principle, the step from x^k to x^{k+1} involves the construction of a local parametrization of $\mathscr{M}$ and the selection of a predicted point w from which a local parametrization algorithm, such as GPHI in Table 1, converges to the desired next point x^{k+1} on $\mathscr{M}$. For the local parametrization at x^k we require a nonzero vector $v^k \in \mathbb{R}^n$ such that (11) holds which here means that

$$v^k \notin \mathrm{rge}\, DF(x^k)^T. \tag{24}$$

It is natural to call $T_x M^\perp = N_x M$ the normal space of $\mathscr{M}$ at x (under the natural inner product of $\mathbb{R}^n$). Thus (24) means that v^k should not be a normal vector of $\mathscr{M}$ at x^k. Once v^k is available, the local parametrization algorithm GPHI requires the solution of the augmented system

$$
\begin{pmatrix} Fx \\ (v^k)^{\mathrm{T}}(x - x^k) - y \end{pmatrix} = 0,
\tag{25}
$$

for given local coordinates $y \in \mathbb{R}^1$.

In summary then, three major choices are involved in the design of a continuation process of this type, namely,

  (i) the coordinate direction v^k at each step,
 (ii) the predicted point z^k at each step,
(iii) the corrector process for solving system (25).

In most cases a linear predictor $y = x^k + hv$ is chosen, whence (ii) subdivides into the choice (ii-a) of the predictor-direction $v_r \in \mathbb{R}^n$, and (ii-b) of the steplength h. The so-called pseudo-arclength method (see [20]) uses for u and v the (normalized) direction of the tangent of $\mathcal{M}$ at x^k while in the PITCON code (see [11]), only the prediction v is along the tangent direction while u is a suitable natural basis vector of $\mathbb{R}^n$. The local iterative process (iii); that is, the corrector, usually is a chord Newton method with the Jacobian evaluated at x^k or y as the iteration matrix. Other correctors include update methods as well as certain multigrid approaches (see e.g. the PLTMG package described in [8] and the references given there).

6. Simplicial approximations of manifolds

As before suppose that $F : \mathbb{R}^n \mapsto \mathbb{R}^m$, $n = m + d$, $d \geqslant 1$, is a C^ρ map, $\rho \geqslant 1$, and a submersion on $\mathcal{M} := F^{-1}(0)$. Then $\mathcal{M}$ is a d-dimensional C^ρ submanifold of $\mathbb{R}^n$ or the empty set, which is excluded. Obviously, for $d \geqslant 2$ we can apply continuation methods to compute paths on $\mathcal{M}$, but, it is certainly not easy to develop a good picture of a multi-dimensional manifold solely from information along some paths on it. This has led in recent years to the development of methods for a more direct approximation of implicitly defined manifolds of dimension exceeding one.

The case of implicitly defined manifolds has been addressed only fairly recently. The earliest work appears to be due to Allgower and Schmidt [6] (see also [7]) and uses a piecewise-linear continuation algorithm to construct a simplicial complex in the ambient space $\mathbb{R}^n$ that encloses the implicitly given d-dimensional manifold $\mathcal{M}$. In other words, this piecewise linear approach does not generate directly a simplicial approximation of $\mathcal{M}$ in the sense of Section 2.2. But, of course, such an approximation can be obtained from it. In fact, by using linear programming tools, points on the intersection of the n-simplices with the manifold can be computed. These points form polytopes which, in turn, can be subdivided to generate d-simplices that form the desired simplicial approximation. A disadvantage of this approach appears to be that the computational complexity is only acceptable for low ambient dimensions n. In fact, the method was mainly intended for surface and volume approximations.

A first method for the direct computation of a d-dimensional simplicial complex approximating an implicitly defined manifold $\mathcal{M}$ in a neighborhood of a given point of $\mathcal{M}$ was developed in [29]. There standardized patches of triangulations of the tangent spaces $T_x M$ of $\mathcal{M}$ are projected onto the manifold by smoothly varying projections constructed by a moving frame algorithm. An implementation of a globalized version of the method for the case $d = 2$ is described in [32] and

Brodzik [10] extended this global algorithm to the case of dimensions larger than two. In [10] also various applications are discussed and a general survey of methods in this area is provided.

A different method was developed in [24] which does not aim at the explicit construction of a simplicial complex on the implicitly defined, two-dimensional manifold, but on tessellating it by a cell-complex. This complex is formed by nonoverlapping cells with piecewise curved boundaries that are obtained by tracing a fish-scale pattern of one-dimensional paths on the manifold. Hence this approach appears to be intrinsically restricted to two-dimensional manifolds.

7. Further topics

There are a number of topics relating to the general area of continuation methods which, in view of space limitations, could not be addressed here.

While we restricted attention to problem in finite-dimensional spaces, many of the results can be extended to an infinite-dimensional setting. But, in that case various additional questions arise. For instance, typically in applications, the nonlinear equations (1) or (6) represent parametrized boundary value problems which must be discretized before we can apply any of the computational procedures. This raises the question how to define and estimate the discretization error. For parametrized equations already the definition of such errors is nontrivial; in fact, the development of a rigorous theory of discretization errors for parametrized nonlinear boundary value problems is of fairly recent origin. The case of a scalar parameter was first studied in the three-part work [9]. There mildly nonlinear boundary value problems were considered and the three parts concerned estimates at different types of points on the solution path. In particular, Part I addressed the case when λ can be used as local variable, while in Parts II and III estimates at simple limit points and simple bifurcation points are presented, respectively. Of course, a principal aspect of the latter two cases is the development of suitable local parametrizations.

For certain discretizations of equations defined by Fredholm operators on Hilbert spaces that involve a finite-dimensional parameter vector, a general theory of discretization errors was developed in [15]. Some applications of this theory to boundary value problems of certain quasilinear partial differential equations and finite-element discretizations are given in [38].

In Sections 4 and 5 we considered only general smooth maps. For more special systems it is, of course, possible to develop further refinements of the methods. In particular, for polynomial systems an extensive literature exists on homotopy methods for computing all zeros (in principle). For references see, e.g., [25,4], as well as the discussion in [41] of a very sophisticated code for polynomial systems that exploits their structure. It may be noted that — except for some early work — these results on polynomial systems formulate the problem in the complex projective space $\mathbf{CP}^n$.

As in other areas of computational mathematics, complexity studies have also become a topic of increasing interest in connection with continuation methods. In a series of five articles Shub and Smale developed a theory on the complexity of Bezout's theorem which concerns homotopy methods for computing all solutions of a system of polynomial equations. As an introduction to the results, we cite here the survey article [35] and, for further references, the fifth part [36] of the series.

Besides our brief comments about bifurcations in Section 3.2, we had to exclude any further discussion of the numerous results on computational methods for bifurcation problems. For some introduction and references see, e.g., [21,3,4]. A survey of methods for computing the simplest type of these points — the so-called *limit* points — was given in [23]. Methods for these and also the *simple bifurcation points* are incorporated in several of the existing packages for continuation problems, including, e.g., ALCON written by Deuflhard, Fiedler, and Kunkel [14], and BIFPACK by Seydel [34].

References

[1] R. Abraham, J.E. Marsden, T. Ratiu, Manifolds, Tensor Analysis, and Applications, 2nd ed., Springer, New York, 1988.

[2] P. Alexandroff, H. Hopf, Topologie, Chelsea, New York, 1965 (original edition, Berlin, Germany, 1935).

[3] E.L. Allgower, K. Georg, Numerical Continuation Methods, Series in Computers and Mathematics, Vol. 13, Springer, New York, 1990.

[4] E.L. Allgower, K. Georg, Numerical path following, in: P.G. Ciarlet, J.L. Lions (Eds.), Handbook of Numerical Analysis, Vol. V, North-Holland, Amsterdam, 1997, pp. 3–207.

[5] E.L. Allgower, K. Georg, Piecewise linear methods for nonlinear equations and optimization, J. Comput. Appl. Math. 124 (2000) 245–261, this issue.

[6] E.L. Allgower, P.H. Schmidt, An algorithm for piecewise linear approximation of an implicitly defined manifold, SIAM J. Numer. Anal. 22 (1985) 322–346.

[7] E.L. Allgower, S. Gnutzmann, Simplicial pivoting for mesh generation of implicitly defined surfaces, Comput. Aided Geom. Design 8 (1991) 305–325.

[8] R.E. Bank, PLTMG: A Software Package for Solving Elliptic Partial Differential Equations, Frontiers in Applied Mathematics, Vol. 15, SIAM, Philadelphia, PA, 1994.

[9] F. Brezzi, J. Rappaz, P.A. Raviart, Finite dimensional approximation of nonlinear problems, Part I: Branches of nonsingular solutions, Numer. Math. 36 (1981) 1–25; Part II: Limit points, Numer. Math. 37 (1981) 1–28; Part III: Simple bifurcation points, Numer. Math. 38 (1981) 1–30.

[10] M.L. Brodzik, Numerical approximation of manifolds and applications, Ph.D. Thesis, Department of Mathematics, University of Pittsburgh, Pittsburgh, PA, 1996.

[11] J. Burkardt, W.C. Rheinboldt, A locally parametrized continuation process, ACM Trans. Math. Software 9 (1983) 215–235.

[12] S.N. Chow, J. Mallett-Paret, J.A. Yorke, Finding zeros of maps: Homotopy applications to continuation methods, Math. Comput. 32 (1978) 887–899.

[13] R.X. Dai, W.C. Rheinboldt, On the computation of manifolds of foldpoints for parameter-dependent problems, SIAM J. Numer. Anal. 27 (1990) 437–446.

[14] P. Deuflhard, B. Fiedler, P. Kunkel, Efficient numerical path-following beyond critical points, SIAM J. Numer. Anal. 24 (1987) 912–927.

[15] J.P. Fink, W.C. Rheinboldt, On the discretization error of parametrized nonlinear equations, SIAM J. Numer. Anal. 20 (1983) 732–746.

[16] J.P. Fink, W.C. Rheinboldt, A geometric framework for the numerical study of singular points, SIAM J. Numer. Anal. 24 (1987) 618–633.

[17] M. Golubitsky, D.G. Schaeffer, Singularities and Groups in Bifurcation Theory, Vol. I, Springer, New York, 1985.

[18] M. Golubitsky, I. Stewart, D.G. Schaeffer, Singularities and Groups in Bifurcation Theory, Vol. II, Springer, New York, 1988.

[19] B. Hong, Computational methods for bifurcation problems with symmetries on the manifold, Ph.D. Thesis, University of Pittsburgh, Department of Mathematics, Pittsburgh, PA, 1991.

[20] H.B. Keller, Global homotopies and Newton methods, in: C. de Boor, G.H. Golub (Eds.), Recent Advances in Numerical Analysis, Academic Press, New York, 1978, pp. 73–94.

[21] H.B. Keller, Lectures on Numerical Methods in Bifurcation Problems, Springer, New York, 1987.

[22] E. Lahaye, Une méthode de résolution d'une catégorie d'équations transcendentes, C.R. Acad. Sci. Paris 198 (1934) 1840–1942.

[23] R. Melhem, W.C. Rheinboldt, A comparison of methods for determining turning points of nonlinear equations, Computing 29 (1982) 201–226.

[24] R. Melville, S. Mackey, A new algorithm for two-dimensional continuation, Comput. Math. Appl. 30 (1995) 31–46.

[25] A.P. Morgan, Solving Polynomial Systems using Continuation for Engineering and Scientific Problems, Prentice-Hall, Englewood Cliffs, NJ, 1987.

[26] J.M. Ortega, W.C. Rheinboldt, Iterative Solutions of Nonlinear Equations in Several Variables, Academic Press, New York, 1970.

[27] P.J. Rabier, W.C. Rheinboldt, On a computational method for the second fundamental tensor and its application to bifurcation problems, Numer. Math. 57 (1990) 681–694.

[28] W.C. Rheinboldt, Solution fields of nonlinear equations and continuation methods, SIAM J. Numer. Anal. 17 (1980) 221–237.

[29] W.C. Rheinboldt, On the computation of multi-dimensional solution manifolds of parametrized equations, Numer. Math. 53 (1988) 165–181.

[30] W.C. Rheinboldt, On the sensitivity of parametrized equations, SIAM J. Numer. Anal. 30 (1993) 305–320.

[31] W.C. Rheinboldt, MANPACK: a set of algorithms for computations on implicitly defined manifolds, Comput. Math. Appl. 27 (1996) 15–28.

[32] W.C. Rheinboldt, Methods for Solving Systems of Nonlinear Equations, SIAM, Philadelphia, PA, 1998.

[33] J. Schwartz, Nonlinear functional analysis (1963/64), Courant Institute of Mathematics and Science, New York, 1964.

[34] R. Seydel, BIFPACK: a program package for continuation, bifurcation, and stability analysis, Technical Report Version 2.3+, University of Ulm, Germany, 1991.

[35] M. Shub, Some remarks on Bezout's theorem and complexity theory, in: M.W. Hirsch, J.E. Marsden, M. Shub (Eds.), From Topology to Computation, Springer, Berlin, 1993, pp. 443–455.

[36] M. Shub, S. Smale, Complexity of Bezout's theorem, Part V. Polynomial time, Theoret. Comput. Sci. 133 (1994) 141–164.

[37] M.J. Todd, The Computation of Fixed Points and Applications, Lecture Notes in Economics and Mathematical Systems, Vol. 124, Springer, New York, 1976.

[38] T. Tsuchiya, A priori and a posteriori error estimates of finite element solutions of parametrized nonlinear equations, Ph.D. Thesis, Department of Mathematics, University of Maryland, College Park, MD, 1990.

[39] L.T. Watson, S.C. Billups, A.P. Morgan, Algorithm 652: HOMPACK: a suite of codes for globally convergent homotopy algorithms, ACM Trans. Math. Software 13 (1987) 281–310.

[40] L.T. Watson, M. Sosonkina, R.C. Melville, A.P. Morgan, H.F. Walker, Algorithm 777: HOMPACK90: a suite of FORTRAN 90 codes for globally convergent homotopy algorithms, ACM Trans. Math. Software 23 (1997) 514–549.

[41] S.M. Wise, A.J. Sommese, L.T. Watson, POLSYS_PLP: a partitioned linear product homotopy code for solving polynomial systems of equations, ACM Trans. Math. Software, to appear.

![ELSEVIER logo]

Journal of Computational and Applied Mathematics 124 (2000) 245–261

JOURNAL OF
COMPUTATIONAL AND
APPLIED MATHEMATICS

www.elsevier.nl/locate/cam

Piecewise linear methods for nonlinear equations and optimization

Eugene L. Allgower [1,*], Kurt Georg [1,2]

Department of Mathematics, Colorado State University, Fort Collins, CO 80523, USA

Received 7 April 1999; received in revised form 18 November 1999

Abstract

Piecewise linear methods had their beginning in the mid-1960s with Lemke's algorithm for calculating solutions to linear complementarity problems. In the 1970s and 1980s activity moved on to computing fixed points of rather general maps and economic equilibria. More recently, they have been used to approximate implicitly defined manifolds, with applications being made to computer graphics and approximations of integral over implicitly defined manifolds. In this paper we present the basic ideas of piecewise linear algorithms and a selection of applications. Further references to the literature on piecewise linear algorithms are indicated. © 2000 Elsevier Science B.V. All rights reserved.

1. Introduction

Piecewise linear algorithms, also referred to in the literature as simplicial algorithms, can be used to generate piecewise linear manifolds which approximate the solutions of underdetermined systems of equations $H(x)=0$, where $H : \mathbb{R}^{N+K} \to \mathbb{R}^N$ may be a mapping having relaxed smoothness properties. Of particular interest and importance is the case $K = 1$, in which case the algorithms produce an approximation of an implicitly defined curve. If the defining map H is itself piecewise linear, then $H^{-1}(0)$ is a polygonal path. More generally, H may also be piecewise smooth, or in some instances, an upper semi-continuous multi-valued map. Intuitively, the approximations which are produced result from traversing through cells of a tiling of $\mathbb{R}^{N+K}$ which intersect $H^{-1}(0)$. Often the tilings which are used are triangulations of $\mathbb{R}^{N+K}$ into simplices and hence the term simplicial algorithms occurs.

Piecewise linear algorithms have been used to find solutions to complementarity problems, fixed points of mappings, and economic equilibria [33]. Many classical theorems of analysis which can

---

* Corresponding author.

E-mail address: allgower@math.colostate.edu (E.L. Allgower).

[1] Partially supported by NSF via grant # DMS-9870274.

[2] http://www.math.colostate.edu/~georg/.

0377-0427/00/$ - see front matter © 2000 Elsevier Science B.V. All rights reserved.
PII: S 0377-0427(00)00427-1

be proven by means of homotopies or degree theory have been re-examined in terms of piecewise linear algorithms. Indeed, these algorithms can be viewed as a constructive approach to the Brouwer degree. Recent applications with $K > 1$ have been made to obtain computer graphical approximations of surfaces, surface and volume integrals, and solutions of differential–algebraic equations. In the first part of the paper we deal with the case $K = 1$, and in Section 9, we consider the more general case $K > 1$.

The first prominent example of a piecewise linear algorithm was designed in [27,26] to calculate a solution of the linear complementarity problem, see Section 7. This algorithm played a crucial role in the development of subsequent piecewise linear algorithms. Scarf [32] gave a numerically implementable proof of the Brouwer fixed point theorem, based upon Lemke's algorithm. Eaves [17] observed that a related class of algorithms can be obtained by considering piecewise linear approximations of homotopy maps. Concurrently, Merrill [28] gave a restart version for fixed points of upper semi-continuous maps. Thus the piecewise linear continuation methods began to emerge as a parallel to the classical embedding or predictor corrector numerical continuation methods (see [2,3]).

Piecewise linear methods require no smoothness of the underlying equations and hence have, at least in theory, a more general range of applicability than classical embedding methods. In fact, they can be used to calculate fixed points of set-valued maps. They are more combinatorial in nature and are closely related to the topological degree. Piecewise linear continuation methods are usually considered to be less efficient than the predictor corrector methods when the latter are applicable, especially for large N. The reasons for this lie in the fact that steplength adaptation and exploitation of special structure are more difficult to implement for piecewise linear methods. Some efforts to overcome this were given, e.g., in [31,35].

Many applications of piecewise linear algorithms for optimization and complementarity problems have recently been superceded by interior point methods which can handle the much larger (but more special) systems frequently occuring in practical problems (see, e.g., [25,29,37]).

We cast the notion of piecewise linear algorithms into the general setting of subdivided manifolds which we will call *piecewise linear manifolds*. Lack of space precludes an extensive bibliography. The older literature on the subject is well documented (see, e.g., [2,3]).

2. Basic facts

A piecewise linear algorithm consists of moving (pivoting) through cells which subdivide the domain of the map H. Let us formally introduce the basic notions.

Let $\mathbf{E}$ denote some ambient finite-dimensional Euclidean space which contains all points arising in the sequel. A *half-space* η and the corresponding *hyperplane* $\partial\eta$ are defined by $\eta = \{y \in \mathbf{E}: x^* y \leqslant \alpha\}$ and $\partial\eta = \{y \in \mathbf{E}: x^* y = \alpha\}$, respectively, for some $x \in \mathbf{E}$ with $x \neq 0$ and some $\alpha \in \mathbb{R}$. A finite intersection of half-spaces is called a *cell*. If σ is a cell and ξ a half-space such that $\sigma \subset \xi$ and $\tau := \sigma \cap \partial\xi \neq \emptyset$, then the cell τ is called a *face* of σ. For reasons of notation we consider σ also to be a face of itself, and all other faces are *proper* faces of σ. The *dimension* of a cell is the dimension of its affine hull. In particular, the dimension of a singleton is 0 and the dimension of the empty set is -1. If the singleton $\{v\}$ is a face of σ, then v is called a *vertex* of σ. If τ is a face of σ such that $\dim \tau = \dim \sigma - 1$, then τ is called a *facet* of σ. The *interior* of a cell σ consists of all points of σ which do not belong to a proper face of σ.

A *piecewise linear manifold* of dimension N is a system $\mathcal{M} \neq \emptyset$ of cells of dimension N such that the following conditions hold:

1. If $\sigma_1, \sigma_2 \in \mathcal{M}$, then $\sigma_1 \cap \sigma_2$ is a common face of σ_1 and σ_2.
2. A cell τ of dimension $N-1$ can be a facet of at most two cells in $\mathcal{M}$.
3. The family $\mathcal{M}$ is *locally finite*, i.e., any relatively compact subset of

$$|\mathcal{M}| := \bigcup_{\sigma \in \mathcal{M}} \sigma \tag{1}$$

meets only finitely many cells $\sigma \in \mathcal{M}$.

A simple example of a piecewise linear manifold is $\mathbb{R}^N$ subdivided into unit cubes with integer vertices.

We introduce the *boundary* $\partial \mathcal{M}$ of $\mathcal{M}$ as the system of facets which are common to exactly one cell of $\mathcal{M}$. Generally, we cannot expect $\partial \mathcal{M}$ to again be a piecewise linear manifold. However, this is true for the case that $|\mathcal{M}|$ is convex. Two cells which have a common facet τ are called *adjacent*. Moving from one cell to another through a common facet is called *pivoting*.

It is typical of piecewise linear path following that at any particular step only one current cell is stored in the computer, along with some additional data, and the pivoting step is performed by calling a subroutine which makes use of the data to determine an adjacent cell which then becomes the new current cell.

A cell of particular interest is a *simplex* $\sigma = [v_1, v_2, \ldots, v_{N+1}]$ of dimension N which is defined as the convex hull of $N+1$ affinely independent points $v_1, v_2, \ldots, v_{N+1} \in \mathbf{E}$. These points are the vertices of σ. If a piecewise linear manifold $\mathcal{M}$ of dimension N consists only of simplices, then we call $\mathcal{M}$ a *pseudo-manifold* of dimension N. Such manifolds are of special importance, see, e.g., [34]. If a pseudo-manifold $\mathcal{T}$ subdivides a set $|\mathcal{T}|$, then we also say that $\mathcal{T}$ *triangulates* $|\mathcal{T}|$. We will use the notions pseudo-manifold and triangulation somewhat synonymously. Some triangulations of $\mathbb{R}^N$ of practical importance were already considered in [11,21]. Eaves [19] gave an overview of standard triangulations.

If σ is a simplex in a pseudo-manifold $\mathcal{T}$ and τ is a facet of σ which is not in the boundary of $\mathcal{T}$, then there is exactly one simplex $\tilde{\sigma}$ in $\mathcal{T}$ which is different from σ but contains the same facet τ, and there is exactly one vertex v of σ which is not a vertex of $\tilde{\sigma}$. We call v the vertex of σ *opposite* τ. There is also exactly one vertex $\tilde{v}$ of $\tilde{\sigma}$ opposite τ. We say that σ is pivoted across τ into $\tilde{\sigma}$, and that the vertex v of σ is pivoted into $\tilde{v}$.

A simple triangulation can be generated by the following pivoting rule (pivoting by reflection; see [12]) if

$$\sigma = [v_1, v_2, \ldots, v_i, \ldots, v_{N+1}]$$

is a simplex in $\mathbb{R}^N$, and τ is the facet opposite a vertex v_i, then σ is pivoted across τ into $\tilde{\sigma} = [v_1, v_2, \ldots, \tilde{v}_i, \ldots, v_{N+1}]$ by setting

$$\tilde{v}_i = \begin{cases} v_{i+1} + v_{i-1} - v_i & \text{for } 1 < i < N+1, \\ v_2 + v_{N+1} - v_1 & \text{for } i = 1, \\ v_N + v_1 - v_{N+1} & \text{for } i = N+1. \end{cases} \tag{2}$$

In fact, a minimal (nonempty) system of N-simplices in $\mathbb{R}^N$ which is closed under the above pivoting rule is a triangulation of $\mathbb{R}^N$. We note that the above-described triangulation maintains a consistent ordering of the vertices of the simplices.

3. Piecewise linear algorithms

Let $\mathcal{M}$ be a piecewise linear manifold of dimension $N+1$. We call $H:|\mathcal{M}| \to \mathbb{R}^N$ a *piecewise linear map* if the restriction $H_\sigma : \sigma \to \mathbb{R}^N$ of H to σ is an affine map for all $\sigma \in \mathcal{M}$. In this case, H_σ can be uniquely extended to an affine map on the affine space spanned by σ. The Jacobian H'_σ is piecewise constant and has the property $H'_\sigma(x-y) = H_\sigma(x) - H_\sigma(y)$ for x, y in this affine space. Note that under a choice of basis H'_σ corresponds to an $(N, N+1)$-matrix which has a one-dimensional kernel in case of nondegeneracy, i.e., if its rank is maximal.

If $\mathcal{M}$ is a pseudo-manifold triangulating a set $X = |\mathcal{M}|$, and if $\tilde{H}:X \to \mathbb{R}^k$ is a map, then the *piecewise linear approximation* of $\tilde{H}$ (with respect to $\mathcal{M}$) is defined as the unique piecewise linear map $H:X \to \mathbb{R}^k$ which coincides with $\tilde{H}$ on all vertices of $\mathcal{M}$, i.e., $\tilde{H}$ is affinely interpolated on the simplices of $\mathcal{M}$.

A piecewise linear algorithm is a method for following a polygonal path in $H^{-1}(0)$. To handle possible degeneracies, we introduce a concept of regularity. A point $x \in |\mathcal{M}|$ is called a *regular point* of H if x is not contained in any face of dimension $< N$, and if H'_τ has maximal rank N for all faces τ containing x. A value $y \in \mathbb{R}^N$ is a *regular value* of H if all points in $H^{-1}(y)$ are regular. By definition, y is vacuously a regular value if it is not contained in the range of H. If a point or value is not regular it is called *singular*.

The following analogue of Sard's theorem holds for piecewise linear maps (see, e.g., [18]). This enables us to confine ourselves to regular values. We note that degeneracies can also be handled via the closely related concept of lexicographical ordering (see [8,16,34]).

Theorem 3.1 (Perturbation Theorem). *Let $H:\mathcal{M} \to \mathbb{R}^N$ be a piecewise linear map where $\mathcal{M}$ is a piecewise linear manifold of dimension $N+1$. Then for any relatively compact subset $C \subset |\mathcal{M}|$ there are at most finitely many $\varepsilon > 0$ such that $C \cap H^{-1}(\vec{\varepsilon})$ contains a singular point of H. Consequently, $\vec{\varepsilon}$ is a regular value of H for almost all $\varepsilon > 0$. Here we use the notation*

$$\vec{\varepsilon} := \begin{pmatrix} \varepsilon \\ \varepsilon^2 \\ \vdots \\ \varepsilon^N \end{pmatrix}.$$

Let 0 be a regular value of H. This implies that $H^{-1}(0)$ consists of polygonal paths whose vertices are always in the interior of some facet. If σ is a cell, then $\sigma \cap H^{-1}(0)$ is a segment (two end points), a ray (one end point), a line (no end point) or empty. The latter two cases are not of interest for piecewise linear path following. A step of the method consists of following the ray or segment from one cell into a uniquely determined adjacent cell. The method is typically started at a point of the boundary or on a ray (coming from infinity), and it is typically terminated at a point of the boundary or in a ray (going to infinity), The numerical linear algebra (*piecewise linear step*) required to perform one step of the method is typical for linear programming and usually involves $O(N^2)$ operations for dense matrices.

On the other hand, even if 0 is not a regular value of H, the above theorem helps us to do something similar. Namely, $\sigma \cap H^{-1}(\vec{\varepsilon})$ is a segment (two end points) for all sufficiently small $\varepsilon > 0$, a ray (one end point) for all sufficiently small $\varepsilon > 0$, a line (no end point) for all sufficiently small $\varepsilon > 0$ or empty for all sufficiently small $\varepsilon > 0$. This leads us to the following

Definition 3.1. We call an N-dimensional facet τ *completely labeled* with respect to an affine map $H : \tau \to \mathbb{R}^N$, if $\tau \cap H^{-1}(\vec{\varepsilon}) \neq \emptyset$ for all sufficiently small $\varepsilon > 0$. We call a cell σ of dimension $> N$ *transverse* with respect to an affine map $H : \sigma \to \mathbb{R}^N$, if $\sigma \cap H^{-1}(\vec{\varepsilon}) \neq \emptyset$ for all sufficiently small $\varepsilon > 0$.

Instead of following the paths $H^{-1}(0)$ for a regular value 0, we now follow more specifically the *regularized paths*

$$\bigcup \{ H^{-1}(0) \cap \sigma : \sigma \text{ transverse} \}.$$

Of course, this set coincides with $H^{-1}(0)$ for the case that 0 is a regular value of H.

For $\varepsilon > 0$ sufficiently small and $\vec{\varepsilon}$ a regular value of H, a node of the polygonal paths $H^{-1}(\vec{\varepsilon})$ corresponds to a completely labeled facet (which is intersected), and hence the piecewise linear algorithm traces such completely labeled facets belonging to the same cell. The method is usually started either on the boundary, i.e., in a completely labeled facet $\tau \in \partial \mathcal{M}$, or on a ray, i.e., in a transverse cell $\sigma \in \mathcal{M}$ which has only one completely labeled facet.

Hence, a piecewise linear algorithm generates a succession of transverse cells σ_i and completely labeled facets τ_i such that σ_i, σ_{i+1} have the common facet τ_i. We are thus led to the following generic version:

Algorithm 3.1 (Piecewise Linear Algorithm).
  1. Start
    (a) start from the boundary
      (i) let $\tau_1 \in \partial \mathcal{M}$ be completely labeled
      (ii) find the unique $\sigma_1 \in \mathcal{M}$ such that $\tau_1 \subset \sigma_1$
    (b) start from a ray
      (i) let $\sigma_0 \in \mathcal{M}$ have precisely one completely labeled facet τ_1
      (ii) pivoting step:
         find $\sigma_1 \in \mathcal{M}, \sigma_1 \neq \sigma_0$ such that $\tau_1 \subset \sigma_1$
  2. for $i = 1, 2, 3, \ldots$
    (a) if τ_i is the only completely labeled facet of σ_i
        then **stop** (ray termination)
    (b) else
        piecewise linear step:
        find the other completely labeled facet τ_{i+1} of σ_i

(c) if $\tau_{i+1} \in \partial\mathcal{M}$
    then **stop** (boundary termination)
(d) else
    pivoting step:
    find $\sigma_{i+1} \in \mathcal{M}$, $\sigma_{i+1} \neq \sigma_i$, such that $\tau_{i+1} \subset \sigma_{i+1}$

4. Numerical considerations

From a numerical point of view, two steps of a piecewise linear algorithm have to be efficiently implemented. Usually, a current cell σ and a completely labeled facet τ of σ are stored via some characteristic data.

A pivoting step consists of finding the adjacent cell $\tilde{\sigma}$ sharing the same facet τ. The implementation of this step is dependent on the special piecewise linear manifold under consideration. Typically, this step is performed by only a few operations. The pivoting rule (2) is a simple example.

A piecewise linear step consists of finding a second completely labeled facet $\tilde{\tau}$ of σ (if it exists, otherwise we have ray termination). This is usually computationally more expensive than the pivoting rule and typically involves some numerical linear algebra.

Let us consider an example. We assume that a cell of dimension $N+1$ is given by

$$\sigma := \{x \in \mathbb{R}^{N+1}: Lx \geqslant c\},$$

where $L: \mathbb{R}^{N+1} \to \mathbb{R}^m$ is a linear map and $c \in \mathbb{R}^m$ is a given value. Furthermore, let us assume that

$$\tau_i := \{x \in \mathbb{R}^{N+1}: Lx \geqslant c, e_i^* Lx = e_i^* c\},$$

for $i = 1, 2, \ldots, m$, is a numbering of all the facets of σ. Here and in the following e_i denotes the ith unit vector, i.e., the ith column of the identity matrix.

On the cell σ, the piecewise linear map $H: \mathcal{M} \to \mathbb{R}^N$ reduces to an affine map, and hence there is a linear map $A: \mathbb{R}^{N+1} \to \mathbb{R}^N$ and a vector $b \in \mathbb{R}^N$ such that the segment of the path in σ can be written as

$$\sigma \cap H^{-1}(0) = \{x \in \mathbb{R}^{N+1}: Ax = b, Lx \geqslant c\}. \tag{3}$$

Let τ_i be completely labeled. This implies that the rank of A is N. If we exclude degeneracies, then $\tau_i \cap H^{-1}(0) = \{x_0\}$ is a singleton, and there is a unique vector t in the one-dimensional kernel $A^{-1}(0)$ such that $e_i^* Lt = -1$. Since x_0 is in the interior of τ_i (by excluding degeneracies), we have

$$e_j^* Lx_0 > e_j^* c \quad \text{for } j = 1, \ldots, m, \ j \neq i$$

and hence $x_0 - \lambda t$ is in the interior of σ for small $\lambda > 0$.

If (3) is a ray, then $e_j^* L(x_0 - \lambda t) > e_j^* c$ for all $\lambda > 0$. Otherwise, we have $e_j^* Lt > 0$ for at least one index j, and since we are excluding degeneracies, the minimization

$$k := \arg\min \left\{ \frac{e_j^*(Lx_0 - c)}{e_j^* Lt} : j = 1, \ldots, m, e_j^* Lt > 0 \right\} \tag{4}$$

yields the unique completely labeled facet τ_k of σ with $k \neq i$. For the minimum

$$\lambda_0 := \frac{e_k^*(Lx_0 - c)}{e_k^* Lt} > 0,$$

we obtain: $\sigma \cap H^{-1}(0) = \{x_0 - \lambda t: \ 0 \leqslant \lambda \leqslant \lambda_0\}$.

Minimizations such as (4) are typical for linear programming, and the numerical linear algebra can be efficiently handled by standard routines. Successive linear programming steps can often make use of previous matrix factorizations via update methods. In the case of a pseudo-manifold $\mathcal{M}$ where the cell σ is a simplex, it is convenient to handle the numerical linear algebra with respect to the barycentric coordinates based on the vertices of σ. Then the equations become particularly simple, (see, e.g., [2, Sections 12.2–12.4] or [34] for details).

We now give some examples of how the piecewise linear path following methods are used.

5. Piecewise linear homotopy algorithms

Let us see how the above ideas can be used to approximate zero points of a map $G: \mathbb{R}^N \to \mathbb{R}^N$ by applying piecewise linear methods to an appropriate homotopy map. In order to also allow for applications to optimization problems or other nonlinear programming problems, we consider the case where G is not necessarily continuous, e.g., G might be a selection of a multi-valued map. For the case that $\bar{x}$ is a point of discontinuity of G, we have to generalize the notion of a zero point in an appropriate way, as described below.

Eaves [17] presented the first piecewise linear homotopy method for computing a fixed point. A restart method based on somewhat similar ideas was developed in [28]. Fixed point problems and zero point problems are obviously equivalent.

As an example of a piecewise linear homotopy algorithm, let us sketch the algorithm of Eaves and Saigal [20]. We consider a triangulation $\mathcal{T}$ of $\mathbb{R}^N \times (0, 1]$ into $(N + 1)$-simplices σ such that every simplex has vertices in adjacent levels $\mathbb{R}^N \times \{2^{-k}, 2^{-k-1}\}$ and a diameter $\leqslant C^k$ for some $k = 0, 1, \ldots$ and some $0 < C < 1$ which is not dependent on k. We call such a triangulation a refining triangulation (with refining factor C). Of course, the main point here is to obtain a triangulation which is easily implemented. The first such triangulation was proposed in [17]. Todd [34] gave a triangulation with refining factor $\frac{1}{2}$. Subsequently, many triangulations with arbitrary refining factors were developed (see the books [14,19]). To ensure success (i.e., convergence) of the algorithms, it is necessary to require a boundary condition.

Let us first introduce some notation. For $x \in \mathbb{R}^N$ we denote by $\mathcal{U}(x)$ the system of neighborhoods of x. By $\overline{\mathrm{co}}(X)$ we denote the closed convex hull of a set $X \subset \mathbb{R}^N$. By $\mathbb{R}_{\Sigma}^N$ we denote the system of compact convex nonempty subsets of $\mathbb{R}^N$. We call the map $G: \mathbb{R}^N \to \mathbb{R}^N$ *asymptotically linear* if the following three conditions hold:

1. G is *locally bounded*, i.e., each point $x \in \mathbb{R}^N$ has a neighbourhood $U \in \mathcal{U}(x)$ such that $G(U)$ is a bounded set.
2. G is *differentiable at* ∞, i.e., there exists a linear map $G_\infty': \mathbb{R}^N \to \mathbb{R}^N$ such that $\|x\|^{-1}\|G(x) - G_\infty' x\| \to 0$ for $\|x\| \to \infty$.
3. G_∞' is nonsingular.

If a map $G : \mathbb{R}^N \to \mathbb{R}^N$ is locally bounded, then we define its *set-valued hull* $G_\Sigma : \mathbb{R}^N \to \mathbb{R}^N_\Sigma$ by setting

$$G_\Sigma(x) := \bigcap_{U \in \mathcal{U}(x)} \overline{\mathrm{co}}(G(U)).$$

It is not difficult to see that G_Σ is upper semi-continuous, and that G is continuous at x if and only if $G_\Sigma(x)$ is a singleton. By using a degree argument on the set-valued homotopy

$$H_\Sigma(x, \lambda) := (1 - \lambda)G'_\infty x + \lambda G_\Sigma(x),$$

it can be seen that G_Σ has at least one zero point, i.e., a point $\bar{x}$ such that $0 \in G_\Sigma(\bar{x})$. Our aim here is to approximate such a solution numerically.

We now construct a piecewise linear homotopy for an asymptotically linear map $G : \mathbb{R}^N \to \mathbb{R}^N$. First we define $\tilde{H} : \mathbb{R}^N \times [0, \infty) \to \mathbb{R}^N$ by setting

$$\tilde{H}(x, \lambda) := \begin{cases} G'_\infty(x - x_1) & \text{for } \lambda = 1, \\ G(x) & \text{for } \lambda < 1. \end{cases}$$

Here x_1 is a chosen starting point of the method. Then we consider a refining triangulation $\mathcal{T}$ of $\mathbb{R}^N \times (0, 1]$ as above, and we use the piecewise linear approximation H of $\tilde{H}$ (with respect to $\mathcal{T}$) to trace the polygonal path in $H^{-1}(0)$ which contains the starting point $(x_1, 1)$.

The boundary $\partial \mathcal{T}$ is a pseudo-manifold which triangulates the sheet $\mathbb{R}^N \times \{1\}$. If we assume that the starting point $u_1 := (x_1, 1)$ is in the interior of a facet $\tau_1 \in \partial \mathcal{T}$, then it is immediately clear that τ_1 is the only completely labeled facet of $\partial \mathcal{T}$. Hence, the piecewise linear algorithm started in τ_1 cannot terminate in the boundary, and since all cells of $\mathcal{T}$ are compact, it cannot terminate in a ray. Hence, it has no termination. Thus, the piecewise linear algorithm generates a sequence $\tau_1, \tau_2, \ldots$ of completely labeled facets of $\mathcal{T}$. Let us also consider the polygonal path generated by the piecewise linear algorithm. This path is characterized by the nodes $(x_1, \lambda_1), (x_2, \lambda_2), \ldots$ such that (x_i, λ_i) is the unique zero point of the piecewise linear homotopy H in τ_i for $i = 1, 2, \ldots$. The resulting algorithm, i.e., applying Algorithm 3.1 to the above homotopy H, is due to Eaves [17] and Eaves and Saigal [20].

We call $\bar{x} \in \mathbb{R}^N$ an *accumulation point* of the algorithm if

$$\liminf_{i \to \infty} \|x_i - \bar{x}\| = 0.$$

The following convergence theorem holds.

Theorem 5.1. *The set A of accumulation points generated by the Eaves–Saigal algorithm is compact, connected and nonempty. Each point $\bar{x} \in A$ is a zero point of G_Σ, i.e., we have $0 \in G_\Sigma(\bar{x})$.*

A proof can be found in [3, p. 153].

As a consequence, if the set-valued hull G_Σ has only isolated zero points, then the sequence x_i generated by the Eaves–Saigal algorithm converges to a zero point of G_Σ.

As a simple example, we consider the situation of the celebrated Brouwer fixed-point theorem. Let $F : C \to C$ be a continuous map on a convex, compact, nonempty subset $C \subset \mathbb{R}^N$ with nonempty interior. We define an asymptotically linear map $G: \mathbb{R}^N \to \mathbb{R}^N$ by setting

$$G(x) := \begin{cases} x - F(x) & \text{for } x \in C, \\ x - x_1 & \text{for } x \notin C. \end{cases}$$

Here, a point x_1 in the interior of C is used as a starting point. The above piecewise linear algorithm generates a point $\bar{x} \in \mathbb{R}^N$ such that $0 \in G_\Sigma(\bar{x})$. If $\bar{x} \notin C$, then $G_\Sigma(\bar{x}) = \{\bar{x} - x_1\}$, but $\bar{x} \neq x_1$ implies that this case is impossible. If $\bar{x}$ is an interior point of C, then $G_\Sigma(\bar{x}) = \{\bar{x} - F(\bar{x})\}$, and hence $\bar{x}$ is a fixed point of F. If $\bar{x}$ is in the boundary ∂C, then $G_\Sigma(\bar{x})$ is the convex hull of $\bar{x} - x_1$ and $\bar{x} - F(\bar{x})$, and hence $\bar{x} = (1 - \lambda)x_1 + \lambda F(\bar{x})$ for some $0 \leqslant \lambda \leqslant 1$. But $\lambda < 1$ would imply that $\bar{x}$ is an interior point of C, and hence we have $\lambda = 1$, and again $\bar{x}$ is a fixed point of F. Hence, the above piecewise linear homotopy algorithm generates a fixed point of F in either case.

Many similar asymptotically linear maps can be constructed which correspond to important nonlinear problems (see e.g., [2, Chapter 13]). In particular, let us mention two examples that relate to nonlinear optimization (see [2, Examples 13.1.17 and 13.1.22]).

Consider the constrained minimization problem

$$\min_x \{\theta(x): \psi(x) \leqslant 0\}, \tag{5}$$

where $\theta, \psi: \mathbb{R}^N \to \mathbb{R}$ are convex. We assume the Slater condition

$$\{x: \psi(x) < 0, \|x - x_0\| < r\} \neq \emptyset$$

and the boundary condition that the problem

$$\min_x \{\theta(x): \psi(x) \leqslant 0, \|x - x_0\| \leqslant r\}$$

has no solution on the boundary $\{x: \|x - x_0\| = r\}$ for some suitable $x_0 \in \mathbb{R}^N$ and $r > 0$. This boundary condition is satisfied, for example, if

$$\{x: \psi(x) \leqslant 0\} \subset \{x: \|x - x_0\| < r\}$$

or more generally, if

$$\emptyset \neq \{x: \psi(x) \leqslant 0\} \cap \{x: \theta(x) \leqslant C\} \subset \{x: \|x - x_0\| < r\}$$

for some $C > 0$. Let us define the map $G: \mathbb{R}^N \to \mathbb{R}^N$ by

$$G(x) \in \begin{cases} \partial \theta(x) & \text{for } \psi(x) \leqslant 0 \text{ and } \|x - x_0\| < r, \\ \partial \psi(x) & \text{for } \psi(x) > 0 \text{ and } \|x - x_0\| < r, \\ \{x - x_0\} & \text{for } \|x - x_0\| \geqslant r, \end{cases}$$

where ∂ indicates the set of subdifferentials of a convex function. G is asymptotically linear with Jacobian $G'(\infty) = \text{Id}$. Hence, by Theorem 5.1, we obtain a zero point $0 \in G_\Sigma(\bar{x})$. It can be shown that $\bar{x}$ solves the minimization problem (5).

As a second example let us consider the nonlinear complementarity problem: Find an $x \in \mathbb{R}^N$ such that

$$x \in \mathbb{R}^N_+, \quad g(x) \in \mathbb{R}^N_+, \quad x^* g(x) = 0, \tag{6}$$

where $g: \mathbb{R}^N \to \mathbb{R}^N$ is a continuous map.

Here $\mathbb{R}_+$ denotes the set of nonnegative real numbers, and in the sequel we also denote the set of positive real numbers by $\mathbb{R}_{++}$. For $x \in \mathbb{R}^N$ we also introduce the positive part $x_+ \in \mathbb{R}^N_+$ by setting $e_i^* x_+ := \max\{e_i^* x, 0\}$ for $i = 1, \ldots, N$ and the negative part $x_- \in \mathbb{R}^N_+$ by $x_- := (-x)_+$. The following formulae are then obvious: $x = x_+ - x_-$, $(x_+)^*(x_-) = 0$.

It can be seen that x is a solution of (6) if and only if $x = \bar{x}_+$ where $\bar{x}$ is a zero point of the map

$$x \mapsto g(x_+) - x_-. \tag{7}$$

We assume the following coercivity condition: There is a bounded open neighborhood $V \in \mathbb{R}^N$ such that

$$x^* g(x) > 0 \quad \text{for all } x \in \partial V \cap R^N_+.$$

We choose

$$\alpha > \max\{\|g(x)\|_\infty + \|x\|_\infty : x \in \bar{V} \cap R^N_+\}$$

and define

$$\Omega := \{x \in \mathbb{R}^N : \|x_-\|_\infty < \alpha, \; x_+ \in V\}.$$

Now define $G : \mathbb{R}^N \to \mathbb{R}^N$ by

$$G(x) = \begin{cases} x & \text{if } x \notin \Omega, \\ g(x_+) - x_- & \text{if } x \in \Omega. \end{cases}$$

Again, G is asymptotically linear and $G'(\infty) = \mathrm{Id}$. Hence, we have a zero point $0 \in G_\Sigma(\bar{x})$. It can be shown that $\bar{x}$ is a zero point of the map (7) and hence a solution of (6).

6. Index and orientation

Nearly all piecewise linear manifolds $\mathcal{M}$ which are of importance for practical implementations, are orientable. If $\mathcal{M}$ is orientable and of dimension $N + 1$, and if $H : \mathcal{M} \to \mathbb{R}^N$ is a piecewise linear map, then it is possible to introduce an index for the piecewise linear solution manifold $H^{-1}(0)$ which has important invariance properties and also yields some useful information. It should be noted that this index is closely related to the topological index which is a standard tool in topology and nonlinear analysis. Occasionally, index arguments are used to guarantee a certain qualitative behavior of the solution path.

We begin with some basic definitions. Let $\mathbf{F}$ be a linear space of dimension k. An *orientation* of $\mathbf{F}$ is a function $\mathbf{or} : F^k \to \{-1, 0, 1\}$ such that the following conditions hold:

1. $\mathbf{or}(b_1, \ldots, b_k) \neq 0$ if and only if $b_1, \ldots, b_k$ are linearly independent.
2. $\mathbf{or}(b_1, \ldots, b_k) = \mathbf{or}(c_1, \ldots, c_k) \neq 0$ if and only if the transformation matrix between $b_1, \ldots, b_k$ and $c_1, \ldots, c_k$ has positive determinant.

It is clear from the basic facts of linear algebra that any finite-dimensional linear space permits exactly two orientations.

Let σ be a cell of dimension k and aff σ its affine hull. We introduce the k-dimensional linear space $\mathrm{tng}\,\sigma := \{x - y : x, y \in \mathrm{aff}\,\sigma\}$ as the *tangent space* of σ. The cell σ is oriented by orienting this tangent space. Such an orientation $\mathbf{or}_\sigma$ of σ *induces an orientation* $\mathbf{or}_{\tau,\sigma}$ on a facet τ of σ by the following convention:

$$\mathbf{or}_{\tau,\sigma}(b_1,\ldots,b_{k-1}) := \mathbf{or}_\sigma(b_1,\ldots,b_k)$$

whenever $b_1,\ldots,b_{k-1}$ is a basis for tng τ, and $b_1,\ldots,b_k$ is a basis for tng σ such that b_k points from τ into the interior of the cell σ. It is routine to check that the above definition of $\mathbf{or}_{\tau,\sigma}$ indeed satisfies the definition of an orientation.

If $\mathcal{M}$ is a piecewise linear manifold of dimension $N+1$, then an *orientation of* $\mathcal{M}$ is a choice of orientations $\{\mathbf{or}_\sigma\}_{\sigma\in\mathcal{M}}$ such that

$$\mathbf{or}_{\tau,\sigma_1} = -\mathbf{or}_{\tau,\sigma_2} \tag{8}$$

for each τ which is a facet of two different cells $\sigma_1, \sigma_2 \in \mathcal{M}$. By making use of the standard orientation

$$\mathbf{or}(b_1,\ldots,b_N) := \mathrm{sign}\,\det(b_1,\ldots,b_N)$$

of $\mathbb{R}^N$, it is clear that any piecewise linear manifold of dimension N which subdivides a subset of $\mathbb{R}^N$ is oriented in a natural way.

If $H:\mathcal{M}\to\mathbb{R}^N$ is a piecewise linear map on a piecewise linear manifold of dimension $N+1$ such that zero is a regular value of H, then it is clear that the system

$$\ker H := \{\sigma\cap H^{-1}(0)\}_{\sigma\in\mathcal{M}}$$

is a one-dimensional piecewise linear manifold which subdivides the solution set $H^{-1}(0)$. For the case that $\mathcal{M}$ is oriented, the orientation of $\mathcal{M}$ and the natural orientation of $\mathbb{R}^N$ induce an orientation of $\ker H$. Namely, for $\xi\in\ker H$, $v\in\mathrm{tng}(\xi)$ and $\sigma\in\mathcal{M}$ such that $\xi\subset\sigma$, the definition

$$\mathbf{or}_\xi(v) := \mathbf{or}_\sigma(b_1,\ldots,b_N,v)\,\mathrm{sign}\,\det(H'_\sigma b_1,\ldots,H'_\sigma b_N) \tag{9}$$

is independent of the special choice of $b_1,\ldots,b_N \in \mathrm{tng}(\sigma)$, provided the $b_1,\ldots,b_N$ are linearly independent. Clearly, an orientation of the one-dimensional manifold $\ker H$ is just a rule which indicates a direction for traversing each connected component of $\ker H$. Keeping this in mind, we now briefly indicate why the above definition indeed yields an orientation for $\ker H$.

Let τ be a facet of $\mathcal{M}$ which meets $H^{-1}(0)$ and does not belong to the boundary $\partial\mathcal{M}$, let $\sigma_1,\sigma_2 \in \mathcal{M}$ be the two cells containing τ, and let $\xi_j := H^{-1}(0)\cap\sigma_j \in \ker H$ for $j=1,2$. If $b_1,\ldots,b_N$ is a basis of $\mathrm{tng}(\tau)$, and if $a_j\in\mathrm{tng}(\xi_j)$ points from τ into σ_j, then from condition (8) it follows that

$$\mathbf{or}_{\sigma_1}(b_1,\ldots,b_N,a_1) = -\mathbf{or}_{\sigma_2}(b_1,\ldots,b_N,a_2)$$

and hence (9) implies that

$$\mathbf{or}_{\xi_1}(a_1) = -\mathbf{or}_{\xi_2}(a_2),$$

which is exactly the right condition in the sense of (8) to ensure that the manifold $\ker H$ is oriented.

7. Lemke's algorithm

The first prominent example of a piecewise linear algorithm was designed in [26,27] to calculate a solution of the linear complementarity problem. Subsequently, several authors have studied complementarity problems from the standpoint of piecewise linear homotopy methods; see the references in [3, Section 38]. Complementarity problems can also be handled via interior point methods (see [29,37]). Linear complementarity problems arise in quadratic programming, bimatrix games, variational inequalities and economic equilibria problems. Hence numerical methods for their solution have been of considerable interest. For further references, see [10].

We present the Lemke algorithm as an example of a piecewise linear algorithm since it played a crucial role in the development of subsequent piecewise linear algorithms. Let us consider the following *linear complementarity problem*: Given an affine map $g : \mathbb{R}^N \to \mathbb{R}^N$, find an $x \in \mathbb{R}^N$ such that

$$x \in \mathbb{R}^N_+, \quad g(x) \in \mathbb{R}^N_+, \quad x^* g(x) = 0. \tag{10}$$

If $g(0) \in \mathbb{R}^N_+$, then $x = 0$ is a trivial solution to the problem. Hence, this trivial case is always excluded and the additional assumption $g(0) \notin \mathbb{R}^N_+$ is made.

It is not difficult to show the following: Define $f : \mathbb{R}^N \to \mathbb{R}^N$ by $f(z) := g(z_+) - z_-$. If x is a solution of the linear complementarity problem, then $z := x - g(x)$ is a zero point of f. Conversely, if z is a zero point of f, then $x := z_+$ solves the linear complementarity problem.

The advantage which f provides is that it is obviously a piecewise linear map if we subdivide $\mathbb{R}^N$ into orthants. This is the basis for our description of Lemke's algorithm. For a fixed $d \in \mathbb{R}^N_{++}$ we define the homotopy $H : \mathbb{R}^N \times [0, \infty) \to \mathbb{R}^N$ by

$$H(x, \lambda) := f(x) + \lambda d. \tag{11}$$

For a given subset $I \subset \{1, 2, \ldots, N\}$ an orthant can be written in the form

$$\sigma_I := \{(x, \lambda) : \lambda \geqslant 0, \ e_i^* x \geqslant 0 \text{ for } i \in I, \ e_i^* x \leqslant 0 \text{ for } i \in I'\}, \tag{12}$$

where I' denotes the complement of I. The collection of all such orthants forms a piecewise linear manifold $\mathcal{M}$ (of dimension $N + 1$) which subdivides $\mathbb{R}^N \times [0, \infty)$. Furthermore, it is clear that $H : \mathcal{M} \to \mathbb{R}^N$ is a piecewise linear map since $x \mapsto x_+$ switches its linearity character only at the co-ordinate hyperplanes.

Let us assume for simplicity that zero is a regular value of H. We note however, that the case of a singular value is treated in the same way by using perturbation techniques. Lemke's algorithm is started on a ray: if $\lambda > 0$ is sufficiently large, then

$$(-g(0) - \lambda d)_+ = 0 \quad \text{and} \quad (-g(0) - \lambda d)_- = g(0) + \lambda d \in \mathbb{R}^N_{++}$$

and consequently

$$H(-g(0) - \lambda d, \lambda) = 0.$$

Hence, the ray defined by

$$\lambda \in [\lambda_0, \infty) \mapsto -g(0) - \lambda d \in \sigma_\emptyset \tag{13}$$

$$\text{for} \quad \lambda_0 := \max_{i=1,\ldots,N} \frac{-e_i^* g(0)}{e_i^* d} \tag{14}$$

is used (for decreasing λ-values) to start the path following. Since the piecewise linear manifold $\mathcal{M}$ consists of the orthants of $\mathbb{R}^N \times [0, \infty)$, it is finite, and there are only two possibilities:

1. The algorithm terminates on the boundary $|\partial \mathcal{M}| = \mathbb{R}^N \times \{0\}$ at a point $(z, 0)$. Then z is a zero point of f, and hence z_+ solves the linear complementarity problem.

2. The algorithm terminates on a secondary ray. Then it can be shown (see [9]), that the linear complementarity problem has no solution, at least if the Jacobian g' belongs to a certain class of matrices.

Let us illustrate the use of index and orientation by showing that the algorithm generates a solution in the sense that it terminates on the boundary under the assumption that all principal minors of the Jacobian g' are positive. Note that the Jacobian g' is a constant matrix since g is affine.

For $\sigma_I \in \mathcal{M}$, see (12), we immediately calculate the Jacobian

$$H'_{\sigma_I} = (f'_{\sigma_I}, d),$$

where

$$f'_{\sigma_I} e_i = \begin{cases} g' e_i & \text{for } i \in I, \\ e_i & \text{for } i \in I'. \end{cases} \tag{15}$$

If $\xi \in \ker H$ is a solution path in σ_I, then formula (9) yields

$$\mathbf{or}_\xi(v) = \text{sign det } f'_{\sigma_I} \, \mathbf{or}_{\sigma_I}(e_1, \dots, e_N, v)$$

and since $\mathbf{or}_{\sigma_I}(e_1, \dots, e_N, v) = \text{sign}(v^* e_{N+1})$ by the standard orientation in $\mathbb{R}^{N+1}$, we have that $\det f'_{\sigma_I}$ is positive or negative if and only if the λ-direction is increasing or decreasing, respectively, while ξ is traversed according to its orientation. It is immediately seen from (15) that $\det f'_{\sigma_I}$ is obtained as a *principal minor* of g', i.e., by deleting all columns and rows of g' with index $i \in I'$ and taking the determinant of the resulting matrix (where the determinant of the "empty matrix" is assumed to be 1). Since we start in the negative orthant $\sigma_\emptyset$ where the principal minor is 1, we see that the algorithm traverses the primary ray against its orientation, because the λ-values are initially decreased. Hence, the algorithm continues to traverse $\ker H$ against its orientation. For the important case that all principal minors of g' are positive, the algorithm must continue to decrease the λ-values and thus it stops at the boundary $|\partial \mathcal{M}| = \mathbb{R}^N \times \{0\}$. Hence, in this case the algorithm finds a solution. Furthermore, it is clear that this solution is unique, since $\ker H$ can contain no other ray than the primary ray.

8. Further aspects of piecewise linear algorithms

Lack of space precludes the presentation of specific details of the extensive activity in piecewise linear methods which took place in the eighties and nineties. In particular, considerable activity took place on variable dimension algorithms, studies were made on the efficiency of triangulations, and on the complexity of piecewise linear methods. Literature of these developments until approximately 1994 can be found in [3]. The Netherlands school which works on piecewise linear methods continues to be active in this field, see, e.g., the recent publications and references cited therein: [15,23,24,36].

Many of the newer developments can be generally described in the following way: Very special piecewise linear manifolds are constructed for special classes of problems, e.g., special economic equilibrium problems or special complementarity problems. The aims are to fit the construction of the manifold to the problem in such a way that a convergence proof, leading to an existence theorem for solutions, can be carried out, and/or the resulting piecewise linear algorithm is easily implemented and becomes very efficient.

9. Approximating manifolds

Let us now consider the case $K > 1$. The ideas of numerical continuation [3] and piecewise linear methods can be extended to the approximation of implicitly defined manifolds $\tilde{H}^{-1}(0)$ where $\tilde{H} : \mathbb{R}^{N+K} \to \mathbb{R}^N$.

For simplicity, we assume in this section that zero is a regular value of the smooth map $\tilde{H} : \mathbb{R}^{N+K} \to \mathbb{R}^N$. Hence $\tilde{\mathcal{M}} := \tilde{H}^{-1}(0)$ is a smooth K-dimensional manifold. Before we discuss the methods for obtaining piecewise linear approximations of $\tilde{\mathcal{M}}$, let us briefly indicate the well-known fact that the Gauss–Newton method can be used to obtain a nonlinear projector P from a neighborhood U of $\tilde{\mathcal{M}}$ onto $\tilde{\mathcal{M}}$.

More precisely, given a point $v_0 \in U$, the sequence

$$v_{i+1} = v_i - \tilde{H}'(v_i)^+ \tilde{H}(v_i), \quad i = 0, 1, \ldots \tag{16}$$

converges quadratically to a point $v_\infty := P(v_0) \in \tilde{\mathcal{M}}$. Here $\tilde{H}'(v_i)^+$ denotes the Moore–Penrose inverse.

Rheinboldt [30], has exploited this idea to project a standard triangulation of the tangent space at a point of $\tilde{\mathcal{M}}$ onto $\tilde{\mathcal{M}}$, which leads to a local triangulation of the manifold in a neighborhood of that point. This method will be discussed elsewhere in this volume. The method is well-suited for approximating smooth manifolds in which the dimension N is large, such as in multiple parameter nonlinear eigenvalue problems (see, e.g., [30]). It has been applied to the calculation of fold curves and to differential–algebraic equations (see [13]).

The approximation of implicit surfaces has also been an active area of research in computer graphics, where $H : \mathbb{R}^3 \to \mathbb{R}^1$ (see [7] for bibliography and references to software).

A global approximation of $\tilde{\mathcal{M}}$ can be obtained via piecewise linear algorithms. This has been developed in [5,6] (see also [2, Chapter 15; 3, Section 40.2]).

Piecewise linear algorithms have been applied to the visualization of body surfaces, and to the approximation of surface and body integrals [4] (see also [3, Section 41]). They can also be used as automatic mesh generators for boundary element methods [22]. For software for surface and volume approximation via piecewise linear methods; see the URL of the second author.

We begin with a description of the underlying ideas. Let us suppose that $\mathcal{T}$ triangulates the space $\mathbb{R}^{N+K}$. An important advantage of the usual standard triangulations is that any simplex can be very compactly stored and cheaply recovered by means of an $(N + K)$-tuple of integers m corresponding to its barycenter. It is also possible to perform the pivoting steps directly on the integer vector m and thereby to save some arithmetic operations.

As in Section 3, let H denote the piecewise linear approximation of $\tilde{H}$ with respect to $\mathcal{T}$. The definitions of regular points and regular values extend analogously to this context. We again obtain a perturbation theorem, i.e., the proof of Theorem 3.1 involving ε-perturbations, generalizes verbatim if 1 is replaced by K.

If zero is a regular value of H, the zero set $H^{-1}(0)$ carries the structure of a K-dimensional piecewise linear manifold. We formulate this last remark more precisely.

Theorem 9.1. *Let zero be a regular value of H. If $\sigma \in \mathscr{T}$ has a non-empty intersection with $H^{-1}(0)$, then $\mathscr{M}_\sigma := \sigma \cap H^{-1}(0)$ is a K-dimensional polytope, and the family*

$$\mathscr{M} := \{\mathscr{M}_\sigma : \sigma \in \mathscr{T},\ \sigma \cap H^{-1}(0) \neq \emptyset\}$$

is a K-dimensional piecewise linear manifold.

The following algorithm describes the fundamental steps for obtaining the piecewise linear manifold $\mathscr{M}$ approximating $\tilde{\mathscr{M}}$. We again make the assumptions that $\tilde{H} : \mathbb{R}^{N+K} \to \mathbb{R}^N$ is a smooth map, $\mathscr{T}$ is a triangulation of $\mathbb{R}^{N+K}$, and zero is a regular value of both $\tilde{H}$ and its piecewise linear approximation H. Analogously to the definitions preceding Algorithm 3.1, we call a simplex $\sigma \in \mathscr{T}$ *transverse* if it contains an N-face which is completely labeled with respect to H; see Definition 3.1. In the algorithm below, the input is one transverse simplex σ, and the output is the maximal set Σ of transverse simplices that meet a certain given compact domain D and are connected to σ. For any transverse simplex $\sigma \in \Sigma$, the dynamically varying set $V(\sigma)$ keeps track of all vertices which remain to be checked in order to find all possible new transverse simplices via pivoting.

Algorithm 9.1 (PL Approximation of a Manifold).
1. input:
   (a) a transverse starting simplex $\sigma \in \mathscr{T}$
   (b) a compact subset $D \subset \mathbb{R}^{N+K}$ for bounding the search
2. initialization:
   $\Sigma := \{\sigma\}$ and $V(\sigma) := $ set of vertices of σ
3. while $V(\sigma) \neq \emptyset$ for some $\sigma \in \Sigma$
   (a) get $\sigma \in \Sigma$ such that $V(\sigma) \neq \emptyset$, and get $v \in V(\sigma)$
   (b) pivot the vertex v into v' to get an adjacent simplex σ'
   (c) if $\sigma' \cap \sigma$ is not transverse or $\sigma' \cap D = \emptyset$
       delete v from $V(\sigma)$
   (d) else if σ' is not new, i.e., $\sigma' \in \Sigma$
       delete v from $V(\sigma)$ and v' from $V(\sigma')$
   (e) else σ' is added to the list Σ, i.e.,
       (i) $\Sigma := \Sigma \cup \{\sigma'\}$
       (ii) $V(\sigma') := $ set of vertices of σ'
       (iii) delete v from $V(\sigma)$ and v' from $V(\sigma')$

For purposes of exposition, we have formulated the above algorithm in a very general way. A number of items remain to be discussed. We will show below how a starting simplex in 1a can be obtained in the neighborhood of a point $x \in \tilde{\mathscr{M}}$. The list Σ can be used to generate a K-dimensional connected piecewise linear manifold

$$\mathscr{M} := \{\mathscr{M}_\sigma\}_{\sigma \in \Sigma}$$

(see Theorem 9.1). This piecewise linear manifold approximates $\tilde{\mathscr{M}}$ quadratically in the mesh size of $\mathscr{T}$, as was shown in [1] (see also [3, Section 40.3]). If $\tilde{\mathscr{M}}$ is compact, the generated piecewise linear manifold will be compact without boundary, provided the mesh of the triangulation is sufficiently small and the bounding set D is sufficiently large. It is not really necessary to perform the pivot in 3(b) if $\sigma' \cap \sigma$ is not transverse, since this will already be known from the current data. In the comparing process 3(d), it is crucial that compact exact storing is possible for standard triangulations. The list searching in 3(a) and 3(d) can be performed via efficient binary tree searching.

The piecewise linear manifold $\mathcal{M}$ furnishes an initial coarse piecewise linear approximation of $\tilde{\mathcal{M}}$. Several improvements are possible. The first is that a Gauss–Newton type method as in (16) can be used to project the nodes of $\mathcal{M}$ onto $\tilde{\mathcal{M}}$. Thus a new piecewise linear manifold $\mathcal{M}_1$ is generated which inherits the adjacency structure of the nodes from $\mathcal{M}$ and has nodes on $\tilde{\mathcal{M}}$.

In many applications (e.g., boundary element methods) it is desirable to uniformize the mesh $\mathcal{M}_1$. A very simple and successful means of doing this is "mesh smoothing". One such possible method consists of replacing each node of the mesh by the average of the nodes with which it shares an edge and by using the resulting point as a starting value for a Gauss–Newton type process to iterate back to $\tilde{\mathcal{M}}$. The edges or nodal adjacencies are maintained as before. Three or four sweeps of this smoothing process over all of the nodes of $\mathcal{M}_1$ generally yields a very uniform piecewise linear approximation of $\tilde{\mathcal{M}}$.

Another step which is useful for applications such as boundary element methods is to locally subdivide the cells of the piecewise linear manifolds $\mathcal{M}$ or $\mathcal{M}_1$ into simplices in such a way that the resulting manifold can be given the structure of the pseudo-manifold $\mathcal{M}_2$.

Once an approximating pseudo-manifold $\mathcal{M}_2$ has been generated, it is easy to refine it by, e.g., the well-known construction of halving all edges of each simplex $\tau \in \mathcal{M}_2$, triangulating it into 2^K subsimplices and projecting the new nodes back onto $\tilde{\mathcal{M}}$.

We have assumed that zero is a regular value of H. In fact, as in the Perturbation Theorem 3.1 and following remarks, $\vec{\varepsilon}$-perturbations and the corresponding general definition "completely labeled" automatically resolves singularities even if zero is not a regular value of H. The situation is similar to the case $K = 1$.

Let us next address the question of obtaining a transverse starting simplex. If we assume that a point x on $\tilde{\mathcal{M}}$ is given, then it can be shown that any $(N + K)$-simplex with barycenter x and sufficiently small diameter is transverse (see [3, Section 40.3]).

Algorithm 9.1 merely generates a list Σ of transverse simplices. For particular purposes such as boundary element methods, computer graphics, etc., a user will wish to have more information concerning the structure of the piecewise linear manifold $\mathcal{M}$, e.g., all nodes of the piecewise linear manifold $\mathcal{M}$ together with their adjacency structure. Hence, to meet such requirements, it is necessary to customize the above algorithm for the purpose at hand.

References

[1] E.L. Allgower, K. Georg, Estimates for piecewise linear approximations of implicitly defined manifolds, Appl. Math. Lett. 1 (5) (1989) 1–7.

[2] E.L. Allgower, K. Georg, Numerical Continuation Methods: An Introduction, Series in Computational Mathematics, Vol. 13, Springer, Berlin, 1990, p. 388.

[3] E.L. Allgower, K. Georg, Numerical path following, in: P.G. Ciarlet, J.L. Lions (Eds.), Handbook of Numerical Analysis, Vol. 5, North-Holland, Amsterdam, 1997, pp. 3–207.

[4] E.L. Allgower, K. Georg, R. Widmann, Volume integrals for boundary element methods, J. Comput. Appl. Math. 38 (1991) 17–29.

[5] E.L. Allgower, S. Gnutzmann, An algorithm for piecewise linear approximation of implicitly defined two-dimensional surfaces, SIAM J. Numer. Anal. 24 (1987) 452–469.

[6] E.L. Allgower, Ph.H. Schmidt, An algorithm for piecewise-linear approximation of an implicitly defined manifold, SIAM J. Numer. Anal. 22 (1985) 322–346.

[7] J. Bloomenthal, Introduction to Implicit Surfaces, Morgan Kaufmann, San Francisco, 1997.

[8] A. Charnes, Optimality and degeneracy in linear programming, Econometrica 20 (1952) 160–170.

[9] R.W. Cottle, Solution rays for a class of complementarity problems, Math. Programming Study 1 (1974) 58–70.

[10] R.W. Cottle, J.-S. Pang, R.E. Stone, The Linear Complementarity Problem, Academic Press, New York, 1992.

[11] H.S.M. Coxeter, Discrete groups generated by reflections, Ann. Math. 6 (1934) 13–29.

[12] H.S.M. Coxeter, Regular Polytopes, 3rd ed., MacMillan, New York, 1963.

[13] R.-X. Dai, W.C. Rheinboldt, On the computation of manifolds of fold points for parameter-dependent problems, SIAM J. Numer. Anal. 27 (1990) 437–466.

[14] C. Dang, Triangulations and Simplicial Methods, Lecture Notes in Economics and Mathematical Systems, Springer, Berlin, 1995.

[15] C. Dang, H. van Maaren, A simplicial approach to integer programming, Math. Oper. Res. 23 (1998) 403–415.

[16] G.B. Dantzig, Linear Programming and Extensions, Princeton University Press, Princeton, NJ, 1963.

[17] B.C. Eaves, Homotopies for the computation of fixed points, Math. Programming 3 (1972) 1–22.

[18] B.C. Eaves, A short course in solving equations with PL homotopies, R.W. Cottle, C.E. Lemke (Eds.), Nonlinear Programming, SIAM-AMS Proceedings, Vol. 9, American Mathematical Society, Providence, RI, 1976, pp. 73–143.

[19] B.C. Eaves, A Course in Triangulations for Solving Equations with Deformations, Lecture Notes in Economics and Mathematical Systems, Vol. 234, Springer, Berlin, 1984.

[20] B.C. Eaves, R. Saigal, Homotopies for computation of fixed points on unbounded regions, Math. Programming 3 (1972) 225–237.

[21] H. Freudenthal, Simplizialzerlegungen von beschränkter Flachheit, Ann. Math. 43 (1942) 580–582.

[22] K. Georg, Approximation of integrals for boundary element methods, SIAM J. Sci. Statist. Comput. 12 (1991) 443–453.

[23] P.J.J. Herings, A.J.J. Talman, Intersection theorems with a continuum of intersection points, J. Optim. Theory Appl. 96 (1998) 311–335.

[24] A. Kirman (Ed.), Elements of General Equilibrium Analysis, Blackwell, Oxford, 1998.

[25] M. Kojima, N. Megiddo, T. Noma, A. Yoshise, A Unified Approach to Interior Point Algorithms for Linear Complementarity Problems, Lecture Notes in Computer Science, Vol. 538, Springer, Berlin, 1991.

[26] C.E. Lemke, Bimatrix equilibrium points and mathematical programming, Management Sci. 11 (1965) 681–689.

[27] C.E. Lemke, J.T. Howson, Equilibrium points of bimatrix games, SIAM J. Appl. Math. 12 (1964) 413–423.

[28] O. Merrill, Applications and extensions of an algorithm that computes fixed points of a certain upper semi-continuous point to set mapping, Ph.D. Thesis, University of Michigan, Ann Arbor, MI, 1972.

[29] Yu.E. Nesterov, A. Nemirovskii, Interior-Point Polynomial Algorithms in Convex Programming, SIAM, Philadelphia, PA, 1993.

[30] W.C. Rheinboldt, On the computation of multi-dimensional solution manifolds of parameterized equations, Numer. Math. 53 (1988) 165–182.

[31] R. Saigal, M.J. Todd, Efficient acceleration techniques for fixed point algorithms, SIAM J. Number. Anal. 15 (1978) 997–1007.

[32] H.E. Scarf, The approximation of fixed points of a continuous mapping, SIAM J. Appl. Math. 15 (1967) 1328–1343.

[33] H.E. Scarf, T. Hansen, The Computation on Economic Equilibria, Yale University Press, New Haven, CT, 1973.

[34] M.J. Todd, The Computation of Fixed Points and Applications, Lecture Notes in Economics and Mathematical Systems, Vol. 124, Springer, Berlin, 1976.

[35] M.J. Todd, Piecewise-linear homotopy algorithms for sparse systems of nonlinear equations, SIAM J. Control Optim. 21 (1983) 204–214.

[36] G. van der Laan, A.J.J. Talman, Z. Yang, Existence and approximation of robust stationary points on polytopes, SIAM J. Control Optim., in preparation.

[37] S.J. Wright, Primal–Dual Interior-Point Methods, SIAM, Philadelphia, PA, 1997.

NH

ELSEVIER

Journal of Computational and Applied Mathematics 124 (2000) 263–280

JOURNAL OF
COMPUTATIONAL AND
APPLIED MATHEMATICS

www.elsevier.nl/locate/cam

Interval mathematics, algebraic equations and optimization

M.A. Wolfe

*School of Mathematical and Computational Sciences, University of St. Andrews, Mathematical Institute,
North Haugh, St. Andrews, Fife KY16 9SS, UK*

Received 20 June 1999; received in revised form 12 December 1999

Abstract

This is a brief survey of some of the applications of interval mathematics to the solution of systems of linear and nonlinear algebraic equations and to the solution of unconstrained and constrained nonlinear optimization problems. © 2000 Elsevier Science B.V. All rights reserved.

1. Introduction

When a computer program using only real floating point machine arithmetic is run, data error, rounding error, truncation error and in–out conversion error combine in an extremely complex manner to produce error in the output which is unknown. A bound on the exact solution of the problem which the program was intended to solve could be determined by the computer itself if floating point machine *interval* arithmetic were to be used, together with appropriate algorithms. A bound which is so obtained can be made arbitrarily sharp to within the limits imposed by the accuracy of the data and of the machine interval arithmetic and of the limit imposed by the allowed CPU time. Furthermore the nonexistence, existence and uniqueness of solutions of problems in given regions can often be determined rigorously by using a now-extensive body of mathematical knowledge called *interval mathematics* which was used effectively for the first time by Moore [30]. A great deal of information about interval mathematics, its researchers, available literature and computational resources may be obtained from the interval mathematics web site `http://cs.utep.edu/interval-comp/main.html`. People in Europe should find it quicker to use the new (recently set up) mirror URL `http://www.lsi.upc.es/ ~robert/interval-comp/main.html`.

E-mail address: michael@dcs.st-and.ac.uk (M.A. Wolfe).

0377-0427/00/$ - see front matter © 2000 Elsevier Science B.V. All rights reserved.
PII: S 0377-0427(00)00421-0

Interval mathematics is playing an increasingly important rôle in scientific computing and is being used to obtain computationally rigorous bounds on solutions of numerical problems from an increasingly large number of subject areas. Survey articles of some applications of interval mathematics are in [22]. The subject areas which have been studied include systems of linear and nonlinear algebraic equations and global optimization: a necessarily brief survey of some of the contributions to these subject areas appears in Sections 3 and 4 of this article.

The number of contributions to the fundamental theory of interval mathematics and to its applications to pure and applied science are now so numerous that it is impossible completely to survey even the title-subjects of the present paper. Therefore one has to rely on a fairly long list of references to avoid an over-lengthy survey.

2. Notation

The fundamental interval mathematics needed in Sections 3 and 4 is in [6,32]; the fundamental real analysis is in [37,34]. This section contains some notation and some basic definitions that are needed to understand statements made in Sections 3 and 4. An excellent introduction to interval arithmetic computation is in [14] which contains PASCAL-XSC software listings for the algorithms described therein, and several references. See also [4].

Definition 1. A real interval $[x] = [\underline{x}, \bar{x}]$ has infimum $\underline{x} \in \mathbb{R}^1$ and supremum $\bar{x} \in \mathbb{R}^1$ with $\underline{x} \leqslant \bar{x}$. If $X \subseteq \mathbb{R}^1$ then $\mathbb{I}(X) = \{ [x] \mid [x] \subseteq X \}$.

Definition 2. The sum, difference, product and quotient of $[x], [y] \in \mathbb{I}(\mathbb{R}^1)$ are defined by

$$[x] + [y] = [\underline{x} + \underline{y}, \bar{x} + \bar{y}],$$

$$[x] - [y] = [\underline{x} - \bar{y}, \bar{x} - \underline{y}],$$

$$[x] \cdot [y] = [\min S, \max S],$$

$$[x]/[y] = [\min T, \max T],$$

where $S = \{\underline{x} \cdot \underline{y}, \underline{x} \cdot \bar{y}, \bar{x} \cdot \underline{y}, \bar{x} \cdot \bar{y}\}$ and $T = \{\underline{x}/\underline{y}, \underline{x}/\bar{y}, \bar{x}/\underline{y}, \bar{x}/\bar{y}\}$.

Definition 3. The midpoint $m([x])$, the magnitude $|[x]|$, the mignitude $\langle [x] \rangle$, the width $w([x])$ and the radius $\mathrm{rad}([x])$ of $[x] \in \mathbb{I}(\mathbb{R}^1)$ are defined by $\check{x} = m([x]) = (\underline{x} + \bar{x})/2$, $|[x]| = \max\{|x| \mid x \in [x]\}$, $\langle [x] \rangle = \min\{|x| \mid x \in [x]\}$, $w([x]) = \bar{x} - \underline{x}$ and $\mathrm{rad}([x]) = w([x])/2$.

Definition 4. A real interval vector $[x] = ([x]_i) = [\underline{x}, \bar{x}] \in \mathbb{I}(\mathbb{R}^n)$ (a box) has infimum $\underline{x} = (\underline{x}_i) \in \mathbb{R}^n$ and supremum $\bar{x} = (\bar{x}_i) \in \mathbb{R}^n$ and if $[A] = ([A]_{i,j}) = ([\underline{A}_{i,j}, \bar{A}_{i,j}]) = [\underline{A}, \bar{A}] \in \mathbb{R}^{m \times n}$ is an interval matrix, where $\underline{A} = (\underline{A}_{i,j}) \in \mathbb{R}^{m \times n}$ and $\bar{A} = (\bar{A}_{i,j}) \in \mathbb{R}^{m \times n}$ with $\underline{A} \leqslant \bar{A}$, then $\check{A} = m([A]) = (\underline{A} + \bar{A})/2$, $|[A]| = \max\{|\underline{A}|, |\bar{A}|\}$, $w([A]) = \bar{A} - \underline{A}$ and $\mathrm{rad}([A]) = w([A])/2$ in which the usual componentwise ordering is assumed.

Definition 5. The interval hull $\square X$ of the set $X \subset \mathbb{R}^{m \times n}$ is defined by $\square X = [\inf(X), \sup(X)]$.

Definition 6. The interval matrix $[A] \in \mathbb{I}(\mathbb{R}^{n \times n})$ is regular if and only if $(\forall A \in [A])\ A^{-1} \in \mathbb{R}^{n \times n}$ exists, and is strongly regular if and only if $\check{A}^{-1}[A]$ is regular. The matrix inverse $[A]^{-1}$ of the regular matrix $[A] \in \mathbb{I}(\mathbb{R}^{n \times n})$ is defined by $[A]^{-1} = \square\{A^{-1} | A \in [A]\}$.

Note that $[A]^{-1} \in \mathbb{I}(\mathbb{R}^{n \times n})$. If $[A] = [\underline{A}, \bar{A}]$ then $[A] \geqslant 0$ means that $\underline{A} \geqslant 0$. If no ambiguity could arise when $[x] \in \mathbb{I}(\mathbb{R}^1)$ and $\underline{x} = \bar{x}$ then one may write $[x] = x$. Similar remarks apply to $[x] \in \mathbb{I}(\mathbb{R}^n)$ and to $[A] \in \mathbb{I}(\mathbb{R}^{m \times n})$.

Recall that $A = (A_{i,j}) \in \mathbb{R}^{n \times n}$ is an M-matrix if and only if $\exists A^{-1} \geqslant 0$ and $A_{i,j} \leqslant 0\ (i, j = 1, \ldots, n)$ $(i \neq j)$.

Definition 7. The matrix $[A] \in \mathbb{I}(\mathbb{R}^{n \times n})$ is an (interval) M-matrix if and only if $(\forall A \in [A])\ A \in \mathbb{R}^{n \times n}$ is an M-matrix.

Furthermore $[A]$ is an M-matrix if and only if $\underline{A}$ and $\bar{A}$ are M-matrices, and if $[A]$ is an M-matrix then $[A]$ is regular and $[A]^{-1} = [\bar{A}^{-1}, \underline{A}^{-1}] \geqslant 0$. More generally, if $\underline{A}$ and $\bar{A}$ are regular, $\underline{A}^{-1} \geqslant 0$ and $\bar{A}^{-1} \geqslant 0$ then $[A]$ is regular and $[\bar{A}^{-1}, \underline{A}^{-1}] \geqslant 0$. For more detail see [32].

Definition 8. The matrix $[A]$ is inverse nonnegative if and only if $[A]$ is regular and $[A]^{-1} \geqslant 0$.

Definition 9. The comparison matrix $\langle [A] \rangle = (\langle [A] \rangle_{i,j}) \in \mathbb{I}(\mathbb{R}^{n \times n})$ of $[A] = ([A]_{i,j}) \in \mathbb{R}^{n \times n}$ is defined by

$$\langle [A] \rangle_{i,j} = \begin{cases} -|[A]_{i,j}| & (i \neq j), \\ \langle [A]_{i,i} \rangle & \text{otherwise.} \end{cases}$$

Definition 10. The interval matrix $[A] \in \mathbb{I}(\mathbb{R}^{n \times n})$ is an (interval) H-matrix if and only if $\langle [A] \rangle$ is an M-matrix.

Definition 11. $[A] \in \mathbb{I}(\mathbb{R}^{n \times n})$ is strictly diagonally dominant if and only if

$$\langle [A]_{i,i} \rangle > \sum_{j \neq i} |[A]_{i,j}| \quad (i = 1, \ldots, n).$$

If $[A]$ is strictly diagonally dominant then $[A]$ is an H-matrix. Every M-matrix is an H-matrix but not conversely.

The expression $[x] \overset{\circ}{\subset} [y]$ where $[x], [y] \in \mathbb{I}(\mathbb{R}^1)$ means that $\underline{y} < \underline{x} \leqslant \bar{x} < \bar{y}$. Similarly $[A] \overset{\circ}{\subset} [B]$ means that $[A]_{i,j} \overset{\circ}{\subset} [B]_{i,j}\ (i = 1, \ldots, m; j = 1, \ldots, n)$ where $[A], [B] \in \mathbb{I}(\mathbb{R}^{m \times n})$. Recall that the spectral radius $\rho(A)$ of $A \in \mathbb{R}^{n \times n}$ is defined by

$$\rho(A) = \max\{|\lambda| \mid \lambda \in \mathbb{C}\ Ax = \lambda x,\ x \in \mathbb{C}^n\}.$$

Definition 12. The function $[f]: \mathbb{I}(D) \subseteq \mathbb{I}(\mathbb{R}^1) \to \mathbb{I}(\mathbb{R}^1)$ is an interval extension of the function $f: D \subseteq \mathbb{R}^1 \to \mathbb{R}^1$ if $f(x) = [f](x)$ $(x \in D)$ and $f(x) \in [f]([x])$ $(\forall x \in [x] \in \mathbb{I}(D))$.

Definition 13. The function $[f]: \mathbb{I}(D) \subseteq \mathbb{I}(\mathbb{R}^1) \to \mathbb{I}(\mathbb{R}^1)$ is an interval enclosure of f if $f(x) \in [f]$ $([x])$ $(\forall x \in [x] \in \mathbb{I}(D))$.

Definition 14. The function $[f]: \mathbb{I}(D) \subseteq \mathbb{I}(\mathbb{R}^1) \to \mathbb{I}(\mathbb{R}^1)$ is an interval evaluation of f if for $(x \in [x] \in \mathbb{I}(D))$ $[f]([x])$ is obtained from the expression $f(x)$ by replacing x with $[x]$ and real arithmetic operations with interval arithmetic operations provided that $[f]([x])$ exists.

See [6,32] for more detail.

Definition 15. The function $[f]: \mathbb{I}(D) \to \mathbb{I}(\mathbb{R}^1)$ is inclusion isotonic if and only if $([x] \subseteq [y] \in \mathbb{I}(D))$ $\Rightarrow ([f]([x]) \subseteq [f]([y]))$.

Definition 16. The function $[f]: \mathbb{I}(D) \subseteq \mathbb{I}(\mathbb{R}^1) \to \mathbb{I}(\mathbb{R}^1)$ is continuous at $[\hat{x}] \in \mathbb{I}(D)$ if and only if, given $\varepsilon > 0$, there exists $\delta > 0$ such that $(q([x], [\hat{x}]) < \delta \wedge [x] \in \mathbb{I}(D)) \Rightarrow (q([f]([x]), [f]([\hat{x}])) < \varepsilon)$ where $q: \mathbb{I}(\mathbb{R}^1) \times \mathbb{I}(\mathbb{R}^1) \to \mathbb{R}^1$ is defined by $q([x], [y]) = \max\{|\underline{x} - \underline{y}|, |\bar{x} - \bar{y}|\}$.

The concepts of interval extension, interval evaluation, interval enclosure, and inclusion isotonicity are defined componentwise for $f: D \subseteq \mathbb{R}^n \to \mathbb{R}^m$ $(n, m > 1)$ and continuity is defined using the metrics $\sigma_n: \mathbb{I}(\mathbb{R}^n) \times \mathbb{I}(\mathbb{R}^n) \to \mathbb{R}^1$ defined by $\sigma_n([x], [y]) = \|q_n([x], [y])\|_\infty$ and $\sigma_{m,n}: \mathbb{I}(\mathbb{R}^{m \times n}) \times \mathbb{I}(\mathbb{R}^{m \times n}) \to \mathbb{R}^1$ defined by $\sigma_{m,n}([A], [B]) = \|q_{m,n}([A], [B])\|_\infty$, in which $q_n([x], [y]) = (q([x]_i, [y]_i)) \in \mathbb{R}^n$ and $q_{m,n}([A], [B]) = (q([A]_{i,j}, [B]_{i,j})) \in \mathbb{R}^{m \times n}$.

3. Algebraic equations

Interval algorithms for bounding the solutions of systems of nonlinear algebraic equations and optimization problems often require interval algorithms for bounding the solutions of systems of linear algebraic equations, and for determining the nonexistence, existence and uniqueness of solutions of systems of linear and nonlinear algebraic equations in given regions. Several interval arithmetic algorithms for bounding zeros of $f: \mathbb{R}^n \to \mathbb{R}^n$ are extensions of well-known real arithmetic algorithms for estimating zeros of f of the kind described in [34].

3.1. Linear algebraic equations

Many significant contributions to the problem of bounding the solutions of systems of linear algebraic equations up to 1990 are described in [6,32,54] and in references therein. Some more recent contributions are mentioned in this section. Relative beginners to this subject should consult [6,32].

The system of linear interval algebraic equations (interval linear system) $[A]x = [b]$ where $[A] \in \mathbb{I}(\mathbb{R}^{n \times n})$ and $[b] \in \mathbb{I}(\mathbb{R}^n)$ is the set of systems of real linear algebraic equations (real linear systems)

$\{Ax = b \mid \exists A \in [A] \wedge \exists b \in [b]\}$. The solution set $\Sigma([A],[b])$ of $[A]x = [b]$ is defined by

$$\Sigma([A],[b]) = \{x \in \mathbb{R}^n \mid \exists A \in [A] \wedge \exists b \in [b], \ Ax = b\}.$$

In general $\Sigma([A],[b])$ is not convex (it is a polytope) [41] and is NP-hard to compute [25,24,44]. If $[A]$ is regular then $\Sigma([A],[b])$ is bounded and the interval hull $[A]^H[b] = \square\Sigma([A],[b])$ of $\Sigma([A],[b])$ is defined: indeed if $[A]$ is regular then [32]

$$[A]^H[b] = \square\{A^{-1}b \mid A \in [A] \wedge b \in [b]\},$$

where $[A]^H$ is called the hull inverse of $[A]$. Note that $[A]^H[b]$ is not an interval matrix-interval vector product; it represents the result of the *mapping* $[A]^H$ applied to $[b]$.

If $[A]$ is merely regular and $n > 2$ then it is difficult sharply to bound $[A]^H[b]$. Fortunately, in many cases of practical importance $[A]$ and $[b]$ have properties which make it easier to bound $[A]^H[b]$ sharply.

Alefeld et al. [7] have considered the set of inequalities which characterize the sets $S = \{x \in \mathbb{R}^n \mid Ax = b, \ A \in [A], \ b \in [b]\}$ and $S_{\mathrm{sym}} = \{x \in \mathbb{R}^n \mid Ax = b, \ A = A^{\mathrm{T}} \in [A], \ b \in [b]\} \subseteq S$. Very little work appears to have been done on bounding the solution set of over-determined linear systems. However, Rohn [43] has considered the problem of bounding $\Sigma([A],[b])$ where $[A] \in \mathbb{I}(\mathbb{R}^{m \times n})$ with $m \geqslant n$, and has described an algorithm for bounding $\Sigma([A],[b])$ which terminates in a finite number of iterations under appropriate conditions.

The interval Gaussian algorithm IGA [6] for bounding $\Sigma([A],[b])$ consists of computing a bound $[x]^G \in \mathbb{I}(\mathbb{R}^n)$ on $\Sigma([A],[b])$ as follows.

Algorithm 1 (IGA).
Set $[A]^{(1)} = [A]$, $[b]^{(1)} = [b]$ *and for* $k = 1,\ldots,n-1$ *compute* $[A]_{i,j}^{(k+1)}$ *and* $[b]_i^{(k+1)}$ *from*

$$[A]_{i,j}^{(k+1)} = \begin{cases} [A]_{i,j}^{(k)} & (1 \leqslant i \leqslant k; 1 \leqslant j \leqslant n), \\ [A]_{i,j}^{(k)} - ([A]_{i,k}^{(k)}[A]_{k,j}^{(k)})/[A]_{k,k}^{(k)} & (k+1 \leqslant i,j \leqslant n), \\ 0 & otherwise \end{cases}$$

and

$$[b]_i^{(k+1)} = \begin{cases} [b]_i^{(k)} & (1 \leqslant i \leqslant k), \\ [b]_i^{(k)} - ([A]_{i,k}^{(k)}[b]_k^{(k)})/[A]_{k,k}^{(k)} & (k+1 \leqslant i \leqslant n). \end{cases}$$

Then for $i = n, \ n-1, \ldots, 1$ *compute* $[x]_i^G$ *from*

$$[x]_n^G = [b]_n^n/[A]_{n,n}^{(n)},$$

$$[x]_i^G = \left([b]_i^{(n)} - \sum_{j=i+1}^n [A]_{i,j}^{(n)}[x]_j^G \right) \bigg/ [A]_{i,i}^{(n)}.$$

The algorithm IGA reduces to the real Gaussian elimination algorithm if $[A]$ and $[b]$ are replaced with $A \in \mathbb{R}^{n \times n}$ and $b \in \mathbb{R}^n$, respectively, and is executable if and only if for $k = 1,\ldots,n$, $0 \notin [A]_{k,k}^{(k)}$ [6]. One writes $[x]^G = \mathrm{IGA}([A],[b]) = [A]^G[b]$ and calls $[A]^G$ the Gauss inverse of $[A]$. As for $[A]^H$, $[A]^G$ is a mapping such that $[A]^G[b]$ is the result of applying the algorithm IGA to $([A],[b])$.

If IGA is executable then $[A]^H[b] \subseteq [A]^G[b]$ [32]. Rohn [45] has illustrated the degree of over-estimation produced by IGA. The algorithm IGA could break down if for some $k \in \{1, \ldots, n\}$, $0 \in [A]_{k,k}^{(k)}$ even when $[A]$ is regular. In order to enhance the applicability of IGA $[A]x = [b]$ is usually pre-conditioned with $\check{A}^{-1}$ to obtain $[M]x = [r]$ where $[M] = \check{A}^{-1}[A]$ and $[r] = \check{A}^{-1}[b]$ so that $[M] = [I] + [-Q, Q]$ where $[I] \in \mathbb{I}(\mathbb{R}^{n \times n})$ is the unit interval matrix and $Q = |\check{A}^{-1}|\mathrm{rad}([A])$ [43].

Hansen [15] has proposed a method for bounding $[M]^H[r]$, noting that $[A]^H[b] \subseteq [M]^H[r]$ [32]. Hansen assumes that $[M]$ is diagonally dominant in order to ensure that $[M]$ is regular and therefore that $[M]^H[r]$ is bounded. Hansen's method is intended to bound $[A]^H[b]$ when $w([A])$ and $w([b])$ are large (i.e. when IGA is likely to break down), but requires the solution of $2n$ real linear systems with the same real matrix.

Rohn [42] has shown that if $\rho(Q) < 1$ where $\rho: \mathbb{R}^{n \times n} \to \mathbb{R}^1$ is as defined in Section 2, then $\exists B = \underline{M}^{-1} = (I - Q)^{-1} \geqslant 0$ and has described an algorithm requiring the solution of only one real linear system. Rohn's algorithm is essentially a reformulation of Hansen's algorithm, and is contained in Theorem 17.

Theorem 17 (Rohn [42]). *If $\rho(Q) < 1$ and $[x] = [M]^H[r]$ then for $i = 1, \ldots, n$*

$$\underline{x}_i = \min\{\hat{x}_i, v_i \hat{x}_i\}$$

and

$$\bar{x}_i = \max\{\tilde{x}_i, v_i \tilde{x}_i\},$$

where

$$\hat{x}_i = -x_i^* + B_{i,i}(\check{r} + |\check{r}|)_i,$$

$$\tilde{x}_i = x_i^* + B_{i,i}(\check{r} - |\check{r}|)_i,$$

$$x_i^* = (B(|\check{r}| + \mathrm{rad}([r])))_i$$

and

$$v_i = 1/(2B_{i,i} - 1) \in (0, 1].$$

Ning and Kearfott [33] have extended the technique of Hansen [15] and Rohn [42] with a formula that bounds $[A]^H[b]$ when $[A]$ is an H-matrix: when $\check{A}$ is diagonal the bound is equal to $[A]^H[b]$. Ning and Kearfott note that $\underline{M}^{-1} \geqslant 0$ implies both that $\rho(Q) < 1$ and that $[M]$ is strongly regular. This leads to the following restatements of Hansen's and Rohn's results.

Theorem 18 (Ning and Kearfott [33]). *Suppose that $\underline{M}^{-1} \geqslant 0$. Let*

$$s^{(i)} = \begin{cases} \bar{r}_i & (j = i), \\ \max\{-\underline{r}_j, \bar{r}_j\} & (j \neq i; j = 1, \ldots, n), \end{cases}$$

$$t^{(i)} = \begin{cases} \underline{r}_i & (j = i), \\ \min\{\underline{r}_j, \bar{r}_j\} & (j \neq i; j = 1, \ldots, n) \end{cases}$$

and

$$c_i = 1/(2B_{i,i} - 1).$$

Then $[M]^H[r] = [\underline{x}, \bar{x}]$, *where*

$$\underline{x}_i = \begin{cases} c_i e_i^{\mathrm{T}} B t^{(i)} & (\underline{x}_i \geqslant 0), \\ e_i^{\mathrm{T}} B t^{(i)} & (\underline{x}_i < 0), \end{cases}$$

$$\bar{x}_i = \begin{cases} e_i^{\mathrm{T}} B s^{(i)} & (\bar{x}_i \geqslant 0), \\ c_i e_i^{\mathrm{T}} B s^{(i)} & (\underline{x}_i < 0), \end{cases}$$

in which e_i^{T} *is the unit vector with ith component unity and all other components zero.*

Theorem 17 is valid with $\underline{M}^{-1} \geqslant 0$ replacing $\rho(Q) < 1$.

Ning and Kearfott [33] have also obtained some interesting results among which are the following.

Theorem 19 (Ning and Kearfott [33]). *Suppose that* $[A] \in \mathbb{I}(\mathbb{R}^{n \times n})$ *and* $[b] \in \mathbb{I}(\mathbb{R}^n)$ *and that* $D \in \mathbb{R}^{n \times n}$ *is diagonal and nonsingular. If* $[M] = D^{-1}[A]$ *and* $[r] = D^{-1}[b]$ *then* $\Sigma([A], [b]) = \Sigma([M], [r])$.

Theorem 20 (Ning and Kearfott [33]). *Let* $[A] \in \mathbb{I}(\mathbb{R}^{n \times n})$ *be an H-matrix, let* $u = \langle [A] \rangle^{-1} |[b]|$ *and for* $i = 1, \ldots, n$ *let* $d_i = (\langle [A] \rangle^{-1})_{i,i}$, *let* $\alpha_i = \langle [A]_{i,i} \rangle - 1/d_i$ *and let* $\beta_i = u_i/d_i - |[b]_i|$. *Then* $[A]^H[b] \subseteq [x]$ *where*

$$[x]_i = \frac{[b]_i + [-\beta_i, \beta_i]}{[A]_{i,i} + [-\alpha_i, \alpha_i]} \quad (i = 1, \ldots, n).$$

Also if $m([A])$ *is diagonal then* $[A]^H[b] = [x]$.

Theorem 21 (Ning and Kearfott [33]). *Suppose that* $[A] \in \mathbb{I}(\mathbb{R}^{n \times n})$ *and* $[b] \in \mathbb{I}(\mathbb{R}^n)$ *are such that* $[A]$ *is inverse positive and* $\underline{b} = -\bar{b} \neq 0$ *so that* $0 \in [b]$ *but* $w([b]) \neq 0$. *Then* $[A]^H[b] = [M]^H[r]$ *where* $[M] = \breve{A}^{-1}[A]$ *and* $[r] = \breve{A}^{-1}[b]$.

Mayer and Rohn [29] have proved necessary and sufficient conditions for the applicability of IGA (Algorithm 1) when $[A]x = [b]$ is preconditioned with $\breve{A}^{-1}$ and when partial pivoting is used. Partial pivoting in Algorithm 1 to produce the algorithm PIGA [29] occurs if either for $k = 1, \ldots, n-1$ two of the rows $k, k+1, \ldots, n$ are interchanged in $[A]^{(k)}$ such that $|[A]_{k,k}^{(k)}| = \max\{|[A]_{i,k}^{(k)}| \mid k \leqslant i \leqslant n, \ 0 \notin [A]_{i,k}^{(k)}\}$ or for $k = 1, \ldots, n-1$ the corresponding columns are permuted so that $|[A]_{k,k}^{(k)}| = \max\{|[A]_{k,j}^{(k)}| \mid k \leqslant j \leqslant n, \ 0 \notin [A]_{k,j}^{(k)}\}$.

Theorem 22 (Mayer and Rohn [29]). *Let* $[A] = [I] + [-Q, Q] \in \mathbb{I}(\mathbb{R}^{n \times n})$ *and* $[b] \in \mathbb{I}(\mathbb{R}^n)$. *Then the following conditions are equivalent:*

(a) *IGA is applicable (i.e., $[x]^G$ exists);*
(b) *PIGA is applicable;*
(c) *$I - Q$ is an M-matrix;*
(d) *$\rho(Q) < 1$;*
(e) *$[A]$ it is regular;*
(f) *$[A]$ is strongly regular;*
(g) *$[A]$ is an H-matrix.*

It should be noted that Theorem 22 is applicable only when $[A] = [I] + [-Q, Q]$.

Shary [49] has described an algorithm for bounding $\Sigma([A], [b])$ which reduces to solving one-point linear system in $\mathbb{R}^{2n}$. A C version of the algorithm is available from the author, who claims, by presenting several examples, that his algorithm seems to be superior to those of Hansen [15], Rohn [42] and Ning and Kearfott [33].

As explained in [32] Krawczyk iteration may be used to bound $[A]^H[b]$ when $[A]$ is strongly regular, but that Gauss–Seidel iteration is superior. Practical details regarding the implementation of IGA and Krawczyk and Gauss–Seidel iteration are given in [21]. A motive for the Krawczyk iteration for bounding the solution of $[A]x = [b]$ discussed in [32] is as follows. If $C, A \in \mathbb{R}^{n \times n}$ and $b \in \mathbb{R}^n$ and $CAx = Cb$ then $x = Cb - (CA - I)x$, giving rise to the Krawczyk iterative procedure

$$[x]^{(0)} = [x],$$

$$[x]^{(k+1)} = \{C[b] - (C[A] - I)[x]^{(k)}\} \cap [x]^{(k)} \quad (k \geq 0),$$

where C is an appropriate pre-conditioner. It is shown in [32] that $(\forall k \geq 0)\ ([A]^H[b] \subseteq [x]) \Rightarrow ([A]^H[b] \subseteq [x]^{(k)})$.

As explained in [32] the interval Gauss–Seidel iterative procedure for bounding $\Sigma([A], [b])$ consists of generating the sequence $([x]^{(k)})$ from

$$[x]_i^{(0)} = [x]_i,$$

$$[x]_i^{(k+1)} = \Gamma\left([A]_{i,i}, [b]_i - \sum_{j<i}[A]_{i,j}[x]_j^{(k+1)} - \sum_{j>i}[A]_{i,j}[x]_j^{(k)}, [x]_i^{(k)}\right), \tag{1}$$

where $i = 1, \ldots, n$ and where if $[u], [v], [w] \in \mathbb{I}(\mathbb{R}^1)$ then

$$\Gamma([u], [v], [w]) = \begin{cases} ([v]/[u]) \cap [w] & (0 \notin [a]), \\ \square([w] \setminus (\underline{v}/\underline{u}, \underline{v}/\overline{u})) & ([b] > 0 \in [a]), \\ \square([w] \setminus (\overline{v}/\overline{u}, \overline{v}/\underline{u})) & ([b] < 0 \in [a]), \\ [x] & (0 \in [a] \wedge 0 \in [b]). \end{cases}$$

If (1) is expressed as $[x]^{(k+1)} = \Gamma([A], [b], [x]^{(k)})$ then [32]

$$\Sigma([A], [b]) \cap [x] \subseteq \Gamma([A], [b], [x]) \subseteq [x],$$

$$([A][x] \cap [b] = \emptyset) \Rightarrow (\Gamma([A], [b], [x]) = \emptyset),$$

$$\Gamma(I, [b], [x]) = [b] \cap [x],$$

from which it follows that the Krawczyk iterates contain the Gauss–Seidel iterates. For more detail see [32]. Frommer and Mayer [13] have presented interval versions of two-stage iterative methods for bounding the solutions of linear systems.

3.2. Nonlinear algebraic equations

The most significant contributions to the problem of bounding the solutions of systems of nonlinear algebraic equations up to 1990 are described in [6,32] and in many references therein. Some recent developments, together with several references and information about Fortran 90 software are in [21]. Some results which have appeared since 1990 are described in this section.

Alefeld [3] has given a survey of the properties of the interval Newton algorithm (IN) for bounding a zero $x^* \in [x] \in \mathbb{I}(D)$ of a function $f : D \subset \mathbb{R}^n \to \mathbb{R}^n$ with $f \in C^1(D)$.

Algorithm 2 (IN).

$[x]^{(0)} = [x]$
for $k = 0, 1, \ldots$
   $x^{(k)} \in [x]^{(k)}$
   $[N]([x]^{(k)}) = x^{(k)} - \text{IGA}([f']([x]^{(k)}), f(x^{(k)}))$
   $[x]^{(k+1)} = [N]([x]^{(k)}) \cap [x]^{(k)}$

In Algorithm 2, $x^{(k)} \in [x]^{(k)}$ is arbitrary, but often one sets $x^{(k)} = m([x]^{(k)})$. As explained in [3] if IGA is not applicable for the interval enclosure $[f']([x]^{(0)})$ of the Jacobian f' of f in $[x]^{(0)}$ and an arbitrary right-hand side, then the problem may be avoided by using the Krawczyk operator [6] $[K] : \mathbb{I}(D) \times \mathbb{I}(\mathbb{R}^n) \times \mathbb{I}(\mathbb{R}^{n \times n}) \to \mathbb{I}(\mathbb{R}^n)$ defined by

$$[K]([x], x, C) = x - Cf(x) + (I - C[f']([x]))([x] - x)$$

in various ways, as, for example, in Algorithm 3.

Algorithm 3.

$[x]^{(0)} = [x]$
for $k = 0, 1, \ldots$
   $x^{(k)} \in [x]^{(k)}$
   $C^{(k)} = (m([f']([x]^{(k)})))^{-1}$
   $[K]([x]^{(k)}, x^{(k)}, C^{(k)}) = x^{(k)} - C^{(k)} f(x^{(k)}) + (I - C^{(k)}[f']([x]^{(k)}))([x]^{(k)} - x^{(k)})$
   $[x]^{(k+1)} = [K]([x]^{(k)}, x^{(k)}, C^{(k)}) \cap [x]^{(k)}$

In Algorithm 3, $x^{(k)} \in [x]^{(k)}$ is arbitrary, but often one sets $x^{(k)} = m([x]^{(k)})$.

If $[K]([x], x, C) \subseteq [x]$ then f has at least one zero $x^* \in [K]([x], x, C)$ [6]. This fact has been used by Alefeld et al. [5] rigorously to bound x^* when $x^{(k)} \to x^*$ ($k \to \infty$) where the real Newton sequence $(x^{(k)})$ is generated from

$$x^{(k+1)} = x^{(k)} - f'(x^{(k)})^{-1} f(x^{(k)}) \quad (k \geq 0).$$

Frommer and Mayer [11] have described modifications of the interval Newton method for bounding a zero $x^* \in [x]$ of $f: D \in \mathbb{R}^n \to \mathbb{R}^n$ where D is an open convex set and $[x] \subseteq D$. These modifications combine two ideas: reusing the same evaluation of the Jacobian $J = f'$ of f $s > 1$ times; approximately solving the Newton linear system by using a 'linear' iterative procedure. It is shown [11] that the R-order[1] of these methods can be $s + 1$. The class of methods which is described in [11] is contained in Algorithm 4.

Algorithm 4.

for $k = 0, 1, \dots$
  $[x]^{(k,0)} = [x]^{(k)}$
for $\ell = 0, \dots, s_k - 1$ {Use $[M]^{(k)}, [N]^{(k)}$ s_k times.}
  $[x]^{(k,\ell,0)} = [x]^{(k,\ell)}$
  for $m = 0, \dots, r_{k,\ell} - 1$ {'Solve' the Newton linear system.}
    $[x]^{(k,\ell,m+1)} = \{x^{(k,\ell)} - \mathrm{IGA}([M]^{(k)}, [N]^{(k)}(x^{(k,\ell)} - [x]^{(k,\ell)}) + f(x^{(k,\ell)}))\} \cap [x]^{(k,\ell,m)}$
  $[x]^{(k,\ell+1)} = [x]^{(k,\ell,r_{k,\ell})}$
$[x]^{(k+1)} = [x]^{(k,s_k)}$

In Algorithm 4, $x^* \in [x]^{(0)}$, $[J]([x]^{(k)}) = [M]^{(k)} - [N]^{(k)}$ is a splitting of the Jacobian $[J]([x]^{(k)}) = [f']([x]^{(k)})$ of f such that IGA is executable for $[M]^{(k)}$, $x^{(k,\ell)} \in [x]^{(k,\ell)}$ is chosen arbitrarily and $k = 0, \dots, s_k$, and $r_{k,\ell}$ ($\ell = 0, \dots, s_k - 1$) are given integers. As explained in [11] Algorithm 4 contains extensions of several known iterative methods using real arithmetic [34] and interval arithmetic [6]. Numerical experience and theoretical results [11] indicate that Algorithm 4 can be more efficient than the algorithms which it contains as special cases. See also [12,26].

Kolev [23] has used an algorithm of Yamamura [65] to describe interval Newton-like algorithms for bounding all of the zeros of nonlinear systems in given boxes. The nonlinear system $f(\tilde{x}) = (f_i(\tilde{x})) = 0 \in \mathbb{R}^{\tilde{n}}$ is transformed into separable form $F(x) = (F_i(x)) = 0 \in \mathbb{R}^n$ where

$$F_i(x) = \sum_{j=1}^{n} F_{i,j}(x_j) \quad (i = 1, \dots, n),$$

$n > \tilde{n}$ and $[b]_{i,j} \in \mathbb{I}(\mathbb{R}^1)$ and $a_{i,j} \in \mathbb{R}^1$ ($i,j = 1, \dots, n$) are determined so that $(\forall x \in [x]_j)$ $f_{i,j}(x_j) \in [b]_{i,j} + a_{i,j}x_j$ ($j = 1, \dots, n$), and $(\forall x \in [x]_j)$ $F_i(x) \in -Ax + [b]_i$ ($i = 1, \dots, n$) where $A = (-a_{i,j}) \in \mathbb{R}^{n \times n}$ and $[b]_i = (\sum_{j=1}^{n} [b]_{i,j}) \in \mathbb{I}(\mathbb{R}^1)$ ($i = 1, \dots, n$). Then $(F(x^*) = 0 \wedge x^* \in [x]) \Rightarrow (x^* \in (A^{-1}[b] \cap [x]))$. Kearfott [20] has shown how a related idea may be used to reduce over-estimation when bounding the solutions of nonlinear systems.

---

[1] R-order is defined in [12].

Yamamura et al. [66] have described a computational test for the nonexistence of a zero of $f : \mathbb{R}^n \to \mathbb{R}^n$ in a convex polyhedral region $X \subset \mathbb{R}^n$ by formulating a linear programming problem whose feasible set contains every zero of f in X. The proposed nonexistence test has been found considerably to improve the efficiency of interval algorithms based on the Krawczyk operator. Furthermore, the proposed test has been found to be very effective if $f(x)$ contains several linear terms and a relatively small number of nonlinear terms as, for example, if $f(x) = Ax + \phi(x)$ where $A \in \mathbb{R}^{n \times n}$ and $\phi : \mathbb{R}^n \to \mathbb{R}^n$ is a diagonal mapping.

Rump [46] has given an extensive 73 page description (with 91 references) of methods for validating zeros of dense and sparse nonlinear systems, and has described an iterative procedure [47] for computing validated solutions of nonlinear systems which is an improvement of that described in [46] and in which epsilon inflation is used.

Epsilon inflation is a valuable tool for verifying and enclosing zeros of a function $f : \mathbb{R}^n \to \mathbb{R}^n$ by expressing $f(x) = 0$ in fixed-point form $g(x) = x$ so that $(f(x^*) = 0) \Leftrightarrow (g(x^*) = x^*)$ and using Brouwer's fixed-point theorem. Mayer [27] has described how epsilon inflation may be used to determine $[x] \in \mathbb{I}(\mathbb{R}^n)$ such that $[g]([x]) \subseteq [x]$ where $[g] : \mathbb{I}(\mathbb{R}^n) \to \mathbb{I}(\mathbb{R}^n)$ is an interval extension of the function $g : \mathbb{R}^n \to \mathbb{R}^n$. Several procedures for applying epsilon inflation and several problems to which epsilon inflation is applicable are described in [28]. The following algorithm, based on one in [28] illustrates how, given an estimate $\tilde{x}$ of x^*, a box $[x]$ may be determined such that $x^* \in [x]$.

Algorithm 5.

Data: $\varepsilon \approx 0.1$, $\tilde{x} \in \mathbb{R}^1$, $k_{MAX} \approx 3$

```
[x] = [x̃, x̃]
outer : do
  k = 0
  inner : do
    k := k + 1
    [y] = [x]_ε   ! Epsilon inflation
    [x] = [g]([y])
  if ([x] ⊆ [y]) then
    exit outer   ! x* ∈ [x]
  end if
  if (k = k_MAX) then
    ε := 5ε   ! The box [x] must be enlarged.
    exit inner
  end if
  end do inner
end do outer
```

Rump [48] has shown that the term $[-\eta, \eta]$ in the epsilon inflation formula

$$[x]_\varepsilon = \begin{cases} [x] + w([x])[-\varepsilon, \varepsilon] & (w[x] \neq 0), \\ [x] + [-\eta, \eta] & \text{otherwise} \end{cases}$$

is necessary, where η is the smallest representable positive machine number.

Wolfe [59,60] has established sufficient conditions for the uniqueness of a zero $x^* \in [x]$ of $f : \mathbb{R}^n \to \mathbb{R}^n$ using the second derivative operator $[L] : \mathbb{I}(\mathbb{R}^n) \times \mathbb{R}^{n \times n} \times \mathbb{R}^n \to \mathbb{I}(\mathbb{R}^n)$ of Qi [36] defined by

$$[L]([x], Y, y) = y - Yf(y) - \tfrac{1}{2}Y[f''] ([x])([x] - y)([x] - y).$$

Theorem 23 (Wolfe [59]). *If* $[L]([x], Y, y) \subseteq [x]$ *and* $w([L]([x], Y, y)) < w([x])$ *where* $y = m([x])$ *and* $Y = f'(y)^{-1}$ *then* $\exists x^* \in [L]([x], Y, y)$ *such that* $f(x^*) = 0$ *and* x^* *is unique in* $[x]$.

It is shown in [59] that under the hypotheses for the convergence of Newton's method given in [64] a box $[x] \in \mathbb{I}(\mathbb{R}^n)$ exists which satisfies the sufficient conditions of Theorem 23. Furthermore, it is shown in [59] how the second derivative operator can be used in a manner similar to that which has been done with the Krawczyk operator in [5].

Shen and Wolfe [53] have given improved forms of the existence and uniqueness tests of Pandian [35] and have related these to results due to Moore and Kioustelidis [31] and to Shen and Neumaier [51].

Recently, Zhang et al. [67] have described a test for the existence and uniqueness of a zero x^* of a continuously differentiable function $f : \mathbb{R}^n \to \mathbb{R}^m$ $(m \leqslant n)$ in a given box $[x] \subset \mathbb{I}(\mathbb{R}^n)$ using the Krawczyk-like operator $[\bar{K}] : \mathbb{I}(\mathbb{R}^n) \to \mathbb{I}(\mathbb{R}^n)$ defined by

$$[\bar{K}]([x]) = x - Yf(x) + (YA - Y[f']([x]))([x] - x),$$

where $x \in [x]$ and $Y \in \mathbb{R}^{n \times m}$ is a (2)-inverse [9] of $A = f'(x)$.

Theorem 24 (Zhang et al. [67]). *If* $[\bar{K}]([x]) \subseteq [x]$ *then* $\exists x^* \in [x] \cap (x + R(Y))$ *where* $R(Y)$ *is the range of* Y *such that* $Yf(x^*) = 0$, *and if also* $w([\bar{K}]([x])) < w([x])$ *then* x^* *is unique in* $[x] \cap (x + R(Y))$.

Zhang, Li and Shen have used the operator $[\bar{K}]$ to construct an extended Krawczyk–Moore algorithm which under appropriate conditions they have shown to converge Q-quadratically to x^*. Thus encouraged Wolfe [62] has determined similar existence and uniqueness tests using the second-derivative operator $[L^+]$ defined by

$$[L^+]([x], Y, y) = y - Yf(y) - \tfrac{1}{2}Y[f'']([x])([x] - y)([x] - y),$$

where $y = m([x])$, and $Y = f'(y)^+$ is the Moore–Penrose generalized inverse [9] of $f'(y)$ and has suggested that the algorithm of Yamamura [65] and the linear programming nonexistence test of Yamamura et al. [66] could be used effectively to extend the usefulness of $[L^+]$.

4. Optimization

The problems of global unconstrained and constrained optimization present great computational difficulties. Interval algorithms for rigorously bounding the solutions of several kinds of such problems now exist. Detailed descriptions of interval optimization algorithms together with several references to earlier work are in [38,40,16,21,22,58] which survey most of what has been done up to 1996.

4.1. Unconstrained optimization

Let $f : \mathbb{R}^n \to \mathbb{R}^1$ be a given function. Then the global unconstrained optimization problem is $\min_{x \in \mathbb{R}^n} f(x)$. In practice, interval methods for unconstrained optimization usually bound the solutions of the so-called *bound-constrained* problem $\min_{x \in [x]} f(x)$ where $[x] \in \mathbb{I}(\mathbb{R}^n)$ [16,40] although $w([x])$ could be limited by twice the largest available machine number. Ratschek and Voller [39] have shown how interval optimization algorithms can be made to work over such domains.

Jansson and Knüppel [19] have described an interval branch-and-bound algorithm for the bound-constrained problem. The algorithm consists of three sub-algorithms and requires no derivatives of the objective function. Numerical results for 22 nontrivial problems are reported and indicate the effectiveness of the algorithm.

4.2. Constrained optimization

Let $f : \mathbb{R}^n \to \mathbb{R}^1$, $c_i : \mathbb{R}^n \to \mathbb{R}^1$ $(i = 1, \ldots, m_1)$ and $h_j : \mathbb{R}^n \to \mathbb{R}^1$ $(j = 1, \ldots, m_E)$ be given functions. Then a global constrained optimization problem is $\min_{x \in X} f(x)$ where

$$X = \{x \in [x] \mid c_i(x) \leqslant 0 \ (i = 1, \ldots, m_1), \ h_j(x) = 0 \ (j = 1, \ldots, m_E)\} \tag{2}$$

and $[x] \in \mathbb{I}(\mathbb{R}^n)$.

It is relatively easy to construct interval algorithms for the bound-constrained problem in which $n = 1$ and $X = [x] \in \mathbb{I}(\mathbb{R}^1)$, and for the inequality-constrained problem in which $n = 1$ $m_I \geqslant 0$ and $m_E = 0$ [56]. Both problems are much more difficult when $n > 1$ [55,57,63] as is the problem in which $n > 1$, $m_I > 0$ and $m_E > 0$ [21,22]. A detailed account of the techniques used in the construction of interval algorithms for constrained global optimization problems up to 1996 is in [21] which contains 247 references, and in Chapter 2 of [22] which contains 67 references. A brief survey of some of the fundamental ideas which are used in interval methods for global optimization is in [58].

If in Eq. (2) $f, c_i, h_j \in C^2(D)$ $(i = 1, \ldots, m_I; j = 1, \ldots, m_E)$ then the Fritz John optimality conditions [8] may be used to solve the constrained optimization problem: details are given in [16–18]. Kearfott [21] has shown how the Fritz John conditions may be used when

$$X = \{x \in [x] \mid c_i(x) \leqslant 0 \ (i = 1, \ldots, m_I)\} \tag{3}$$

as well as when X is defined by Eq. (2). The system of nonlinear algebraic equations corresponding to the Fritz John conditions depends upon the constraint set X. The set X defined by (3) gives rise to a simpler nonlinear system than the system corresponding to X defined by (2) as explained in [21], in which appropriate software is discussed.

Recently, Adjiman et al. [2] have described the so-called alpha branch-and-bound (αBB) method for the constrained optimization problem with X defined by Eq. (2) when $f, c_i, h_j \in C^2(D)$ $(i = 1, \ldots, m_I; \ j = 1, \ldots, m_E)$. The αBB method depends upon the ability to generate a sharp convex underestimate L of the objective function f of the form

$$L(x) = f(x) + \sum_{i=1}^{n} \alpha_i (\underline{x}_i - x_i)(\bar{x}_i - x_i).$$

Methods for the determination of the parameters $\alpha_i > 0$ $(i = 1, \ldots, n)$ which ensure that L is convex $(\forall x \in [x])$ are described in [2]. One method in particular has been found to be very effective [1] and has been obtained by noting that if $H_L : \mathbb{R}^n \to \mathbb{R}^{n \times n}$ and $H_f : \mathbb{R}^n \to \mathbb{R}^{n \times n}$ are the Hessians of L and f, respectively, then

$$H_L(x) = H_f(x) + 2\Delta,$$

where $\Delta = \mathrm{diag}(\alpha_i)$ and using the following theorems.

Theorem 25. *The function L is convex if and only if the Hessian $H_L(x)$ of L is positive semi-definite $(\forall x \in [x])$.*

Theorem 26. *If $[d] \in \mathbb{I}(\mathbb{R}^n)$, $[A] \in \mathbb{I}(\mathbb{R}^{n \times n})$ is symmetric and*

$$\alpha_i = \max \left\{ 0, -\frac{1}{2} \left(\underline{a}_{i,i} - \sum_{j \neq i} |a|_{i,j} d_j / d_i \right) \right\} \quad (i = 1, \ldots, n),$$

where $|a|_{i,j} = \max\{ |\underline{a}_{i,j}|, |\bar{a}_{i,j}| \}$ then $(\forall A \in [A])$ $A_L = A + 2\Delta$ with $\Delta = \mathrm{diag}(\alpha_i)$ is positive semi-definite.

Using the theoretical results described in [2] Adjiman et al. [1] have implemented αBB and have obtained computational results for various test problems [10], chemical engineering design problems, generalized geometric programming problems and batch process design problems.

Theoretical devices other than the Fritz John conditions have been used to bound the solutions of various minimax problems. Shen et al. [52] have described an interval algorithm for bounding the solutions of the minimax problem

$$\min_{z \in [\hat{z}]} \max_{y \in [\hat{y}]} f(y, z),$$

where $f : D \subseteq \mathbb{R}^m \times \mathbb{R}^n \to \mathbb{R}^1$ is a given function with $f \in C^2(D)$, and $[\hat{z}] \in \mathbb{I}(\mathbb{R}^m)$ and $[\hat{y}] \in \mathbb{I}(\mathbb{R}^n)$ are such that $[\hat{z}] \times [\hat{y}] \subseteq \mathbb{I}(D)$. If $[\hat{x}] = ([\hat{y}], [\hat{z}]) \in \mathbb{I}(D)$ then the algorithm systematically sub-divides $[\hat{x}]$ into sub-boxes. If $[x] \subseteq [\hat{x}]$ and $[x]$ does not contain a minimax point x^* then $[x]$ is discarded using the following rules. If $f'(x) = (f'_y(x), f'_z(x))$ where $f'_y(x) = (\partial/\partial y_i f(y, z)) \in \mathbb{R}^n$ and $f'_z(x) = (\partial/\partial z_i f(y, z)) \in \mathbb{R}^m$ and for some $i \in \{1, \ldots, m\}$, $0 \notin [f'_z]([x])$ then $x^* \notin [x]$. If $f''(x) = (f''_y(x), f''_z(x))$ where $f''_y(x) = (\partial^2/\partial y_i \partial y_j f(y, z)) \in \mathbb{R}^{n \times n}$ and $f''_z(x) = (\partial^2/\partial z_i \partial z_j f(y, z)) \in \mathbb{R}^{m \times m}$ and for some $i \in \{1, \ldots, m\}$, $[f''_z]_{i,i} > 0$ then $x^* \notin [x]$. If $(\forall x \in ([y], [z]))$

$$f(x) > f_U \geq \min_{z \in [\hat{z}]} \max_{y \in [\hat{y}]} f(y, z),$$

where f_U is a continually updated upper bound on $f(x^*)$ then $x^* \notin ([\hat{y}], [z])$ so that the entire 'strip', $([\hat{y}], [z])$ of $[\hat{x}]$ may be discarded.

If $x^* \in [x] \overset{\circ}{\subset} [\hat{x}]$ then $f'_y(x^*) = 0$ and an interval Newton method is used to bound x^*. Points $x \in [x]$ such that

$$f(\check{x}) + (x - \check{x})^{\mathrm{T}} f'_x(\check{x}) + \tfrac{1}{2}(x - \check{x})^{\mathrm{T}} f''_x(x)(x - \check{x}) \leq f_U$$

are retained in $[x]$ as explained in [52].

Shen et al. [50] have described an interval algorithm for bounding the solutions of the discrete minimax problem $\min_{x\in[x]} \max_{1\leqslant i\leqslant m} f_i(x)$ where $f_i : D\subseteq\mathbb{R}^n \rightarrow \mathbb{R}^1$ are functions with $f_i\in C(D)$ $(i=1,\ldots,m)$. The algorithm depends on the facts that if $F_p : D\subseteq\mathbb{R}^n \rightarrow \mathbb{R}^1$ is defined by

$$F_p(x) = \frac{1}{p}\log\left\{\sum_{i=1}^m \exp(pf_i(x))\right\}, \tag{4}$$

then $(1\leqslant p\leqslant q) \Rightarrow (F_q(x)\leqslant F_p(x)\,(\forall x\in[x]))$ and $(\forall x\in[x])$

$$f(x)\leqslant F_p(x)\leqslant f(x) + (\log m)/p, \tag{5}$$

where $f(x) = \max_{i=1}^m f_i(x)$, so that $F_p(x)\downarrow f(x)$ $(p\rightarrow\infty)$ $(\forall x\in[x])$.

Wolfe [61] has described interval algorithms for bounding solutions of the discrete minimax problems $\min_{x\in[a,b]} \max_{1\leqslant i\leqslant m_f} f_i(x)$ and $\min_{x\in X} \max_{1\leqslant i\leqslant m_f} f_i(x)$ with

$$X = \{x\in[a,b]\,|\,c_j(x)\leqslant 0 \quad (j=1,\ldots,m_c)\}, \tag{6}$$

where $f_i : [a,b]\subset\mathbb{R}^1 \rightarrow \mathbb{R}^1$ are given functions with $f_i\in C^1([a,b])$ $(i=1,\ldots,m_f)$ and $c_j : [a,b]\subset\mathbb{R}^1 \rightarrow \mathbb{R}^1$ are given functions with $c_j\in C^1([a,b])$ $(j=1,\ldots,m_c)$. The algorithms depend on the following ideas. From Eqs. (4) and (5) with $n=1$, if $[F_p] : \mathbb{I}(\mathbb{R}^1) \rightarrow \mathbb{I}(\mathbb{R}^1)$ is an interval extension of $F_p : \mathbb{R}^1 \rightarrow \mathbb{R}^1$, $x^*\in[x]\subseteq[a,b]$ and $[F_p]([x]) = [\underline{F}_p, \bar{F}_p]$ then

$$\underline{F}_p - (\log m)/p \leqslant \min_{x\in[x]} f(x)\leqslant\bar{F}_p,$$

so if $x^*\in[x]^{(i)}\subseteq[a,b]$ for at least one $i\in\{1,\ldots,\ell\}$ then

$$f(x^*)\in\square\{[F_p]([x]^{(i)}) + [-(\log m)/p, 0]\,|\,i=1,\ldots,\ell\}. \tag{7}$$

Result (7) is also true if X replaces $[a,b]$. If $X=[a,b]$ and an upper bound F_U on $F_p(x^*)$ is known then sub-intervals $[x]^{(i)}$ of $[a,b]$ which could contain x^* are obtained using an idea of Hansen [16] by bounding the set

$$[x]^{(1)} \cup [x]^{(2)} = \{t\in\mathbb{R}^1\,|\,(F_p(\check{x}) - F_U) + [F_p']([x])(t - \check{x})\leqslant 0\}\cap[x], \tag{8}$$

where $\check{x} = m([x])$. If X is defined by Eq. (6) and $[x]\neq\emptyset$ is of unknown feasibility then the same idea which is used in (7) allows one to determine $\{x\in[x]^{(i)}\,|\,c_i(x)\leqslant 0\}\subseteq[x]_1^{(i)}\cup[x]_2^{(i)}$ where

$$[x]_1^{(i)} \cup [x]_2^{(i)} = \{t\in\mathbb{R}^1\,|\,c_i(\check{x}) + [c_i']([x])(t - \check{x})\leqslant 0\}\cap[x].$$

5. The future

The principal obstacles to the application of interval mathematics to problems in pure and applied science appear to have nothing to do with its applicability. It would appear that the most serious obstacle to the acceptance of interval mathematics in the past has been associated with the lack of widely available appropriate computational facilities. Serial and parallel machines, programming languages and compilers suitable for interval computation are slowly becoming more accessible. Even at present interval mathematics has effectively been used to solve problems in, for example, structural engineering, mechanical engineering, electrical engineering, chemical engineering, computer-assisted proof, control theory, economics, robotics and medicine. Most, if not all, of these applications need linear or nonlinear systems or optimization problems to be solved. Often algorithms using only real arithmetic may be used to obtain approximate solutions which may then be validated using interval algorithms.

One objection to the use of interval algorithms is that interval arithmetic is slow compared with real arithmetic, although for some problems the converse has been found to be true. The increasing use of parallel machines and interval algorithms designed to use them may help to remove the objection, especially for very large-scale problems.

Automatic differentiation has proved to be invaluable when implementing algorithms for solving nonlinear systems and nonlinear optimization problems. It is to be hoped that *all* compilers for programming languages such as Fortran and C++ for example will in future contain facilities for rigorous interval arithmetic and for automatic differentiation: this would have a profound effect on both the efficiency and the design of algorithms.

Acknowledgements

The author is grateful to the referees for their helpful remarks which have helped to improve the clarity of the manuscript.

References

[1] C.S. Adjiman, I.P. Androulakis, C.A. Floudas, A global optimization method, αBB, for general twice-differentiable constrained NLPs – II, Implementation and computational results, Comput. Chem. Eng. 22 (1998) 1159–1179.

[2] C.S. Adjiman, S. Dallwig, C.A. Floudas, A. Neumaier, A global optimization method, αBB, for general twice-differentiable constrained NLPs – I, Theoretical advances, Comput. Chem. Eng. 22 (1998) 1137–1158.

[3] G. Alefeld, Inclusion methods for systems of nonlinear equations – the interval Newton method and modifications, in: J. Herzberger (Ed.), Topics in Validated Computations, Elsevier, Dordrecht, 1994, pp. 7–26.

[4] G. Alefeld, D. Claudio, The basic properties of interval arithmetic, its software realizations and some applications, Comput. Struct. 67 (1998) 3–8.

[5] G. Alefeld, A. Gienger, F. Potra, Efficient numerical validation of solutions of nonlinear systems, SIAM J. Numer. Anal. 31 (1994) 252–260.

[6] G. Alefeld, J. Herzberger, Introduction to Interval Computations, Academic Press, New York, 1983.

[7] G. Alefeld, V. Kreinovich, G. Mayer, The shape of the symmetric solution set, in: R.B. Kearfott, V. Kreinovich (Eds.), Applications of Interval Computations, Kluwer Academic Publishers, Dordrecht, 1996, pp. 61–79.

[8] M.S. Bazarraa and C.M. Shetty, Foundations of Optimization, Lecture Notes in Economics and Mathematical Systems, Vol. 122, Springer, Berlin, 1976.

[9] A. Ben-Israel, T.N.E. Greville, Generalized Inverses: Theory and Applications, Wiley, New York, 1974.

[10] C.A. Floudos, P.M. Pardolos, A Collection of Test Problems for Constrained Optimization, Lecture Notes in Computer Science, Vol. 455, Springer, Berlin, 1990.

[11] A. Frommer, G. Mayer, Efficient methods for enclosing solutions of systems of nonlinear equations, Computing 44 (1990) 221–235.

[12] A. Frommer, G. Mayer, On the R-order of Newton-like methods for enclosing solutions of nonlinear equations, SIAM J. Numer. Anal. 27 (1990) 105–116.

[13] A. Frommer, G. Mayer, Two-stage interval iterative methods, preprint, 1993.

[14] R. Hammer, M. Hocks, U. Kulisch, D. Ratz, Numerical Toolbox for Verified Computing I, Springer, Berlin, 1993.

[15] E.R. Hansen, Bounding the solution of interval linear equations, SIAM J. Numer. Anal. 29 (1992) 1493–1503.

[16] E. Hansen, Global Optimization using Interval Analysis, Marcel Dekker, New York, 1992.

[17] E.R. Hansen, G.W. Walster, Bounds for Lagrange multipliers and optimal points, Comput. Math. Appl. 25 (1993) 59–69.

[18] E.R. Hansen, G.W. Walster, Nonlinear equations and optimization, Comput. Math. Appl. 25 (1993) 125–145.

[19] C. Jansson, O. Knüppel, A branch and bound algorithm for bound constrained optimization problems without derivatives, J. Global Optim. 7 (1995) 297–331.

[20] R.B. Kearfott, Decomposition of arithmetic expressions to improve the behaviour of interval iteration for nonlinear systems, Computing 47 (1991) 169–191.

[21] R.B. Kearfott, Rigorous Global Search: Continuous Problems, Kluwer Academic Publishers, Dordrecht, 1996.

[22] R.B. Kearfott, V. Kreinovich (Eds.), Applications of Interval Computations, Kluwer Academic Publishers, Dordrecht, 1996.

[23] L. Kolev, A new method for global solution of systems of non-linear equations, Reliab. Comput. 4 (1998) 125–146.

[24] V. Kreinovich, A. Lakeyev, Linear interval equations: computing enclosures with bounded relative or absolute overestimation is NP-hard, Reliab. Comput. 2 (1996) 341–350.

[25] V. Kreinovich, A. Lakeyev, S. Noskov, Optimal solution of interval linear systems is intractible (NP-hard), Interval Comput. 1 (1993) 6–14.

[26] G. Mayer, Some remarks on two interval-arithmetic modifications of the Newton method, Computing 48 (1992) 125–128.

[27] G. Mayer, Success in epsilon-inflation, in: G. Alefeld, A. Frommer (Eds.), Scientific Computing and Validated Numerics, Akademie Verlag, Berlin, 1996, pp. 98–104.

[28] G. Mayer, Epsilon inflation in verification algorithms, J. Comput. Appl. Math. 60 (1995) 147–169.

[29] G. Mayer, J. Rohn, On the applicability of the interval Gaussian algorithm, Reliab. Comput. 4 (1998) 205–222.

[30] R.E. Moore, Interval arithmetic and automatic error analysis in digital computing, Ph.D. Thesis, Stanford University, 1962.

[31] R.E. Moore, J.B. Kioustelidis, A simple test for accuracy of approximate solutions to nonlinear (or linear) systems, SIAM J. Numer. Anal. 17 (1980) 521–529.

[32] A. Neumaier, Interval Methods for Systems of Equations, Cambridge University Press, Cambridge, 1990.

[33] S. Ning, R.B. Kearfott, A comparison of some methods for solving linear interval equations, SIAM J. Numer. Anal. 34 (1997) 1289–1305.

[34] J.M. Ortega, W.C. Rheinboldt, Iterative Solution of Nonlinear Equations in Several Variables, Academic Press, New York, 1970.

[35] M.C. Pandian, A convergence test and componentwise error estimates for Newton methods, SIAM J. Numer. Anal. 22 (1985) 779–791.

[36] L. Qi, A note on the Moore test for nonlinear systems, SIAM J. Numer. Anal. 19 (1982) 851–857.

[37] L.B. Rall, Computational Solution of Nonlinear Operator Equations, Wiley, New York, 1969.

[38] H. Ratschek, J. Rokne, New Computer Methods for Global Optimization, Ellis Horwood, Chichester, 1988.

[39] H. Ratschek, R.L. Voller, Global optimization over unbounded domains, SIAM J. Control Optim. 28 (1990) 528–539.

[40] H. Ratschek, J. Rokne, Interval methods, in: R. Horst, P.M. Pardolos (Eds.), Handbook of Global Optimization, Kluwer Academic Publishers, Dordrecht, 1995, pp. 751–828.

[41] J. Rohn, On nonconvexity of the solution set of a system of linear interval equations, BIT 30 (1989) 161–165.

[42] J. Rohn, Cheap and tight bounds: the recent result by E. Hansen can be made more efficient, Interval Comput. 4 (1993) 13–21.

[43] J. Rohn, Enclosing solutions of overdetermined systems of linear interval equations, Reliab. Comput. 2 (1996) 167–171.

[44] J. Rohn, Linear interval equations: computing enclosures with bounded relative overestimation is NP-hard, in: R.B. Kearfott, V. Kreinovich (Eds.), Applications of Interval Computations, Kluwer Academic Publishers, Dordrecht, 1996, pp. 81–89.

[45] J. Rohn, On overestimation produced by the interval Gaussian algorithm, Reliab. Comput. 3 (1997) 363–368.

[46] S.M. Rump, Verification methods for dense and sparse systems of equations, in: J. Herzberger (Ed.), Topics in Validated Computations, Elsevier, Amsterdam, 1994, pp. 63–135.

[47] S.M. Rump, Improved iteration schemes for validation algorithms for dense and sparse nonlinear systems, Computing 57 (1996) 77–84.

[48] S.M. Rump, A note on epsilon inflation, Reliab. Comput. 4 (1998) 371–375.

[49] S.P. Shary, Algebraic approach in the "outer problem" for interval linear equations, Reliab. Comput. 3 (1997) 103–135.

[50] Z. Shen, Z. Huang, M.A. Wolfe, An interval maximum entropy method for a discrete minimax problem, Appl. Math. Comput. 87 (1997) 49–68.

[51] Z. Shen, A. Neumaier, A note on Moore's interval test for zeros of nonlinear systems, Computing 40 (1988) 85–90.

[52] Z. Shen, A. Neumaier, M.C. Eiermann, Solving minimax problems by interval methods, BIT 30 (1990) 742–751.

[53] Z. Shen, M.A. Wolfe, On certain computable tests and componentwise error bounds, Computing 50 (1993) 353–368.

[54] M.A. Wolfe, Interval methods for linear systems, in: E. Spedicato, M. Vespucci (Eds.), NATO Advanced Study Institute Computer Algorithms for Solving Linear Algebraic Equations: the State of the Art, Universita' Degli Studi di Bergamo, 1990, pp. 128–140.

[55] M.A. Wolfe, An interval algorithm for constrained global optimization, J. Comput. Appl. Math. 50 (1994) 605–612.

[56] M.A. Wolfe, On global optimization in $\mathbb{R}$ using interval arithmetic, Optim. Methods Software 3 (1994) 60–76.

[57] M.A. Wolfe, An interval algorithm for bound-constrained global optimization, Optim. Methods Software 6 (1995) 145–159.

[58] M.A. Wolfe, Interval methods for global optimization, Appl. Math. Comput. 75 (1996) 179–206.

[59] M.A. Wolfe, On a second derivative test due to Qi, Reliab. Comput. 4 (1998) 223–234.

[60] M.A. Wolfe, A note on a uniqueness theorem for the second-derivative test of Qi, to appear.

[61] M.A. Wolfe, On discrete minimax problems in $\mathbb{R}^1$ using interval arithmetic, Reliab. Comput. 5 (1999) 371–383.

[62] M.A. Wolfe, On bounding solutions of underdetermined systems, to appear.

[63] M.A. Wolfe, L.S. Zhang, An interval algorithm for nondifferentiable global optimization, Appl. Math. Comput. 63 (1994) 101–122.

[64] T. Yamamoto, Error bounds for Newton's iterates derived from the Kantorovich theorem, Numer. Math. 48 (1986) 91–98.

[65] K. Yamamura, An algorithm for representing functions of many variables by superpositions of functions of one variable and addition, IEEE Trans. Circuits Systems 1 (43) (1996) 338–340.

[66] K. Yamamura, H. Kawata, A. Tokue, Interval solution of nonlinear equations using linear programming, BIT 38 (1) (1998) 186–199.

[67] Zhang Danqing, Li Weiguo, Shen Zuhe, Solving underdetermined systems with interval methods, Reliab. Comput. 5 (1999) 23–33.

ELSEVIER

Journal of Computational and Applied Mathematics 124 (2000) 281–302

JOURNAL OF
COMPUTATIONAL AND
APPLIED MATHEMATICS

www.elsevier.nl/locate/cam

Interior-point methods

Florian A. Potra[a], Stephen J. Wright[b],*

[a]Department of Mathematics and Statistics, University of Maryland, Baltimore County, Baltimore, MD 21250, USA
[b]Division of Mathematics and Computer Science, Argonne National Laboratory, 9700 South Cass Avenue, Argonne, IL 60439-4844, USA

Received 18 November 1999; received in revised form 10 February 2000

Abstract

The modern era of interior-point methods dates to 1984, when Karmarkar proposed his algorithm for linear programming. In the years since then, algorithms and software for linear programming have become quite sophisticated, while extensions to more general classes of problems, such as convex quadratic programming, semi-definite programming, and nonconvex and nonlinear problems, have reached varying levels of maturity. We review some of the key developments in the area, and include comments on the complexity theory and practical algorithms for linear programming, semi-definite programming, monotone linear complementarity, and convex programming over sets that can be characterized by self-concordant barrier functions. © 2000 Elsevier Science B.V. All rights reserved.

1. Introduction

In their survey article [6], Freund and Mizuno wrote

Interior-point methods in mathematical programming have been the largest and most dramatic area of research in optimization since the development of the simplex method...Interior-point methods have permanently changed the landscape of mathematical programming theory, practice and computation... .

Although most research in the area was devoted to linear programming, the authors claimed that

semidefinite programming is the most exciting development in mathematical programming in 1990s.

Although various interior-point methods had been considered one way or another from the 1950s, and investigated quite extensively during the 1960s [5], it was the publication of the seminal paper

---

* Corresponding author.
E-mail addresses: potra@math.umbc.edu (F.A. Potra), wright@mcs.anl.gov (S.J. Wright).

0377-0427/00/$ - see front matter © 2000 Elsevier Science B.V. All rights reserved.
PII: S 0377-0427(00)00433-7

of Karmarkar [11] that placed interior-point methods at the top of the agenda for many researchers. On the theoretical side, subsequent research led to improved computational complexity bounds for linear programming (LP), quadratic programming (QP), linear complementarity problems (LCP) semi-definite programming (SDP) and some classes of convex programming problems. On the computational side, high-quality software was eventually produced, much of it freely available. The general performance of computational tools for linear programming improved greatly, as the sudden appearance of credible competition spurred significant improvements in implementations of the simplex method.

In the first years after Karmarkar's initial paper, work in linear programming focused on algorithms that worked with the primal problem, but were more amenable to implementation than the original method or that had better complexity bounds. A particularly notable contribution from this period was Renegar's algorithm [21], which used upper bounds on the optimal objective value to form successively smaller subsets of the feasible set, each containing the solution, and used Newton's method to follow the analytic centers of these subsets to the primal optimum. A new era was inaugurated with Megiddo's paper [13], originally presented in 1987, which described a framework for primal–dual framework algorithms. The primal–dual viewpoint proved to be extremely productive. It yielded new algorithms with interesting theoretical properties, formed the basis of the best practical algorithms, and allowed for transparent extensions to convex programming and linear complementarity. In 1989, Mehrotra described a practical algorithm for linear programming that remains the basis of most current software; his work appeared in 1992 [14]. Meanwhile, Nesterov and Nemirovskii [16] were developing the theory of self-concordant functions, which allowed algorithms based on the primal log-barrier function for linear programming to be extended to wider classes of convex problems, particularly semi-definite programming and second-order cone programming (SOCP). Nesterov and Todd [17,18] extended the primal–dual approach along similar lines to a more restricted class of convex problems that still included SDP and SOCP. Other work on interior-point algorithms for SDPs, which have a wide variety of applications in such areas as control and structural optimization, was already well advanced by this point. Work on these algorithms gained additional impetus when it was recognized that approximate solutions of NP-hard problems could thereby be obtained in polynomial time.

We now outline the remainder of the paper. Section 2 discusses linear programming, outlining the pedigree of the most important algorithms and various computational issues. In Section 3, we discuss extensions to quadratic programming and linear complementarity problems, and compare the resulting algorithms with active-set methods. Semi-definite programming is the topic of Section 4. Section 5 contains some elements of the theory of self-concordant functions and self-scaled cones. Finally, we present some conclusions in Section 6.

There are many other areas of optimization in which areas in which interior-point approaches have made an impact, though in general the state of the art is less mature than for the areas mentioned above. General convex programming problems of the form

$$\min_x \ f(x) \quad \text{s.t.} \ g_i(x) \leqslant 0, \ i = 1, 2, \ldots, m$$

(where f and g_i, $i = 1, 2, \ldots, m$, are convex functions) can be solved by extensions of the primal–dual approach of Section 3. Interestingly, it is possible to prove superlinear convergence of these primal–dual algorithms without assuming linear independence of the active constraints at the solution. This observation prompted recent work on improving the convergence properties of other algorithms, notably sequential quadratic programming. A number of researchers have used interior-point methods in algorithms for combinatorial and integer programming problems. (In some cases, the interior-point method is used to find an inexact solution of related problems in which the integrality constraints are relaxed.) In decomposition methods for large linear and convex problems, such as Dantzig–Wolfe/column generation and Benders' decomposition, interior-point methods have been used to find inexact solutions of the large master problems, or to approximately solve analytic center subproblems to generate test points. Additionally, application of interior-point methodology to non-convex nonlinear programming has occupied many researchers for some time now. The methods that have been proposed to date contain many ingredients, including primal–dual steps, barrier and merit functions, and scaled trust regions.

For references to work mentioned in the previous paragraph, and for many other results discussed but not cited in this paper, please see the bibliography of the technical report in [28].

A great deal of literature is available to the reader interested in learning more about interior-point methods. A number of recent books [27,29,23] give overviews of the area, from first principles to new results and practical considerations. Theoretical background on self-concordant functionals and related developments is described in [16,22]. Technical reports from the past five years can be obtained from the Interior-Point Methods Online Web site at www.mcs.anl.gov/otc/InteriorPoint.

2. Linear programming

We consider first the linear programming problem, which is undoubtedly the optimization problem solved most frequently in practice. Given a cost vector $c \in \mathbb{R}^n$, m linear equality constraints defined by a matrix $A \in \mathbb{R}^{m \times n}$ and a vector $b \in \mathbb{R}^m$, the linear programming problem can be stated in its standard form as

$$\min_{x} c^T x \quad \text{s.t. } Ax = b, \ x \geqslant 0. \tag{2.1}$$

The restriction $x \geqslant 0$ applies componentwise, that is, all components of the vector $x \in \mathbb{R}^n$ are required to be nonnegative.

The simplex method developed by Dantzig between 1947 and 1951 has been the method of choice for linear programming. While performing very well in practice, its worst-case computational complexity is exponential, as shown by the example of Klee and Minty from 1972. The problem of existence of a (weakly) polynomial algorithm for solving linear programs with integer data was solved by Khachiyan in 1979. He proved that the ellipsoid method solves such programs in $O(n^2L)$ iterations, requiring a total of $O(n^4L)$ bit operations, where L is the length of a binary coding of the input data, that is

$$L = \sum_{i=0}^{m} \sum_{j=0}^{n} \lceil \log_2(|a_{ij}| + 1) + 1 \rceil$$

with $a_{i0} = b_i$ and $a_{0j} = c_j$.

There are no known implementations of the ellipsoid method for linear programming that are remotely competitive with existing practical codes. The merit of the celebrated paper of Karmarkar [11] consisted not so much in lowering the bound on the computational complexity of LP to $O(nL)$ iterations, requiring a total of $O(n^{3.5}L)$ bit operations, as in the fact that it was possible to implement his algorithm with reasonable efficiency. The theoretical computational complexity of interior-point methods for LP was eventually lowered to $O(\sqrt{n}L)$ iterations, requiring a total of $O(n^3L)$ bit operations by a number of authors. Goldfarb and Todd [8] provide a good reference for these complexity results. By using fast matrix multiplication techniques, the complexity estimates can be reduced further. Quite recently, Anstreicher [1] proposed an interior-point method, combining partial updating with a preconditioned gradient method, that has an overall complexity of $O(n^3/\log n)$ bit operations. The paper [1] contains references to recent complexity results for LP.

The best of these complexity results, all of which are of major theoretical importance, are obtained as a consequence of global linear convergence with factor $1 - c/\sqrt{n}$. In what follows we will describe a simple interior algorithm that achieves this rate. We assume that the linear program (2.1) has a strict interior, that is, the set

$$\mathscr{F}^0 \stackrel{\text{def}}{=} \{x \mid Ax = b,\; x > 0\}$$

is nonempty, and that the objective function is bounded below on the set of feasible points. Under these assumptions, (2.1) has a (not necessarily unique) solution.

By using a logarithmic barrier function to account for the bounds $x \geq 0$, we obtain the parameterized optimization problem

$$\min_x\; f(x; \mu) \stackrel{\text{def}}{=} \frac{1}{\mu} c^{\mathrm{T}} x - \sum_{i=1}^{n} \log x_i, \quad \text{s.t. } Ax = b, \tag{2.2}$$

where log denotes the natural logarithm and $\mu > 0$ denotes the barrier parameter. Because the logarithmic function requires its arguments to be positive, the solution $x(\mu)$ of (2.2) must belong to $\mathscr{F}^0$. It is well known (see, for example, [26, Theorem 5]) that for any sequence $\{\mu_k\}$ with $\mu_k \downarrow 0$, all limit points of $\{x(\mu_k)\}$ are solutions of (2.1).

The traditional SUMT approach [5] accounts for equality constraints by including a quadratic penalty term in the objective. When the constraints are linear, as in (2.1), it is simpler and more appropriate to handle them explicitly. By doing so, we devise a *primal barrier algorithm* in which a projected Newton method is used to find an approximate solution of (2.2) for a certain value of μ, and then μ is decreased. Note that

$$\nabla_{xx}^2 f(x; \mu) = -X^{-2}, \quad \nabla_x f(x; \mu) = (1/\mu)c + X^{-1}e,$$

where $X = \mathrm{diag}(x_1, x_2, \ldots, x_n)$ and $e = (1, 1, \ldots, 1)^{\mathrm{T}}$. The projected Newton step Δx from a point x satisfies the following system:

$$\begin{bmatrix} -\mu X^{-2} & A^{\mathrm{T}} \\ A & 0 \end{bmatrix} \begin{bmatrix} \Delta x \\ \lambda^+ \end{bmatrix} = -\begin{bmatrix} c + \mu X^{-1}e \\ Ax - b \end{bmatrix}, \tag{2.3}$$

so that Eq. (2.3) are the same as those that arise from a sequential quadratic programming algorithm applied to (2.2), modulo the scaling by μ in the first line of (2.3). A line search can be performed along Δx to find a new iterate $x + \alpha \Delta x$, where $\alpha > 0$ is the step length.

The prototype primal barrier algorithm can be specified as follows:

primal barrier algorithm
Given $x^0 \in \mathscr{F}^0$ and $\mu_0 > 0$;
Set $k \leftarrow 0$;
repeat
      Obtain x^{k+1} by performing one or more Newton steps (2.3),
              starting at $x = x^k$, and fixing $\mu = \mu_k$;
      Choose $\mu_{k+1} \in (0, \mu_k)$; $k \leftarrow k + 1$;
until some termination test is satisfied.

A *short-step* version of this algorithm takes a single Newton step at each iteration, with step length $\alpha = 1$, and sets

$$\mu_{k+1} = \mu_k \left/ \left(1 + \frac{1}{8\sqrt{n}}\right)\right. .\tag{2.4}$$

It is known (see, for instance, [22, Section 2.4]) that if the feasible region of (2.1) is bounded, and x^0 is sufficiently close to $x(\mu_0)$ in a certain sense, then we obtain a point x^k whose objective value $c^{\mathrm{T}}x^k$ is within ε of the optimal value after

$$O\left(\sqrt{n}\log\frac{n\mu_0}{\varepsilon}\right) \quad \text{iterations,}\tag{2.5}$$

where the constant factor disguised by the $O(\cdot)$ depends on the properties of (2.1) but is independent of n and ε. For integer data of bitlength L, it is known that if $\varepsilon \leqslant 2^{-2L}$ then x^k can be rounded to an exact solution in $O(n^3)$ arithmetic operations. Moreover, provided we can choose the initial point such that $\mu_0 \leqslant 2^{\beta L}$ for some positive constant β, the iteration complexity will be $O(\sqrt{n}L)$.

The rate of decrease of μ in short-step methods is too slow to allow good practical behavior, so *long-step* variants have been proposed that decrease μ more rapidly, while possibly taking more than one Newton step for each μ_k and also using a line search. Although long-step algorithms have better practical behavior, the complexity estimates associated with them typically are no better than estimate (2.5) for the short-step approach. In fact, a recurring theme of worst-case complexity estimates for linear programming algorithms is that no useful relationship exists between the estimate and the practical behavior of the algorithm. Indeed, as we have seen above, the best-known iteration complexity bound is obtained from a rather slow linear convergence rate. Good practical performance is obtained by algorithms that are superlinearly convergent.

Better practical algorithms are obtained from the primal–dual framework. These methods recognize the importance of the path of solutions $x(\mu)$ to (2.2) in the design of algorithms, but differ from the approach above in that they treat the dual variables explicitly in the problem, rather than as adjuncts to the calculation of the primal iterates. The dual problem for (2.1) is

$$\max_{(\lambda,s)} b^{\mathrm{T}}\lambda \quad \text{s.t. } A^{\mathrm{T}}\lambda + s = c, \quad s \geqslant 0,\tag{2.6}$$

where $s \in \mathbb{R}^n$ and $\lambda \in \mathbb{R}^m$, and the optimality conditions for x^* to be a solution of (2.1) and (λ^*, s^*)

to be a solution of (2.6) are that $(x, \lambda, s) = (x^*, \lambda^*, s^*)$ satisfies

$$Ax = b, \tag{2.7a}$$

$$A^{\mathrm{T}}\lambda + s = c, \tag{2.7b}$$

$$X Se = 0, \tag{2.7c}$$

$$(x, s) \geqslant 0, \tag{2.7d}$$

where $X = \mathrm{diag}(x_1, x_2, \ldots, x_n)$ and $S = \mathrm{diag}(s_1, s_2, \ldots, s_n)$. Primal–dual methods solve (2.1) and (2.6) simultaneously by generating a sequence of iterates (x^k, λ^k, s^k) that in the limit satisfies conditions (2.7). As mentioned above, the *central path* defined by the following perturbed variant of (2.7) plays an important role in algorithm design:

$$Ax = b, \tag{2.8a}$$

$$A^{\mathrm{T}}\lambda + s = c, \tag{2.8b}$$

$$X Se = \mu e, \tag{2.8c}$$

$$(x, s) > 0, \tag{2.8d}$$

where $\mu > 0$ parameterizes the path. Note that these conditions are simply the optimality conditions for the problem (2.2): If $(x(\mu), \lambda(\mu), s(\mu))$ satisfies (2.8), then $x(\mu)$ is a solution of (2.2). We have from (2.8c) that a key feature of the central path is that

$$x_i s_i = \mu \quad \text{for all } i = 1, 2, \ldots, n, \tag{2.9}$$

that is, the pairwise products $x_i s_i$ are identical for all i.

   In primal–dual algorithms, steps are generated by applying a perturbed Newton methods to the three equalities in (2.8), which form a nonlinear system in which the number of equations equals the number of unknowns. We constrain all iterates (x^k, λ^k, s^k) to have $(x^k, s^k) > 0$, so that the matrices X and S remain positive diagonal throughout, ensuring that the perturbed Newton steps are well defined. Supposing that we are at a point (x, λ, s) with $(x, s) > 0$ and the feasibility conditions $Ax = b$ and $A^{\mathrm{T}}\lambda + s = c$ are satisfied, the primal–dual step $(\Delta x, \Delta \lambda, \Delta s)$ is obtained from the following system:

$$
\begin{bmatrix} 0 & A & 0 \\ A^{\mathrm{T}} & 0 & I \\ 0 & S & X \end{bmatrix}
\begin{bmatrix} \Delta\lambda \\ \Delta x \\ \Delta s \end{bmatrix}
= -\begin{bmatrix} 0 \\ 0 \\ X Se - \sigma\mu e + r \end{bmatrix}, \tag{2.10}
$$

where $\mu = x^{\mathrm{T}}s/n$, $\sigma \in [0, 1]$, and r is a perturbation term, possibly chosen to incorporate higher-order information about the system (2.8), or additional terms to improve proximity to the central path.

Using the general step (2.10), we can state the basic framework for primal–dual methods as follows:

primal–dual algorithm
Given (x^0, λ^0, s^0) with $(x^0, s^0) > 0$;
Set $k \leftarrow 0$ and $\mu_0 = (x^0)^{\mathrm{T}} s^0 / n$;
repeat
      Choose σ_k and r^k;
      Solve (2.10) with $(x, \lambda, s) = (x^k, \lambda^k, s^k)$ and $(\mu, \sigma, r) = (\mu_k, \sigma_k, r^k)$
            to obtain $(\Delta x^k, \Delta \lambda^k, \Delta s^k)$;
      Choose step length $\alpha_k \in (0, 1]$ and set
            $(x^{k+1}, \lambda^{k+1}, s^{k+1}) \leftarrow (x^k, \lambda^k, s^k) + \alpha_k (\Delta x^k, \Delta \lambda^k, \Delta s^k)$;
            $\mu_{k+1} \leftarrow (x^{k+1})^{\mathrm{T}} s^{k+1} / n$; $k \leftarrow k + 1$;
until some termination test is satisfied.

The various algorithms that use this framework differ in the way that they choose the starting point, the centering parameter σ_k, the perturbation vector r^k, and the step α_k. the simplest algorithm – a short-step path-following method similar to the primal algorithm described above – sets

$$r^k = 0, \quad \sigma_k \equiv 1 - \frac{0.4}{\sqrt{n}}, \quad \alpha_k \equiv 1 \tag{2.11}$$

and, for suitable choice of a feasible starting point, achieves convergence to a feasible point (x, λ, s) with $x^{\mathrm{T}} s / n \leqslant \varepsilon$ for a given ε in

$$O\left(\sqrt{n} \log \frac{\mu_0}{\varepsilon} \right) \text{ iterations.} \tag{2.12}$$

Note the similarity of both the algorithm and its complexity estimate to the corresponding primal algorithm. As in that case, algorithms with better practical performance but not necessarily better complexity estimates can be obtained through more aggressive, adaptive choices of the centering parameter (that is, σ_k closed to zero). They use a line search to maintain proximity to the central path. The proximity requirement dictates, implicitly or explicitly, that while condition (2.9) may be violated, the pairwise products must not be too different from each other. For example, some algorithms force the iterates to remain in l_2-neighborhoods of the central path of the form

$$\mathcal{N}(\beta) \stackrel{\text{def}}{=} \{(x, \lambda, s) \mid (x, s) > 0, \ \|Xs - \mu e\|_2 \leqslant \beta\}. \tag{2.13}$$

A very interesting algorithm of this type is the Mizuno–Todd–Ye predictor corrector method which can be described as follows:

predictor–corrector algorithm
Given $(x^0, \lambda^0, s^0) \in \mathcal{N}(0.25)$
Set $k \leftarrow 0$ and $\mu_0 = (x^0)^{\mathrm{T}} s^0 / n$;
repeat
    Set $(x, \lambda, s) \leftarrow (x^k, \lambda^k, s^k)$ and $(\mu, \sigma, r) \leftarrow (\mu_k, 0, 0)$;
    Solve (2.10) and set $(u, w, v) \leftarrow (\Delta x, \Delta \lambda, \Delta s)$;
        to obtain $(\Delta x^k, \Delta \lambda^k, \Delta s^k)$;
    Choose step length α_k as the largest $\alpha_k \in (0, 1]$ such that:
        $(x, \lambda, s) + \alpha(u, w, v) \in \mathcal{N}(0.25)$
    Set $(x, \lambda, s) \leftarrow (x, \lambda, s) + \alpha_k(u, w, v)$ and $(\mu, \sigma, r) \leftarrow (\mu_k, (1 - \alpha_k), 0)$;
    Solve (2.10) and set
        $(x^{k+1}, \lambda^{k+1}, s^{k+1}) \leftarrow (x, \lambda, s) + (\Delta x, \Delta \lambda, \Delta s)$;
        $\mu_{k+1} \leftarrow (x^{k+1})^{\mathrm{T}} s^{k+1} / n$; $k \leftarrow k + 1$;
until some termination test is satisfied.

It can be proved that the above algorithm has the iteration complexity bound (2.12), the same as the short-step algorithm defined by (2.11). We note that the predictor–corrector method requires the solution of two linear systems per iteration (one in the predictor step and another one in the corrector step), while the short-step algorithm requires only the solution of one linear system per iteration. However, numerical experiments show that the predictor–corrector algorithm is significantly more efficient than the short-step algorithm. This is explained by the fact that while with the short-step algorithm μ_k decreases by a fixed factor at each step, i.e.,

$$\mu_{k+1} = \left(1 - \frac{0.4}{n} \right) \mu_k, \quad k = 0, 1, 2, \ldots \tag{2.14}$$

the predictor–corrector algorithm, by its adaptive choice of σ_k, allows μ_k to decrease faster, especially close to the solution. Ye et al. [30] proved that the predictor–corrector algorithm is quadratically convergent in the sense that

$$\mu_{k+1} \leqslant B\mu_k^2, \quad k = 0, 1, 2, \ldots \tag{2.15}$$

for some constant B independent of k. This constant may be large, so that (2.15) ensures a better decrease of μ_k that (2.14) only if μ_k is sufficiently small (specifically, $\mu_k < (1 - 0.4/n)/B$). There are examples in which quadratic convergence cannot be observed until quite late in the algorithm — the last few iterations. Even in these examples, the linear decrease factor in μ_k in early iterations is much better than $(1 - 0.4/n)$, because of the adaptive choice of σ_k.

Even better reductions of μ_k in the early iteration can be obtained by considering larger neighborhoods of the central path than the l_2-neighborhoods (2.13). The worst-case complexity bounds of the resulting algorithms deteriorates — $\mathrm{O}(nL)$ instead of $\mathrm{O}(\sqrt{n}L)$ — but the practical performance is better.

Quadratic convergence, or, more generally, superlinear convergence is also important for the following reason. The condition of the linear systems to be solved at each iteration often worsens as μ_k becomes small, and numerical problems are sometimes encountered. Superlinearly convergent algorithms need to perform only a couple of iterations with these small μ_k. When μ_k is small enough, a projection can be used to identify an exact solution. A finite-termination strategy can also be implemented by using the Tapia indicators to decide which components of x and s are zero at the solution [4]. The use of a finite-termination strategy in conjunction with superlinearly convergent algorithms for linear programming is somewhat superfluous, since the domain range of μ_k values for which superlinear convergence is obtained appears to be similar to the range on which finite termination strategies are successful. Once the iterates enter this domain, the superlinear method typically converges in a few steps, and the savings obtained by invoking a finite termination strategy are not great.

In the above algorithms we assumed that a starting point satisfying exactly the linear constraints and lying in the interior of the region defined by the inequality constraints is given. In practice, however, it may be difficult to obtain such a starting point, so many efficient implementations of interior-point methods use starting points that lie in the interior of the region defined by the inequality constraints but do not necessarily satisfy the equality constraints. Such methods are called infeasible-interior-point methods, and they are more difficult to analyze. The first global convergence result for such methods was obtained by Kojima, Megiddo and Mizuno, while the first polynomial complexity result was given by Zhang [32]. The computational complexity of the infeasible-interior-point algorithms typically is worse than in the feasible case. An advantage is that these algorithms can solve problems for which no strictly feasible points exist. They also can be used to detect the infeasibility of certain linear programming problems.

A different way of dealing with infeasible starting points was proposed by Ye et al. [31]. Starting with a linear programming problem in standard form and with a possibly infeasible starting point whose x and s components are strictly positive, they construct a homogeneous self-dual linear program for which a strictly feasible starting point is readily available. The solution of the original problem is obtained easily from the solution of the homogeneous program. When the original linear program is infeasible, this fact can be ascertained easily from the solution of the homogeneous problem.

The practical performance of a numerical algorithm is explained better by a probabilistic complexity analysis than by a worst-case complexity analysis. For example, the probabilistic computational complexity of the simplex method is strongly polynomial (that is, a polynomial in the dimension n of the problem only), which is closer to practical experience with this method than the exponential complexity of the worst-case analysis (see [3] and the literature cited therein). As mentioned above, the worst-case complexity of interior-point methods is weakly polynomial, in the sense that the iteration bounds are polynomials in the dimension n and the bitlength of the data L. In [2], it is shown that from a probabilistic point of view the iteration complexity of a class of interior-point methods is $O(\sqrt{n} \ln n)$. Thus the probabilistic complexity of this class on interior-point methods is strongly polynomial, that is, the complexity depends only on the dimension of the problem and not on the binary length of the data.

Most interior-point software for linear programming is based on Mehrotra's predictor–corrector algorithm [14], often with the higher-order enhancements described in [9]. This approach uses an adaptive choice of σ_k, selected by first solving for the pure Newton step (that is, setting $r = 0$ and $\sigma = 0$ in (2.10)). If this step makes good progress in reducing μ, we choose σ_k small so that the step actually taken is quite close to this pure Newton step. Otherwise, we enforce more centering and calculate a conservative direction by setting σ_k closer to 1. The perturbation vector r^k is chosen to improve the similarity between system (2.10) and the original system (2.8) that it approximates. Gondzio's technique further enhances r^k by performing further solves of the system (2.10) with a variety of right-hand sides, where each solve reuses the factorization of the matrix and is therefore not too expensive to perform.

To turn this basic algorithmic approach into a useful piece of software, we must address many issues. These include problem formulation, presolving to reduce the problem size, choice of the step length, linear algebra techniques for solving (2.10), and user interfaces and input formats.

Possibly, the most interesting issues are associated with the linear algebra. Most codes deal with a partially eliminated form of (2.10), either eliminating Δs to obtain

$$\begin{bmatrix} 0 & A \\ A^\mathrm{T} & -X^{-1}S \end{bmatrix} \begin{bmatrix} \Delta\lambda \\ \Delta x \end{bmatrix} = - \begin{bmatrix} 0 \\ -X^{-1}(X\,Se - \sigma\mu e + r) \end{bmatrix} \tag{2.16}$$

or eliminating both Δs and Δx to obtain a system of the form

$$A(S^{-1}X)A^\mathrm{T}\Delta\lambda = t, \tag{2.17}$$

to which a sparse Cholesky algorithm is applied. A modified version of the latter form is used when dense columns are present in A. These columns may be treated as a low-rank update and handled via the Sherman–Morrison–Woodbury formula or, equivalently, via a Schur complement strategy applied to a system intermediate between (2.16) and (2.17). In many problems, the matrix in (2.17) becomes increasingly ill-conditioned as the iterates progress, eventually causing the Cholesky process to break down as negative pivot elements are encountered. A number of simple (and in some cases counterintuitive) patches have been proposed for overcoming this difficulty while still producing useful approximate solutions of (2.17) efficiently.

Despite many attempts, iterative solvers have not shown much promise as means to solve (2.17), at least for general linear programs. A possible reason is that, besides its poor conditioning, the matrix lacks the regular spectral properties of matrices obtained from discretizations of continuous operators. Some codes do, however, use preconditioned conjugate gradient as an alternative to iterative refinement for improving the accuracy, when the direct approach for solving (2.17) fails to produce a solution of sufficient accuracy. The preconditioner used in this case is simply the computed factorization of the matrix $A(S^{-1}X)A^\mathrm{T}$.

A number of interior-point linear programming codes are now available, both commercially and free of charge. Information can be obtained from the World-Wide Web via the URL mentioned earlier. It is difficult to make blanket statements about the relative efficiency of interior-point and simplex methods for linear programming, since significant improvements to the implementations of both techniques continue to be made. Interior-point methods tend to be faster on large problems and can better exploit multiprocessor platforms, because the expensive operations such as Cholesky factorization of (2.17) can be parallelized to some extent. They are not able to exploit "warm start" information — a good prior estimate of the solution, for instance, — to the same extent as simplex methods. For this reason, they are not well suited for use in contexts such as branch-and-bound or branch-and-cut algorithms for integer programming, which solve many closely related linear programs.

Several researchers have devised special interior-point algorithms for special cases of (2.1) that exploit the special properties of these cases in solving the linear systems at each iteration. One algorithm for network flow problems uses preconditioned conjugate–gradient methods for solving

(2.17), where the preconditioner is built from a spanning tree for the underlying network. For multicommodity flow problems, there is an algorithm for solving a version of (2.17) in which the block-diagonal part of the matrix is used to eliminate many of the variables, and a preconditioned conjugate–gradient method is applied to the remaining Schur complement. Various techniques have also been proposed for stochastic programming (two-stage linear problems with recourse) that exploit the problem structure in performing the linear algebra operations.

3. Extensions to convex quadratic programming and linear complementarity

The primal–dual algorithms of the preceding section are readily extended to convex quadratic programming (QP) and monotone linear complementarity problems (LCP), both classes being generalizations of linear programming. Indeed, many of the convergence and complexity properties of primal–dual algorithm were first elucidated in the literature with regard to monotone LCP rather than linear programming.

We state the convex QP as

$$\min_x \ c^T x + \tfrac{1}{2} x^T Q x \quad \text{s.t. } Ax = b, \quad x \geqslant 0, \tag{3.18}$$

where Q is a positive-semi-definite matrix. The monotone LCP is defined by square matrices M and N and a vector q, where M and N satisfy a monotonicity property: all vectors y and z that satisfy $My + Nz = 0$ have $y^T z \geqslant 0$. This problem requires us to identify vectors y and z such that

$$My + Nz = q, \quad (y,z) \geqslant 0, \quad y^T z = 0. \tag{3.19}$$

With some transformations, we can express the optimality conditions (2.7) for linear programming, and also the optimality conditions for (3.18), as a monotone LCP. Other problems fit under the LCP umbrella as well, including bimatrix games and equilibrium problems. The central path for this problem is defined by the following system, parametrized as in (2.8) by the positive scalar μ:

$$My + Nz = q, \tag{3.20a}$$

$$YZe = \mu e, \tag{3.20b}$$

$$(y,z) > 0 \tag{3.20c}$$

and a search direction from a point (y,z) satisfying (3.20a) and (3.20c) is obtained by solving a system of the form

$$\begin{bmatrix} M & N \\ Z & Y \end{bmatrix} \begin{bmatrix} \Delta y \\ \Delta z \end{bmatrix} = - \begin{bmatrix} 0 \\ YZe - \sigma \mu e + r \end{bmatrix}, \tag{3.21}$$

where $\mu = y^T z / n$, $\sigma \in [0,1]$, and, as before, r is a perturbation term. The corresponding search direction system for the quadratic program (3.18) is identical to (2.10) except that the $(2,2)$ block in the coefficient matrix is replaced by Q. The primal–dual algorithmic framework and the many variations within this framework are identical to the case of linear programming, with the minor difference that the step length should be the same for all variables. (In linear programming, different step lengths usually are taken for the primal variable x and the dual variables (λ, s).)

Complexity results are also similar to those obtained for the corresponding linear programming algorithm. For an appropriately chosen starting point (y^0, z^0) with $\mu_0 = (y^0)^T z^0 / n$, we obtain convergence to a point with $\mu \leqslant \varepsilon$ in

$$O\left(n^\tau \log \frac{\mu_0}{\varepsilon}\right) \quad \text{iterations,}$$

where $\tau = \frac{1}{2}$, 1, or 2, depending on the algorithm. Fast local convergence results typically require an additional strict complementarity assumption that is automatically satisfied in the case of linear programming. Some authors have proposed superlinear algorithms that do not require strict complementarity, but these methods require accurate identification of the set of degenerate indices before the fast convergence becomes effective.

The LCP algorithms can, in fact, be extended to a wider class of problems involving the so-called sufficient matrices. Instead of requiring M and N to satisfy the monotonicity property defined above, we require there to exist a nonnegative constant κ such that

$$y^T z \geqslant -4\kappa \sum_{i \mid y_i z_i > 0} y_i z_i \quad \text{for all } y, z \text{ with } My + Nz = 0.$$

The complexity estimates for interior-point methods applied to such problems depends on the parameter κ, so that the complexity is not polynomial on the whole class of sufficient matrices. Potra and Sheng [19] propose a large-step infeasible-interior-point method for solving $P_*(\kappa)$-matrix linear complementarity problems with a number of strong properties. The algorithm generates points in a large neighborhood of an infeasible central path, and each iteration requires only one matrix factorization. If the problem has a solution, the algorithm converges from an arbitrary positive starting point. The computational complexity of the algorithm depends on the quality of the starting point. If a well centered starting point is feasible or close to being feasible, it has $O((1+\kappa)\sqrt{n}L)$-iteration complexity. In cases in which such a starting point is not readily available, a modified version of the algorithm terminates in $O((1+\kappa)^2 nL)$ steps either by finding a solution or by determining that the problem is not solvable. Finally, high-order local convergence is proved for problems having a strictly complementary solution. We note that while the properties of the algorithm (e.g. computational complexity) depend on κ, the algorithm itself does not.

Primal–dual methods have been applied to many practical applications of (3.18) and (3.19), including portfolio optimization, optimal control, and ℓ_1 regression (see [28] for references).

The interior-point approach has a number of advantages over the active-set approach from a computational point of view. It is difficult for an active-set algorithm to exploit any structure inherent in both Q and A without redesigning most of its complex linear algebra operations: the operations of adding a constraint to the active set, deleting a constraint, evaluating Lagrange multiplier estimates, calculating the search direction, and so on. In the interior-point approach, on the other hand, the only complex linear algebra operation is solution of the linear system (3.21) — and this operation, though expensive, is relatively straightforward. Since the structure and dimension of the linear system remain the same at all iterations, the routines for solving the linear systems can exploit fully the properties of the systems arising from each problem class or instance. In fact, the algorithm can be implemented to high efficiency using an object-oriented approach, in which the implementer of each new problem class needs to supply only code for the factorization and solution of the systems (3.21), optimized for the structure of the new class, along with a number of simple operations such as inner-product calculations. Code that implements upper-level decisions (choice of parameter σ, vector r, steplength α) remains efficient across the gamut of applications of (3.19) and can simply be reused by all applications.

We note, however, that active-set methods would still require much less execution time than interior-point methods in many contexts, especially when "warm start" information is available and when the problem is generic enough that not much benefit is gained by exploiting its structure.

The extension of primal–dual algorithms from linear programming to convex QP is so straight-forward that a number of the interior-point linear programming codes have recently been extended to handle problems in the class (3.18) as well. In their linear algebra calculations, most of these codes treat both Q and A as general sparse matrices, and hence are efficient across a wide range of applications. By contrast, implementations of active-set methods for (3.18) that are capable of handling even moderately sized problems have not been widely available.

4. Semi-definite programming

As mentioned in the introduction, semi-definite programming (SDP) has been one of the most active areas of optimization research in the 1990s. SDP consists in minimizing a linear functional of a matrix subject to linear equality and inequality constraints, where the inequalities include membership of the cone of positive-semi-definite matrices. SDP is a broad paradigm; it includes as special cases linear programming, (linearly constrained) QP, quadratically constrained QP and other optimization problems (see [16,25]). Semi-definite programming has numerous applications in such diverse areas as optimal control, combinatorial optimization, structural optimization, pattern recognition, trace factor analysis in statistics, matrix completions, etc. See the excellent survey paper [25] for some instances. It was only after the advent of interior-point methods, however, that efficient solution methods for SDP problems were available. During the past few years an impressive number of interior-point methods for SDP have been proposed. Some of them have been successfully implemented and used to solve important application problems. However the theory and practice of interior-point methods for SDP has not yet reached the level of maturity of interior-point methods for LP, QP, and LCP. One reason that the study of interior-point methods for SDP is extremely important is that while LP, QP, and LCP can also be solved by other methods (e.g. the simplex method or Lemke's method), interior-point methods appear to be the only efficient methods for solving general SDP problems presently known.

To define the SDP, we introduce the notation $\mathscr{S}\mathbb{R}^{n \times n}$ to represent the set of $n \times n$ symmetric matrices, and the inner product $X \bullet Z$ of two matrices in this set, which is defined as

$$X \bullet Z = \sum_{i=1}^{n} \sum_{j=1}^{n} x_{ij} z_{ij}.$$

The SDP in standard form is then

$$\min_{X} C \bullet X \quad \text{s.t. } X \succcurlyeq 0, \ A_i \bullet X = b_i, \ i = 1, 2, \ldots, m, \tag{4.22}$$

where $X \in \mathscr{S}\mathbb{R}^{n \times n}$, and its associated dual problem is

$$\max_{y, S} \ b^T \lambda \quad \text{s.t.} \ \sum_{i=1}^{m} \lambda_i A_i + S = C, \ S \succcurlyeq 0, \tag{4.23}$$

where $S \in \mathscr{S}\mathbb{R}^{n \times n}$ and $\lambda \in \mathbb{R}^m$.

In what follows, we will consider only primal–dual interior-point methods that simultaneously solve the primal and dual problems. Points on the central path for (4.22), (4.23) are defined by the following parametrized system:

$$\sum_{i=1}^{m} \lambda_i A_i + S = C, \tag{4.24a}$$

$$A_i \bullet X = b_i, \quad i = 1, 2, \ldots, m, \tag{4.24b}$$

$$XS = \mu I, \tag{4.24c}$$

$$X \succcurlyeq 0, \quad S \succcurlyeq 0, \tag{4.24d}$$

where as usual μ is the positive parameter. Unlike the corresponding equations for linear programming, system (4.24a), (4.24b), (4.24c) is not quite "square", since the variables reside in the space $\mathscr{S}\mathbb{R}^{n \times n} \times \mathbb{R}^m \times \mathscr{S}\mathbb{R}^{n \times n}$ while the range space of the equations is $\mathscr{S}\mathbb{R}^{n \times n} \times \mathbb{R}^m \times \mathbb{R}^{n \times n}$. In particular, the product of two symmetric matrices (see (4.24c)) is not necessarily symmetric. Before Newton's method can be applied the domain and range have to be reconciled. The various primal–dual algorithms differ partly in the manner in which they achieve this reconciliation.

The paper of Todd [24] is witness to the intensity of research in SDP interior-point methods: It describes 20 techniques for obtaining search directions for SDP, among the most notable being the following:

(1) the AHO search direction proposed by Alizadeh, Haeberly and Overton;
(2) the KSH/HRVW/M search direction independently proposed by Kojima, Shindoh and Hara; Helmberg, Rendl, Vanderbei and Wolkowicz; and later rediscovered by Monteiro;
(3) the NT direction introduced by Nesterov and Todd.

Most of the search directions for SDP are obtained by replacing Eq. (4.24c) by a "symmetric" one whose range lies in $\mathscr{S}\mathbb{R}^{n \times n}$

$$\Theta(X, S) = 0. \tag{4.25}$$

Primal–dual methods are then derived as perturbed Newton's methods applied to (4.24a), (4.24b), (4.25). Examples of symmetrizations (4.25) include the Monteiro–Zhang family [15], in which

$$\Theta(X, S) = H_P(XS),$$

where

$$H_P(M) = \tfrac{1}{2}[PMP^{-1} + (PMP^{-1})^T]$$

(with a given a nonsingular matrix $P \in \mathbb{R}^{n \times n}$) is the symmetrization operator of Zhang. The search directions (1)–(3) mentioned above are obtained by taking P equal to I, $S^{1/2}$, and $[S^{1/2}(S^{1/2} X S^{1/2})^{-1/2} S^{1/2}]^{1/2}$, respectively.

Even if the SDP has integer data, its solution cannot in general be expressed in terms of rational numbers, so that the exact solution cannot be obtained in a finite number of bit operations. We say that an interior-point method for SDP "is polynomial" if there is a positive constant ω such that the distance to optimum (or the duality gap) is reduced by a factor of $2^{-O(L)}$ in at most $O(n^{\omega}L)$ iterations. In this case, we will say that the interior-point method has $O(n^{\omega}L)$ iteration complexity. The iteration complexity appears to be dependent on the choice of search direction. The best results obtained to date show that some feasible interior-point methods based on small neighborhoods for the central path have $O(\sqrt{n}L)$ iteration complexity for all three search directions mentioned above.

Monteiro and Zhang [15] proved that algorithms acting in large neighborhoods of the central path have $O(nL)$ iteration complexity if based on the NT direction and $O(n^{3/2}L)$ if based on the KSH/HRVW/M search direction. They also gave iteration complexity bounds (which depend on the condition number of matrices J_x and J_s defined by $P^{\mathrm{T}}P = X^{-1/2}J_xX^{1/2} = S^{-1/2}J_sS^{1/2}$) for algorithms acting in the large neighborhood that are based on the MZ^* family of directions. This family is a subclass of the MZ family that contains the NT and the KSH/HRVW/M directions but not the AHO direction. So far, no complexity results are known for algorithms based on the large neighborhood and the AHO direction.

The analysis of infeasible interior-point algorithms for SDP is considerably more difficult than that of their feasible counterparts. The first complexity result in this respect was obtained by Kojima, Shindoh, and Hara, who showed that an infeasible-interior-point potential reduction method for SDP has $O(n^{5/2}L)$ iteration complexity. Subsequently, Zhang analyzed an infeasible-interior-point method, based on the KSH/HRVW/M search direction, that has $O(n^2L)$ iteration complexity when acting in the semi-large neighborhood and $O(n^{5/2}L)$ iteration complexity in the large neighborhood of the central path. The analysis of the Mizuno–Todd–Ye predictor–corrector method for infeasible starting points was performed independently by Kojima, Shida and Shindoh and Potra and Sheng. The analysis in the latter paper shows that the iteration complexity depends on the quality of the starting point. If the problem has a solution, then the algorithm is globally convergent. If the starting point is feasible or close to feasible, the algorithms finds an optimal solution in at most $O(\sqrt{n}L)$ iterations. If the starting point is large enough according to some specific criteria, then the algorithm terminates in at most $O(nL)$ steps either by finding a strictly complementary solution or by determining that the primal–dual problem has no solution of norm less than a specified size.

Superlinear convergence is especially important for SDP since no finite termination schemes exist for such problems. As predicted by theory and confirmed by numerical experiments, the condition number of the linear systems defining the search directions increases like $1/\mu$, so that the respective systems become quite ill conditioned as we approach the solution. As we observed in the case of linear programming, an interior-point method that is not superlinearly convergent is unlikely to obtain high accuracy in practice. On the other hand, superlinearly convergent interior-point methods often achieve good accuracy (duality measure of 10^{-10} or better) in substantially fewer iterations than indicated by the worse-case global linear convergence rate indicated by the analysis.

The local convergence analysis for interior-point algorithms for SDP is much more challenging than for linear programming. Kojima, Shida and Shindoh [12] established superlinear convergence of the Mizuno–Todd–Ye predictor–corrector algorithm based on the KSH/HRVW/M search direction under the following three assumptions:

(A) SDP has a strictly complementary solution;

(B) SDP is nondegenerate in the sense that the Jacobian matrix of its KKT system is nonsingular;
(C) the iterates converge tangentially to the central path in the sense that the size of the neighborhood containing the iterates must approach zero namely,

$$\lim_{k\to\infty} ||(X^k)^{1/2}S^k(X^k)^{1/2} - (X^k \bullet S^k/n)I||_F/(X^k \bullet S^k/n) = 0.$$

Assumptions (B) and (C) are quite restrictive; similar conditions are not required for the superlinear convergence of interior-point methods for linear programming or QP. Potra and Sheng [20] proved superlinear convergence of the same algorithm under assumption (A) together with the following condition:

(D) $\lim_{k\to\infty} X^k S^k/\sqrt{X^k \bullet S^k} = 0,$

which is clearly weaker than (C). Of course both (C) and (D) can be enforced by the algorithm, but the practical efficiency of such an approach is questionable. From a theoretical point of view, however, it is known from [20] that a modified version of the algorithm of [12] that uses several corrector steps in order to enforce (C) has polynomial complexity and is superlinearly convergent under assumption (A) only. It is well known that assumption (A) is necessary for superlinear convergence of interior-point methods that take Newton-like steps even in the QP case. (However, there are methods for convex QP and monotone LCP that attain superlinear convergence by making explicit guesses of the set of degenerate indices.)

Kojima et al. [12] also gave an example suggesting that interior-point algorithms for SDP based on the KSH/HRVW/M search direction are unlikely to be superlinearly convergent without imposing a condition like (C) or (D). In a later paper they showed that a predictor–corrector algorithm using the AHO direction is quadratically convergent under assumptions (A) and (B). They also proved that the algorithm is globally convergent, but no polynomial complexity bounds have yet been found. It appears that the use of the AHO direction in the corrector step has a strong effect on centering. This property is exploited in a recent paper of Ji et al. [10] who proved that the Mizuno–Todd–Ye algorithm, based on the MZ-family is superlinear under assumptions (A) and (D). They also showed that under assumptions (A) and (B) the algorithm has Q-order 1.5 if scaling matrices in the corrector step have bounded condition number, and Q-order 2 if the scaling matrices in both predictor and corrector step have bounded condition number. In particular, these results apply for the AHO direction, where the scaling matrix is the identity matrix. References to the results cited above can be found in [10].

Over the past several years we have witnessed an intense research effort on the use of SDP for finding approximate solution of (NP-hard) combinatorial optimization problems. In what follows, we will describe the technique of Goemans and Williamson, which yields an approximate solution whose value is within 13% of optimality for the MAX CUT problem [7].

In MAX CUT, we are presented with an undirected graph with N whose edges w_{ij} have nonnegative weights. The problem is choose a subset $\mathscr{S} \subset \{1, 2, \ldots, N\}$ so that the sum of weights of the edges that cross from $\mathscr{S}$ to its complement is minimized. In other words, we aim to choose $\mathscr{S}$ to maximize the objective

$$w(\mathscr{S}) \overset{\text{def}}{=} \sum_{i\in\mathscr{S}, j\notin\mathscr{S}} w_{ij}.$$

This problem can be restated as an integer quadratic program by introducing variables y_i, $i = 1, 2, \ldots, N$, such that $y_i = 1$ for $i \in \mathcal{S}$ and $y_i = -1$ for $i \notin \mathcal{S}$. We then have

$$\max_y \frac{1}{2} \sum_{i<j} w_{ij}(1 - y_i y_j) \quad \text{s.t. } y_i \in \{-1, 1\} \text{ for all } i = 1, 2, \ldots, N. \tag{4.26}$$

This problem is NP-complete. Goemans and Williamson replace the variables $y_i \in \mathbb{R}$ by vectors $v_i \in \mathbb{R}^N$ and consider instead the problem

$$\max_{v_1, v_2, \ldots, v_N} \frac{1}{2} \sum_{i<j} w_{ij}(1 - v_i^{\mathrm{T}} v_j) \quad \text{s.t. } \|v_i\| = 1 \text{ for all } i = 1, 2, \ldots, N. \tag{4.27}$$

This problem is a relaxation of (4.26) because any feasible point y for (4.26) corresponds to a feasible point

$$v_i = (y_i, 0, 0, \ldots, 0)^{\mathrm{T}}, \quad i = 1, 2, \ldots, N$$

for (4.27). Problem (4.27) can be formulated as an SDP by changing variables $v_1, v_2, \ldots, v_N$ to a matrix $Y \in \mathbb{R}^{N \times N}$, such that

$$Y = V^{\mathrm{T}} V \quad \text{where } V = [v_1, v_2, \ldots, v_N].$$

The constraints $\|v_i\| = 1$ can be expressed simply as $Y_{ii} = 1$, and since $Y = V^{\mathrm{T}} V$, we must have Y semi-definite. The transformed version of (4.27) is then

$$\max \frac{1}{2} \sum_{i<j} w_{ij}(1 - Y_{ij}) \quad \text{s.t. } Y_{ii} = 1, \ i = 1, 2, \ldots, N \quad \text{and} \quad Y \succcurlyeq 0,$$

which has the form (4.22) for appropriate definitions of C and A_i, $i = 1, 2, \ldots, N$. We can recover V from Y by performing a Cholesky factorization. The final step of recovering an approximate solution to the original problem (4.26) is performed by choosing a random vector $r \in \mathbb{R}^N$, and setting

$$y_i = \begin{cases} 1 & \text{if } r^{\mathrm{T}} v_i > 0, \\ -1 & \text{if } r^{\mathrm{T}} v_i \leqslant 0. \end{cases}$$

A fairly simple geometric argument shows that the expected value of the solution so obtained has objective value at least 0.87856 of the optimal solution to (4.26).

Similar relaxations have been obtained for many other combinatorial problems, showing that is possible to find good approximate solutions to many NP-complete problems by using polynomial algorithms. Such relaxations are also useful if we seek *exact* solutions of the combinatorial problem by means of a branch-and-bound or branch-and-cut strategy. Relaxations can be solved at each node of the tree (in which some of the degrees of freedom are eliminated and some additional constraints are introduced) to obtain both a bound on the optimal solution and in some cases a candidate feasible solution for the original problem. Since the relaxations to be solved at adjacent nodes of the tree are similar, it is desirable to use solution information at one node to "warm start" the SDP algorithm at a child node.

5. Convex programming

One of the most surprising results in interior-point methods is the fact that interior-point algorithms from LP can be extended to general convex programming problems, at least in a theoretical sense. The key to such an extension was provided in [16]. These authors explored the properties of self-concordant functions, and described techniques in which the inequality constraints in a convex programming problem are replaced by self-concordant barrier terms in the objective function. They derived polynomial algorithms by applying Newton-like methods to the resulting parametrized reformulations.

The fundamental property of self-concordant functions is that their third derivative can be bounded by some expression involving their second derivative at each point in their domain. This property implies that the second derivative does not fluctuate too rapidly in a relative sense, so that the function does not deviate too much from the second-order approximation on which Newton's method is based. Hence, we can expect Newton's method to perform reasonably well on such a function.

Given a finite-dimensional real vector space $\mathscr{V}$, an open, nonempty convex set $\mathscr{S} \subset \mathscr{V}$, and a closed convex set $\mathscr{T} \subset \mathscr{V}$ with nonempty interior, we have the following formal definition.

Definition 1. The function $F: \mathscr{S} \to \mathbb{R}$ is *self-concordant* if it is convex and if the following inequality holds for all $x \in \mathscr{S}$ and all $h \in \mathscr{V}$:

$$|D^3 F(x)[h, h, h]| \leqslant 2(D^2 F(x)[h, h])^{3/2}, \tag{5.28}$$

where $D^k F[h_1, h_2, \ldots, h_k]$ denotes the kth differential of F along the directions $h_1, h_2, \ldots, h_k$.

F is called *strongly self-concordant* if $F(x_i) \to \infty$ for all sequences $x_i \in \mathscr{S}$ that converge to a point on the boundary of $\mathscr{S}$.

F is a *ϑ-self-concordant barrier* for $\mathscr{T}$ if it is a strongly self-concordant function for int $\mathscr{T}$, and the parameter

$$\vartheta \stackrel{\text{def}}{=} \sup_{x \in \text{int } \mathscr{T}} F'(x)^{\mathrm{T}} [F''(x)]^{-1} F'(x) \tag{5.29}$$

is finite.

Note that the exponent $\frac{3}{2}$ on the right-hand side of (5.28) makes the condition independent of the scaling of the direction h. It is shown in [16, Corollary 2.3.3], that if $\mathscr{T} \neq \mathscr{V}$, then the parameter ϑ is no smaller than 1.

It is easy to show that log-barrier function of Section 2 is an n-self-concordant barrier for the positive orthant $\mathbb{R}^n_+$ if we take

$$\mathscr{V} = \mathbb{R}^n, \quad \mathscr{T} = \mathbb{R}^n_+, \quad F(x) = -\sum_{i=1}^{n} \log x_i.$$

where $\mathbb{R}^n_+$ denotes the positive orthant. Another interesting case is the second-order cone (or "ice-cream cone"), for which we have

$$\mathscr{V} = \mathbb{R}^{n+1}, \qquad \mathscr{T} = \{(x, t) \,|\, ||x||_2 \leqslant t\}, \qquad F(x, t) = -\log(t^2 - ||x||^2), \tag{5.30}$$

where $t \in \mathbb{R}$ and $x \in \mathbb{R}^n$. In this case, F is a two-self-concordant barrier. Second-order cone programming consists in minimizing a linear function subject to linear equality constraints together with inequality constraints induced by second-order cones. Convex quadratically constrained quadratic programs can be posed in this form, along with sum-of-norms problems and many other applications.

A third important case is the cone of positive-semi-definite matrices, for which we have

$\mathcal{V} = n \times n$ symmetric matrices,

$\mathcal{T} = n \times n$ symmetric positive-semi-definite matrices,

$F(X) = -\log \det X$

for which F is an n-self-concordant barrier. This barrier function can be used to model the constraint $X \succeq 0$ in (4.22).

Self-concordant barrier functions allow us to generalize the primal barrier method of Section 2 to problems of the form

$$\min \langle c, x \rangle \quad \text{s.t. } Ax = b, \ x \in \mathcal{T}, \tag{5.31}$$

where $\mathcal{T}$ is a closed convex set, $\langle c, x \rangle$ denotes a linear functional on the underlying vector space $\mathcal{V}$, and A is a linear operator. Similarly to (2.2), we define the barrier subproblem to be

$$\min_x f(x; \mu) \overset{\text{def}}{=} \frac{1}{\mu} \langle c, x \rangle + F(x) \quad \text{s.t. } Ax = b, \tag{5.32}$$

where $F(x)$ is a self-concordant barrier and $\mu > 0$ is the barrier parameter. Note that by the Definition 1, $f(x; \mu)$ is also a strongly self-concordant function. The primal barrier algorithm for (5.31) based on (5.32) is as follows:

primal barrier algorithm
Given $x^0 \in \text{int } \mathcal{T}$ and $\mu_0 > 0$;
Set $k \leftarrow 0$;
repeat
      Obtain $x^{k+1} \in \text{int } \mathcal{T}$ by performing one or more projected Newton steps
         for $f(\cdot; \mu_k)$, starting at $x = x^k$;
      Choose $\mu_{k+1} \in (0, \mu_k)$;
until some termination test is satisfied.

As in Sections 2–4, the worst-case complexity of algorithms of this type depends on the parameter ϑ associated with F but not on any properties of the data that defines the problem instance. For example, we can define a short-step method in which a single full Newton step is taken for each value of k, and μ is decreased according to

$$\mu_{k+1} = \mu_k \left/ \left(1 + \frac{1}{8\sqrt{\vartheta}} \right) \right. .$$

Given a starting point with appropriate properties, we obtain an iterate x^k whose objective $\langle c, x^k \rangle$ is within ε of the optimum in

$$O\left(\sqrt{\vartheta} \log \frac{\vartheta \mu_0}{\varepsilon} \right) \quad \text{iterations.}$$

Long-step variants are discussed in [16]. The practical behavior of the methods does, of course, depend strongly on the properties of the particular problem instance.

The primal–dual algorithms of Section 2 can also be extended to more general problems by means of the theory of self-scaled cones developed in [17,18]. The basic problem considered is the conic programming problem

$$\min \langle c, x \rangle \quad \text{s.t. } Ax = b, \ x \in K, \tag{5.33}$$

where $K \subset \mathbb{R}^n$ is a closed convex cone, that is, a closed convex set for which $x \in K \Rightarrow tx \in K$ for all nonnegative scalars t, and A denotes a linear operator from $\mathbb{R}^n$ to $\mathbb{R}^m$. The dual cone for K is denoted by K^* and defined as

$$K^* \stackrel{\text{def}}{=} \{s \mid \langle s, x \rangle \geqslant 0 \text{ for all } x \in K\}$$

and we can write the dual instance of (5.33) as

$$\max \langle b, \lambda \rangle \quad \text{s.t. } A^*\lambda + s = c, \ s \in K^*, \tag{5.34}$$

where A^* denotes the adjoint of A. The duality relationships between (5.33) and (5.34) are more complex than in linear programming, but if either problem has a feasible point that lies in the interior of K or K^*, respectively, the strong duality property holds. This property is that when the optimal value of either (5.33) or (5.34) is finite, then both problems have finite optimal values, and these values are the same.

K is a self-scaled cone when its interior $\operatorname{int} K$ is the domain of a self-concordant barrier function F with certain strong properties that allow us to define algorithms in which the primal and dual variables are treated in a perfectly symmetric fashion and play interchangeable roles. The full elucidation of these properties is quite complicated. It suffices to note here that the three cones mentioned above – the positive orthant $\mathbb{R}^n_+$, the second-order cone (5.30), and the cone of positive-semi-definite symmetric matrices – are the most interesting self-scaled cones, and their associated barrier functions are the logarithmic functions mentioned above.

To build algorithms from the properties of self-scaled cones and their barrier functions, the Nesterov–Todd theory defines a *scaling point* for a given pair $x \in \operatorname{int} K$, $s \in \operatorname{int} K^*$ to be the unique point w such that $H(w)x = s$, where $H(\cdot)$ is the Hessian of the barrier function. In the case of linear programming, it is easy to verify that w is the vector in $\mathbb{R}^n$ whose elements are $\sqrt{x_i/s_i}$. The Nesterov–Todd search directions are obtained as projected steepest descent direction for the primal and dual barrier subproblems (that is, (5.32) and its dual counterpart), where a weighted inner product involving the matrix $H(w)$ is used to define the projections onto the spaces defined by the linear constraints $Ax = b$ and $A^*\lambda + s = c$, respectively. The resulting directions satisfy the following linear system:

$$\begin{bmatrix} 0 & A & 0 \\ A^* & 0 & I \\ 0 & H(w) & I \end{bmatrix} \begin{bmatrix} \Delta\lambda \\ \Delta x \\ \Delta s \end{bmatrix} = - \begin{bmatrix} 0 \\ 0 \\ s + \sigma\mu\nabla F(x) \end{bmatrix}, \tag{5.35}$$

where $\mu = \langle x, s \rangle / \vartheta$. (The correspondence with (2.10) is complete if we choose the perturbation term to be $r = 0$.) By choosing the starting point appropriately, and designing schemes to choose the parameters σ and step lengths along these directions, we obtain polynomial algorithms for this general setting. The NT direction in the previous section is the specialization of the above search directions for semi-definite programming.

6. Conclusions

Interior-point methods remain an active and fruitful area of research, although the frenetic pace that characterized the area has slowed in recent years. Interior-point codes for linear programming codes have become mainstream and continue to undergo development, although the competition from the simplex method is stiff. Semi-definite programming has proved to be an area of major impact. Applications to quadratic programming show considerable promise, because of the superior ability of the interior-point approach to exploit problem structure efficiently. The influence on nonlinear programming theory and practice has yet to be determined, even though significant research has already been devoted to this topic. Use of the interior-point approach in decomposition methods appears promising, though no rigorous comparative studies with alternative approaches have been performed. Applications to integer programming problems have been tried by a number of researchers, but the interior-point approach is hamstrung here by competition from the simplex method with its superior warm-start capabilities.

Acknowledgements

This work was supported in part by NSF under grant DMS-9996154 and by the Mathematical, Information, and Computational Sciences Division subprogram of the Office of Advanced Scientific Computing Research, U.S. Department of Energy, under Contract W-31-109-Eng-38.

We are grateful to an anonymous referee for a speedy but thorough review.

References

[1] K.M. Anstreicher, Linear programming in $O([n^3/\ln n]L)$ operations, CORE Discussion Paper 9746, Université Catholique de Louvain, Louvain-la-Neuve, Belgium, January 1999, SIAM J. Optim., in preparation.

[2] K.M. Anstreicher, J. Ji, F.A. Potra, Y. Ye, Average performance of a self-dual interior-point algorithm for linear programming, in: P. Pardalos (Ed.), Complexity in Numerical Optimization, World Scientific, Singapore, 1993, pp. 1–15.

[3] K.H. Borgwardt, The Simplex Method: A Probabilistic Analysis, Springer, Berlin, 1987.

[4] A.S. El-Bakry, R.A. Tapia, Y. Zhang, A study of indicators for identifying zero variables in interior-point methods, SIAM Rev. 36 (1) (1994) 45–72.

[5] A.V. Fiacco, G.P. McCormick, Nonlinear Programming: Sequential Unconstrained Minimization Techniques, Wiley, New York, 1968 (reprinted by SIAM, Philadelphia, PA, 1990).

[6] R.M. Freund, S. Mizuno, Interior point methods: current status and future directions, Optima 51 (1996) 1–9.

[7] M.X. Goemans, D.P. Williamson, Improved approximation algorithms for maximum cut and satisfiability problems using semidefinite programming, J. Assoc. Comput. Mach. 42 (6) (1995) 1115–1145.

[8] D. Goldfarb, M.J. Todd, Linear programming, in: G.L. Nemhauser, A.H.G. Rinnooy Kan, M.J. Todd (Eds.), Optimization, North-Holland, Amsterdam, 1989, pp. 73–170.

[9] J. Gondzio, Multiple centrality corrections in a primal–dual method for linear programming, Comput. Optim. Appl. 6 (1996) 137–156.

[10] J. Ji, F.A. Potra, R. Sheng, On the local convergence of a predictor–corrector method for semidefinite programming, SIAM J. Optim. 10 (1999) 195–210.

[11] N. Karmarkar, A new polynomial-time algorithm for linear programming, Combinatorica 4 (1984) 373–395.

[12] M. Kojima, M. Shida, S. Shindoh, Local convergence of predictor–corrector infeasible-interior-point algorithms for SDPs and SDLCPs, Math. Programming, Ser. A 80 (2) (1998) 129–160.

[13] N. Megiddo, Pathways to the optimal set in linear programming, in: N. Megiddo (Eds.), Progress in Mathematical Programming: Interior-Point and Related Methods Springer, New York, 1989, pp. 131–158 (Chapter 8).

[14] S. Mehrotra, On the implementation of a primal–dual interior point method, SIAM J. Optim. 2 (1992) 575–601.

[15] R.D.C. Monteiro, Y. Zhang, A unified analysis for a class of long-step primal–dual path-following interior-point algorithms for semidefinite programming, Math. Programming Ser. A 81 (3) (1998) 281–299.

[16] Yu.E. Nesterov, A.S. Nemirovskii, Interior Point Polynomial Methods in Convex Programming: Theory and Applications, SIAM, Philadelphia, PA, 1994.

[17] Yu.E. Nesterov, M.J. Todd, Self-scaled barriers and interior-point methods for convex programming, Math. Oper. Res. 22 (1997) 1–42.

[18] Yu.E. Nesterov, M.J. Todd, Primal–dual interior-point methods for self-scaled cones, SIAM J. Optim. 8 (1998) 324–362.

[19] F.A. Potra, R. Sheng, A large-step infeasible-interior-point method for the P_*-matrix LCP, SIAM J. Optim. 7 (2) (1997) 318–335.

[20] F.A. Potra, R. Sheng, Superlinear convergence of interior-point algorithms for semidefinite programming, J. Optim. Theory Appl. 99 (1) (1998) 103–119.

[21] J. Renegar, A polynomial-time algorithm, based on Newton's method, for linear programming, Math. Programming 40 (1988) 59–93.

[22] J. Renegar, A mathematical view of interior-point methods in convex optimization, unpublished notes, June 1999.

[23] C. Roos, J.-Ph. Vial, T. Terlaky, Theory and Algorithms for Linear Optimization: An Interior Point Approach, Wiley-Interscience Series in Discrete Mathematics and Optimization, Wiley, New York, 1997.

[24] M.J. Todd, A study of search directions in primal–dual interior-point methods for semidefinite programming, Technical Report, School of Operations Research and Industrial Engineering, Cornell University, Ithaca, NY, February 1999.

[25] L. Vandenberghe, S. Boyd, Semidefinite programming, SIAM Rev. 38 (1) (1996) 49–95.

[26] M.H. Wright, Interior methods for constrained optimization, in: Acta Numer. 1992, Cambridge University Press, Cambridge, 1992, pp. 341–407.

[27] S.J. Wright, Primal–Dual Interior-Point Methods, SIAM, Philadelphia, PA, 1997.

[28] S.J. Wright, Recent developments in interior-point methods, preprint ANL/MCS-P783-0999, Mathematics and Computer Science Division, Argonne National Laboratory, Argonne, IL, September 1999.

[29] Y. Ye, Interior Point Algorithms: Theory and Analysis, Wiley-Interscience Series in Discrete Mathematics and Optimization, Wiley, New York, 1997.

[30] Y. Ye, O. Güler, R.A. Tapia, Y. Zhang, A quadratically convergent $O(\sqrt{n}L)$-iteration algorithm for linear programming, Math. Programming Ser. A 59 (1993) 151–162.

[31] Y. Ye, M.J. Todd, S. Mizuno, An $O(\sqrt{n}L)$-iteration homogeneous and self-dual linear programming algorithm, Math. Oper. Res. 19 (1994) 53–67.

[32] Y. Zhang, On the convergence of a class of infeasible-interior-point methods for the horizontal linear complementarity problem, SIAM J. Optim. 4 (1994) 208–227.

Journal of Computational and Applied Mathematics 124 (2000) 303–318

JOURNAL OF
COMPUTATIONAL AND
APPLIED MATHEMATICS

www.elsevier.nl/locate/cam

Complementarity problems

Stephen C. Billups[a,1], Katta G. Murty[b,*]

[a]*Department of Mathematics, University of Colorado, Denver, CO 80217-3364, USA*
[b]*Department of Industrial and Operations Engineering, University of Michigan, Ann Arbor, MI 48109-2117, USA*

Received 7 May 1999; received in revised form 17 November 1999

Abstract

This paper provides an introduction to complementarity problems, with an emphasis on applications and solution algorithms. Various forms of complementarity problems are described along with a few sample applications, which provide a sense of what types of problems can be addressed effectively with complementarity problems. The most important algorithms are presented along with a discussion of when they can be used effectively. We also provide a brief introduction to the study of matrix classes and their relation to linear complementarity problems. Finally, we provide a brief summary of current research trends. © 2000 Elsevier Science B.V. All rights reserved.

Keywords: Complementarity problems; Variational inequalities; Matrix classes

1. Introduction

The distinguishing feature of a complementarity problem is the set of *complementarity conditions*. Each of these conditions requires that the product of two or more nonnegative quantities should be zero. (Here, each quantity is either a decision variable, or a function of the decision variables). Complementarity conditions made their first appearance in the optimality conditions for continuous variable nonlinear programs involving inequality constraints, which were derived by Karush in 1939. But the significance of complementarity conditions goes far beyond this. They appear prominently in the study of equilibria problems and arise naturally in numerous applications from economics, engineering and the sciences. There is therefore a great deal of practical interest in the development of robust and efficient algorithms for solving complementarity problems.

The early motivation for studying the Linear Complementarity Problem (LCP) was because the KKT optimality conditions for linear and Quadratic Programs (QP) constitute an LCP of the form

---

* Corresponding author.
E-mail address: murty@umich.edu (K.G. Murty).
[1] Research partially supported through NSF grant DMS-9973321.

0377-0427/00/$ - see front matter © 2000 Elsevier Science B.V. All rights reserved.
PII: S 0377-0427(00)00432-5

(1), or a mixed LCP of the form (2) (see next section for statements of these problems). However, the study of LCP really came into prominence in 1963 when Howson, in his Ph.D. thesis, and Lemke and Howson [10] showed that the problem of computing a *Nash equilibrium point* of a *bimatrix game* can be posed as an LCP of the form (1), and developed an elegant and efficient constructive procedure (the *complementary pivot method*) for solving it. In 1968, the unification of linear and quadratic programs and bimatrix games under the LCP format by Cottle and Dantzig was seen as a fundamental breakthrough, and the study of complementarity problems suddenly blossomed.

The Nonlinear Complementarity Problem (NCP) was introduced by Cottle in his Ph.D. thesis in 1964, and the closely related Variational Inequality Problem (VIP) was introduced by Hartman and Stampacchia in 1966, primarily with the goal of computing stationary points for nonlinear programs. While these problems were introduced soon after the LCP, most of the progress in developing algorithms for these problems did not begin until the late 1970s.

Well over a thousand articles and several books have been published on the subject of complementarity problems. We limit the scope of this paper. First we describe what complementarity problems are and try to give a sense of what types of problems can be addressed effectively within this framework. This includes a description of the various types of complementarity problems in Section 2, as well as a discussion of applications in Section 3. In Sections 4 and 5 we describe the most important computational algorithms for solving complementarity problems and discuss when these methods are most likely to be successful. In Section 6, we give a brief introduction to the study of matrix classes, which represents a very rich field within LCP. Finally, in Section 7, we discuss some of the current trends in complementarity research. For a detailed comprehensive treatment of the LCP, we refer the reader to the books by Murty [12] and Cottle et al. [3]. For a general treatment of NCP and VIP we recommend Ref. [8]. Additional references are given in [2], as well as in the references mentioned above.

2. The various complementarity problems

The simplest and most widely studied of the complementarity problems is the LCP, which has often been described as a *fundamental problem* because the first order necessary optimality conditions for QP involving inequality constraints in nonnegative variables form an LCP: given $M \in R^{n \times n}$, $q \in R^n$, find $w = (w_j) \in R^n$, $z = (z_j) \in R^n$ satisfying

$$w - Mz = q; \quad w, z \geqslant 0; \quad w^T z = 0. \tag{1}$$

We denote this LCP by the symbol (q, M). The name comes from the third condition, the *complementarity condition* which requires that at least one variable in the pair (w_j, z_j) should be equal to 0 in the solution of the problem, for each $j = 1$ to n. This pair is therefore known as the *jth complementary pair* in the problem, and for each j, the variable w_j is known as the *complement* of z_j and vice versa. The LCP (q, M) is said to be *monotone* if the matrix M is *positive semidefinite* (PSD).

A slight generalization of the LCP is the *mixed LCP* (mLCP): given $A \in R^{n \times n}$, $B \in R^{m \times m}$, $C \in R^{n \times m}$, $D \in R^{m \times n}$, $a \in R^n$, $b \in R^m$, find $u \in R^n$, $v \in R^m$ satisfying

$$a + Au + Cv = 0, \quad b + Du + Bv \geqslant 0, \quad v \geqslant 0, \quad v^T(b + Du + Bv) = 0. \tag{2}$$

The mLCP is a mixture of the LCP and a system of linear equations, which correspond to the unrestricted variables u. The first order necessary optimality conditions for a quadratic program involving some equality and some inequality constraints are in this form. In (2), if A is nonsingular, then u can be eliminated from it using $u = -A^{-1}(a + Cv)$ and then (2) becomes the LCP $(b - DA^{-1}a, \ B - DA^{-1}C)$. This mLCP is said to be monotone if the matrix

$$\begin{pmatrix} A & C \\ D & B \end{pmatrix}$$

in (2) is PSD.

Another generalization of the LCP is the *horizontal LCP* or hLCP: given $N \in R^{n \times n}$, $M \in R^{n \times n}$, $q \in R^n$ find $w \in R^n$, $z \in R^n$ satisfying

$$Nw - Mz = q; \quad w, z \geqslant 0; \quad w^T z = 0. \tag{3}$$

The hLCP (3) becomes the standard LCP if $N = I$. Also, if N is nonsingular, then (3) is equivalent to the LCP $(N^{-1}q, N^{-1}M)$. The hLCP is said to be monotone if for any two pairs of points (w^1, z^1) and (w^2, z^2) satisfying $Nw - Mz = q$ we have $(w^1 - w^2)^T(z^1 - z^2) \geqslant 0$. Note that if $N = I$, this is equivalent to the matrix M being PSD.

For each $i = 1$ to n let m_i be a positive integer, and $m = \sum_{i=1}^n m_i$. The *Vertical LCP* or VLCP is another generalization of the LCP for which the input data are $M \in R^{m \times n}$, $q \in R^m$ partitioned as follows:

$$M = \begin{pmatrix} M^1 \\ \vdots \\ M^n \end{pmatrix}, \quad q = \begin{pmatrix} q^1 \\ \vdots \\ q^n \end{pmatrix}$$

where for each $i = 1$ to n, $M^i \in R^{m_i \times n}$, $q^i \in R^{m_i}$. Given this data, the VLCP is to find $z = (z_i) \in R^n$ satisfying

$$q + Mz \geqslant 0, \quad z \geqslant 0, \quad z_i \prod_{j=1}^{m_i} (q^i + M^i z)_j = 0, \quad i = 1, \ldots, n. \tag{4}$$

If $m_i = 1$ for all i, then this VLCP becomes the standard LCP.

In the spirit of the VLCP, we can define a *general horizontal linear complementarity problem* (HLCP) involving a vector $q \in R^n$, a square matrix $N \in R^{n \times n}$, a rectangular matrix $M \in R^{n \times m}$ where $m \geqslant n$, and a partition of the vector of variables $z = (z^1, \ldots, z^n)^T \in R^m$ where each z^i is again a vector consisting of one or more variables. Given this data, the problem is to find a $w = (w_i) \in R^n$ and a $z = (z^1, \ldots, z^n)^T \in R^m$ satisfying $Nw - Mz = q$, $w \geqslant 0$, $z \geqslant 0$, and for each i at least one variable among $\{w_i, z^i\}$ is 0. Clearly (3) is a special case of this problem.

Then there is the *generalized LCP* (GLCP) with data $A, B \in R^{m \times n}$, $C \in R^{m \times d}$ and $q \in R^m$. The problem is to find $(x \in R^n, s \in R^n, z \in R^d)$ satisfying $Ax + Bs + Cz = q$, $(x, s, z) \geqslant 0$, $x^T s = 0$.

Now we will present some nonlinear generalizations of the LCP. The most important of these is the NCP: given a mapping $F(z) = (F_i(z)) : R^n \to R^n$, find a $z \in R^n$ satisfying

$$z \geqslant 0, \quad F(z) \geqslant 0, \quad z^T F(z) = 0. \tag{5}$$

If $F(z)$ is the affine function $q + Mz$, then (5) becomes the LCP (q, M).

A further generalization of the NCP is the VIP: given a mapping $F(z) = (F_i(z)) : R^n \to R^n$, and $\emptyset \neq K \subset R^n$, find a $z^* \in K$ satisfying

$$(y - z^*)^{\mathrm{T}} F(z^*) \geqslant 0 \quad \text{for all } y \in K \tag{6}$$

denoted by VI(K,F). If $K = \{z: z \geqslant 0\}$, then the z^* solving (6) also solves (5). Also, if K is polyhedral and F is affine, it can be verified that VI(K,F) is an LCP. When K is a rectangular region defined by $K := \prod_{i=1}^{n} [l_i, u_i], -\infty \leqslant l_i < u_i \leqslant \infty$, $i = 1, \ldots, n$, this is called the Box Constrained VIP (BVIP), which is also commonly referred to as the (nonlinear) Mixed Complementarity Problem (MCP).

For any subset $K \subset R^n$, its polar cone, denoted by K^*, is defined to be $\{y \in R^n : x^{\mathrm{T}} y \geqslant 0$ for all $x \in K\}$. Another generalization of the NCP is the *complementarity problem over a cone*: given a mapping $F(z) = (F_i(z)) : R^n \to R^n$ and a cone K in R^n, find a $z \in K$ satisfying $F(z) \in K^*$, and $z^{\mathrm{T}} F(z) = 0$. This problem, denoted by CP(K,F), reduces to NCP (5) if $K = R^n_+$. Also, since K is a cone, CP(K,F) and VI(K,F) have the same solution set.

3. Example applications

Complementarity problems arise naturally in the study of many phenomena in economics and engineering. A comprehensive and excellent treatment of applications of complementarity problems is provided in [6]. Additionally, a large collection of problems from a variety of application areas can be found in the MCPLIB library of test problems [4]. Applications of complementarity from the field of economics include general Walrasian equilibrium, spatial price equilibria, invariant capital stock, and game-theoretic models. In engineering, complementarity problems arise in contact mechanics, structural mechanics, obstacle and free boundary problems, elastohydrodynamic lubrication, and traffic equilibrium.

As a rule of thumb, the complementarity framework should be considered whenever the system being studied involves complementary pairs of variables (that is, where one or the other member of each pair must be at its bound). For example, in contact mechanics, the force between two objects is complementary to the distance between the two objects; there is no force unless the distance between the objects is zero. As another example, in Walrasian equilibrium problems, the price of a commodity is complementary to excess supply of the commodity; if there is excess supply, the price will fall either until the demand rises to eliminate the excess supply, or until the price is zero.

3.1. Piecewise linear equations

Consider the LCP (q, M). For each $z \in R^n$ define $h_i(z) = \min\{z_i, (q + Mz)_i\}$, and let $h(z) = (h_i(z))$. Then $h(z) : R^n \to R^n$ is a piecewise linear concave function, and clearly, solving the LCP (q, M) is equivalent to solving the system of piecewise linear equations $h(z) = 0$. Conversely, under a mild nonsingularity assumption, any piecewise linear system of equations can be reformulated as a linear complementarity problem.

In the same way, the VLCP (4) is equivalent to the system of piecewise linear equations $H(z) = 0$ where $H(z) = (H_i(z))$ and $H_i(z) = \min\{z_i, (q^i + M^i z)_1, \ldots, (q^i + M^i z)_{m_i}\}$ for $i = 1$ to n.

3.2. An application involving a smallsize convex QP model

Since the optimality conditions for convex QP form an LCP or mLCP, any application involving convex QP offers an application for LCP or mLCP. We present a recent application, described by Murty [13], in supply chain management, which is becoming increasingly popular.

An important issue in supply chain management is to forecast the demand for each item and to determine when to place orders for it and the order quantities. Classical analysis in the inventory management literature assumes that the distribution of demand is known and typically assumes this distribution to be normal. This assumption confers many theoretical advantages, the principal among which is the fact that the normal distribution is fully characterized by only two parameters, the mean and the standard deviation. So, when the distribution changes, one just has to change the values of these two parameters in the models.

In recent times, in the computer and electronics industries and many other manufacturing industries, the rapid rate of technological change is resulting in new products replacing the old periodically. The result is that product life cycles are shortened. The short life cycle itself is partitioned into three distinct periods. A *growth period* at the beginning of life sees the demand for the item growing due to gradual market penetration, reaching its peak by the end of this period. This is followed by a short *stable period* during which the demand for the item is relatively stable. This is followed by the final *decline period* during which the demand for the item undergoes a steady decline until it is replaced at the end by a technologically superior one. Because of this constant change, classical models based on a single stable demand distribution are not suitable. We need to use models that frequently and periodically update the demand distribution based on recent data.

Approximating demand distributions by something like the normal or gamma distributions, which are characterized by two or fewer parameters, allows us the freedom to change only those few parameters when updating the demand distribution. This appears quite inadequate to capture all the dynamic changes occurring in the shapes of demand distributions. A better strategy is to approximate the demand distribution by its histogram from past data. In this approximation, called the *discretized demand distribution*, the range of variation of the demand is divided into a convenient number (about 10 to 25 in practice) demand intervals, and the probability associated with each interval in the initial distribution is taken to be its relative frequency among historical data.

Let $I_1, \ldots, I_n$ be the demand intervals and $p = (p_1, \ldots, p_n)^{\mathrm{T}}$ the vector of probabilities associated with them in the present distribution. For $i = 1$ to n, let r_i be the relative frequency in I_i over the most recent k periods (if the period is a day for example, k could be about 50) at the time of updating. Let $x = (x_1, \ldots, x_n)^{\mathrm{T}}$ denote the unknown current (i.e., updated) probability vector. $r = (r_1, \ldots, r_n)^{\mathrm{T}}$ is an estimate of x, but it is based on too few (only k) observations. We can take an estimate of x to be the $\bar{y}$ which is the optimum solution of the following quadratic program. This $\bar{y}$ will be used in place of p in the next planning period.

$$\min \quad \beta \sum_{i=1}^{n} (p_i - y_i)^2 + (1 - \beta) \sum_{i=1}^{n} (r_i - y_i)^2$$

$$\text{subject to} \quad \sum_{i=1}^{n} y_i = 1 \quad \text{and all } y_i \geqslant 0,$$

where $0 < \beta < 1$ is a weight. Typically $\beta = 0.9$ works well. The reason for choosing the weight of the second term in the objective function to be small is because the relative frequency vector r is based on a small number of observations. Since the quadratic objective function in this model is the weighted sum of squared forecast errors over all demand intervals, it has the effect of tracking gradual changes in the demand distribution when used at every ordering period. Optimal ordering policies based on discretized demand distributions are discussed in [13].

The optimum solution for the quadratic program above is $\bar{y} = (\bar{z}_1, \ldots, \bar{z}_{n-1}, \bar{w}_n)^{\mathrm{T}}$ where $\bar{w} = (\bar{w}_j)^{\mathrm{T}}$, $\bar{z} = (\bar{z}_j)^{\mathrm{T}}$ is the solution of the LCP with data

$$
q = \begin{pmatrix} -1 - \beta(p_1 - p_n) - (1 - \beta)(r_1 - r_n) \\ -1 - \beta(p_2 - p_n) - (1 - \beta)(r_2 - r_n) \\ \vdots \\ -1 - \beta(p_{n-1} - p_n) - (1 - \beta)(r_{n-1} - r_n) \\ 1 \end{pmatrix}, \quad M = \begin{pmatrix} 2 & 1 & \ldots & 1 & 1 \\ 1 & 2 & \ldots & 1 & 1 \\ \vdots & \vdots & & \vdots & \vdots \\ 1 & 1 & \ldots & 2 & 1 \\ -1 & -1 & \ldots & -1 & 0 \end{pmatrix}.
$$

This LCP is of small order, and it can be solved very conveniently using the complementary pivot algorithm, or any other pivoting algorithm for the LCP mentioned above. Simple codes for the complementary pivot algorithm using the explicit inverse of the basis, available from several sources, work very well on problems of this size. In supply chain management, such a model is solved for each item at every ordering period, providing an excellent source of application for the pivoting algorithms for the LCP.

3.3. Traffic equilibrium

The following traffic equilibrium example from MCPLIB [4] illustrates the connection between equilibrium and complementarity.

The problem involves five cities, numbered 1 through 5, connected by a network of one-way roads called links, (see Fig. 1). Each city i must ship a quantity d_i of a commodity to the third city clockwise from itself. For example, city 1 ships to city 4, city 2 to city 5, and so on. Naturally, the goal is to ship the commodity in the shortest time possible. However, the time to ship along a given path is determined by the total flow of traffic on the links making up that path.

From the figure, it is clear that for each city, there are only two possible shortest paths: shipping counterclockwise along the outside loop or clockwise along the inside loop. Let x_i represent the amount shipped from city i to city $i + 3$ (modulo 5) along the outside path, and let y_i represent the amount shipped along the inside path. We say that a set of flows $x = (x_i)$, $y = (y_i)$ is feasible if it satisfies the demands $x + y \geq d$ and $x, y \geq 0$, where $d = (d_i)$.

From the flow vectors x and y, it is possible to determine the traffic on each link of the network. For example, the outside link between cities 3 and 4 will have flow given by $x_1 + x_2 + x_3$. The delay on a link k is determined by the total traffic on the link and is assumed to be a convex function of the traffic flow. The delay for a given path is then the sum of the delays of all the links making up that path. From this discussion, it follows that for each city i, the delay O_i along the outside path is determined by the flow vector x, and the delay I_i along the inside path is determined by the flow vector y. This can be encapsulated by defining the two functions $O(x) = (O_i(x))$ and $I(y) = (I_i(y))$.

We define the *effective delay* between two cities to be the maximum delay among paths with nonzero flow between the two cities. Each city chooses a shipping strategy in order to minimize its

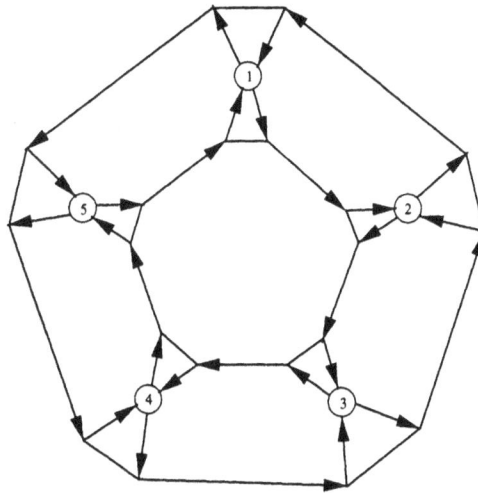

Fig. 1. A traffic network.

effective delay subject to the shipping strategies of the other cities remaining constant. This minimum is achieved either when the city ships everything along the path with shortest delay, or when both the inside and the outside paths have equal delay. To see this, note that if the delay for the inside path is less than the delay for the outside path, the shipper can improve the effective delay by shipping more along the inside path. This reduces the traffic on the outside path, which reduces the delay on the outside path thereby reducing the effective delay.

An equilibrium traffic pattern emerges when all five cities are shipping optimally subject to the shipping strategies of the other cities remaining constant. From the above discussion, this is equivalent to the complementarity conditions

$$0 \leqslant O(x) - u, \quad x \geqslant 0, \quad x^{\mathrm{T}}(O(x) - u) = 0,$$

$$0 \leqslant I(y) - u, \quad y \geqslant 0, \quad y^{\mathrm{T}}(I(y) - u) = 0,$$

where we have introduced the additional variable $u \in \mathbb{R}^5$ to represent the effective delay. Notice that u is complementary to the demand constraint. In particular, there can only be excess supply if the effective delay is zero.

The conditions described above lead to the NCP (5), with $z := (x, y, u)$ and

$$F \begin{pmatrix} x \\ y \\ u \end{pmatrix} := \begin{pmatrix} O(x) - u \\ I(y) - u \\ x + y - d \end{pmatrix}.$$

It should be noted, however, that this problem can be solved more efficiently using a generalization of the NCP. In particular, this problem can be reformulated as a BVIP involving only five variables, instead of the fifteen used above. Details of this are provided in [4].

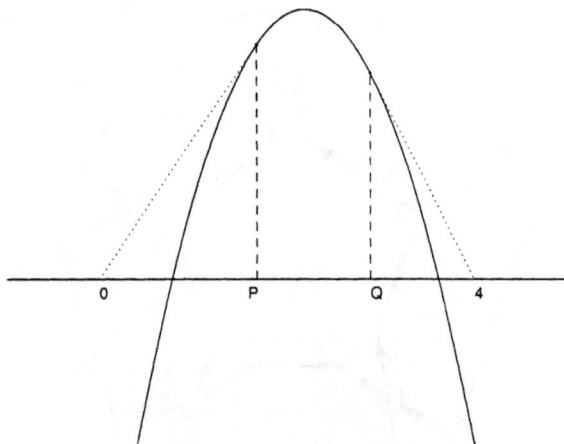

Fig. 2. An elastic string stretched over an obstacle.

3.4. Obstacle and free boundary problems

The obstacle problem consists of finding the equilibrium position of an elastic membrane that is held at a fixed position on its boundary and which lies over an obstacle.

Consider stretching an elastic string fixed at the endpoints $(0,0)$, and $(4,0)$ over an obstacle defined by a function f (in this example, we use $f(x) = 1 - (x - 2.2)^2$ – see Fig. 2). Notice that the position of the string will be defined by $f(x)$ for x between the unknown points P and Q, and that in the intervals $0 \leqslant x \leqslant P$, and $Q \leqslant x \leqslant 4$, the string will lie along straight line segments connecting $(0,0)$ to $(P, f(P))$ and $(Q, f(Q))$ to $(4,0)$, respectively. If we represent the equilibrium position of the string by the function u, then u must satisfy the following conditions:

$$u(0) = 0, \quad u(4) = 0,$$
$$u'(P) = f'(P), \quad u'(Q) = f'(Q),$$
$$u(x) = f(x) \quad \text{for } P \leqslant x \leqslant Q,$$
$$u''(x) = 0 \quad \text{for } 0 < x < P \text{ or } Q < x < 4.$$

This representation of the problem is complicated by the presence of the free boundaries P and Q. The complementarity framework allows a simpler representation, which does not require free boundaries. First, note that since there is no downward force on the string, $u''(x) \leqslant 0$ for all x, except possibly at $x = P$ or $x = Q$ where u'' may be discontinuous. Also, note that $u(x) \geqslant f(x)$ everywhere. Finally, at each point x, either $u''(x) = 0$ or $u(x) = f(x)$. Thus, if we ignore momentarily the discontinuity of u'' at P and Q, we see that u must satisfy the conditions

$$u(0) = u(4) = 0,$$
$$u(x) \geqslant f(x), \qquad 0 \leqslant x \leqslant 4.$$
$$u''(x) \leqslant 0,$$
$$(u(x) - f(x))u''(x) = 0,$$

This system can be solved numerically using a finite difference or finite element scheme. For example, using a central difference scheme on a regular mesh with step size $h = 4/n, u$ is approximated by the vector $u = (u_0, u_1, \ldots, u_n)$, where $u_i := f(x_i)$, $x_i := x_0 + ih$, $i = 0, \ldots, n$, and $x_0 = 0$. The above system is then approximated by

$$u_0 = u_n = 0,$$

$$\frac{u_{i-1} - 2u_i + u_{i+1}}{h^2} \leqslant 0,$$

$$u_i - f(x_i) \geqslant 0, \qquad \qquad i = 1, \ldots, n-1.$$

$$(u_i - f(x_i))\frac{u_{i-1} - 2u_i + u_{i+1}}{h^2} = 0,$$

Using the simple change of variables $z_i := u_i - f(x_i)$, this system is equivalent to the linear complementarity problem (q, M), where M is an $(n-1) \times (n-1)$ matrix and q is an $(n-1)$-vector defined by

$$M = \begin{pmatrix} 2 & -1 & 0 & 0 & \cdots & 0 \\ -1 & 2 & -1 & 0 & \cdots & 0 \\ 0 & -1 & 2 & -1 & \cdots & 0 \\ \vdots & & \ddots & \ddots & \ddots & \vdots \\ 0 & \cdots & 0 & -1 & 2 & -1 \\ 0 & \cdots & & 0 & -1 & 2 \end{pmatrix}, \quad q := \begin{pmatrix} -2f(x_1) + f(x_2) \\ f(x_1) - 2f(x_2) + f(x_3) \\ \vdots \\ f(x_{n-3}) - 2f(x_{n-2}) + f(x_{n-1}) \\ f(x_{n-2}) - 2f(x_{n-1}) \end{pmatrix}.$$

The solution $z = (z_1, \ldots, z_{n-1})$ of this LCP then gives the discrete approximation to u at the interior grid points by the relation $u_i = z_i + f(x_i)$, $i = 1, \ldots, n-1$.

4. Algorithms for the LCP

The fascination of the subject stems from the fact that it exhibits enormous diversity. Depending on the properties of the data matrix M, the LCP can be so nice at one end that it admits an extremely simple greedy type algorithm for its solution, or be an intractable NP-hard problem at the other end.

Chung (published in 1989, but result discussed in [12] also) has shown that the LCP (q, M) with general integer data is NP-hard. The only known algorithms that are guaranteed to process the LCP (q, M) with no restrictions on the data are enumerative algorithms.

A *complementary vector* of variables in (1) is a vector $y = (y_1, \ldots, y_n)^\mathrm{T}$ where $y_j \in \{w_j, z_j\}$ for all $j = 1$ to n. A complementary vector is said to be *basic* if the set of column vectors associated with them in (1) form a nonsingular matrix, and it is said to be *basic feasible* if the basic solution of (1) associated with it is nonnegative. Clearly, the basic solution associated with a basic feasible complementary vector is a solution of the LCP.

4.1. Pivotal methods

The first class of methods to be developed for the LCP are the pivotal methods which try to obtain a basic feasible complementary vector through a series of pivot steps. These methods are variants of the complementary pivot method of Lemke and Howson [10]. The mathematical principle used by Lemke and Howson in the complementary pivot method has been applied by Scarf in 1967 to develop a method for computing fixed points, and for solving systems of nonlinear equations, using partitions of R^n into sets called primitive sets. In the 1970s and early 1980s a lot of researchers have extended this work and developed a variety of methods for computing fixed points and for solving nonlinear equations and complex equilibrium problems using triangulations of R^n. These methods are now called *simplicial methods*, or *piecewise linear methods* because they employ piecewise linear approximations of maps, or also *complementary pivot methods*. Prime candidates among these are the methods of Merrill, and Eaves and Saigal, developed in 1972.

The simplicial methods start at a solution of an artificially set up simple system, and trace a path through the n-dimensional simplices of the triangulation, which, when the method works, terminates with a simplex that contains an easily computed approximate solution of the original system. Using a homotopy interpretation of this path, a variety of other homotopy and path tracing algorithms have been developed for solving systems of nonlinear equations.

Almost all the pivotal methods are guaranteed to process the LCP (q,M) when the matrix M is PSD; this class of LCPs is equivalent to the class of convex quadratic programs. All these algorithms are finite procedures when applied to the classes of problems for which they are guaranteed to work, and in practice these algorithms are quite efficient on problems of reasonable size. However, in 1978, Murty showed that in the worst case, two of the most important among them, Lemke's method and Murty's least-index method, require $O(2^n)$ pivot steps to solve monotone LCPs of order n; this same result has been extended to other pivotal methods for the LCP.

The most famous among the pivotal methods for the LCP is Lemke's method, which we now describe in some detail. Given an LCP (q,M), define the feasible region to be the set

$$\mathscr{F} := \{z \mid w := Mz + q \geqslant 0,\ z \geqslant 0\}.$$

Elements of $\mathscr{F}$ are called *feasible solutions*, and extreme points (or vertices) of $\mathscr{F}$ are called *basic feasible solutions*. A feasible solution z is said to be *complementary* if $z_i w_i = 0$ for all i and *almost complementary* if $z_i w_i = 0$ for all but one i, where $w := Mz + q$. Clearly, a point z is a complementary feasible solution if and only if z solves (q,M).

For simplicity of discussion, we assume that the LCP is *nondegenerate*, which means that for every basic feasible solution z, the vector (z,w) has exactly n nonzero components. Under this nondegeneracy assumption, every complementary feasible solution is an extreme point of $\mathscr{F}$, and every almost complementary solution lies on an edge of $\mathscr{F}$, where we define an edge to be the intersection of $n-1$ linearly independent hyperplanes of the form $z_i = 0$ or $w_j = 0$. If every point on the edge is almost complementary, we call the edge an *almost complementary edge*. If the edge is unbounded, it is called an *almost complementary ray*.

It can be shown that every almost complementary (but not complementary) extreme point (satisfying, for example, $w_j z_j = 0$, for $j = 2,\dots,n$) is incident with exactly two almost complementary edges of $\mathscr{F}$, every point on which also satisfies the same conditions ($w_j z_j = 0$, for $j = 2,\dots,n$). Hence, the set of all such solutions is a collection of paths. Lemke's method traces exactly one such path in this collection, beginning with an almost complementary extreme point incident to an almost complementary ray. Such a path must terminate, either with a complementary basic feasible solution, or with a secondary almost complementary ray. In the first case, a solution to the LCP is found. In the second case, the method has failed to produce a solution to the problem.

To get the algorithm started, it is necessary to construct an almost complementary extreme point that is incident to an almost complementary ray. Since this may be difficult to find for the original LCP (q, M), we instead solve the augmented LCP $(\tilde{q}, \tilde{M})$, where $\tilde{M} \in \mathbb{R}^{(n+1) \times (n+1)}$ and $\tilde{q} \in \mathbb{R}^{n+1}$ are defined by

$$\tilde{M} := \begin{pmatrix} 1 & 0 \\ e & M \end{pmatrix}, \quad \tilde{q} = \begin{pmatrix} 0 \\ q \end{pmatrix},$$

where $e \in \mathbb{R}^n$ is the vector of all ones. Notice that a point (z_0, z) is a solution to $(\tilde{q}, \tilde{M})$ if and only if $z_0 = 0$ and z is a solution to (q, M). Notice further that by choosing $\delta = \max\{|q_i| \,|\, q_i < 0\}$, the point $(z_0, z) = (\delta, 0)$ is an almost complementary extreme point of $\mathcal{F}$ that is incident to the almost complementary ray $\{(z_0, z) = (\delta, 0) + t(1, 0) \,|\, t \geqslant 0\}$. Thus, a starting point for Lemke's method can easily be generated for this augmented LCP.

When the LCP is degenerate, the above method can be modified using a degeneracy resolution technique such as lexicographical ordering. For details of such techniques, we refer the reader to the monographs [12,3].

4.2. Interior point methods

The other important class of methods for the LCP are the interior point methods. Interior point methods originated from an algorithm introduced by Karmarkar in 1984 for solving linear programs. The most successful interior point methods follow a path in $\mathcal{F}^0 = \{(w, z) : w - Mz = q, w > 0, z > 0\}$ (hence the name interior point methods) in an effort to reduce $w^{\mathrm{T}} z$ to 0. One such method defines this path as the set of solutions to the following parameterized system

$$w - Mz = q,$$

$$w_i z_i = \mu, \quad i = 1, \ldots, n, \tag{7}$$

$$w > 0, \quad z > 0,$$

where each choice of the parameter μ yields a different point along the path. This path is followed by generating a sequence of iterates $\{(w^k, z^k)\}$, starting from a feasible point (w^0, z^0). Each step is calculated by solving the system

$$\begin{pmatrix} -I & M \\ Z^k & W^k \end{pmatrix} \begin{pmatrix} \Delta w^k \\ \Delta z^k \end{pmatrix} = \begin{pmatrix} w^k - Mz^k - q \\ -W^k Z^k e + \mu_k e \end{pmatrix},$$

where W^k and Z^k are the diagonal matrices whose diagonal components are defined by $W_{ii}^k = w_i^k$ and $Z_{ii}^k = z_i^k$, and $\mu_k = (z^k)^{\mathrm{T}} w^k / n$. Note that this system is just Newton's method applied to (7).

The next iterate (w^{k+1}, z^{k+1}) is then determined by

$$(w^{k+1}, z^{k+1}) = (w^k, z^k) + \alpha_k (\Delta w^k, \Delta z^k),$$

where the steplength α_k is chosen to ensure that the iterates do not get too close to the boundary of the positive orthant.

The mLCP and hLCP are solved using essentially a similar strategy. In fact, in a 1995 paper, Guler showed that a monotone hLCP can be reduced to a monotone LCP in the same variables. This allows any interior point method for monotone LCPs to be generalized to solve monotone hLCPs. Moreover, the iterates generated by the interior point method in solving the hLCP correspond directly to iterates that would be produced by applying the interior point method to the corresponding LCP. Thus, any convergence results that hold true for the monotone LCP also hold true for the monotone hLCP.

Interior point methods for convex QP essentially use the above methods on the KKT conditions, which form an LCP or mLCP. These methods have polynominal time worst-case complexity for monotone complementarity problems, which correspond to convex quadratic programs (Convex QPs).

4.3. Computational applications and limitations

Convex QP models like the one discussed in Section 3 involving not too large a number of variables and constraints, appear very commonly in many sciences and are a major source of application for the pivoting algorithms for the LCP. These are the algorithms of choice when the number of variables + the number of constraints in a convex QP is of the order of 100 or less. However, when a convex QP is a large-scale problem (i.e., when the number of variables + the number of constraints is $\gg 100$) computational tests indicate that the active set methods of nonlinear programming are much better suited to solve it than the pivot methods applied to the LCP or mLCP formed by its first-order optimality constraints. The newly developed interior point methods for convex QP also compete with active set methods for solving large-scale problems.

Convex QPs appear very prominently in the Sequential (or Recursive) QP Approach (SQP) for solving nonlinear programs. The SQP approach solves nonlinear programs (either convex or nonconvex) using a series of steps; each step consisting of solving a convex QP to find a search direction and then a line search to find an optimum step length in that direction. Most publicly distributed nonlinear programming software packages based on SQP do not use the LCP based algorithms for solving the convex QP in each step because the authors of these codes assume that the users will apply them to solve large scale nonlinear programs. The convex QP solvers in these codes are usually based on some type of active set approach.

Methods based on the LCP or mLCP formed by the first-order necessary optimality conditions (i.e., the KKT conditions) are not suitable for handling nonconvex QPs. This is because all these methods focus only on finding a KKT point for the problem and never even compute the objective value in any step or track how it is changing over the steps. Even if these methods obtain a KKT point, there are no efficient methods known to check whether that KKT point is even a local minimum for the original nonconvex QP as shown by Murty and Kabadi in 1987. For handling nonconvex QP, descent methods, which try to decrease the objective value in each step, are definitely to be preferred in practice.

In the same way, the complementary pivot or simplicial methods for computing fixed points and solving nonlinear equations, using triangulations of R^n, do not have any measure to track the progress of the algorithm from one step to the next. For this reason, these methods are currently not popular.

5. Methods for the NCP

We now turn our focus to methods for solving NCPs. For simplicity, we discuss only the standard NCP defined by (5); although it is straightforward to implement these methods in the context of some more general formulations.

The first methods we consider are sequential LCP methods. These methods generate a sequence of iterates $\{z^k\}$, such that z^{k+1} is a solution to a linear complementarity problem (q^k, M^k), where q^k and M^k are chosen to approximate F near x^k. Depending on the choice of M^k and q^k, various algorithms can be generated, each of which is analogous to a standard iterative method for solving nonlinear systems of equations. Among these are Newton, quasi-Newton, Jacobi, successive overrelaxation, symmetrized Newton, and projection methods. Details of how to choose M^k and q^k for each of these methods can be found in [8]. Here we focus on Newton's method, which corresponds to the choices $M^k := \nabla F(z^k)$ and $q^k = F(z^k) - M^k z^k$. Notice here that $F^k(z) := M^k z + q^k$ is the first order Taylor approximation to F, making this method analogous to Newton's method for nonlinear equations. Josephy showed in 1979 that, in a neighborhood of a solution z^* to the NCP, the iterates produced by this method are well-defined and converge quadratically to z^* provided that ∇F is locally Lipschitzian at z^* and that a certain strong regularity assumption is satisfied.

As with Newton's method for nonlinear equations, it is desirable to employ a globalizing strategy to increase the domain of convergence of the method. One such strategy for nonlinear equations is a backtracking linesearch, in which a step is chosen along the Newton direction $d^k := z^{k+1} - z^k$ so as to ensure sufficient descent of a merit function at every iteration. Unfortunately, attempts to apply this strategy to the NCP have been largely unsuccessful and have generally required very strong assumptions. The difficulty lies in the fact that for reasonable choices of merit functions, the Newton direction is not guaranteed to be a descent direction, even when z^k is not a stationary point. Instead, a backtracking strategy that is not restricted to the Newton direction is needed.

Such a strategy was proposed by Ralph in 1994. He devised a path search algorithm in which global convergence is achieved by searching along a piecewise linear path connecting z^k to the solution of the LCP (q^k, M^k). This path is generated by a complementary pivot algorithm, similar to Lemke's method, which is used to solve the LCP subproblem. This path search strategy is the basis for the highly successful PATH algorithm developed by Dirkse and Ferris in 1995.

Another class of methods for NCP involves reformulating the problem as a system of nonlinear equations. This involves constructing a function $H : \mathbb{R}^n \to \mathbb{R}^n$ with the property that zeros of H correspond to solutions of the NCP. Such a function H is called an NCP-function. Perhaps the simplest example is to define H by

$$H_i(z) := \min(F_i(z), z_i)$$

Many other NCP functions have been studied in the literature [7]. Interestingly, while smooth NCP functions exist, they are generally not in favor computationally since they have singular Jacobian matrices at degenerate solutions (a degenerate solution is a point z^* such that $F_i(z^*)$ and z_i are both zero for some index i). Thus, the NCP functions of interest are usually only piecewise differentiable.

Once the NCP function H has been constructed, a generalized Newton method can be used to find a solution to $H(z) = 0$. Generalized Newton methods are similar to Newton's method except that the Newton equation

$$\nabla F(z^k) d^k = -F(z^k),$$

which is used to calculate the search direction d^k, is replaced by the equation

$$V^k d^k = -F(z^k),$$

where V^k is an element of the Clarke subdifferential or the B-subdifferential of $F(z^k)$.

An alternative approach to applying a generalized Newton method is to approximate the nonsmooth system by a family of smooth functions. This is the fundamental idea behind the so-called *smoothing methods*. An excellent review of these techniques can be found in [7]. The basic idea of these techniques is to approximate the function H by a family of smooth approximations H_μ parameterized by the scalar μ. Under suitable assumptions, the solutions $z(\mu)$ to the perturbed systems $H_\mu(z) = 0$ form a smooth trajectory parameterized by μ, leading to a zero of H as $\mu \downarrow 0$. The smoothing methods generate a sequence of iterates that follow this trajectory, and in that sense are similar to interior point methods. Numerous smoothers have been proposed in the literature for complementarity problems.

Another class of algorithms for monotone NCP are interior point algorithms. In similar fashion to the interior point techniques for LCP, these methods follow the central path defined by

$$w = F(z), \quad (w, z) > 0, \quad w_i z_i = \mu, \tag{8}$$

which leads to a solution as $\mu \downarrow 0$.

The final class of methods we discuss are continuation or homotopy methods. Like the smoothing and interior point algorithms, these methods work by introducing an additional variable μ and then following a path which leads to a solution. However, unlike the smoothing and interior point methods, the continuation methods do not assume that μ decreases monotonically along this path.

6. The geometry of LCP, matrix classes

For any matrix $D = (d_{ij})$ of order $m \times n$ we let $D_{.j}$ denote its jth column; for any $J \subset \{1, \ldots, n\}$ we let $D_{.J}$ denote the $m \times |J|$ matrix consisting of columns $D_{.j}$ for $j \in J$; and for any $P \subset \{1, \ldots, m\}$, $J \subset \{1, \ldots, n\}$ we let D_{PJ} denote the $|P| \times |J|$ matrix $(d_{ij}: i \in P, j \in J)$.

Consider the LCP (1). Let $y = (y_1, \ldots, y_n)^{\mathrm{T}}$ be a complementary vector for it. Let $A_{.j}$ be the column associated with y_j in (1). Hence $A_{.j} \in \{I_{.j}, -M_{.j}\}$ for each $j = 1$ to n. The cone $\mathrm{Pos}\{A_{.1}, \ldots, A_{.n}\} = \{x \in R^n : x = \sum_{j=1}^n \alpha_j A_{.j}, \ \alpha_j \geqslant 0 \ \forall j\}$ is known as the *complementary cone* associated with y for (1). Clearly, there are 2^n complementary cones, and the LCP (1) has a solution iff q belongs to some complementary cone. Hence the LCP (1) is equivalent to finding a complementary cone containing q. The geometric study of the LCP using complementary cones has been initiated by Murty in 1968 in his Ph.D. thesis.

It can be verified that if the matrix $M = I$, the unit matrix of order n, the complementary cones become the orthants of R^n. In an earlier paper not focused on the LCP, Samelson, Thrall, and Wesler defined in 1958 the complementary cones with respect to a square matrix M as a generalization of orthants, and investigated the question of what conditions on the matrix M would guarantee that these complementary cones form a partition of R^n. They established that the required condition is that M must be a *P-matrix*, i.e., a square matrix all of whose principal subdeterminants are > 0. For the LCP (1), this result leads to the theorem that (1) has a unique solution for all $q \in R^n$ iff M is a P-matrix.

The geometric study of the LCP has been the object of enduring study in the literature ever since. This study is purely mathematical in nature, and not motivated by immediate practical application.

We briefly summarize some of the other major results in this geometric investigation. This research has identified a wide variety of classes of square matrices that correspond to certain properties related to the LCP. Let $S(q,M) = \{z : (w := Mz + q, z)$ is a solution of the LCP $(q,M)\}$.

The w-part of the solution of the LCP (q,M) is unique $\forall q \in R^n$ iff M is a *column adequate matrix* (this requires that M must be a P_0-*matrix* (i.e., all its principal subdeterminants are $\geqslant 0$), and for any $J \subset \{1,\ldots,n\}$ M_{JJ} is singular iff the columns of $M_{\cdot j}$ are themselves linearly dependent).

For all $q \in R^n$, every solution of the LCP (q,M) is locally unique iff M is a *nondegenerate matrix* (M belongs to this class iff all its principal subdeterminants are nonzero).

The LCP (q,M) has a unique solution $\forall q > 0$ iff M is a *semimonotone matrix* (class denoted by E_0) (M belongs to this class if $\forall J \subset \{1,\ldots,n\}$, the system $M_{JJ}x_J < 0$, $x_J \geqslant 0$ has no solution).

The LCP (q,M) has a unique solution $\forall q \geqslant 0$ iff M is a *strictly semimonotone matrix* (class denoted by E) (M belongs to this class if $\forall J \subset \{1,\ldots,n\}$, the system $M_{JJ}x_J \leqslant 0$, $x_J \geqslant 0$, $x_J \neq 0$ has no solution).

These are some of the main ones, but there are so many other classes of matrices identified, we refer the reader to [3,12,1] for a summary and some recent work in the area. Theoretical studies on the geometry of complementary cones continues to be very actively pursued.

Given a square matrix M of order n, to check whether M is PSD or positive definite (PD) requires at most n Gaussian pivot steps, and hence can be carried out very efficiently. However, for many of the matrix classes defined above, checking membership is intractable, as shown in [12].

The study of the mathematical aspects of the geometry of LCP, matrix classes, and establishing connections between LCP and other branches of mathematics such as degree theory through the study of the degree of piecewise linear equation formulations of the LCP continue to be pursued very actively.

7. Generalizations and current trends

Complementarity problems and variational inequalities remain a vigorous area of research. While excellent algorithms have been developed, there is still much attention being devoted to developing new algorithms. Much of this interest lies in expanding the classes of functions for which algorithms can be proven effective. In the realm of LCPs, this, in part, has motivated the study of matrix classes. In the NCP and VI arenas, new algorithms are constantly being introduced. Some of these are based on new merit or NCP functions [5] leading to variants of damped Newton-type methods. Others are variants of path-following algorithms, including homotopy, smoothing [5], interior point [9], and, recently, noninterior point methods. Also various globalizing strategies, such as regularizing methods [7], tunneling and filled functions, trust region methods and proximal perturbation strategies have recently been explored.

Another trend is focused on developing algorithms that do not require Jacobian evaluations. These include projection methods, quasi-Newton methods, and derivative-free methods [7].

Finally, we mention mathematical programs with equilibrium constraints (MPECs). These can be defined in the form

$$\min \quad f(x, y)$$

$$\text{subject to} \quad (x, y) \in Z \subset \mathbb{R}^{n+m},$$
$$y \text{ solves } VI(C(x), F(x, \cdot))$$

where $x \in \mathbb{R}^n$, $y \in \mathbb{R}^m$ are the decision variables, Z is a closed set, C is a set-valued mapping, and $f : \mathbb{R}^{m+n} \to \mathbb{R}$ and $F : \mathbb{R}^{m+n} \to \mathbb{R}^{m+n}$ are given functions. Here the constraints state that the variable y must be a solution to a variational inequality that is parameterized by x. Numerous applications of MPECs have been identified. These include misclassification minimization in machine learning; robotics; continuous network design; discrete transit planning; optimal design of mechanical structures; and Stackelberg leader-follower games, which have numerous applications in economics, such as oligopolistic market analysis.

The MPEC is an extremely difficult problem to solve. This is due largely to the fact that the feasible region $\{(x, y) \subset Z \,|\, y \text{ solves } VI(C(x), F(x, \cdot))\}$ is not convex, and in some cases, is not even closed. Nevertheless, a number of reasonable algorithms exist for solving MPECs, and research in this area remains vigorous. The reader is referred to [11] for a detailed treatment on MPECs.

References

[1] S. Bandyopadhyay, On Lipschitzian, INS and connected matrices in LCP, Ph.D. Thesis, Indian Statistical Institute, Delhi, 1998.

[2] S.C. Billups, K.G. Murty, Complementarity problems, UCD/CCM Report No. 147, Department of Mathematics, University of Colorado at Denver, Denver, CO, 1999.

[3] R.W. Cottle, J.-S. Pang, R.E. Stone, The Linear Complementarity Problem, Academic Press, Boston, 1992.

[4] S.P. Dirkse, M.C. Ferris, MCPLIB: a collection of nonlinear mixed complementarity problems, Optim. Methods Software 5 (1995) 319–345.

[5] M.C. Ferris, J.-S. Pang (Eds.), Complementarity and Variational Problems: State of the Art, SIAM, Philadelphia, PA, 1997.

[6] M.C. Ferris, J.-S. Pang, Engineering and economic applications of complementarity problems, SIAM Rev. 39 (1997) 669–713.

[7] M. Fukushima, L. Qi (Eds.), Reformulation – Nonsmooth, Piecewise Smooth, Semismooth and Smoothing Methods, Kluwer, Dordrecht, 1998.

[8] P.T. Harker, J.-S. Pang, Finite-dimensional variational inequality and nonlinear complementarity problems: a survey of theory, algorithms and applications, Math. Program. 48 (1990) 161–220.

[9] M. Kojima, N. Megiddo, T. Noma, A. Yoshise, A unified approach to interior point algorithms for linear complementarity problems, Lecture Notes in Computer Science, Vol. 538, Springer, Berlin, 1991.

[10] C.E. Lemke, J.T. Howson, Equilibrium points of bimatrix games, SIAM J. Appl. Math. 12 (1964) 413–423.

[11] Z.-Q. Luo, J.-S. Pang, D. Ralph, Mathematical Programs with Equilibrium Constraints, Cambridge University Press, Cambridge, 1996.

[12] K.G. Murty, Linear Complementarity, Linear and Nonlinear Programming, Heldermann Verlag, Berlin, 1988, Free public access at http://www-personal.engin.umich.edu/~murty/book/LCPbook/index.html.

[13] K.G. Murty, Supply chain management in the computer industry, Manuscript, Department of IOE, University of Michigan, Ann Arbor, 1998.

N·H

ELSEVIER

Journal of Computational and Applied Mathematics 124 (2000) 319–340

JOURNAL OF
COMPUTATIONAL AND
APPLIED MATHEMATICS

www.elsevier.nl/locate/cam

Evolution and state-of-the-art in integer programming

Hanif D. Sherali[a,*], Patrick J. Driscoll[b]

[a] *Department of Industrial and Systems Engineering, Virginia Polytechnic Institute and State University, Blacksburg, VA 24061, USA*
[b] *Department of Mathematical Sciences, U.S. Military Academy, West Point, NY 10996, USA*

Received 2 July 1999; received in revised form 9 September 1999

Abstract

Under a unifying theme of exploiting both algebraic and polyhedral special structures present in integer linear programming problems, we discuss the evolution of both technique and philosophy leading to the current state-of-the-art for modeling and solving this challenging class of problems. Integrated throughout the discussion are insights into the rationale and motivation that have contributed in a large part to the past and present direction of research in this fascinating field. © 2000 Elsevier Science B.V. All rights reserved.

1. Modeling concepts

Integer programming (IP) problems involve the optimization of a linear objective function subject to a set of linear constraints, as in linear programming, but with the added complexity that a subset of variables are required to take on certain discrete values. In most interesting applications, these discrete values are 0 or 1, leading to the class of 0–1 *mixed integer programming problems* (MIP). Although limited progress has been made to-date in nonlinear integer programming, we focus here exclusively on the linear case and consequently use IP and MIP as synonymous with this class of problems.

It is well known in discrete optimization literature that in order to be able to solve reasonably sized instances of challenging classes of problems, two particular features must come into play. First, a good model of the problem must be constructed in the sense that it affords a tight underlying linear programming representation, and second, any inherent special structures must be exploited, both in the process of model formulation and in algorithmic developments. The intent is to try to construct *good* models, rather than simply *mathematically correct* models (see Jeroslow and Lowe, 1984, 1985; [14]). The much sought after characteristic of good, or *tight* models is that

---

* Corresponding author. Tel.: +1-540-231-6656; fax: +1-540-231-3322.
E-mail address: hanifs@vt.edu (H.D. Sherali).

0377-0427/00/$ - see front matter © 2000 Elsevier Science B.V. All rights reserved.
PII: S 0377-0427(00)00431-3

they more closely approximate the convex hull of integer feasible solutions, at least in the vicinity of optimal solutions, because all commercially available software use the LP relaxation within their solution methods as a primary means of avoiding direct confrontation with the inherent combinatorial challenge these problems introduce. Often adding the "right" type of constraints to integer programs, even at the expense of increasing problem size, enhances rather than inhibits solvability. Nemhauser and Wolsey (1988; [11], and Wolsey (1998; [16]) provide detailed expositions on various modeling and algorithmic issues related to this class of problems.

To illustrate the concept of tight model representations, consider the class of fixed-charge location problems, in which up to m potential supply facilities having capacities $s_1, s_2, \ldots, s_m$ can be constructed to serve n customers having corresponding demands $d_1, d_2, \ldots, d_n$. The objective function seeks to minimize the annualized sum of the fixed costs of construction and the variable costs of distributing supply to meet customer demands. Let x_{ij} denote the shipping quantity from facility i to customer j, for $i = 1, \ldots, m$ and $j = 1, \ldots, n$. Let y_i be a binary variable that equals 1 if facility i is constructed and 0 otherwise. Then the following constraint is sufficient to insure that whenever $y_i = 0$, we must have $x_{ij} = 0$ for all $j = 1, \ldots, n$ and that when $y_i = 1$, we have the full supply s_i available for facility i, $i = 1, \ldots, m$:

$$\sum_{j=1}^{n} x_{ij} \leqslant s_i y_i \quad \text{for } i = 1, \ldots, m. \tag{1}$$

Although mathematically correct, the model's solvability can be substantially improved by incorporating the constraints

$$0 \leqslant x_{ij} \leqslant y_i \min\{s_i, d_j\} \quad \text{for } i = 1, \ldots, m, \ j = 1, \ldots, n, \tag{2}$$

which are redundant in the discrete sense but strengthen the continuous relaxation by enforcing variable upper bounds on each individual x_{ij}, as opposed to the aggregated bound embodied in (1). As a consequence, restrictions (2) are referred to as *disaggregated* constraints. For sparse distribution solutions in which $0 \leqslant x_{ij} = d_j \leqslant s_i$ for some j, (2) implies that $d_j \leqslant d_j y_i$, or that $y_i = 1$ (in concert with $y_i \leqslant 1$).

Likewise, if the coefficient of y_i in (2) can be reduced via some logical deductions based on the structure and data of the problem without eliminating possible integer solutions, the LP relaxation would be further tightened, thereby reducing the prospect of fractionating y_i in any solution to the LP relation. Such a strategy of altering constraint coefficients in order to tighten the LP relaxation's feasible region to more closely represent the convex hull of integer feasible solutions is known as *coefficient reduction*. Barnhart et al. (1993; [3]) present a wide variety of methods for improving the solvability of mixed-integer programs through a careful formulation of the initial problem.

Often, there exists a natural symmetry inherent in the problem itself that, if propagated to the model, can force a solution process to explore and eliminate alternative symmetric solutions, thus needlessly miring the process (see Sherali and Smith, 1999; [14]). Consider a machine procurement and scheduling problem in which the binary variable $y_{ij} = 1$ if machine i of type j is purchased, and 0 otherwise. Suppose that there are $n_j = 10$ machines of some type j available. If the desired solution requires three of these machines to be purchased, a traditional formulation might admit $\binom{10}{3}$ possible symmetric combinations of solutions just for the procurement portion of the problem dealing with this machine type. Compounded with alternative dispatch strategies, it is easy to imagine the combinatorial explosion that will result. However, providing a decision hierarchy that establishes an identity to the procurement of these machines (and similarly to their dispatchment), defeats such symmetric reflections in the model. One possible technique imposes an additional set of constraints $y_{ij} \geqslant y_{i+1,j}$ for $i = 1, \ldots, n_j - 1$, for each machine type j, which places a priority on selecting lower index machines first. In the foregoing example, this would require the selection of the first three machines to be the only admissible combination. Similar constraints can be introduced to impose dispatchment hierarchies in the model.

This paper traces the evolution of several fundamental advances in linear integer programming with specific emphasis on the insights provided by such developments. In addition, we present an overview of several state-of-the-art developments that define current research tracks in integer programming that hold promise for efficient problem solution in the future.

2. Preliminary developments

2.1. Branch-and-bound

Prior to the 1970s when researchers began to employ significant elements of polyhedral study, two primary approaches for solving combinatorial optimization problems dominated the literature. The first was a *branch-and-bound* approach pioneered by Land and Doig (1960; [14]) and later extended by Dakin (1966; [14]), an enumerative divide-and-conquer technique in which the feasible region is successively partitioned based on enforcing a finite set of alternative logical restrictions that must be satisfied at any stage. The second was the *cutting plane* approach, a procedure that iteratively generates valid inequalities to delete fractional vertices of the region of feasible solutions to the linear programming relaxation while preserving integer feasible solutions and inducing convergence toward a discrete optimum. Although both techniques held promise for the eventual solution of successively larger and larger problem instances, each possessed inherent shortcomings that would not be overcome until several years later.

Although branch-and-bound performed reasonably well on limited-sized problem formulations (see Lawler and Wood, 1966; Mitra, 1973; [14]), the number of subproblems that were required to be solved during the execution of the algorithm increased exponentially with the size of the problem, thus inhibiting the solution of even moderate-sized practical integer problems.

Strategies were consequently developed to limit the growth of the search tree created by the partitioning mechanism of the general branch-and-bound algorithm, and hence, reduce the computational burden imposed by any sizable problem instance. Early on, these strategies focused exclusively on algorithmic aspects of branch-and-bound, such as composing the partitioning strategy, and choosing which subproblem to solve, among others (Beale, 1979; [14]). However, none of the commercial and academic codes developed during this period attempted to resolve the one inherent weakness indigenous to branch-and-bound: to compute the lower (or upper) bound for each subproblem, the procedure had to solve a naive linear programming (LP) relaxation in which the integer requirements on the variables are simply relaxed. The *integrality gap* that existed between the value of this naive relaxation, $v(\text{LP})$, and that for the integer problem, $v(\text{IP})$, defined by $|v(\text{LP}) - v(\text{IP})|$, was frequently far too large relative to $v(\text{IP})$, requiring extensive branchings to resolve the values of the integer variables. Two possible options were evident to overcome this handicap: reduce the size of the gap prior to handling the problem over to a branch-and-bound algorithm (reformulation), or reduce the size of the gap during the run-time to branch-and-bound (cut generation).

2.2. Cutting planes

Seeking to close the integrality gap during the branch-and-bound process, researchers looked to a strategy for generating cutting plane constraints as needed at particular nodes of the branching tree. An early class of cutting planes, introduced by Gomory (1960; [14]), served to validly separate the current subproblem LP relaxation solution $\bar{x}$ from the convex hull of feasible integer solutions for all subsequent subproblems, thus achieving a certain tightening of the problem's feasible domain. Although certain problems, such as set covering problems, were solved more effectively using cuts like Gomory's, cutting planes such as these possessed several shortcomings that detracted from their appeal. They behaved very differently on quite similar problems, tended to cause numerical ill-conditioning, and proved to be weak in the sense that they did little to reduce the size of the integrality gap for most problems (Gondran, 1979; [14]).

An additional source of difficulty arose when cutting planes introduced for a particular subproblem during the execution of branch-and-bound maintained validity only for subsequent domain divisions on the same branch. Localizing constraint validity in this manner increases the amount of temporary storage required by a branch-and-bound algorithm that incorporates such cuts, and increases the overall computational burden by having to generate node-customized cuts. This is more of a burden on algorithms that use a breadth-first search scheme than on those based principally on a depth-first search from the viewpoint of data storage.

As it turns out, a part of the difficulty associated with Gomory's cuts is the numerical ill-conditioning caused by generating several constraints nearly parallel to the objective function contours. Furthermore, the true merit of this technique is realized best when used in concert with branch-and-bound, rather than as a stand-alone procedure. With the advent of increased computational power, Balas et al. (1996; [14]) and Ceria et al. (1998; [14]) have recently resurrected Gomory cuts to reexamine their use in conjunction with branch-and-bound, demonstrating that cuts of this type still hold promise for solving problems with general integer variables.

It became apparent early on that any cuts generated either prior to handling the problem to a branch-and-bound procedure or during its execution had to be deeper, imposing a more significant structural tightening of the LP feasible region. The integrality gap should be significantly reduced by any constraints that are added to the problem. The most ideal situation, of course, would occur when each additional constraint represents a facet of the convex hull of feasible solutions.

An interesting marriage of propositional logic and integer programming cutting planes evolved from the work of Granot and Hammer (1972; [14]) into what is now called *constraint logic programming* (CLP). Though initially championed through the efforts of Williams (1977; [14]) and Wilson (1977; [14]), various researchers have continued to explore the boundary between logical constraint satisfiability and integer programming, particularly 0–1 IP's, afforded by CLP (see Hooker, 1998; [14]). Whereas an IP formulated problem within a branch-and-bound framework experiences progressive partitioning of its initial variable domain through altered variable bounds, CLP seeks to alter the domains of variables directly by exploiting logical inferences induced by the interaction of constraints. For example, suppose that x, y, and z are integer variables defined on $[1,10]$ with constraints $y < z$ and $x = y + z$. The restriction $y < z$ thus implies $y < 10$ and $z > 1$ directly, and, together with $x = y + z$ imply $x > 2$, $y < 9$, and $z < 10$. The logically altered domains are thus $x \in [3, 10]$, $y \in [1, 8]$, and $z \in [2, 9]$, thus affording a tightening of the original domain of definition. In other applications, particularly in the context of scheduling, solution methods based on deriving logical inferences implied by the operational constraints of the problem can be more effective in complex situations than examining mathematical formulations of such constraints. Hooker (1988, 1998; [14]) and Williams (1995; [14]) further linked the resolution afforded by CLP with cutting plane approaches, demonstrating that this approach might offer a better chance of success for certain limited problem sizes than traditional IP solution methods mentioned previously (see Brailsford et al., 1995; [14]).

3. Polyhedral developments

In the 1970s, the focus of interest for combinatorial optimization – originally tending exclusively to algorithmic aspects of these problems – shifted to an investigation of the facial structure of *polyhedra* associated with these problems. Research into *polyhedral theory* was motivated by the desire to obtain tighter integer programming formulations. New theorems and algorithms emerged that identified important classes of constraints that could be generated as needed within a branch-and-bound environment. These cuts possessed an important property distinguishing them from earlier cutting planes such as Gomory's: they frequently defined *facets* of the problem's underlying polyhedral structure or substructure. *Facets*, or *facet-defining* inequalities, are valid inequalities that intersect the integer polyhedron P_I, defined by the constraints of the problem, in a face of dimension one less than the dimension of P_I. Facets belong to a system of linear inequalities that minimally define the convex hull of feasible solutions to an integer programming problem.

This new shift in approach pioneered by Padberg (1973; [14]) recognized the value of exploiting special structural properties present in combinatorial optimization problems in order to derive classes of strong valid inequalities that are tailored to the particular type of problem. Although a complete polyhedral description for most integer programming problems proved elusive, branch-and-bound procedures incorporating even *partial* convex hull descriptions provided by facetial constraints tended to perform better, generating fewer active nodes in the enumeration process and terminating faster because of improved linear programming relaxation bounds.

The success of this polyhedral approach created a spark of excitement as researchers recognized the value of examining a given problem via various alternative viewpoints, such as graph-theoretic representations, in order to derive strong valid inequalities. The ensuing research extended both an understanding of polyhedra for specially structured problems, and of classes of problems for which particular types of facetial cuts could be readily identified and generated (Grötchel and Padberg, 1975; [14]).

Apart from identifying classes of facets, another algebraic approach for seeking tighter formulations arose in which "equivalent" inequalities could be generated that have exactly the same 0–1 feasible solutions as the original one, but are "stronger" than the original inequality in that they delete a portion of the fractional continuous region (Bradley, Hammer, Wolsey, 1974; [14]). These "minimum equivalent inequalities" possessed integer coefficients, and separated feasible and infeasible points in a defined "strongest" possible way. Sherali and Shetty (1980; [14]) introduced several criteria for generating deep disjunctive cuts, and Ceria et al. (1998; [14]) used similar constructs to derive strong valid inequalities. Boyd (1994; [14]) generated a class of deep cutting planes known as *Fenchel cuts* to enhance the *separation* problem associated with using Lagrangian relaxations to guide integer programming problems. Computational results indicated that with respect to the number of cuts generated and the extent to which the linear programming–integer programming gap is closed, Fenchel cuts appear to be as effective as facets of the individual knapsack polyhedra.

3.1. Lifting lower-dimensional facets

Paralleling efforts to identify polyhedral characteristics of integer programming problems during the 1970s were attempts to extract and exploit *implicit* structures inherent in these problems. These attempts were extensions of an observation first noted by Pollatschek (1970; [14]) with regard to the structure afforded by independence systems: it is possible to obtain facets (valid inequalities) of the convex hull of binary integer feasible solutions to a general 0–1 program by *lifting* facets and valid inequalities of lower dimensional polytopes. Specifically, this process commences with a restricted or projected feasible region in which several of the binary variables are fixed at zero or one values, and for which a facet defining inequality is available that is also valid for the original integer polytope. In a procedure known as *sequential lifting*, the restricted variables are released, or "lifted", from their fixed binary values one at a time in some predetermined order, while modifying the current inequality so that it remains valid and continues to be a facet of the one higher dimensional polytope. If this lifting is done in a manner such that the coefficients of the released binary variables are determined simultaneously, then the process is referred to as one of *simultaneous lifting*.

Padberg (1973; [14]) introduced a procedure for calculating the sequentially lifted coefficients in a manner that proved easy to implement within branch-and-bound methods. His procedure was first applied to the set packing polytope, and then extended to any 0–1 linear program having a nonnegative coefficient matrix. The same procedure was used for other 0–1 problems containing sufficient explicit structure to facilitate this approach. It is worth noting that the lifted variable coefficients obtained in this sequential lifting were integer valued, and the resulting facets depended upon the sequence in which the variables were introduced. Also, as pointed out by Padberg (1979; [14]), packing, covering, and knapsack problems can ultimately be viewed as optimization problems over undirected graphs. This observation facilitated insights into the underlying structure of these problems that might have gone unnoticed. Gu et al. (1995; [14]) discuss various aspects related to the derivation, complexity, and implementation of lifting procedures.

A generalized procedure in which sets of variables from the subset $N \setminus N'$ were simultaneously introduced was proposed by Zemel (1978; [14]). This *simultaneous lifting* was found to generate additional facets unobtainable by sequential lifting, i.e., facets having *fractional* coefficients. However, because the number of integer programs one has to solve to generate all the facets of a given 0–1 polytope was prohibitively large, it was questionable whether this generalized approach would prove computationally advantageous. Zemel (1978; [14]) observed that the procedure becomes more effective when it is applied to problems whose special structure helps to cut down the number of integer programs required to be solved, or simplifies the solution. The necessity to exploit special structures, when present, again became apparent. Indeed, even the sequential lifting process for GUB-constrained 0–1 problems as studied by Sherali and Lee (1992; [14]) involves lifting groups of variables from individual GUB sets in a simultaneous fashion.

4. Algorithmic advances

The scope of integer programming applications increased steadily since the 1970s, but not dramatically so because the size of an integer programming problem measured in numbers of constraints, variables, and nonzero coefficients remained a poor guide as to its difficulty. Unless the number of integer variables was very small ($n \leqslant 20$), it was important that $v(\text{LP})$ should provide a fair guide to $v(\text{IP})$, or at least, the LP solution should approximate the best integer solution (Beale, 1979; [14]). However, even the incorporation of deep cuts within a branch-and-bound methodology failed to remove some impediments. Sometimes, cuts would destroy the matrix structure present in the problem prior to their addition, hence inducing difficulties. Miliotis (1976; [14]) showed that for the symmetric traveling salesman problem, cuts such as Gomory's tend to be dense, having very few zero coefficients. Efforts to preserve sparsity while incorporating these cuts would therefore be self-defeating. Additionally, whereas weaker cuts like Gomory's were discarded when they exhibited slack, facetial cut constraints tended to be retained in storage throughout calculations. Depending upon the number of cuts generated for specific subproblems of a branching tree, this additional storage requirement presented a significant obstacle in tackling large-scale problems on machines employing either a single memory system, or a parallel environment possessing global shared memory architecture. This point underscores the importance of finding an approach that generates facetial cuts valid for the *complete* polytope and not simply a subset of the original feasible region, both from a standpoint of minimizing storage requirements during problem execution, and providing an initial problem formulation that is as tight as possible prior to execution.

Miliotis (1978; [14]) proposed an algorithmic approach for the traveling salesman problem that is typical of the class of algorithms that incorporates cutting planes. Starting with the linear assignment constraints, the LP relaxation is solved repeatedly, but instead of branching upon some fractionating variable, cutting planes are used to drive a fractional solution to an integer one. When an integer solution is found, it is tested for feasibility by checking if it yields a feasible Hamiltonian circuit. If so, then this solution is optimal and the algorithm terminates. Otherwise, cut constraints are generated, added to the current set of constraints, and the process is reiterated. Interestingly, the computations for this particular algorithm were performed in integer arithmetic. This was achieved by storing the determinant and the cofactor matrix of each linear programming basis separately. Martin (1966; [14]) had previously used a slightly different starting formulation with upper bounds of 1 on the relaxed binary variables, permitting two city subtours to also occur. Apart from this approach was the type of problem preprocessing advocated by Brearley et al. (1975; [14]) that would later prove to be of fundamental importance.

5. Hybrid algorithms

Enthusiasm over identifying and exploiting underlying polyhedral structures of combinatorial optimization problems showed no signs of waning in the 1980s. This period ushered in several seminal papers that presented comprehensive computational strategies incorporating ideas of exploiting special structures present in problem formulations. Most, if not all, of the large-scale mathematical programming systems designed to solve 0–1 mixed-integer programs, such as MPSX-MIP/370, APEX-III, UMPIRE, FMPS, LP 6000, etc., used branch-and-bound algorithms based on exploring a sequence of

LP relaxations in which only a subset of the 0–1 variables were fixed, with all other variables being treated as continuous. However, for large-scale problems having many 0–1 variables and a highly structured coefficient matrix, run-times in finding and proving optimality were often excessive.

Hybrid algorithms that combined branch-and-bound and cutting plane generation schemes were introduced to solve large-scale 0–1 integer programming problems. Injected within this hybrid approach was an *implicit enumeration* concept in which various 0–1 completions of partial solutions and logical tests were used to prove that certain binary variables must take on specific 0 or 1 values to yield improving, feasible solutions. This type of approach is particularly attractive for problem formulations in which the number of 0–1 variables greatly exceeds the number of continuous variables.

For general 0–1 mixed-integer programs, Johnson and Suhl (1980; [14]) applied the principles of implicit enumeration using Benders' (1962; [14]) cuts that essentially provide reduced-cost-based objective representations in the projected space of simply the binary variables. Lougee–Heimer (1998; [14]) proposed a clarification of, and an extension to, the SOS-based methods of Johnson et al. (1985; [14]), showing that their results can be viewed as adjusting the coefficient on a single binary variable, while exploiting the SOS restrictions.

The disadvantages of using explicit Benders' cuts mirror those noted earlier for Gomory's cuts: (a) the cuts are usually dense, and adding them alters any special structure previously present in the formulation; and, (b) the number of cuts generated for a general MIP is large, imposing a memory storage burden on the algorithm. Primarily for these reasons, Johnson and Suhl (1980; [14]) used implicit Benders' cuts via reduced costs of LP relaxations in which several binary variables were also treated as continuous and were projected out of the derived cut. This produced valid inequalities in terms of certain key identified binary variables at specific nodes of the enumerative tree. Their numerical results indicated that the more frequently the LP relaxation is solved, the fewer the number of nodes that need to be generated to prove optimality. This further reinforces the importance of having a tight LP relaxation in-hand for use in estimating the appropriate lower/upper bounds.

Crowder and Padberg (1980; [14]) utilized another hybrid algorithm that combined LP-based branch-and-bound and specialized facetial cutting planes derived from the special structure provided by the zero-one integer programming formulation of the symmetric traveling salesman problem (TSP). Using this approach, they solved to optimality 10 problems ranging from 48 to 318 cities. Although several formulations of the symmetric traveling salesman program exist, the specific 0–1 programming formulation used by Crowder and Padberg (1980; [14]) incorporated constraints requiring the degree of every node in the solution to be two, along with generalized subtour elimination constraints known as *comb* inequalities. These latter constraints were initially relaxed, and specific members of this class were iteratively generated as needed to delete the current fractional linear programming solution. In addition, reduced-cost-based objective expressions were used to deduce fixed values for the binary variables (i.e., to select an edge or not). To their surprise, the entire procedure never iterated more than three times in order to find an optimal tour. The largest problem tested, involving 318 cities, only 216 nodes within the two executed branch-and-bound procedures. Their results emphasized the suitability of facet-defining cutting planes for the purpose of proving optimality in difficult combinatorial optimization problems. Moreover, it underscored the importance of exploiting special structures within a polyhedral environment in a manner that either directly provides facetial constraints, or, as later will prove equally as important, affords access to such facetial constraints. It is worthwhile noting that the cutting planes generated were appended to the *main* LP relaxation problem, and not simply a branch-specific subproblem, thereby insuring that they were globally valid for the entire branch-and-bound decision tree. This provides a computational advantage for a serial computing machine, but not necessarily for a parallel architecture. The adaption of a scheme to generate only globally valid constraints could introduce additional complexity within a parallel implementation using distributed memory, since the amount of necessary communication between processors to pass cutting plane information might increase substantially. Thus, although pursuing globally valid cutting planes for the purposes of generating tighter initial formulations retains its importance, their pursuit during run-time might sacrifice computational efficiency in a parallel environment.

5.1. Cutting planes for 0–1 problems

Crowder, Johnson and Padberg (1983; [5]) extended the general approach of Crowder and Padberg (1980; [14]) to the case of large-scale 0–1 linear programming problems. Their focus was on exploiting problem sparsity with the motivation that if a tight representation is constructed for each individual knapsack polytope defined by the problem constraints, then this would yield a tight representation for the convex hull of feasible solutions for the overall problem as well.

Accordingly, several novel ad hoc preprocessing procedures that keyed on various algebraic characteristics of the individual constraints were used to improve the associated formulation. A given problem formulation was first preprocessed to: (a) identify variables that could be fixed at zero or one and to check for blatant infeasibility, (b) tighten the constraints through coefficient reductions, and (c) determine constraints of the problem that are rendered inactive by previous manipulations. These preprocessing steps have become standard routines in current mixed-integer programming and pure integer programming software packages. Often, special structured constraints are highlighted to assist in this preprocessing step, as well as in branching strategies. Hoffman and Padberg (1991; [9]) describe a typical schema of this type in which they classify a zero-one problem's constraints into appropriate sets such as:

1. Specially ordered sets (SOS) of constraints of the type $\sum_{j \in L} x_j + \sum_{j \in H} \bar{x}_j \leqslant 1$, where $\bar{x}_j \equiv (1 - x_j)$ and x_j, $j \in L \cup H$ are binary variables.
2. Invariant knapsack constraints having nonzero coefficients of values ± 1 but which cannot be transformed into SOS constraints.
3. Plant location constraints of the type $\sum_{j \in P} x_j \leqslant \theta x_p$, where x_j, $j \in P \cup \{p\}$ are binary variables, and θ is some supply type of coefficient.

The cutting planes generated by Crowder et al. (1983; [5]) were globally valid for the entire enumeration tree, and exploited the inherent structure of individual problem constraints. The method of generating these constraints represented a union of the polyhedral research of Padberg (1973; [14]) and Balas and Zemel (1978; [14]) with computational algorithms. Two types of valid inequalities were derived in attempting to delete a fractional solution to the current LP relaxation. These were lifted minimal cover inequalities, and a more general, although less easily identifiable, class of $1 - k$ configuration constraints.

At the time this particular approach was introduced, no technically good algorithms existed to identify the facetial versions of these two types of inequalities. Consequently, Crowder et al. (1983; [5]) implemented a *continuous* version of identifying and lifting the desired minimal cover and

$(1, k)$-configuration constraints within their algorithm, hence providing strong valid inequalities, although not necessarily facets of the underlying knapsack polytopes. Later, Zemel (1989; [14]) demonstrated how actual lifted facets could be computed in polynomial time. Problems having up to 2756 0–1 variables were solved to optimality using this approach. Crowder et al. (1983; [5]) also demonstrated that if the cutting plane routine was switched off, the more difficult test cases remained unsolved after several hours of cpu time, while previously being solvable in 1–3 min of cpu time.

The success of Crowder et al. (1983; [5]) provided momentum to improving the techniques used to strengthen the initial linear programming formulation relaxation prior to attempting to solve the integer optimization problem. Johnson et al. (1985; [14]) applied a similar methodology as that of Crowder et al. (1983; [5]) to 0–1 integer programming problems arising from large-scale planning models possessing specially structured constraints. They enhanced the preprocessing step, in particular, to include the concept of *probing* in which a binary variable x_j is set either to zero or one, and logical tests are then conducted on the resulting problem. If this problem turns out to be infeasible, then the original variable can be fixed at the complement value. For the largest problems, they used probing on only the most important variables. Savelsbergh (1994; [14]) describes many such preprocessing and probing techniques for mixed-integer programs. These have been incorporated into the commercial software MINTO, for example.

Continuing to focus on improving problem formulation to reduce the size of the integrality gap, Martin and Schrage (1985; [14]) introduced a method to generate an initial set of cuts designed to tighten both pure 0–1 and mixed-integer 0–1 problems. Their method was based on identifying an implied inequality that contains only a subset of variables from a given parent constraint, but one that permits a reduction in its coefficients leading to a tightened cutting plane. This was called a *subset coefficient reduction* procedure.

No single research emphasized the point of using hybrid algorithms more than the seminal papers by Hoffman and Padberg (1985 [14]; [9]) that addressed an LP-based combinatorial approach for 0–1 mixed-integer programming problems. These papers assimilated the existing body of knowledge into a framework that combined preprocessing, heuristics, and cutting plane generation. The emphasis in this philosophy of approach was clear: reduce the integrality gap as much as possible before handing the problem over to a branch-and-bound procedure.

Padberg and Rinaldi (1987; [14]) successfully implemented a slight modification of this generic approach to solve to optimality a 532-city symmetric TSP involving 141,246 0–1 variables. The constraints generated for use as cutting planes consisted only of subtour elimination constraints and comb constraints, similar to those used by Crowder and Padberg (1980; [14]). However, Padberg and Rinaldi (1987; [14]) used a different manner of generating these cuts which rendered them valid for the entire branch-and-bound tree, leading to what is now widely referred to as a *branch-and-cut* scheme.

Capitalizing on the apparent success of this approach, researchers extended its application to several other classes of problems in the ensuing years. Van Roy and Wolsey (1987; [14]) extended the practice of automatically reformulating a given combinatorial optimization problem via preprocessing to mixed-integer programming, noting that reformulation using strong valid inequalities was necessary for solving or finding good optimal solutions. They also suggested that this technique could easily be improved to handle generalized upper bound constraints. Goemans (1989; [14]) did exactly that, developing a family of facet-defining valid inequalities for the class of 0–1 mixed-integer programming problems having a single knapsack constraint along with variable upper bounds on the continuous variables of the problem. Within this same category, Dietrich and Escudero (1990; [14]) extended the notion of coefficient reduction for 0–1 knapsack constraints in the presence of variable bounding constraints. They demonstrated that stronger reductions can be obtained by exploiting the bounding relationship between variables, and aspect not considered by Crowder et al. (1983; [5]). Later Dietrich and Escudero (1992; [14]) applied a similar strategy to tighten cover-induced inequalities using 0–1 knapsack constraints, and, if available, cliques whose variables were included in the cover.

Padberg and Rinaldi (1991; [14]) provided a more in-depth treatment of branch-and-cut for large-scale symmetric traveling salesman problems, noting, among other results, the '*tailing-off*' effect on the best LP relaxation objective function value, $v(\mathrm{LP})$, obtained at node 0 via cut generation. For example, in one particular problem involving 263 cities, the value $v(\mathrm{LP})$ increased quickly in the first 10 iterations, then it increased by less than 9 units in the following 30 iterations, and finally it took 71 iterations to increase by less than half a unit. This '*tailing off*' effect revealed the inability of the cut generator to extricate the current optimal LP solution out of the corner of the polytope where it was trapped, while the integrality gap was still significant. In general, it is important to detect the onset of this phenomenon, and to resort to branching when it is observed to occur.

For some specially structured problems, a variable redefinition technique was proposed by Martin (1987; [14]) for mixed-integer linear programming problems. Whereas the efforts of Crowder and Padberg (1980; [14]), Crowder et al. (1983; [5]), and Hoffman and Padberg (1985; [14]) strived to close the integrality gap by characterizing either fully or partially the facial structure of conv(X), the foundation of this approach rested on the idea that by dropping certain complicating constraints, a specially structured subproblem would result that could be reformulated using an entirely different set of decision variables, or a subset of the original variables plus some new auxiliary variables. This reformulation process yields an equivalent problem formulation that tightens the LP relaxation by constructing a partial convex hull for a subset of constraints. The selection of a linear transformation to relate different mixed-integer formulations is a natural one because the image of a polytope under a linear transformation is also a polytope. Eppen and Martin (1987; [14]) reported encouraging results applying this methodology to solve multi-item capacitated lot-sizing problems. In five out of the seven problems tested, the optimal LP relaxation solution was identical to the optimal solution to the integer programming problem. For the other two problems, the integrality gap was closed to 1.9% and 0.15% of integer optimality. Martin (1991; [14]) also introduced a new method for automatically generating auxiliary variable reformulations for problems solvable by a class of cutting planes, in which the latter are implicitly modeled into the problem via their associated separation program. Although encouraging computational results were obtained for such embedded separation problem models, perhaps equally as interesting was the open research question posed by Martin motivated by the insights provided by this reformulation strategy: can a polynomial size reformulation be generated in this manner for any problem that has a polynomial separation algorithm for finding valid cuts?

5.2. Column generation

In the context of solving large-scale linear programming problems, column generation methods were proposed in the 1960s by Dantzig and Wolfe for block-angular structured problems, and by Gilmore and Gomory for the cutting stock problem (see Lasdon (1970; [14]) for a general discussion on such procedures). These methods decompose a problem formulation into a master problem and appropriately defined subproblem(s), and then proceed to iterate between the two problems, exchanging variable values and reduced cost information as necessary to advance toward an optimal solution to the original problem. More recently, these concepts have arisen in the context of mixed-integer programming problems with the motivation of deriving specially structured formulations for some classes of problems that possess tight LP relaxations.

For example, when deploying synchronous optical rings (SONETS) in telecommunication applications, there arises the problem of allocating customer nodes to rings and routing traffic between pairs of such nodes subject to capacity restrictions, so as to minimize total supporting equipment costs (see Sutter et al., 1998; [14]). A traditional formulation defines binary variables to represent the assignment of nodes to rings along with an accompanying set of dependent traffic routing variables. Due to the inherent symmetry present among the rings, this formulation can produce very weak LP relaxations because of the essentially identical symmetric reflections it admits for any given solution. However, a column-generation-based model can be formulated that defines and associates an integer variable with each possible feasible configuration of customer nodes assigned to a ring, with these variables representing the number of times such a configuration is implemented. In this manner, the column generation formulation automatically circumvents the aforementioned difficulty by suppressing such a replication of symmetric solutions. Moreover, it tends to have a tighter LP relaxation because it eliminates several types of fractional extremal solutions admitted by the traditional formulation.

Of course, such a revised formulation now produces an enormous amount of columns (primal variables), not all of which are explicitly available. This prompts the application of column generation methods to solve the underlying LP relaxation. The key step is that of *pricing* new nonbasic columns that need to be introduced into the overall master program via a suitable subproblem. The process iterates as described above until convergence is obtained in solving the underlying relaxation of the problem. When such a column generation scheme is embedded within a branch-and-bound framework, the resultant procedure is called a *branch-and-price* algorithm.

Aside from exploiting the problem's structure in defining the master and pricing subproblems, another key aspect that requires particular attention is the branching mechanism. Depending upon the application, this must be carefully designed so as to preserve the structure of the subproblems, since branching inherently excludes certain configurations of solutions from being considered. The subproblem should then be capable of shifting through the admissible configurations in order to propose new columns to the master problem. Barnhart et al. (1998; [4]) provide an overview and additional insights into branch-and-price approaches.

6. Reformulation-linearization technique (RLT)

Permeating throughout all these developments was a sense that perhaps there might exist some unifying theory that, taken as a whole, would either identify commonly shared characteristics of the underlying polytopes, or define a structure for tightening LP relaxations via various classes of facetial and non-facetial deep cuts. However, the missing component in this, and other approaches of a similar nature, was the specification of an entire hierarchy based upon a single approach that has the potential to recover the complete convex hull representation, if one was willing to expend a sufficient amount of algebraic and computational effort. The reformulation-linearization technique (RLT) introduced by Sherali and Adams (1990,1994; [14]) did precisely this. Ironically, such a unifying hierarchy was always obtainable, but was effectively concealed behind the non-intuitive idea that a combinatorial optimization problem must first be moved into a higher-dimensional space, transferring integer complexity to nonlinear complexity. By subsequently re-linearizing the problem, and projecting it back into the space of original variables, any desired degree of tightening could then be achieved. The RLT is a method that generates tight LP relaxations in this manner for not only constructing exact solution algorithms, but also to design powerful heuristic procedures for large classes of discrete combinatorial and continuous nonconvex programming problems. Its development initially focused on 0–1 and mixed 0–1 linear and polynomial programs, and later branched into the more general family of continuous, nonconvex polynomial programming problems. The book by Sherali and Adams (1999; [13]) provides a more comprehensive treatment of this subject, along with a historical perspective of related developments.

For the family of mixed 0–1 linear (and polynomial) programs with n 0–1 variables, the RLT generates an n-level hierarchy, with the nth level providing an explicitly algebraic characterization of the convex hull of integer feasible solutions. The RLT consists of two basic steps – a *reformulation* step in which additional nonlinear valid inequalities are automatically generated, and a *linearization* step in which each product term is replaced by a single continuous variable. The level of the hierarchy directly corresponds to the degree of the polynomial terms produced during the reformulation step.

The basic approach is as follows. Suppose the initial problem formulation is composed of both binary variables x_j, $j \in J$, and continuous variables y_k, $k \in K$. In the reformulation step, given a selected level $d \in \{1, \ldots, n\}$, RLT first constructs various nonnegative polynomial factors of degree d comprised of the product of some d binary variables x_j or their complements $(1 - x_j)$, but does not employ both in the same product term. These factors, referred to as *bound-factors*, are then used to multiply each of the defining constraints in the problem (including variable bounding restrictions), to create a nonlinear polynomial mixed-integer 0–1 programming problem. Suitable additional constraint–factor products could also be used to further enhance the procedure. Here, for a structural inequality of the form $\sum_{j \in J} \alpha_j x_j \geqslant \beta$, for example, the nonnegative expression $(\sum_{j \in J} \alpha_j x_j - \beta)$ is referred to as a *constraint factor*, and could likewise be used to generate product constraints. Next, using the relationship $x_j^2 = x_j$ for each binary variable x_j, $j \in J$, which in effect accounts for the tightening of the LP relaxation, the linearization step substitutes a single variable w_J and v_{Jk}, respectively, in place of each nonlinear term of the type $\prod_{j \in J} x_j$ and $y_k \prod_{j \in J} x_j$. The resulting linearized problem defines a higher-dimensional polyhedral set X_d in terms of the original variables (x, y) and the new variables (w, v). Denoting the conceptual projection of X_d onto the space of original (x, y)-variables as X_{Pd}, Sherali and Adams (1990,1994; [14]) showed that as d varies from 1 to n, the underlying LP relaxation polyhedron is progressively tightened via a hierarchy of relaxations leading to the convex hull of integer feasible solutions $(\text{conv}(X))$ to the initial problem:

$$X_{P0} \supseteq X_{P1} \supseteq X_{P2} \supseteq \cdots \supseteq X_{Pd} \supseteq \cdots \supseteq X_{Pn} \equiv \text{conv}(X). \tag{3}$$

The hierarchy of higher-dimensional representations produced in this manner markedly strengthens the LP relaxation, as is evidenced not only by the fact that $\text{conv}(X)$ is obtained at the highest level, but that in computational studies on many classes of problems, even the first level representation helps design algorithms that significantly dominate existing procedures in the literature.

Based on a special case of the RLT process that employs the bound-factors for only a single variable at a time, Balas, Ceria, and Cornuéjols (1993; [14]) described a *lift-and-project* cutting plane algorithm that was shown to produce encouraging results. More recently, Balas et al. (1994; [14]) have applied this cutting plane approach to the maximum clique problem, demonstrating a high degree of closure of the integrality gap when using cuts based on the special structure afforded by the clique constraints.

At its inception, RLT was designed to employ factors involving 0–1 variables stemming from 0–1 mixed-integer programs. By generalizing the concept of factors to include Lagrange inter-polating polynomials, Sherali and Adams (1999; [13]) have extended this development to derive parallel results for the case of general integer problems. Also, in the context of unconstrained quadratic pseudo-Boolean programming problems, Boros et al. (1989; [14]) demonstrated indepen-dently how a hierarchy of relaxations leading to the convex hull representation could be generated. Likewise, Lovasz and Shrijver (1991; [14]) independently developed a similar existential hierarchy for 0–1 linear programs using a succession of pairwise constraint products followed by projection operations.

The RLT process was extended and enhanced by Sherali et al. (1998; [14]) through the use of more generalized constraint factors that imply the bounding restrictions $0 \leqslant x_j \leqslant 1$, for all $j \in J$. As a result, it not only subsumed the previous development, but provided the opportunity to exploit frequently arising special structures such as generalized and variable upper bounds, covering, partitioning, and packing constraints, as well as structural sparsity, and to identify special cases where lower level RLT applications could produce the convex hull representation. More importantly, a new concept was introduced in this paper which combined *conditional logic* based deductions with the generation of RLT constraints in order to derive tighter RLT representations at lower levels of the process. This is analogous to the concept of domain refining *constraint propagation* (Puget, 1995; [14]) in CLP. Sherali and Driscoll (1998; [14]) used this conditional logic approach to generate significantly tightened relaxations for the asymmetric traveling saleman problem and its precedence constrained counterpart based on the ordering structure of Miller–Tucker–Zemlin subtour elimination constraints along with any specified explicit procedence structures. Lougee-Heimer and Adams (1999; [14]) applied this conditional logic approach within a framework that strategically computes quadratic RLT inequalities and then suitably surrogates these nonlinear constraints in order to yield strong valid linear inequalities. They also demonstrated that this conditional logic construct subsumes all of the aforementioned preprocessing techniques.

The RLT also provides information that directly bridges the gap between discrete and continuous sets. Since the level-n formulation characterizes the convex hull, all valid inequalities in the original variable space must be obtainable via a suitable projection. Thus, such a projection operation serves as an all-encompassing tool for generating valid inequalities. Adams et al. (1998; [14]) examined related *persistency* issues for certain classes of constrained and unconstrained pseudo-Boolean problems, whereby variables that take on 0–1 values at an optimum to an RLT relaxation would persist to take on these same values at an optimum to the original problem. A different class of relaxations based on certain generalized bound-factor products is also shown by Adams and Lassiter (1998; [14]) to possess the persistency property. This class of problems subsumes others known in the literature (such as the vertex packing problem) to share this property.

7. Facial disjunctive programming

The class of 0–1 mixed-integer programs is subsumed by an important generalized class of problems known as *facial disjunctive programs* (FDP), which can be stated as follows:

$$\text{FDP}: \text{Minimize } \{cx: x \in X \cap Y\}, \tag{4}$$

where X is a nonempty polytope in $\mathbb{R}^n$, $c \in \mathbb{R}^n$, and where Y is a conjunction of some $\hat{h}$ disjunctions given in the so-called *conjunctive normal form* (conjunction of disjunctions)

$$Y = \bigcap_{h \in H} \left[\bigcup_{i \in Q_h} \{x: a_i^h x \geqslant b_i^h\} \right]. \tag{5}$$

Here, $H = \{1, \dots, \hat{h}\}$, and for each $h \in H$, there is a specified disjunction that requires at least one of the inequalities $a_i^h x \geqslant b_i^h$, for $i \in Q_h$, to be satisfied. The terminology *facial* conveys the feature that $X \cap \{x: a_i^h x \geqslant b_i^h\}$ defines a face of X for each $i \in Q_h$, $h \in H$.

For example, in the context of 0–1 mixed-integer problems, the set X represents the LP relaxation of the problem, H represents the index set for binary variables, and for each x_h, $h \in H$, the corresponding disjunction in (5) states that $x_h \leqslant 0$ or $x_h \geqslant 1$ should hold true (where $0 \leqslant x_h \leqslant 1$ is included within X). Balas (1998; [2]) showed that for FDPs, the convex hull of feasible solutions can be constructed inductively by starting with $K_0 = X$ and then determining

$$K_h = \text{conv} \left[\bigcup_{i \in Q_h} \left(K_{h-1} \bigcap \{x: a_i^h x \geqslant b_i^h\} \right) \right] \quad \text{for } h = 1, \dots, \hat{h}, \tag{6}$$

where $K_{\hat{h}}$ produces $\text{conv}(X \cap Y)$. Supporting this construction, Balas (1985; [14]) had earlier demonstrated that a hierarchy of relaxations $K_0, K_1, \dots, K_{\hat{h}}$ could be generated for FDPs that spans the spectrum from the linear programming to the convex hull representation. Each member in this hierarchy can also be viewed as being obtained by representing the feasible region of the original problem as the intersection of the union of certain polyhedra, and then taking the hull-relaxation for this representation. Here, for a set $D = \bigcap_j D_j$, where each D_j is the union of certain polyhedra, the *hull-relaxation* of D is defined as $h - \text{rel}(D) = \bigcap_j \text{conv}(D_j) \supseteq \text{conv}(D)$.

Interestingly, the RLT construct introduced earlier can be specialized to derive K_h defined by (6) for 0–1 mixed-integer programs. In this case,

$$K_h = \text{conv} \left[(K_{h-1} \cap \{x: x_h \leqslant 0\}) \bigcup (K_{h-1} \cap \{x: x_h \geqslant 1\}) \right] \tag{7}$$

can be obtained by multiplying the implicitly defined constraints of K_{h-1} by x_h and $(1 - x_h)$ and then linearizing the resulting problem. This RLT approach, along with the cutting plane concept of Jeroslow (1977; [14]), is used by Balas et al. (1993; [14]) in their *lift-and-project* hierarchy of relaxations and cutting plane algorithm. The more general RLT process generates tighter relaxations at each level which can be viewed as hull-relaxations produced by the intersection of the convex hull of the union of certain specially constructed polyhedra (see Sherali and Adams (1994; [14])). No direct realization of (7) can produce these relaxations. Following a similar concept of adopting the second-level RLT relaxation while imposing binary restrictions on variables taken two at a time, Balas (1997; [14]) presented an enhanced procedure that considers two variables at a time to define the disjunctions.

8. Post-solution analysis

The approach of post-solution analysis, which encompasses post-optimality analysis and includes de-bugging a scenario when it yields inconsistent or anomalous information, or exhibits infeasibility or unboundedness, is highly well developed for linear programming problems. In the context of integer programming, the field is not as mature, although there exists an ever widening body of related literature that attempts to parallel the development for linear programs (see Greenberg, 1998; [8]). A majority of the literature deals with post-optimality analysis that investigates the effect of data changes on the optimum, or on how other data parameters must vary to compensate for a change in some data parameter, as well as on *stability* issues involved with determining *stability regions* over which variations in specific parameters retain the optimality of the current solution.

Schrage and Wolsey (1985; [14]) have examined the type of information that needs to be stored when implementing branch-and-bound/cut procedures in order to investigate sensitivity analysis issues concerned with the effect of right-hand side variations on the optimum, or concerned with the range of values of the cost coefficient of a new 0–1 activity that would prevent perturbing the current optimum. Hooker (1996; [14]) contrasts this traditional approach to that of using *inference duality*, whereby sensitivity analysis is conducted by viewing the role that each constraint plays in inferring a bound on the optimal value, and thereby, in constituting the optimal solution. The case of parametric variation in the objective function has been treated by Jenkins (1987; [14]) using a cutting plane approach.

An interesting device that plays a useful role in sensitivity analysis for integer programs is that of a *Gröbner basis*. For a family of integer programs defined by varying the right-hand side, the (*reduced*) Gröbner basis is a minimal set of discrete perturbation directions (called a *test set*) such that for each nonoptimal solution to a given program in this family, there is at least one element in the test set which will provide an improved solution. Thomas (1995; [14]) describes a geometric Buchberger algorithm for computing such a basis, and shows how members of the underlying family of integer programs can then be solved using this construct. Sturmfels and Thomas (1997; [14]) also show how Gröbner bases can be used to determine equivalence classes of objective vectors, where two objective vectors are said to be *equivalent* if they yield the same optimal solutions for every right-hand side vector. Hosten and Sturmfels (1995; [14]) describe a software package **GRIN** for computing Gröbner bases for integer programs using Buchberger's algorithm. This is a tool designed for use in both combinatorial optimization and computational algebra contexts.

For an excellent comprehensive survey on this growing field of post-solution analysis in integer programming, we refer the reader to Greenberg (1998; [8]).

9. Stochastic integer programming

The field of stochastic integer programming is a relatively new and budding area of research which injects complicating discrete decisions into an already complex area of stochastic optimization (see Birge, 1997; [14]). The types of distributions considered might be general, or discrete and finite, leading, respectively, to simulation–optimization methods such as quasi-gradient algorithms (see Ermoliev, 1983; [14]), or to decomposition methods (see Birge, 1985; [14]). The principal type of problem that has been widely considered by researchers in the context of stochastic integer programming is the linear two-stage problem with integer recourse (which is generalizable, at least in theory, to multi-stage problems). Here, a first-stage decision is made in a "here-and-now" context, then some random realization manifests itself, following which, a second-stage recourse action is taken. The objective is to minimize the total cost associated with the *non-anticipative* first-stage decision and the expected value of the related, consequent second-stage recourse decision. While both the stages might involve discrete decisions, the main complication arises in the context of second-stage decisions being integer valued, leading to generally nonconvex and nondifferentiable (possibly discontinuous) optimal value functions as viewed in the projected space of the first-stage decisions.

Caroe and Tind (1998; [14]) describe a generalization of the L-shaped method that was developed originally for stochastic linear programming to this class of problems, combining concepts of feasibility and optimality cuts from the generalized Benders' decomposition method with branch-and-bound concepts and with Gomory's fractional cutting plane algorithm. An alternative dual decomposition scheme that layers the model through the definition of auxiliary duplicated variables, and then applies Lagrangian dual/relaxation techniques is described in Caroe and Schultz (1999; [14]).

In the more general context of continuous or discrete stochastic optimization, Norkin et al. (1998; [14]) describe a novel branch-and-bound approach that employs stochastic lower and upper bounds to guide the partitioning and fathoming strategies in an infinite process that converges with probability one. For a more detailed survey of this field, we refer the reader to Schultz et al. (1996; [12]), and Stougie and van der Vlerk (1997; [15]).

10. Meta-heuristics

The class of integer programming problems is well known to be NP-hard (see Garey and Johnson, 1979; [14]), although there exist many special cases that admit polynomial-time algorithms. While even among classes of NP-hard special cases of integer programs, there exist problems for which practical-sized instances can be solved rather effectively, resorting to heuristic solution approaches in general is inevitable for most large-scale applications. It should be pointed out that even exact branch-and-cut algorithms need to actively search for good quality solutions in order to be effective by way of enhancing the fathoming efficiency of the procedure. The pivot-and-complement heuristic of Balas and Martin (1980; [14]) is a popular procedure that is frequently implemented to achieve this purpose. Here, we briefly focus on stand-alone meta-heuristics for solving discrete or combinatorial optimization problems.

Glover (1994; [14]) describes the principal concepts and advancements of the *Tabu-Search* method which he introduced in the 1980s (see also Glover and Laguna, 1997; [7]). In this procedure, a neighborhood structure of any given solution is defined, and the search process moves from a current solution to the best solution in the neighborhood, even if its value is non-improving. In order to avoid cycling, however, a *tabu list* of forbidden solutions, or forbidden types of solution modifications, is maintained and is continually updated based on observed recent trends. In concept, this methodology possesses a great deal of flexibility in accommodating the history of the search process and permitting an exploitation of problem structure to adaptively customize the procedure to the particular type of problem being faced.

Simulated annealing is another meta-heuristic, but one that is less attuned to problem structure.

The fundamental idea here is to randomly generate a neighboring solution and accept it with a probability of 1 if it is an improving solution, and with a probability of $0 < e^{-\delta/T} < 1$ otherwise, where δ is the increase in the (minimization) objective value, and T is a *temperature* parameter. This latter parameter is gradually decreased as the method progresses according to a prescribed *cooling* schedule, so that worsening solutions are accepted with an ever diminishing probability. The concept for this approach is motivated by an analogy in metallurgy wherein a heated metal is cooled at an appropriately tuned rate in order to permit its molecules to settle in a desirable configuration that imparts it certain good structural properties. Aarts and Korst (1989; [1]) and Hajek (1985; [14]) provide a review of the theory, convergence properties, and applications of this procedure.

Genetic algorithms, or more generally, *evolutionary heuristics*, attempt to mimic nature's process in evolving better endowed specimens of a given population. In the algorithmic framework for such methods, the procedure begins with some desirable, finite population of solutions, as opposed to a single solution. Each solution is represented via some coded binary or integer string of numbers and its fitness is evaluated based on a merit function (which might be a form of a penalty function). A pair of *parent* solutions are randomly selected, with a higher probability of selection being ascribed to superior solutions. The two parents are combined using a crossover scheme that attempts to merge the strings representing them in a suitable fashion to produce an *offspring* solution. Offsprings can also be modified by some random *mutation* perturbation. Based on the fitness of the offsprings thus produced versus that of the original population, a new population of the same cardinality as before is composed, and the process is repeated. For further reading on this subject, we refer the reader to Davis (1991; [6]) and Goldberg (1989; [14]).

Two other cutting-edge metaheuristic approaches are worth mentioning because of their potential to radically alter the landscape of discrete computations if they prove successful. Deoxyribonucleic acid, or *DNA computing* employs a direct biochemical manipulation of specially selected DNA strands whose combination will result in the solution of some discrete mathematics problem. This has been used to solve a seven-city TSP (Poole, 1996; [4]). *Quantum* computing, a largely theoretical endeavor, is a hypothetical machine that uses quantum mechanics to perform computations. Both of these approaches are in embryonic stages of development. For further reading on this subject, we refer the reader to Gramss (1998; [14]).

11. Parallel processing

There has been an increasing amount of research performed in recent years focusing on adapting the branch-and-bound algorithm to take advantage of parallel architectures of computing machines involving more than a single processor (see Gendrion and Crainic (1994; [14]) and Eckstein (1994; [14])). In addition to differences in hardware design, the method of controlling operations in each environment differentiates implementation approaches. The main modes of computation can be grouped as *control driven*, *data driven*, and *demand driven* (Treleaven et al. (1982; [14])). Control driven computations rely on the user to specify the exact type and order of operations to be performed. In a data-driven model, operations can be performed as soon as all the necessary operands are available. The demand-driven model waits to perform an operation until some required outcome is needed. All sequential computers essentially use the control-driven method, whereas parallel machines appear to operate most effectively in a data-driven, or *data-flow* environment.

11.1. Parallel aspects of branch-and-bound

The branch-and-bound process contains elements that are amenable to both *coarse-grained* parallelism, and *fine-grained* parallelism, each of which can be employed when executing the algorithm. Coarse-grained parallelism occurs when a program contains certain statements that can be executed in parallel (e.g., a FORTRAN FOR loop). A sequential list of statements that are independent of one another is an example of fine-grained parallelism. Whether one type of parallelism is preferred over the other is dependent upon the specific configuration of memory modules and processors.

As Gendrion and Crainic (1994; [14]) note, the development of parallel branch-and-bound research followed a chronology of early experiments (1975–1982), theoretical studies (1983–1986), and experiments on actual parallel systems (since 1987), with many contributions being made in the process. It is interesting to note that the first use of parallel processing to solve a combinatorial optimization problem by branch-and-bound is credited to Pruul (1975; [14]). Simulating a shared memory system with p processors, $1 \leqslant p \leqslant 5$, because of the lack of parallel hardware, Pruul (1975) nonetheless applied the new methodology to ten 25-city aysmmetric traveling salesman problems, incorporating a bounding mechanism based on the assignment problem, along with subtour elimination branching rules, and a parallel depth-first branching rule. The results of Pruul's study were of little practical value at the time, although this conclusion is clearly not the case today.

Three salient aspects of the branch-and-bound algorithm lend themselves to parallelism. The first accommodates parallelism in the process of performing operations on subproblems generated via branching. For example, after a number of subproblems are generated at the onset, the independent processors could be used to evaluate both the lower bounds available via the subproblems' LP relaxations and the best upper bounds available for each subproblem. The ideal number of subproblems to generate are as many as would be sufficient to avoid *processor starvation* due to a small number of available tasks, which typically hampers speed-up in the early stages of execution.

It is also possible to build the branch-and-bound binary search tree in parallel by performing operations on several subproblems simultaneously. Of course, this implies that each of the processors must have a sufficient capability to execute all the necessary operations of branch-and-bound independently, as in a so-called coarse-grained MIMD (multiple instruction, multiple data) system. Using such a scheme, Boehning et al. (1988; [14]) introduced a parallel branch-and-bound algorithm for integer linear programming problems that was able to achieve superlinear speedup for a subset of the test problems. Each parallel processor selected a subproblem to work on from an available pool of problems shared by one or more processors. The same basic sequence of instructions were executed by each processor: (a) request a problem, (b) add a down-row, (c) add an up-row, (d) analyse the down-node, (e) check for fathoming, (f) put the down-node in the pool of subproblems, (g) analyse the up-node, (h) check for fathoming, (i) put the up-node in the pool of subproblems, and repeat. Their algorithm also made use of cutting planes parallel to the objective function to delete fractional linear programming solutions obtained for the different node subproblems.

Miller and Pekny (1989; [14]) and Pekny and Miller (1992; [14]) implemented a version of branch-and-bound based on a *processor shop model* for the asymmetric traveling salesman problem in which processors examine the list of tasks available and choose one based on some priority rules. They then process the chosen task, place the results in the correct memory location, and extract another task. Within this environment, problems ranging from 50 to 3000 cities were successfully solved on a 14 processor BBN Butterfly Plus computer. Kudva and Pekny (1993; [14])

experienced similar successes testing this approach using various-sized instances of the multiple resource constrained sequencing problem. All of the foregoing implementations achieved success relying on coarse-grained MIMD systems. Although this aspect of branch-and-bound has received some attention on fine-grained SIMD (single instruction, multiple data) systems (Kindervater and Trienekens (1988; [14]), difficulties were encountered trying to effectively use the SIMD architecture – a mismatch of algorithms with hardware.

The last aspect of parallelism for branch-and-bound algorithms arises in the ability to construct different branching trees in parallel by performing operations on several subproblems simultaneously. This type of decomposition is reminiscent of performing sensitivity analysis in linear programming for simultaneous variations in right-hand side values. Each of the branching trees execute different branching, bounding, and evaluation rules, and the information generated in one tree can possibly contribute to the construction of another. Miller and Pekny (1993; [14]) implemented this strategy by varying only the branching rules. Kumar and Kanal (1984; [14]) allowed each of the processors to execute the usual lower bounding technique, but implemented an upper bounding strategy that optimistically diminishes the known best solution value. Janakiram et al. (1988; [14]) experimented with adding a stochastic character to the algorithm by randomizing the selection of the next subproblem to be evaluated by each processor. Their efforts were motivated by the reasoning that the mapping of randomized algorithms onto multiprocessors involves very little scheduling or communications overhead. To avoid possible duplication of work by processors, their technique maintained a global listing of the status of the subproblems at the first k levels of the enumeration tree.

Recognizably, all of the implementations of branch-and-bound strategies are dependent upon the type of parallel architecture used by the computing device. Independent of this differentiation, results from the aforementioned studies are very encouraging, and efforts into this particular vein of research for attacking hard combinatorial optimization problems are expected to blossom in the future. This will especially become the case if the current trend of increased availability of multiprocessor desktop computers continues, along with an accompanying decrease in cost.

12. Final comments

With the advancement in computer technology, refinements in numerical implementation techniques, and the proliferation of computers, there have emerged stable, robust, commercial and research-oriented software for solving mixed-integer programming problems. CPLEX, Inc. (1990) distributes one of the most popular and effective software systems (its most recent version is CPLEX 6.5). This software incorporates preprocessing, heuristic, and lifted cover inequalities generation techniques within an LP-based branch-and-cut approach. The optimization subroutine library (OSL) system distributed by the IBM Corporation (1990; [14]) is a similar, alternative commercial solver.

For the purposes of conducting research development, the software MINTO (Nemhauser et al., 1994; [14]) is a very useful tool. As a stand-alone methodology, MINTO implements far more effort in probing, finding feasible solutions, and generating cuts than CPLEX does. As a result, it is able to solve larger and more difficult problem instances more effectively (see Johnson et al. (1997; [10])). MINTO also permits users to implement their own cut generation schemes and preprocessing routines, thereby facilitating research development.

Cordier et al. (1997; [14]) have also developed a prototype branch-and-cut code (BC-OPT) for solving mixed-integer programming problems. This routine incorporates a variety of cut generation routines based on flow cover, surrogate knapsack, integer knapsack, and Gomory mixed-integer cuts. Another branch-and-cut system known as ABACUS has been developed by Junger and Thienel (1998; [14]). This is an object-oriented framework for implementing customized branch-and-cut or branch-and-price algorithms. All of these foregoing three procedures employ CPLEX to solve the various linear programming relaxations.

Perhaps the most comprehensive and widely accepted set of test problems for researchers desiring to benchmark the performance of newly designed algorithms and tightened problem formulations in integer programming was created by Bixby et al. (1996; [14]) and is maintained on a primary server at Rice University, Texas. Openly distributed at no cost, the files can be obtained over the Internet from the Rice University Software Distribution Center at *http://softlib.rice.edu/softlib/*.

Lastly, a closing comment. Although several classes of optimization problems mentioned in this paper appear to have succumbed to improved algorithms, clever reformulation techniques, and advances in computing technologies, the class of 0–1 mixed-integer programs (MIP) in discrete optimization remains steadfast in its challenge. Despite the fact that individual cases have been studied from the perspective of tightening LP relaxations are exploiting special structures for which success has been noted, the "holy grail" has yet to be discovered. The area of heuristic development is perhaps the most aggressive in its efforts to cross-fertilize the study of integer programming with ideas imported from novel, and sometimes unusual disciplines. This is not to say that progress is not being made on all other fronts as well, for it is clear from this exposition that it is. It is simply that, in the face of failing to solve practically sized instances of mixed-integer programs arising in many applications, and with the enduring conjecture that $P \neq NP$ looming in the background, one is frequently left to wonder whether the tools we are developing are appropriate to eventually defeat the steadfast challenge posed by the underlying combinatorics of such problems.

Acknowledgements

Partially supported under NSF Grant #DMI-9812047, and the Mathematical Sciences Center of Excellence, USMA, West Point, NY.

References

[1] A. Aarts, J. Korst, Simulated Annealing and Boltzmann Machines, Wiley, New York, 1989.

[2] E. Balas, Disjunctive programming: properties of the convex hull of feasible points, Discrete Appl. Math. 89 (1998) 3–44.

[3] C. Barnhart, E.L. Johnson, G.L. Nemhauser, G. Sigismondi, P. Vance, Formulating a mixed integer programming problem to improve solvability, Oper. Res. 41 (6) (1993) 1013–1019.

[4] C. Barnhart, E.L. Johnson, G.L. Nemhauser, M.W.P. Savelsbergh, Branch and price: column generation for solving huge integer programs, Oper. Res. 46 (1998) 316–329.

[5] H. Crowder, E.L. Johnson, M. Padberg, Solving large-scale zero-one linear programming problems, Oper. Res. 5 (1983) 803–834.

[6] L. Davis, A Handbook of Genetic Algorithms, Van Nostrand Reinhold, New York, 1991.

[7] F. Glover, M. Laguna, Tabu Search, Kluwer Academic Publishers, Boston, MA, 1997.

[8] H.J. Greenberg, An annotated bibliography for post-solution analysis in mixed integer and combinatorial optimization, in: D.L. Woodruff (Ed.), Advances in Computational and Stochastic Optimization, Logic Programming and Heuristic Search, Kluwer Academic Publishers, Dordrecht, 1998.

[9] K. Hoffman, M. Padberg, Improving LP-representations of zero-one linear programs for branch-and-cut, Oper. Res. 3 (1991) 121–134.

[10] E.L. Johnson, G. Nemhauser, M.W.P. Savelsbergh, Perspectives on discrete optimization, Proceedings of the EURO/INFORMS Conference, Barcelona, Spain, 1997, pp. 33–55.

[11] G.L. Nemhauser, L.A. Wolsey, Integer and Combinatorial Optimization, Wiley, New York, Mathematical Programming 6 (1988) 48–61.

[12] R. Schultz, L. Stougie, M.H. van der Vlerk, Two-stage stochastic integer programming: a survey, Stat. Neerlandica 50 (1996) 404–416.

[13] H.D. Sherali, W.P. Adams, A Reformulation-Linearization Technique for Solving Discrete and Continuous Nonconvex Problems, Kluwer Academic Publishers, Dordrecht, 1999.

[14] H.D. Sherali, P.J. Driscoll, A bibliography for the evolution and state-of-the-art in integer programming, Technical Report #99-004, Department of Mathematical Sciences, U.S. Military Academy, West Point, NY, 1999.

[15] L. Stougie, M.H. van der Vlerk, Stochastic integer programming, in: M. Dell'Amico, F. Maffioli, S. Martello (Eds.), Annotated Bibliographies in Combinatorial Optimization, Wiley, New York, 1997.

[16] L.A. Wolsey, Integer Programming, Wiley, New York, 1998.

![N·H ELSEVIER logo]

Journal of Computational and Applied Mathematics 124 (2000) 341–360

JOURNAL OF
COMPUTATIONAL AND
APPLIED MATHEMATICS

www.elsevier.nl/locate/cam

Combinatorial optimization: Current successes and directions for the future

Karla L. Hoffman [1]

School of Information Technology and Engineering, George Mason University, Mail Stop 4A6, Fairfax, VI 22030, USA

Received 4 November 1999; received in revised form 31 January 2000

Abstract

Our ability to solve large, important combinatorial optimization problems has improved dramatically in the past decade. The availability of reliable software, extremely fast and inexpensive hardware and high-level languages that make the modeling of complex problems much faster have led to a much greater demand for optimization tools. This paper highlights the major breakthroughs and then describes some very exciting future opportunities. Previously, large research projects required major data collection efforts, expensive mainframes and substantial analyst manpower. Now, we can solve much larger problems on personal computers, much of the necessary data is routinely collected and tools exist to speed up both the modeling and the post-optimality analysis. With the information-technology revolution taking place currently, we now have the opportunity to have our tools embedded into supply-chain systems that determine production and distribution schedules, process-design and location-allocation decisions. These tools can be used industry-wide with only minor modifications being done by each user. © 2000 Elsevier Science B.V. All rights reserved.

Keywords: Problem formulation; Cutting planes; Column-generation; Heuristics; Hybrid algorithms; Parallel processing; Modeling languages and stochastic optimization; Solution analysis

1. Introduction[2]

The versatility of the combinatorial optimization model stems from the fact that in many practical problems, activities and resources, such as machines, airplanes and people are indivisible. Also, many problems (e.g., scheduling) have rules that define a finite number of allowable choices and

---

E-mail address: khoffman@gmuvax.gmu.edu (K.L. Hoffman).

[1] This research has been supported by a grant from the Office of Naval Research.

[2] We note that each section of this paper will include only a limited number of survey references. These survey papers contain the references to the much larger body of work in each area. We have chosen this approach because of the editorial policy of this volume. An alternative copy of this paper with all of the references detailed in the text can be downloaded from the author's homepage at: http://iris.gmu.edu/ ~khoffman.

0377-0427/00/$ - see front matter © 2000 Elsevier Science B.V. All rights reserved.
PII: S 0377-0427(00)00430-1

consequently can appropriately be formulated using procedures that transform the logical alternatives descriptions to linear constraint descriptions where some subset of the variables are required to take on certain discrete values. Such problems are labeled *mixed-integer linear optimization problems.*

This paper will consider problems whereby both the function to be optimized and the functional form of the constraints restricting the possible solutions are linear functions. Although this linear restriction might seem overly constraining, the wealth of real-world problems that either naturally assume this form or can be acceptably transformed, possibly by adding many more variables and constraints, into this mathematical structure is extraordinarily large. Thus, the general linear integer model that we will consider is:

$$\max \quad \sum_{j \in B} c_j x_j + \sum_{j \in I} c_j x_j + \sum_{j \in C} c_j x_j$$

$$\text{subject to} \quad \sum_{j \in B} a_{ij} x_j + \sum_{j \in I} a_{ij} x_j + \sum_{j \in C} a_{ij} x_j \sim b_i \quad (i = 1, \ldots, m),$$

$$l_j \leqslant x_j \leqslant u_j \quad (j \in I \cup C),$$

$$x_j \in \{0, 1\} \quad (j \in B),$$

$$x_j \in \text{integers} \quad (j \in I),$$

$$x_j \in \text{reals} \quad (j \in C)$$

where B is the set of zero–one variables, I is the set of integer variables, C is the set of continuous variables, and the $\sim$ symbol in the first set of constraints denotes the fact that the constraint $I = 1, \ldots, m$ can be either $\leqslant, =,$ or $\geqslant$. The data l_j and u_j are the lower and upper bound values, respectively, for variable x_j. As we are discussing the integer case, there must be some variable in $B \cup I$. If $C = I = \emptyset$, then the problem is referred to as a pure 0–1 linear-programming problem; if $C = \emptyset$, the problem is called a pure integer (linear) programming problem. Otherwise, our problem is a mixed integer (linear) programming problem. Throughout this discussion, we will call the set of points satisfying all constraints S, and the set of points satisfying all but the integrality restrictions, S'.

While linear optimization belongs to the class of problems for which provably good algorithms exist – i.e., algorithms for which the running time is bounded by a polynomial in the size of the input – combinatorial optimization belongs to the class of problems (called *NP-hard problems*) for which provably efficient algorithms do not exist. Even so, when one is careful in choosing among *mathematically correct alternative models* and when one takes advantage of the specific structure of the problem, many very large and important combinatorial problems have been solved in reasonable times. Thus, we begin the discussion by highlighting some of the formulation issues that determine the solvability of the problem.

2. Formulation issues

Since there are often different ways of mathematically representing the same problem, and, since obtaining an optimal solution to a large integer programming in a reasonable amount of computing time may well depend on the way it is "formulated", much recent research has been directed toward the reformulation of combinatorial optimization problems. In this regard, it is sometimes advantageous to increase (rather than decrease) the number of integer variables, the number of constraints, or both. When we discuss the notion of a "good" formulation, we normally think about creating an easier problem to solve that approximates well, the objective function value of the original problem. Since it is the integrality restrictions on the decision variables that destroys the convexity of the feasible region, the most widely used approximation removes this restriction; Such an approximation is known as the *Linear-Programming (LP) relaxation*. However, merely removing these integrality restrictions can alter the structure so significantly that the LP solution is far from the integer solution. One might therefore consider adding additional restrictions to the problem so that, at least in the vicinity of the optimal solution, the linear programming polytope closely approximates the polyhedron described by the *convex hull* of all feasible points to the original combinatorial optimization problem. When one considers adding such constraints to the LP-relaxation iteratively within an overall algorithm, the algorithm is called a *cutting plane algorithm*. More will be said about this in the section on solution approaches.

An example of two very different formulations for the same problem is the *machine-shop-scheduling problem*. Early formulations of this problem took a straight-forward approach of defining the decision variables to be the time at which job i started on machine j, while an alternative formulation might consider providing feasible schedules for each machine and then combining these schedules to form feasible solutions for each job. The first of these formulations has as its linear programming relaxation, an objective function value that is far from the true objective value. The second requires the generation of feasible schedules as input to the formulation and the number of such possible schedules can be enormous. However, the second formulation, although appearing far more work for the modeler and far larger, is the one that allows solvability with current computing technologies. For more on how column generation handles this problem, see the section on Column Generation in this paper.

To illustrate the importance of careful formulations, we include some examples of formulation alternatives that make a difference in the length of time it will take to solve a combinatorial problem. One will obtain a better formulation is one performs, for example, *constraint disaggregation*: In this case, one removes the constraint $\sum_{j=1,\dots,m} x_j = mx_0$ and replaces it with the m-constraints: $x_j = x_0$ for $j = 1,\dots,m$. Similarly, whenever a collection of variables is indistinguishable (e.g. one has k identical machines in a machine scheduling problems), one must provide a decision hierarchy that prioritizes among the identical objects. Otherwise, the LP relaxation will continue to interchange the identical machines and provide alternative fractional solutions with the same objective function value.

Finally, the user must supply bounds that are as tight as they can be, or have the solution procedure attempt to search for tight bounds by examining individual constraints, by probing – a term used to consider the implications of fixing a single variable on all other variables in the problem, or by solving related optimization problems. Without tight bounds, coefficients in the formulation are likely to be large and the resulting LP relaxation weak.

Recently, reformulating these problems as either set-covering or set-partitioning problems, having an extraordinary number of variables, allowed the solution of a variety of difficult problems. Because, for even small instances of the problem, the problem size cannot be explicitly solved, techniques known as *column generation*, which began with the seminal work of Gilmore and Gomory on the cutting stock problem, are employed.

Bramel and Simchi-Levi [9] have shown that the set-partitioning formulation for the vehicle routing problem with time windows is a tight formulation, i.e., the relative gap between the fractional linear programming solution and the global integer solution are close. Similar results have been obtained for the bin-packing problem and the machine-scheduling problem.

Because formulation has such a significant impact on the solvability of the problem, most software packages now contain "automatic" reformulation or *preprocessing* procedures.

For discussions of alternative formulation approaches, see Williams [37] and for automatic preprocessing techniques the papers by Hoffman and Padberg [25], Anderson and Anderson [1] and Brearly et al. [10]. Such preprocessing includes the elimination of variables (by fixing them to their only feasible value deduced through logical implications), the elimination of redundant or non-binding constraints, the tightening of bounds on the variables, coefficient improvement within a row, deducing additional constraint restrictions (adding of cover or clique inequalities), and using ideas from disjunctive programming to strengthen the formulation. Current general-purpose software packages continue to expand the use of such automatic reformulation strategies.

3. Exact solution strategies

Solving combinatorial optimization problems, i.e., finding an optimal solution to such problems can be a difficult task. The difficulty arises from the fact that unlike linear programming, the feasible region of the combinatorial problem is not a convex set. Thus, we must, instead, search a lattice of feasible points, or in the case of the mixed integer case, a set of disjoint half-lines or line segments to find an optimal solution. In linear programming, due to the convexity of the problem, we can exploit that fact that any local solution is a global optimum. In integer programming, problems have many local optima and finding a global optimum to the problem requires one to prove that a particular solution dominates all feasible points by arguments other than the calculus-based derivative-approaches of convex programming.

There are a number of quite different approaches for solving integer-programming problems, and currently, they are frequently combined into "hybrid" solutions that try to exploit the benefits of each. We will highlight the attributes of enumerative techniques, relaxation and decompositions approaches, and of cutting planes. We will then indicate how these have been powerfully combined to tackle very difficult problems.

3.1. Enumerative approaches

The simplest approach to solving a *pure* integer-programming problem is to enumerate all finitely many possibilities. However, due to the "combinatorial explosion" resulting from the parameter "size", only the smallest instances could be solved by such an approach. Sometimes one can *implicitly* eliminate many possibilities by domination or feasibility arguments. Besides straight-forward or implicit enumeration, the most commonly used enumerative approach is called *branch and bound*, where the "branching" refers to the enumeration part of the solution technique and bounding refers to the fathoming of possible solutions by comparison to a known upper or lower bound on the solution value. To obtain an upper bound on the problem (we presume a maximization problem), the problem is relaxed in a way which makes the solution to the relaxed problem, relatively easy to solve.

All commercial branch-and-bound codes relax the problem by dropping the integrality conditions and solve the resultant continuous linear programming problem over the set S'. If the solution to the relaxed linear programming problem satisfies the integrality restrictions, the solution obtained is optimal. If the linear program is infeasible, then so is the integer program. Otherwise, at least one of the integer variables is fractional in the linear programming solution. One chooses one or more such fractional variables and "branches" to create two or more subproblems each of which exclude the prior solution but do not eliminate any feasible integer solutions. These new problems constitute "nodes" on a branching tree, and a linear programming problem is solved for each node created. Nodes can be fathomed if the solution to the subproblem is infeasible, satisfies all of the integrality restrictions, or has an objective function value worse than a known integer solution. A variety of strategies that have been used within the general branch-and-bound framework is described by Linderoth and Savelsbergh [26].

3.2. Lagrangian relaxation and decomposition methods

Relaxing the integrality restriction is not the only approach to relaxing the problem. An alternative approach to the solution to integer programming problems is to take a set of "complicating" constraints into the objective function in a Lagrangian fashion (with fixed multipliers that are changed iteratively). This approach is known as *Lagrangian relaxation*. By removing the complicating constraints from the constraint set, the resulting sub-problem is frequently considerably easier to solve. The latter is a necessity for the approach to work because the subproblems must be solved repetitively until optimal values for the multipliers are found. The bound found by Lagrangian relaxation can be tighter than that found by linear programming, but only at the expense of solving subproblems in *integers*, i.e., only if the subproblems do not have the *integrality property*. (A problem has the integrality property if the solution to the Lagrangian problem is unchanged when the integrality restriction is removed). Lagrangian relaxation requires that one understand the structure of the problem being solved in order to then relax the constraints that are "complicating". A related approach that attempts to strengthen the bounds of Lagrangian relaxation is called *Lagrangian decomposition*. This approach consists of isolating sets of constraints so as to obtain separate, easy problems to solve over each of the subsets. Creating linking variables, which link the subsets, increases the dimension of the problem. All Lagrangian approaches are problem-structure dependent and no underlying general theory – applicable to, for example, arbitrary zero–one problems – has evolved.

Most Lagrangian-based strategies provide approaches, which deal with special row structures. Other problems may possess special column structure, such that when some subset of the variables is assigned specific values, the problem reduces to one that is easy to solve. Benders' decomposition algorithm fixes the complicating variables, and solves the resulting problem iteratively. Based on the problem's associated dual, the algorithm must then find a cutting plane (i.e. a linear inequality) which "cuts off" the current solution point but no integer feasible points. This cut is added to the collection of inequalities and the problem is re-solved. The texts by Nemhauser and Wolsey [31] and Martin [27] provide excellent discussions of relaxation and decomposition methods.

Finally, recent work on algorithms for solving the continuous semi-definite programming problem – a generalization of linear programming – are leading researchers to formulations that consider a semi-definite relaxation of the combinatorial optimization problem. Specifically, Goemans and Williamson [19] have shown that such relaxations provide very strong bounds for the MAX 2SAT (proven to be within 0.931 of optimality), MAX 3SAT (proven to be within 7/8 of the optimal solution), and the maximum cut and MAX DICUT problems. The satisfiability problem of propositional logic is to determine whether or not an assignment of truth values (or the negation) to the variables exists such that the conjunction of all clauses in a truth statement can be satisfied by that assignment. One can transform each clause into a string of quadratic inequalities that significantly tighten the formulation. With the appearance of semidefinite programming software, we can expect to see many important graph-theoretic problems being reformulated in this manner.

Since each of the decomposition approaches described above provide a bound on the integer solution, they can be incorporated into a branch and bound algorithm, instead of the more commonly used linear programming relaxation. However, these algorithms are special-purpose algorithms in that they exploit the "constraint pattern" or special structure of the problem.

3.3. Cutting plane algorithms based on polyhedral combinatorics

Significant computational advances in exact optimization have taken place. Both the size and the complexity of the problems solved have been increased considerably when *polyhedral theory*, developed over the past twenty-five years, was applied to numerical problem solving. The underlying idea of polyhedral combinatorics is to replace the constraint set of an integer-programming problem by an alternative convexification of the feasible points and extreme rays of the problem.

In 1935, Weyl established the fact that a convex polyhedron can alternatively be defined as the intersection of finitely many half-spaces *or* as the convex hull plus the conical hull of some finite number of vectors or points. If the data of the original problem formulation are *rational* numbers, then Weyl's theorem implies the existence of a finite system of linear inequalities whose solution set coincides with the *convex hull* of the mixed-integer points in S which we denote conv(S). Thus, if we can list the set of linear inequalities that completely define the *convexification of* S, then we can solve the integer-programming problem by linear programming. Gomory [18] derived a "cutting plane" algorithm for integer programming problems, which can be viewed as a *constructive* proof of Weyl's theorem, in this context.

Although Gomory's algorithm converges to an optimal solution in finite number of steps, the convergence to an optimum is extraordinarily slow due to the fact that these algebraically derived cuts are "weak" in the sense that they frequently do not even define supporting hyperplanes to the convex hull of feasible points. Worse yet, when many Gomory cuts are added to a problem, the cuts generated may be nearly parallel and thereby cause serious ill conditioning in the basis-matrix-requiring factorization. Finally, an additional problem with these cutting planes was that, if generated within a branch-and-bound tree, the cut was not valid throughout the tree, since the basis representation used to generate these cuts, assumed that certain variables were fixed. Recent work by Balas et al. [4] has suggested approaches to overcome the ill-conditioning problem (by carefully considering when to branch and when to cut). Similarly, they have adopted lifting techniques originally derived for polyhedral-based cuts to force the validity of the cuts throughout the tree. We will first introduce the concepts of polyhedral-based cutting planes and then come back to the promise of Gomory cuts for mixed-integer programming.

Since one is interested in a linear constraint set for conv(S) which is as small as possible, one is led to the consider *minimal* systems of linear inequalities such that each inequality defines a *facet* of the polyhedron conv(S). When viewed as cutting planes for the original problem then the linear inequalities that define facets of the polyhedron conv(S) are "best possible" cuts – they cannot be made "stronger" in any sense of the word without losing some feasible integer or mixed-integer solutions to the problem. Considerable research activity has focused on identifying part (or all) of those linear inequalities for specific combinatorial optimization problems – problem-dependent implementations, of course, that are however derived from an underlying *general* theme due to Weyl's theorem, which applies generally. Since for most interesting integer-programming problems the minimal number of inequalities necessary to describe this polyhedron is exponential in the number of variables, one is led to wonder whether such an approach could ever be computationally practical. It is therefore all the more remarkable that the implementation of cutting plane algorithms based on polyhedral theory has been successful in solving problems of sizes previously believed intractable. The numerical success of the approach can be explained, in part, by the fact that we are interested in *proving* optimality of a *single extreme point* of conv(S). We therefore do not require the *complete* description of F but rather only a partial description of F in the *neighborhood* of the optimal solution.

Thus, a general cutting plane approach relaxes in a first step the integrality restrictions on the variables and solves the resulting linear program over the set S'. If the linear program is unbounded or infeasible, so is the integer program. If the solution to the linear program is integer, then one has solved the integer program. If not, then one solves a *facet-identification problem* whose objective is to find a linear inequality that "cuts off" the fractional linear programming solution while assuring that all feasible integer points satisfy the inequality – i.e. an inequality that "separates" the fractional point from the polyhedron conv(S).

Most of the polyhedral-theory requires one to identify specific sub-structures of the original problem and then based on such structures generate polyhedral cuts (or approximations to such cuts) that separate the hyperplane added from the fractional point. Clearly, we want to generate strong cuts – i.e. cuts that approximate well the convex hull of the integer points around the optimal solution point, and one wishes to generate as few of them as necessary. The separation problem, therefore, is an optimization problem that determines the coefficients of the separating hyperplane such that the distance between this inequality and the fractional point are maximized. Many such formulations have been proposed. Most polyhedral cuts employ algorithms that generate, among all possible, the one that has the maximum geometric distance. This approach has also been proposed for *Fenchel cuts* disjunctive cuts and for general mixed-integer cuts. References and further description can be found in Nemhauser and Wolsey [31], Padberg [32], Martin [27] and Wolsey [38].

Since most of these cuts are based on some substructure of the original problem, the cuts generated will often include only a subset of the entire variable set. The idea is "cut lifting" is quite simple – assume a cutting plane on some subset of the variables has been generated. All other zero–one variables had been assumed to be either at zero or one. We now examine the consequences of having that variable no longer restricted to remain at that bound. It is precisely this lifting procedure that allows one to take Gomory cuts and make them valid throughout the enumeration tree. The ideas related to lifting originated with Padberg [32] and Wolsey [38].

A further approach to determining the convex hull of the integer points considered the role that disjunctions play in zero–one optimization problems. Disjunctions are logical conditions involving the operators "and", "or", and "not". Clearly, zero–one variables are natural disjunctions since these variables can only take on the two values, either zero or one. Using disjunctive arguments, Balas [3] showed that one could incorporate all of the restrictions of a pure zero–one linear programming problem in an equivalent linear programming problem in a much higher dimensional space. Sherali and Adams [33] provided an alternative formulation that also provides the convex hull of all integer points in a nonlinear programming formulation in a higher dimensional space. Upon first glance, these approaches may seem to provide formulations that are too enormous to be practical to consider. However, when one uses either of these formulations, and projects back into the original space of variables, one can obtain a tighter formulation through both variable substitutions and the addition of tightening cutting planes. The "lift and project" algorithm of Ceria et al. is based on these ideas and those of "lifting" back variables not in the generated cut. Separation algorithms based on these ideas require the solution of linear programming problems. Since the process is based on a given fractional solution to an LP relaxation – and not based on any specific structure of the problem, a violated inequality can always be found. For textbooks describing in detail polyhedral cuts, as well as disjunctive cuts, Gomory cuts and Fenchel cuts, see Padberg [32], Wolsey [38], Martin [27], and Nemhauser and Wolsey [31].

A cutting-plane algorithm terminates when: (1) an integer solution is found (we have successfully solved the problem); (2) the linear program is infeasible and therefore the integer problem is infeasible; or (3) no cut is identified by the facet-identification procedures either because a full description of the facial structure is not known or because the facet-identification procedures are inexact, i.e., one is unable to *algorithmically* generate cuts of a known form, or (4) the last few rounds of cut generation has not improved the objective function value sufficiently to warrant continuing the generation process. If we terminate the cutting plane procedure because of either the third or fourth possibilities, then, in general, the process has "tightened" the linear programming formulation so that the resulting linear programming solution value is much closer to the integer solution value.

Thus, cutting planes can be used as a reformulation technique. However, we consider that the overall cutting plane approach is best if incorporated into a bounding algorithm, that allows one to generate cuts not only at the top of the tree, but also throughout the tree search. This method is called "branch and cut". However, before we provide an overall description of such a hybrid algorithm, we must return to our discussions of branch-and-bound. The power of such an algorithm is dependent on the strength of the bounding arguments – when the lower bound equals the upper bound, optimality is proven. Cutting plane procedures provide a mechanism for tightening the bound produced by the relaxation. We must also have another bound – namely, we must have a good feasible solution to the optimization problem. One can wait and hope that one finds this bound within the tree search, or one can use heuristics to generate good bounds early in the process.

4. Heuristics

Operations research analysts have routinely considered using heuristics to obtain good solutions for problems considered too complex to be able to obtain optimal solutions. However, the situation has drastically changed in the past few years. Now, commercial codes whose purpose is to either prove optimality or to terminate once the solution is proven to be within a specified tolerance of optimality, apply heuristic algorithms routinely throughout the procedure so that good bounds are obtained early in the algorithm. Thus, heuristics now serve two very important purposes: to provide good solutions to problems for which current algorithms are incapable of proving optimality within reasonable times and to help in the fathoming efficiency of exact algorithm.

The research in heuristics began with concepts of *local search* whereby one constructs a feasible solution and then iteratively improve that solution by performing local moves, or "swaps".

Constructive algorithms for finding the original feasible solution may be as simple as attempting to construct such a solution greedily, i.e. picking the best single move without any look-ahead, to considering the impact of both rounding up and rounding down a given variable in an linear programming solution. Improving heuristics, similarly, can consider simple neighborhoods of a current solution, or can consider more complicated moves, such as those proposed in the Lin-Kerningham algorithm for the traveling salesman problem.

Alternatives to these construction/improvement procedures became popular in the 1980s when algorithms were proposed that allowed moves that degraded the solution in an attempt to avoid becoming stuck at local solutions. Much of this research applies techniques based on analogies from the natural world – properties of materials, natural selection, neural processing, or properties of learning found in animals.

Simulated annealing algorithms are based on the properties from statistical mechanics whereby an annealing process requires the slow cooling of metals to improve their strength. The analogy is that one will slowly converge to a feasible solution by inserting a randomization component. With a given probability, the algorithm allows moves that degrade the solution. As the algorithm progresses, however, the probability that such moves will be taken decreases. See Hansen [24] for an overview and history of such algorithmic applications to combinatorial optimization.

Similarly, genetic or evolutionary algorithms are based on properties of natural mutation. The analogy here is more obvious, in that every feasible solution to the combinatorial optimization problem is equivalent to a DNA string and each such string is given a value. One then chooses to evolve future generations of the population with "good" attributes. The likelihood that two individuals (parents) mate is dependent upon their objective function value. The mating of two individuals creates a new solution whose attributes are a combination of attributes of each parent. However, an offspring might also contain a mutation – i.e. an attribute that neither parent possessed. One is less likely to generate the same local solutions because the combining process does not center entirely on the best current solution. Goldberg [20] provides a good overview of the research in this area.

Finally, neural networks are based on models of brain function. Artificial neural network algorithms have, as their essential goal, to recognize patterns and to learn "good" responses to a given pattern. In essence, a neural network consists of a set of nodes (neurons) that are capable of receiving information from neighboring nodes and then responding to such neighbors. Since each of these nodes processes the information it receives simultaneously, the idea is that these nodes serve as a powerful parallel processor of information. Eventually, the neural networks "learns" to identify good and bad attributes. There are many alternative approaches to determining the learning strategy – there are self-organizing maps, elastic nets, back-propagation algorithms, feed-forward algorithms, etc. Also, linear-programming and steepest descent algorithms are being used to help "train" nodes in the network more quickly. At the current time, neural nets have not been shown to be competitive with other heuristics. However, the rapid evolution of neural network technology may well make these algorithms effective in the future. For a review of research in neural networks, see the entire issue of J. Comput. 5(4).

Glover and Laguna [17] have generalized many of the attributes of these methods into a method called *tabu-search*. Tabu-search is a meta-heuristic that classifies the attributes that one would wish for in an algorithm. In order to avoid returning to a known local solution too often, the algorithm keeps a list of recent moves and makes such moves forbidden for a given period of time. Thus, at each step, the algorithm must choose among moves that are feasible. The algorithm will choose a move that degrades the solution if no improving moves are possible. Other concepts built into tabu-search, include *diversification* (similar to mutations, these moves force the algorithm into a different parts of the feasible region), *long-term memory* (labeling of moves so that one prevents the repetition of the same series of moves from occurring), and *aspiration rules* (which specify when one can overlook the tabu criteria because, for example, the resulting solution is guaranteed to be better than any solution seen so far). Randomization of algorithms – including randomizing the tabu rules themselves – is easily incorporated into this framework, as is the inclusion of very sophisticated sub-algorithms. For more on meta-heuristics, see J. Comput. 11(4) (1999).

One approach to obtaining good solutions to combinatorial optimization problems – used often for difficult scheduling problems – has evolved within the computer science community. Constraint programming is a language built around concepts of tree-search and logical implications. Various tools are provided to allow the user to easily explore the search space, thereby allowing users to determine the order in which variables are given specific values and the order in which such variables are specified. One language that supports such tree-search is OPL (Optimization Programming Language) and descriptions of the language can be found in Van Hentenryck [36] while the underlying strategies can be found in MacAloon and Trekoff [28].

When considering exact approaches to solving general mixed-integer programming problems, one would like to have a heuristic that employs approaches that are used for other parts of the algorithm, as well. Thus, heuristics that can exploit some or all of the information obtained from the linear-programming relaxation of the problem are most widely used. One can see how to take many of the concepts described above, and apply them to such a heuristic. The simplest approach is to consider a "dive and fix heuristic", whereby we fix some subset of the integer variables to fixed integer values, perform all implied fixing and preprocessing, and again solve the resulting linear programming problem. This process continues until either the LP comes back with an integer feasible solution (considered a success) or stops because there are no feasible solutions to the current LP relaxation. If the latter occurs, one can either stop the algorithm and hope to find a solution at some other iteration, or one can try back-tracking (i.e., unfixing the most recently fixed variables, and fix them to their other bound). Similarly, once only a small subset of the variables remains unfixed, one can enumerate that subset thereby allowing more likelihood of finding a feasible solution. One of the first LP-based heuristics implemented into general IP-software packages was the Pivot and Complement heuristic of Balas and Martin [5].

5. Column generation

One of the recurring themes in many of the approaches to solving combinatorial optimization problems is to examine the structure of the problem and find a relaxation or decomposition of the problem that is easier to solve. One then attempts to strengthen this approximation by either adding constraints, columns or by altering the coefficients in either the constraints or the objective function. One decomposition – often referred to as *column-generation* or *branch-and-price* – that has been extraordinary successful in recent years, is that of Dantzig–Wolfe decomposition. The theory rests on the fact that any feasible point can be represented as a linear combination of the extreme points of the feasible region. Thus, if the constraint set can be divided into two segments (one with nice special structure, for example, a set-partitioning structure), and the other with a structure that allows us to generate extreme points feasible to that structure. We write the problem as

Max cx

subject to $Ax \leqslant b,$

$\qquad\qquad x \in S,$

$\qquad\qquad x$ integer.

The procedure rests on the fact that given a set $S^* = \{x \in S: S \text{ is a bounded set}, x \text{ integer}\}$ then S^* can also be represented as a finite set of points $S^* = \{y_1, y_2, \ldots, y_p\}$. Thus, any point $y \in S^*$ can be represented as $y = \sum_{1 \leqslant k \leqslant p} \lambda_k y_k$ subject to the convexity constraint $\sum_{1 \leqslant k \leqslant p} \lambda_k = 1$ and $\lambda_k \in \{0, 1\}$, $k = 1, 2, \ldots, p$. Thus, one can formulate the problem as:

Max $\displaystyle\sum_{1 \leqslant k \leqslant p} (c y_k) \lambda_k$

subject to $\displaystyle\sum_{1 \leqslant k \leqslant p} (A y_k) \lambda_k \leqslant b,$

$\qquad\qquad \displaystyle\sum_{1 \leqslant k \leqslant p} \lambda_k = 1,$

$\qquad\qquad \lambda_k \in \{0, 1\}, \quad k = 1, 2, \ldots, p.$

For most practical problems the set S^* is too large to enumerate. Instead, one begins by generating sufficient columns so that the "master problem" is guaranteed to have a feasible solution (at least in the LP relaxation to the problem). One then performs a "pricing problem", to identify additional columns that will improve the LP solution to the master problem. This pricing algorithm uses the dual information from the master problem to generate new columns. The master problem is re-solved and the process continues until no column exists that improves the LP. Branching is performed once the LP optimum is found. This approach is especially useful when the resulting master problem has a structure, such as set partitioning that is well known to have a tight LP objective function value and whose polyhedral structure has been well-studied. In addition, this structure may remove symmetries that existed in the compact formulation, and may allow for branching on constraints, referred to as *strong branching*. Finally, there are problems for which the column formulation is the only choice (e.g., crewscheduling problems – For these problems, the rules determining a "feasible" schedule for a crew are so complicated that one cannot write a linear constraint set that describes all the characteristics of the problem.)

Problems that have been successfully solved using this re-formulation include the generalized assignment problem, bin-packing, graph coloring, vehicle routing with time windows, and other complicated delivery problems. For each of these problems, the resulting optimization problem has a set-partitioning, packing or covering structure. Given this structure, one carefully designs the overall algorithm so that symmetries that occur because there are identical machines, trucks or crews can be identified in the generation process. A reformulation is then done that combines the convexity constraints in such a way as to remove the symmetry. Similarly, branching strategies are employed, similar to those in constraint logic, which determines if specific trucks, machines, crews must handle specific types of customers, tasks or flights. Thus, the cutting planes, branching and re-formulation are all strengthened because one better understands the problem characteristics. For an excellent overview of column generation techniques, see the works of Barnhart et al. [6] and of Sol [35].

6. Hybrid algorithms

We next explain how much of the research and development of integer programming methods can be incorporated into a super-algorithm, which uses all that is known about the problem. This method is called "branch-and-cut".

Current software packages include many of the features described above. The major components of these hybrid algorithms consist of automatic reformulation procedures, heuristics which provide "good" feasible integer solutions, and cutting plane procedures which tighten the linear programming relaxation to the combinatorial problem under consideration – all of which is embedded into a tree-search framework as in the branch-and-bound approach to integer programming. Whenever possible, the procedure permanently fixes variables (by reduced cost implications and logical implications) and does comparable conditional fixing throughout the search tree. These four components are combined so as to guarantee optimality of the solution obtained at the end of the calculation. However, the algorithm may also be stopped *early* to produce sub-optimal solutions along with a bound on the remaining error. The cutting planes generated by the algorithm are facets of the convex hull of feasible integer solutions or good polyhedral approximations thereof and as such they are the "tightest cuts" possible. Lifting procedures assure that the cuts generated are valid throughout the search-tree that aids the search process considerably.

Mounting empirical evidence indicates that both pure and mixed integer programming problems can be solved to *proven* optimality in economically feasible computation times by methods based on the polyhedral structure of integer programs. A direct outcome of these research efforts is that similar preprocessing and constraint generation procedures can be found in commercial software packages for combinatorial problems (see [13] and [15] for software implementations of preprocessing and cutting planes).

Finally, we are now seeing algorithms that expand not only the constraint set but also the column set. These algorithms begin by creating a "master problem" and a "pricing problem". It allows the use of all of that we have learned about constraint generation for set-covering and packing structures, allows strong branching and includes heuristics to be used to both generate columns and find feasible solutions to the master problem. There are many issues, however, that are still little understood. When one designs such algorithms, one must consider when to generate columns, when to generate additional cuts, when to search for a better feasible solution and when to branch. When one generates more columns and more constraints, the resulting LP-relaxations become harder to solve. However, the overall time spent solving the problem is likely to be reduced because the number of nodes on the branching tree is reduced substantially. Similarly, spending time finding good feasible solutions allows greater fathoming of the branching tree. It is also important to realize that the successes of these hybrid algorithms are not due to a single component but rather to the interactions and symbiotic relationship among these components. Good upper and lower bounds allow the fixing of variables. The fixing of variables changes the structure of the overall problem, implying new constraints, allowing the heuristic to find new solutions, and altering the rules for searching the tree or generating new columns. Much more testing need to be done to better understand the interactions among these procedures.

The computational successes for difficult combinatorial optimization problems reflect the intense effort devoted to developing the underlying structure of these problems. These approaches may expand the dimensionality of the problem, expand the size of the constraint set, and may require sophisticated heuristic procedures to be embedded in such algorithms. A variety of search techniques might be considered within the mega-procedure. It should be stated, however, that we would not have been able to consider applying such complicated strategies had the underlying "engine" – *linear programming* – not been able to solve the subproblems generated so efficiently. Work on linear programming in the past ten years has substantially altered our strategies toward solving combinatorial problems. See papers in this volume on the changes in this technology. Other breakthroughs may come about because of breakthroughs in our ability to solve efficiently – to global optimality – non-linear programming problems with special structure, such as semi-definite programming problems. We would then begin to use quite different relaxations, which will then alter the cutting-plane and heuristic techniques employed. Thus, successes in one optimization technology naturally bring successes in other very different structures and problems. See Wolsey [38], Nemhauser and Wolsey [31], Padberg [32] and Martin [27] for a detailed discussion of branch-and-cut and polyhedral approaches to solving many important classes of 0–1 programming problems.

7. Parallel implementations

A significant amount of research has taken place recently related to parallel implementations of combinatorial and linear programming algorithms. For linear programming, parallel factorization and pricing schemes have proven extraordinarily successful in shortening the time it takes to solve linear programming problems having millions of variables and thousands of constraints. These algorithms will play a very important role as we expand both the constraint set and the column set of the linear programming relaxations. Again, see other papers in this volume that discuss these important breakthroughs.

When considering how to alter an algorithm so that computations are done across a variety of machines, there are many alternative approaches to consider. One can provide each machine a single node of the branching tree and allow that processor to perform all work associated with that node. Alternatively, one can require that a single machine take on all work associated with a collection of branches. Similarly, one can have machines dedicated to column generation, constraint generation (possibly having many machines each devoted to generating cuts of specific type), and machines dedicated to generating feasible solutions through one or more heuristic schemes.

Parallelization of the search tree has, naturally, seen more study than any of the other approaches, since the subproblems associated with each node are completely independent. However, even in such simple approaches to parallelization, one wishes to share information among nodes as quickly as possible. Cannon and Hoffman [12] designed an algorithm whereby, whenever a processor found a feasible solution better than any previously known, that solution was broadcast to all other processors. Since the only information broadcast was the *value* of the objective function value, such broadcasting was easy to perform. Knowing a better solution value allows fathoming and formulation strengthening to take place instantaneously on all nodes. In addition, these authors stored all constraints in a central pool – a file readable by all processors – so that many nodes could share structural information and not incur the expense of regeneration. The branch-and-cut algorithm is especially suited to this approach, since the cuts generated are applicable throughout the tree.

However, one must store these cuts in a way that avoids serious contention and latency problems. The Cannon–Hoffman approach stored cuts generated from each given row in a separate file so that various processors could be reading different files simultaneously. The file was only accessed if the processor found that the row in question identified a fractional variable. Each cut in the file had a unique identifier so that any cut in the file that were in the existing problem were not re-examined. Each cut also had a key structure that allowed one to also calculate the overlap between that cut and the fractional variables in the current LP solution quickly. In this way, one could examine more closely only cuts likely to be useful to that processor. Having designed a parallel version of a branch-and-cut code to exploit the characteristics of the machines being used (distributed workstations with *no* shared memory), Cannon and Hoffman were capable of achieving superlinear speedups on a set of difficult optimization problems. This approach did not have a master–slave relationship among processors, but rather used a file system again to maintain the list of all tasks still needing work. Whenever a processor completed its work, it would return to this work file and both add new tasks to the file and extract a new task from the file.

Column-generation algorithms have similar challenges to overcome. Decisions about how to share columns among processors are essential. Since each column is generated from some subset of the entire structure one can provide a flag that indicates the structure from which it came. Alternatively, one can store columns in files based on whether that column covers a specific row. In either approach, a processor will only examine files when needing a column having some specific structure.

Much additional research needs to be done to better understand how the many subalgorithms now existing within an overall hybrid algorithm interact. Parallel optimization algorithms may help us "learn" when alternative approaches work best. Appelgate et al. [2] used many distributed workstations to prove optimality to traveling salesmen problems having over 10 000 variables. The algorithm employed required substantial work at each node of the tree. They therefore wanted to carefully choose the variable to branch on before beginning such work. Such considerations resulted in a pivoting strategy to choose the branching variable that is now incorporated as an option in the single processor version of CPLEX (a widely used software package for integer linear programming problems).

Parallel processors may also serve another very important role: currently optimization is used mostly in planning situations. Scheduling algorithms are often used to determine the optimal machine to use to accomplish specific tasks, to determine the announced schedules for crews two months prior to flying, or to determine the schedule of machines before the day begins. However, when the situation changes during the day, users require that the schedule be changed in "real-time". Our algorithms are often not fast enough to supply such answers. Parallel implementations may be able to re-optimize a schedule as complicated as that of an airline when a major storm or maintenance situation causes the existing schedule to no longer be feasible.

8. New developments in modeling and problem generation

Much of this paper has been concerned with the *solution* of difficult and important combinatorial optimization problems. This presumes that the task of correctly modeling the problem and then providing that mathematical model to a solver is a simple task. A major breakthrough in our ability to quickly solve many important problems has been in our ability to model quickly such problems and to provide to other modelers and algorithm developers language that can be quickly understood and whose structure can be readily identified. Modeling languages such as AIMMS [8], AMPL [16], GAMS [11], MIMI, MPL and OPL [36] have allowed analyst to express their problem in languages that directly supports a natural (i.e., more word-like) statement of the problem. All of the above-mentioned languages except MIMI present the problem from a row orientation. MIMI looks at the problem from a process-oriented perspective, and formulates the model in terms of activities (columns). There are also language extensions to many of these that allow one to discuss networks in a natural arc/node descriptive form. Clearly, what is natural for one modeler may not be for another, so flexibility in the ability to describe the model has much value.

A nice attribute of the row-oriented languages is that they allow the user to state the general form of a constraint-set and have the language generate the sequence of constraints that have that form. Since the language allows long naming, as well as constructs such as "while" and "for all" statements, the model is far more readable and changeable quickly. Some of these languages allow the user to separate the model from specific data instances thereby allowing the same model to be used for many alternative instances. Some allow the automatic linking to databases eliminating the need for the extraction of data into new tables solely for the use of an optimization code. Some have Graphical User Interfaces (GUIs) that allow users to present their output in charts and graphs that help explain the results obtained. Some of these languages allow the solving of a string of optimization problems thereby providing a more natural and automatic mechanism for doing sensitivity analysis. Since all data is stored together, the results of this analysis can be displayed in a variety of intuitive, graphical ways.

One of these languages, OPL, has now incorporated language that allows constraint programming to be linked with mathematical optimization tools into a single overall modeling tool. All other languages treat the optimizer as a black box, accessible only through well-defined parameters. OPL, on the other hand, now allows the user to link concepts of user-directed tree search with concepts of optimization relaxation. This new package is a first step in bridging the gap between modelers who treat optimization as black-box solvers and code developers who need to test new algorithmic concepts.

In one sense, MINTO [30] can be considered a pre-cursor (from the optimization-communities' perspective) to OPL. This software package allowed optimizers to use pieces of a general optimization package, and test their own sub-algorithms within this overall package. However, that package was designed specifically for algorithmic developers and did not have the higher-level modeling language tools of the packages discussed above.

OPL, on the other hand, is the first language to attempt to provide higher-level modeling tools and to link these with language constructs specifically designed to help direct tree-search activities. Specifically, constraint programming provides to the optimization community many of the constraint reasoning tools i.e., provides nondeterministic constructs that relieve a modeler from the many mundane implementation aspects of tree-search procedures. Since constraint programming is mostly concerned with proposing software architectures to simplify search algorithms, such methods are likely to be useful in quickening the modelers ability to generate feasible solutions to difficult optimization problems.

The real strength of merging concepts of constraint programming with those of combinatorial optimization, is that we may both better understand and preserve the structure of the underlying problem and we may be able to quickly develop hybrid, meta-algorithms far more powerful than any algorithms we employ today.

Currently, the user of combinatorial optimization algorithms must transform many logical restrictions into a set of linear constraints. Such transformations – as presented in textbooks on linear and integer programming – often destroy an underlying structure and, when linear programming is used as the relaxation, often provide bounds that are far from the optimal integer solution. We believe that a better approach is to have the user supply the problem using logical operators and have the optimization procedure determine the best way to approximate the problem. Thus, instead of requiring the user to transform logical constructs, (such as "A not equal to B", "A only if B", "always choose A before B"), the user supplies these restrictions in the natural form. The modeling language then makes whatever transformations are best for the algorithm used. Similarly: modeling languages should, in the future, allow the user to supply fixed charges, piecewise-linear approximations and graph-related concepts (such as paths, cycles, etc.) in a natural way. The user should also be able to tell the optimizer any information that might help the tree-search or the constraint generation. We do not yet understand how best to perform these tasks, but future versions of modeling languages are likely to allow the user to maintain a transparent descriptions of the underlying problems and allow the optimizer to exploit the underlying structure of such problems far more easily.

With the structure transparent, new algorithms are likely to emerge. Such meta-algorithms will allow all procedures (re-formulation, constraint generation, heuristics, column-generation, and tree-search), to choose sub-algorithms that are most useful for the problem structure exhibited.

9. Understanding the solution

The discussion so far has concentrated on the issues associated with the initial formulation of the problem and current algorithms for solving the problem. However, users want more than a solution vector or objective function value. Users need an understanding of *why* the problem was infeasible. Much progress has been made recently in determining an irreducible infeasible set (IIS) of constraints (see [23]). That is, a subset of constraints defining the overall program that is itself infeasible, but for which any proper subset is feasible.

Similarly, if the problem is feasible, one wants to know the set of constraints that force the optimal solution, and also know the set of constraints that are redundant (or play little role in the solution obtained). One would also like to know whether bounds on specific variables are most restrictive, and, most importantly, which variables were "driving" the problem – i.e., as soon as the value of these variables is known, the problem becomes "easy" to solve.

Current software has incorporated techniques that inform the optimizer this information. We need to develop ways of presenting this information back to the modelers so that they learn far more about the underlying process than is currently provided by the solution vector itself. One software package that provides some of this information is ANALYZE, developed by Greenberg [21] for analyzing linear and integer programs and their solutions.

As the demand for more complex modeling increases, the demand for computer-assisted modeling and analysis will increase. New approaches include the use of artificial-intelligence queries to the model and its outputs, visualization tools to understand the structure of the problem, and a variety of model management tools [34]. One can find a complete bibliographic listing to work on modeling languages, analysis tools and data management tools in the works of Greenberg [22].

10. Stochastic and robust optimization

When our ability to solve large, complex combinatorial optimization problems seemed quite limited, users were satisfied with strictly the solution to the problem posed. But, with our successes has come demands for far more challenging problems to be solved. Although this paper has focused exclusively on the solution of deterministic problems, we acknowledge that demand is growing for solution approaches to the more difficult (but far more realistic) problem – acknowledging that all data is not known with certainty. For such instances, a variety of approaches have been proposed: stochastic integer programming, chance-constrained programming, dynamic programming, and robust optimization.

The simplest approach to handling uncertainty is to estimate the mean value of each parameter and solve a deterministic problem. Then, for those values that have most variability, perform sensitivity analysis on the respective values. Of course, sensitivity analysis in integer programming requires far more effort than for the linear case, so only very small perturbations are usually considered.

Another approach – one commonly used in portfolio optimization and capital budgeting – is to force a diversification of the portfolio (i.e. add constraints that force the portfolio to choose a variety of different types of investments). A second approach adds a penalty to the objective function for the likely event that a constraint will be violated because of variability in the data. A third approach adds new constraints that provide a measurement of risk and then enforces that one does not allow more risk than a given amount. In each of these cases, one has transformed the problem to a deterministic problem. Along these same lines, one can evaluate a reward to risk curve by solving a variety of deterministic optimization problems and having the user determine where along the curve he feels most comfortable.

We now present methods that address the stochasticity directly. One such method is called robust optimization. In this case, stochasticities are addressed via a set of discrete scenarios. Here, one needs to not only specify the scenarios that are likely to occur, but also the utility of the outcomes that occur under each scenario. Here, models either incorporate risk by incorporating variance measures into the objective function or by incorporating expected utility functions. However, in either case, the transformed objective function becomes nonlinear, making the problem more difficult to solve, especially when integrality conditions are imposed. Other reasons for their lack of use are that it is often difficult to obtain the users utility function and/or variance and covariance measures. Also, the resulting solutions are less intuitive to the user. See Mulvey et al. [29] for a discussion of such methods.

Stochastic optimization takes a similar approach to that of robust optimization, but instead of using an expected utility function, it incorporates a penalty for deviation from feasibility for any of the given scenarios (weighted by the expected value of the scenario occurring). The absence of general efficient methods for solving stochastic linear integer problems reflects the fact that, unlike the linear case, very few general properties are known, and what is known is discouraging. One encouraging note is that, when the random variables are appropriately described by a finite distribution, one can obtain approximation algorithms that provide bounds on the solutions obtained (for details see the textbook by Birge [7]).

Thus, although currently, there is little commercial software that incorporates these ideas, it is likely that as our technology for solving deterministic integer linear programming problems improves, we are far more likely to examine ways of incorporating risk issues into our models. We hope that future research will also address the issue of how to incorporate "fuzzy" data, i.e., data for which even the mean value is not known and for which one only has range estimates of its value. Research in the stochastic optimization must also address mechanisms for explaining the suggested results to users in a far more intuitive and understandable fashion. These are extremely difficult problems, and yet, those of most interest to the industrial community that has so benefited from our successes.

11. Where can these successes take us?

Until recently, only large corporations could afford to use combinatorial optimization because the costs of data collection, expensive computer machinery, analyst's time, and the training of employees to use such sophisticated tools were simply too high. Now, computing costs are no longer an issue (every small company has PCs that are capable of running extraordinarily large optimization problems). The data collection have been mostly eliminated because of sophisticated database technology (automated inventory systems, order fulfillment packages, automatic storage of customer requests, etc.). Modeling languages make the time to develop and test models for shorter.

With the growth of the Internet, more people have access to sophisticated tools and information that ever before. Now organizations are faced with an environment marked by increasing complexity, economic pressure and customer expectations. The need for cost reduction and the need for fast product development is imperative. Customers have come to expect high product reliability and sophisticated functionality at a low cost. The need to accomplish these new demands has required that companies focus on their internal business processes and to create relationships with suppliers and their customers so as to achieve maximum efficiency and integration along the entire supply chain. Clearly, optimization can play an important role in these activities. Cost savings can occur by limiting inventory, by continually evaluating all of the logistics costs, and by examining how to minimize the capital tied up in the supply chain. By reducing the cumulative time between product development and delivery to the customer and by elimination of duplication of effort in the supply chain, one can obviously increase long-term profitability.

Enterprise Resource Planning (ERP) information systems are designed to "optimize" across the extended enterprise. Many such systems are in the process of embedding sophisticated combinatorial optimization models within their systems. Once they have been successfully integrated into these systems, entire industries will be using optimization tools routinely for infrastructure design, facility location and sizing, synchronization resources and material flow, resource allocation, transportation and logistics, inventory control and pricing modeling. The impact that such modeling might have on the long-term viability of enterprises could be staggering.

Another exciting challenge for the optimization community is to consider how to provide our tools over the Internet on an as-needed basis. As the software industry moves from having individuals and corporations buy software to individuals leasing software for as short as a few minutes over the Internet, optimization tools can play an important role. Conceivably, someone with a specific scheduling problem would go to a website, provide the data specific to the problem, and nearly-instantly receive a solution. That individual may use such software routinely or only once per year. One of the early entries into this market that allows users to solve optimization applications on the web is the NEOS Project [14].

To achieve these goals a number of issues must be resolved. We must provide intuitive graphical-user-interfaces so that less-technical users will be able to use our tools. We must continue to improve the tools available. Far more research on the mixed-integer problem needs to take place. Considerations of stochasticity, robustness, adjusting of solutions to small data changes (e.g., re-scheduling when something alters the availability of resources), must be considered. Being able to handle simple nonlinearities must be addressed. Similarly, as we continue to improve our ability to solve larger and more complex problems, we are likely to be asked to take on even greater challenges. A very interesting collection of papers on the opportunities for optimization on the world wide web can be found in J. Comput. 10 (1998).

These needs in no way degrade the achievements already made. It is precisely the past successes that have highlighted the need to take on even greater challenges. Happily, our ability to solve more of the real-world problems appears to be accelerating as we have begun to bring divergent lines of research together into mega-algorithms. We must always remember, however, the looming in the shadows is the conjecture that $P \neq NP$, making it unlikely that we will ever be able to solve *all* of the challenges posed. The advances in information technology, make our existing tools much more useful, and provide us with a far greater set of opportunities than we could have hoped for even five years ago. We hope that many in the modeling and algorithmic community will step up to these challenges.

References

[1] E.D. Anderson, K.D. Anderson, Pre-solving in linear programming, Math. Programming 71 (1995) 221–245.

[2] D. Applegate, R. Bixby, W. Cook, Finding cuts in the traveling salesman problem, Rice University Technical Report, 1996.

[3] E. Balas, Disjunctive programming: Properties of the convex hull of feasible points, Discrete Appl. Math. 89 (1998) 3–44; orginally appeared as a Management Science Research Report 348, GSIA, Carnegie Mellon University, 1974.

[4] E. Balas, G. Cornuejols, N. Natraj, Gomory cuts revisited, Operations Res. Lett. 19 (1996) 1–9.

[5] E. Balas, R.K. Martin, Pivot and complement: A heuristic for 0–1 programming, Management Sci. 26 (1980) 86–96.

[6] C. Barnhart, E.L. Johnson, G.L. Nemhauser, M.W.P. Savelsbergh, P.H. Vance, Branch and price column generation for solving huge integer programs, Operations Res. 71 (1996) 221–245.

[7] J.R. Birge, Introduction to Stochastic Programming, Springer, Berlin, 1997.

[8] J. Bisschop, R. Entriken AIMMS The modeling system. Paragon Decision Technology, 1993.

[9] J. Bramel, D. Simchi-Levi, On the effectiveness of set covering formulations for the vehicle routing probelm with time windows, Operations Res. 45 (1997) 295–301.

[10] A.L. Brearly, G. Mitra, H.P. Williams, Analysis of mathematical programming problems prior to applying the simplex method, Math. Programming 8 (1975) 54–83.

[11] A. Brooke, D. Kendrick, A. Meeraus, GAMS, A User's Guide, The Scientific Press, Redwood City, 1988.

[12] T. Cannon, K.L. Hoffman, Large-scale 0–1 programming on distributed workstations, Ann. Oper. Res. 22 (1990) 181–217.

[13] CPLEX Optimization, Inc. (1998) Using the CPLEX Callable Library and CPLEX Mixed Integer Programming Library, Version 5.0.

[14] J. Czyzyk, T. Wisniewski, S. Wright, Optimization case studies in the NEOS Guide, SIAM Rev. 41 (1999) 148–163.

[15] Dash Associates, XPRESS-MP User's Manual, 1994.

[16] R. Fourer, D.M. Gay, B.W. Kernighan, AMPL A Modeling Language for Mathematical programming, The Scientific Press, Palo Alto, CA, 1993.

[17] F. Glover, M. Laguna, Tabu Search, Kluwer Academic, Boston, 1997.

[18] R. Gomory, Solving linear programming problems in integers, R.E. Bellman and M. Hall Jr. (Eds.), Combinatorial Analysis, American Mathematical Society, Providence, RI, pp. 211–216.

[19] M.X. Goemans, D.P. Williamson, Improved approximation algorithms for maximum cut and satisfiability problems using semi-definite programming, J. ACM 42 (6) (1996) 1113–1145.

[20] D.E. Goldberg, Genetic Algorithms in Search Algorithms and Machine Learning, Addison-Wesley, Reading, MA, 1989.

[21] H.J. Greenberg, A computer Assisted Anaysis System for Mathematical Programming Models and Solutions: A User's Guide to Analyze, Kluwer, Hingham, MA, 1993.

[22] H.J. Greenberg, An annotated bibiliography for post-solution analyis in mixed integer and combinatorial optimization, in: D.L. Woodruff (Ed.), Advances in Computational and Stochastic Optimization, Logic Programming and Heuristic Search, Kluwer, Hingham, MA, 1998.

[23] O. Guieu, J. Chinneck, Analyzing infeasible mixed integer and integer linear programs, INFORMS J. Comput. 11 (1999) 63–77.

[24] P. Hansen, The steepest ascent mildest descent heuristic for combinatorial programming, Proceedings of Congress on Numerical Methods in Combinatorial Optimization, Capri, Italy, 1986.

[25] K.L. Hoffman, M. Padberg, Improving the LP-representation of zero-one linear programs for branch-and-cut, ORSA J. Comput. 3 (1991) 121–134.

[26] J.T. Linderoth, M.W. Savelsbergh, A computational study of search strategies for mixed integer programming, INFORMS J. Comput. 11 (1998) 173–187.

[27] R.K. Martin, Large Scale Linear and Integer Optimization, Kluwer, Hingham, MA, 1998.

[28] K. McAloon, C. Tretkoff, Optimization and Computational Logic, Wiley Interscience Series in Mathematics and Optimization, Wiley, New York, 1996.

[29] J. Mulvey, H. Vladimirou, Stochastic network programming for financial planning models, Management Sci. 38 (1992) 1642–1664.

[30] G.L. Nemhauser, M.W.P. Savelsbergh, G. Sigismondi, MINTO: A mixed integer optimizer, Operations Res. Lett. 15 (1994) 47–58.

[31] G.L. Nemhauser, L.A. Wolsey, Integer and Combinatorial Optimization, Wiley, New York, 1988.

[32] M. Padberg, Linear Optimization and Extensions, Springer, Heidelberg, 1995.

[33] H.D. Sherali, W.P. Adams, A hierarchy of relaxation and convex hull characterizations for mixed-integer zero-one programming problems, Discrete Appl. Math. 52 (1994) 83–106.

[34] B. Shetty, H. Bhargava, R. Krishnan (Eds.), Annals of Operations Research 38 (1992) Volume on "Model Management in Operations Research".

[35] M. Sol, Column generation techniques for pickup and delivery problems, Dissertation, Eindhoven University of Technology, Eindhoven, The Netherlands, 8 November, 1994.

[36] P. Van Hentenryck, The OPL Optimization Programming Language, MIT Press, Cambridge, MA, 1999.

[37] H.P. Williams, Model Building in Mathematical Programming, 4th ed., Wiley, New York, 1998.

[38] L.A. Wolsey, Integer Programming, Wiley, New York, 1998.

N·H

ELSEVIER

Journal of Computational and Applied Mathematics 124 (2000) 361–371

JOURNAL OF
COMPUTATIONAL AND
APPLIED MATHEMATICS

www.elsevier.nl/locate/cam

Optimal control

R.W.H. Sargent *

Centre for Process Systems Engineering, Imperial College, Prince Consort Road, London SW7 2BY, UK

Received 20 August 1999; received in revised form 25 November 1999

Abstract

This paper gives a brief historical survey of the development of the theory of the calculus of variations and optimal control, and goes on to review the different approaches to the numerical solution of optimal control problems. © 2000 Elsevier Science B.V. All rights reserved.

Keywords: Historical review; Optimal control theory; Calculus of variations; Numerical solutions

1. Introduction

Optimal control theory is an outcome of the calculus of variations, with a history stretching back over 360 years, but interest in it really mushroomed only with the advent of the computer, launched by the spectacular successes of optimal trajectory prediction in aerospace applications in the early 1960s.

Fortunately, Goldstine [27] has written an excellent treatise on the early history, and there have been three later publications [10,59,64] carrying the story up to the present day. We therefore give only an outline of the main steps in the historical development of the theory, then focus on the development of numerical techniques for solution.

2. A brief history of the theory

Some geometrical optimization problems were known and solved in classical times, such as the line representing the shortest distance between two points, or the "isoperimetric problem": the shape of the plane curve of given length enclosing the largest area. However our story really begins with Galileo, who in 1638 posed two shape problems: the shape of a heavy chain suspended between

---

* Tel.: +0171-594-6604; fax: +0171-594-6606.
E-mail address: r.w.h.sargent@ic.ac.uk (R.W.H. Sargent).

0377-0427/00/$ - see front matter © 2000 Elsevier Science B.V. All rights reserved.
PII: S 0377-0427(00)00418-0

two points (the catenary), and the shape of a wire such that a bead sliding along it under gravity traverses the distance between its end-points in minimum time (the brachistochrone). Later, in 1662, Fermat postulated the principle that light always chooses the path through a sequence of optical media such that it traverses them in minimum time.

Galileo's conjectures on the solutions of his two problems were incorrect, and Newton in 1685 was the first to solve a shape problem — the nose shape of a projectile providing minimum drag — though he did not publish the result until 1694.

In 1696 Johann Bernoulli challenged his contemporaries to solve the brachistochrone problem by the end of the year. Five mathematicians responded to the challenge: Johann's elder brother Jakob, Leibnitz, l'Hopital, Tschirnhaus and Newton. Bernoulli published all their solutions, together with his own, in April 1697.

The competition aroused interest in this type of problem and there followed a period of activity by a number of mathematicians. The resulting ideas were collected in a book [25] published in 1744 by Euler, a student of Bernoulli working in Basel, who remarked "nothing at all takes place in the universe in which some rule of maximum or minimum does not appear".

In essence, Euler formulated the problem in general terms as one of finding the curve $x(t)$ over the interval $a \leqslant t \leqslant b$, with given values $x(a), x(b)$, which minimizes

$$J = \int_a^b L(t, x(t), \dot{x}(t)) \, dt \qquad (1)$$

for some given function $L(t, x, \dot{x})$, where $\dot{x} \equiv dx/dt$, and he gave a necessary condition of optimality for the curve $x(\cdot)$

$$\frac{d}{dx} L_{\dot{x}}(t, x(t), \dot{x}(t)) = L_x(t, x(t), \dot{x}(t)), \qquad (2)$$

where the suffix x or $\dot{x}$ implies the partial derivative with respect to x or $\dot{x}$.

Up to this point the solution techniques had been essentially geometric, but in a letter to Euler in 1755, Lagrange described an analytical approach, based on perturbations or "variations" of the optimal curve and using his "undetermined multipliers", which led directly to Euler's necessary condition, now known as the "Euler–Lagrange equation". Euler enthusiastically adopted this approach, and renamed the subject "the calculus of variations".

The Euler–Lagrange equation, based on first-order variations, yields only a stationarity condition, and it was Legendre in 1786 (see [33]), who studied the second variation and produced a second-order necessary condition of optimality. Legendre derived it for the scalar case, but it was later extended to the vector case by Clebsch and is now known as the Legendre–Clebsch condition

$$L_{\dot{x}\dot{x}}(t, x(t), \dot{x}(t)) \geqslant 0, \quad t \in [a, b], \qquad (3)$$

interpreted as requiring the matrix to be nonnegative definite along the optimal trajectory.

Meanwhile Hamilton [28], through his "principle of least action", had been reformulating the equations of mechanics as a variational principle. He introduced the function, now known as the "Hamiltonian function"

$$H(t, y, x) = \langle y, \dot{x} \rangle - L(t, x, \dot{x}), \qquad (4)$$

where

$$y(t, x, \dot{x}) = L_{\dot{x}}(t, x, \dot{x}) \qquad (5)$$

and in (4) $\dot{x}$ is obtained as a function of (t, y, x) by solving Eq. (5). It follows immediately that

$$\dot{x}(t) = H_y(t, y(t), x(t)), \quad \dot{y}(t) = -H_x(t, y(t), x(t)) \tag{6}$$

if and only if the Euler–Lagrange Eq. (2) is satisfied.

Much later Caratheodory [16] showed that (5) is indeed always solvable for $\dot{x}$ in a neighbourhood defined by weak variations (those for which perturbations δx and $\delta \dot{x}$ are both small) of a regular optimal trajectory (one for which the strict Legendre–Clebsch condition holds). Under these conditions H is well defined and twice continuously differentiable, and $H_{yy}(t, y, x)$ is positive definite if and only if $L_{\dot{x}\dot{x}}(t, x, \dot{x})$ is positive definite.

Hamilton expressed his principle in terms of a pair of partial differential equations, but in 1838 Jacobi showed that it could be more compactly written in terms of what is now known as the Hamilton–Jacobi equation:

$$\phi_t(t, x) + H(t, \phi_x(t, x), x(t)) = 0. \tag{7}$$

If $\phi(t, x)$ is a twice continuously differentiable solution of this equation, then Eq. (6) defines a regular optimal trajectory, provided that

(a) The strict Legendre–Clebsch condition (i.e. $H_{yy}(t, y(t), x(t)) > 0$) is satisfied.
(b) There are no points conjugate [1] to a along the trajectory (the Jacobi condition).

The next step was taken by Weierstrass [66], who considered strong variations (with δx small but no restriction on $\delta \dot{x}$). He considered the special case where $L(t, x, \dot{x})$ is a positively homogeneous function not depending explicitly on t, but with no loss of generality, since any $L(t, x, \dot{x})$ can be transformed to this form. He introduced the "excess function":

$$E(t, x, \dot{x}, u) = L(t, x, u) - L(t, x, \dot{x}) - (u - \dot{x})L_{\dot{x}}(t, x, \dot{x}) \tag{8}$$

with $x(t), \dot{x}(t)$ evaluated along the optimal trajectory.

Again Caratheodory [16] showed that $E > 0$ if and only if $H_{yy} > 0$, thus confirming the sufficiency of the Hamilton–Jacobi solution even under strong variations.

Given the basic problem (1), it was natural to require $x(\cdot)$ to be differentiable on (a, b), and to consider minimization over all such curves, but the development of measure theory allowed the interpretation of (1) as a Lebesgue integral and the relaxation of $x(\cdot)$ to be absolutely continuous on (a, b), incidently providing a closure property for the family of functions J defined for different $\dot{x}(\cdot)$.

Caratheodory was well aware of the need to establish existence of optimal trajectories, and established this through the rather strong sufficient conditions for the existence of the requisite solutions of the Hamilton–Jacobi equation. However, the new viewpoint allowed Tonelli [61] to address directly the problem of existence of optimal trajectories for problem (1). His proof required the convexity of the function $L(t, x, \cdot)$ for all t, x, and a growth condition of the type

$$|L(t, x, \dot{x})| \geqslant \alpha.\|\dot{x}\|^2 - \beta \tag{9}$$

for some positive constants α and β.

---

[1] Conjugacy is defined with respect to solutions of an associated partial differential equation.

The next step was to consider the restriction of the class of admissible functions $\dot{x}(\cdot)$ to a subset of R^n, specifically so that they also satisfy the set of equations

$$g(t,x(t),\dot{x}(t)) = 0, \quad t \in [a,b], \tag{10}$$

which we would today recognize as a general set of differential–algebraic equations. However, sufficient conditions were imposed to ensure that there exist functions $\dot{x}(t) = f(t,x(t))$ satisfying (10) with sufficient degrees of freedom to ensure the existence of neighbouring functions also satisfying (10).

The resulting problem is known as the problem of Lagrange, since it was solved by the use of Lagrange multipliers, and the complete solution can be found in [6,42].

This set the scene for parameterizing the degrees of freedom implicit in the Lagrange problem by considering constraints of the form

$$\dot{x}(t) = f(t,x(t),u(t)), \tag{11}$$

where the parameters $u(t)$ or "controls" can be chosen at each $t \in (a,b)$, possibly restricted to some fixed subset $\Omega \subset \mathbb{R}^m$, yielding the "optimal control problem":

Find $u(\cdot)$ on (a,b) to minimize

$$J = \int_a^b L(t,x(t),\dot{x}(t))\,\mathrm{d}t$$

subject to

$$\dot{x}(t) = f(t,x(t),u(t)), \quad t \in (a,b), \quad u(t) \in \Omega \subset \mathbb{R}^m, \quad t \in (a,b), \quad x(a) \text{ and } x(b) \text{ given.} \tag{12}$$

The necessary conditions of optimality for this problem were established by Pontryagin [54] in his famous "maximum principle", which can be expressed in the form

$$\dot{x}(t) = H'_y(t,y(t),x(t),\dot{x}(t)), \qquad \dot{y}(t) = -H'_x(t,y(t),x(t),\dot{x}(t)),$$
$$H'(t,y(t),x(t),\dot{x}(t)) = \max_{u\in\Omega} H'(t,y(t),x(t),u). \tag{13}$$

The function H' in (13) is not the classical Hamiltonian, but what Clarke [18] later termed the "pseudo-Hamiltonian", still defined by (4) but with $\dot{x}$ as an independent argument instead of being defined as a solution of (5). However, it is easy to see that the Hamilton equations (6) still hold for the pseudo-Hamiltonian if and only if the Euler–Lagrange equation (2) holds, though it is now more natural to express the Legendre–Clebsch condition in the form

$$H'_{\dot{x}\dot{x}}(t,y(t),x(t),\dot{x}(t)) \leqslant 0, \quad t \in (a,b). \tag{14}$$

In fact, Pontryagin et al. [54] considered several variants and extensions of the basic problem (12), and from this point there was an increasing avalanche of publications.

Many workers tackled the problem of pure state inequalities along the trajectory of the form

$$g(t,x(t)) \geqslant 0, \quad t \in (a,b) \tag{15}$$

and this work is well summarized in the recent review in [29], which also describes some open questions remaining in this area. More recently, Sargent [56] has presented necessary conditions of optimality for systems described by a mixed set of general differential–algebraic equations and inequalities.

Roxin [55] extended the classical Tonelli existence theorem to more general problems like (12), and at about the same time Warga [65] showed that convexity of the "extended velocity set" could be dropped by extending the class of admissible controls to include "relaxed" or chattering controls, which allow rapid oscillation between two points of this set, thus effectively replacing the set by its convex hull.

Recently, Bell et al. [4] extended Roxin's result to infinite horizon problems, using the "strong optimality criterion", in the ordinary sense that the integral in (12) remains finite as $b \to \infty$, and is minimized by the optimal control. However, in many cases the integral can become infinite, and successive weakenings of the optimality criteria have been proposed.

Overtaking optimality:

$$\limsup_{b \to \infty} \; [J_b(\hat{x}, \hat{u}) - J_b(x, u)] \leqslant 0.$$

Weakly overtaking optimality:

$$\liminf_{b \to \infty} \; [J_b(\hat{x}, \hat{u}) - J_b(x, u)] \leqslant 0.$$

Finite optimality:

$$J_b(\hat{x}, \hat{u}) - J_b(x, u) \leqslant 0$$

for all (x, u) defined on (a, b) such that $x(b) = \hat{x}(b)$, for all $b > a$.

All these optimality criteria are treated in some detail in the book in [17].

Two more general formulations of the problem have been much studied, mainly by Rockafellar, Clarke, Loewen, Mordukhovich and Vinter:

(a) *The differential inclusion problem*:

$$\text{Minimize } J = l(x(a), x(b)) + \int_a^b L(t, x(t), \dot{x}(t)) \, dt$$

over absolutely continuous functions $x(\cdot)$ on (a, b) subject to:

$$\dot{x}(t) \in F(t, x(t)), \quad t \in (a, b) \text{ a.e.}, \quad (x(a), x(b)) \in C \subset \mathbb{R}^{2n}.$$

Here F is a multifunction, mapping $[a, b] \times \mathbb{R}^n$ into subsets of $\mathbb{R}^n$.

(b) *The generalized Bolza problem*:

$$\text{Minimize } J = l(x(a), x(b)) + \int_a^b L(t, x(t), \dot{x}(t)) \, dt$$

over all absolutely continuous functions $x(\cdot)$ on $[a, b]$.

This deceptively simple form can subsume a wide variety of problems by allowing l and L to take infinite values and defining them appropriately, for example by setting l or L to infinity whenever their arguments fail to satisfy the constraints.

Study of these problems is accompanied by a relaxation of the differentiability conditions necessary in the classical results, requiring the techniques of nonsmooth analysis. To state specific results would take us too far into technicalities, and reference should be made to the textbooks in [18,19,64], and to the survey paper in [20] for further details.

However, in two recent papers [36,37] Loewen and Rockafellar give the latest results on necessary conditions for the generalized Bolza problem, first for fixed values of a and b, and in the later paper for variable values.

Another important issue has been the quest for conditions under which the sufficient conditions obtained via the Hamilton–Jacobi equation are also necessary. Here Clarke and Vinter [21] have shown in the context of the differential inclusion problem that the weakest condition is "local calmness", which is implied by "strong normality", so that the existence of a generalized solution of the Hamilton–Jacobi equation is both necessary and sufficient for optimality for all "reasonable" problems.

Finally, Zeidan [67] has given a complete treatment of second-order necessary and sufficient conditions for problem (12) with additional mixed state-control inequalities along the trajectory. The paper includes some interesting insights on the concept of strong normality, and a connection between the Jacobi condition and solutions of a Riccati equation.

3. Numerical solution techniques

Compared with the intricacies and subtleties of the theoretical development, the history of numerical solution techniques is relatively straightforward. There are essentially three approaches to solve these problems:

1. Solution of the two-point boundary value problem given by the necessary conditions, with solution of the local Hamiltonian optimization problem at each time-step.
2. Complete discretization of the problem, converting it into a finite-dimensional nonlinear program.
3. Finite parameterization of the control trajectory, again converting the problem into a nonlinear program, but with the objective and constraint functions evaluated by integration of the system equations, and their gradients with respect to the control parameters by integration of the adjoint equations or sensitivity equations.

The early numerical methods tackled problems without control or end-point constraints, which of course still yielded a two-point boundary value problem, with given initial values for the system equations and given terminal values for the adjoint system. Bryson [12] and Breakwell [8] used the "shooting method" for solving this problem, guessing the unknown initial values of the adjoint variables, integrating both system and adjoint equations forward, and re-estimating the initial guesses from residuals at the end-point.

Difficulties arose from the extreme sensitivity, even instability, of the solutions to the initial guesses, which led to the "multiple-shooting" technique of Bulirsch and his co-workers [14,58], who subdivided the time interval and re-estimated starting values for each subinterval from the mismatches. An alternative approach, due to Miele and co-workers [43,44] was quasilinearization, in which the system equations were linearized about the current trajectory and integrated forward in time (along with the adjoint equations) for sets of initial conditions spanning the space of the unknown initial values. A linear combination of these solutions was then computed to satisfy the conditions at both initial and final times, and the system re-linearized about the new trajectory.

At almost the same time as these early attempts, Kelley [31] proposed a gradient method, essentially the simplest member the class of control parameterization techniques. He estimated control values on a closely spaced fixed grid, and used this grid to integrate the system equations. He then integrated the adjoint system backwards from the known values at the final time and used the results to obtain the gradients with respect to the control values and hence a correction to the estimated control values.

Of course steepest descent methods have slow final convergence rate, and to speed this up several workers used techniques based on second variations [9,30,32]. At the same time others used numerical acceleration techniques from the parallel developments in nonlinear programming: Lasdon et al. [34] incorporated conjugate gradients while Pollard and Sargent [53] used quasi-Newton approximations. The latter authors also pointed out that the parameterization of the controls need not be tied to the integration step and used piecewise-constant or piecewise-linear controls on a coarser grid, yielding a much smaller optimization problem.

In all these methods, control and terminal constraints were at first dealt with using penalty functions, but control constraints (usually simple bounds) were later treated by projection. A special difficulty arises with end-point equality constraints, because there is then a corresponding number of undetermined terminal values for the adjoint variables and of course the terminal $x(b)$ values from the integration do not necessarily satisfy the terminal equality constraints. Again, the problem was dealt with by penalty functions, but not very effectively. Obviously, the quasilinearization approach provides a general technique for dealing with an arbitrary mix of initial and final equations and inequalities, and Bryson and Ho [11] describe a similar technique for generating a set of adjoint systems, subsequently linearly combined to reduce the residuals of the end-point equality constraints. They also describe "min-H" algorithms which determine the correction to the controls by performing a local constrained optimization of the Hamiltonian at each time-step of the shooting methods. In fact, this textbook [11] provides an excellent summary of the state of the art at the end of the 1960s, except that it completely ignores the development of the complete discretization approach.

Of course, nonlinear programming techniques were also in their infancy over this period, so complete discretization was late on the scene. Early proposals were made in [41,52], and the state of the art was probably summed up by the textbook of Canon et al. [15].

From this point, all three approaches had their followers and were improved using advances in the enabling technologies of nonlinear programming and integration of ordinary differential equations and differential–algebraic equations.

Early methods of complete discretization used finite differences, but Tsang et al. [62] introduced collocation methods, while Biegler and co-workers [22,38,60] developed this approach, showing how to incorporate error measures as constraints to decide on the number and placing of additional nodes in the "collocation on finite elements" method. A parallel development was carried out in [2,5] in the aerospace arena, and recently, Dennis et al. [23] described an interior-point SQP method using a trust-region algorithm.

Sargent and co-workers [57,49,63,51] developed the control parameterization approach, extending it to deal with multistage systems, high-index DAEs and state inequalities, while Bock and his school [7,35] have similarly developed a hybrid approach, with some elements of all three approaches. He starts with a control parameterization, then generates a finite-dimensional nonlinear programme by integrating the system independently over each control subinterval, as in multiple shooting.

Bulirsch and his school [14,58,50,13] and Maurer [39,40,1] have continued to develop the multiple shooting approach to solving the two-point boundary value problem, particularly in relation to dealing with inequality state constraints along the trajectory, while Miele and co-workers have pursued the quasilinearization approach [43–47] linking it with multiple shooting (which they call the "multipoint" approach). Dixon and Bartholomew-Biggs [24] proposed an adjoint-control transformation to help with the sensitivity of the shooting technique, and recently, Fraser-Andrews [26] has combined this with multiple shooting.

Efforts to deal with state inequalities along the trajectory were hampered in the early days by inadequate understanding of the theory, but it is still a formidable problem because of the possibility of discontinuities in the adjoint variables at the "junction-points", where the inequalities become active or cease to be active. The complete discretization approach appears to deal with the problem, since these constraints become ordinary nonlinear constraints of the nonlinear programme, but the discretization destroys the hidden additional constraints arising from differentiations of the original constraints for high-index problems, so a fine discretization is necessary to be sure of satisfying them. Pantelides et al. [51] achieved some success in the control parameterization approach by converting these constraints into end-point constraints by integrating the violations, coupled with a finite set of interior-point constraints, but strictly the problem is still nonsmooth, and the two-point boundary value approach remains the only safe method.

Maurer [39,40,1] has pioneered the judicious use of some pre-analysis of the problem to determine the structure of junction-points, coupled with multiple shooting, and the power of this approach is well demonstrated by the successful solution of the difficult "wind-shear" problem by Bulirsch et al. [50]. However, such solutions require real expertise.

Very recently, Bell and Sargent [3] have sought to avoid the difficulties by conversion of the inequalities to equalities using slack variables and elimination of the resulting bounds by an interior-point approach, converting the optimal control problem into a smooth two-point boundary problem for a DAE system. Preliminary results indicate surprising success in obtaining close approximation of this smooth problem to the original nonsmooth one.

Finally, although the emphasis in this survey has been on optimal control for general nonlinear systems, it would not be complete without a mention of the spectacular successes of the use of discrete-time linear models with quadratic objective function in on-line control applications. This formulation particularly suits the process industries, and requires the solution of a small finite-dimensional quadratic programme on-line, which can easily be solved within the requisite time. Fortunately, a comprehensive review of the relevant issues has recently been published [48], to which the reader is referred for further details.

4. Concluding comments

As already noted, this survey has been limited to optimal control of general nonlinear systems, essentially described by ordinary differential equations or differential–algebraic equations. Although there has been a recent upsurge of interest in control of distributed-parameter systems, it would widen the scope too far to attempt to include this area. Similarly, little has been said of the extensive work on linear systems, and nothing at all on the effects of uncertainty. Each of these would require a separate review of at least the same length. Even so the scope is still enormous and many facets have been left unexplored. It is virtually impossible to keep abreast of all developments, and the selection of topics must inevitably reflect a personal viewpoint.

References

[1] D. Augustin, H. Maurer, Stability and sensitivity analysis of optimal control problems subject to pure state constraints, presented at the 16th International Symposium on Mathematical Programming, Lausanne, August, 1997.

[2] T. Bauer, J.T. Betts, W. Hallman, W.P. Huffman, K. Zondervan, Solving the optimal control problem using a nonlinear programming technique, Parts I and II, AIAA Papers 84-2037 and 84-2038, 1984.

[3] M.L. Bell, R.W.H. Sargent, Optimal control of inequality-constrained DAE systems, Centre for Process Systems Engineering Report No. 98-B , Imperial College, 1998. Comput. Chem. Eng., accepted for publication.

[4] M.L. Bell, R.W.H. Sargent, R.B. Vinter, Existence of optimal controls for continuous-time infinite horizon problems, Int. J. Control 68 (1997) 887–896.

[5] J.T. Betts, W.P. Huffman, Application of sparse nonlinear programming to trajectory optimization, J. Guidance Control Dynamics 15 (1) (1992) 198–206.

[6] G.A. Bliss, The problem of Lagrange in the calculus of variations, Am. J. Math. 52 (1930) 673–744.

[7] H.G. Bock, K.J. Plitt, A multiple-shooting algorithm for direct solution of optimal control problems, Proceedings of the Nineth IFAC World Congress, Budapest, Pergamon Press, Oxford, 1984.

[8] J.V. Breakwell, The optimization of trajectories, SIAM J. 7 (1959) 215–247.

[9] J.V. Breakwell, J. Speyer, A.E. Bryson, Optimization and control of nonlinear systems using the second variation, SIAM J. Control Ser. A 1 (2) (1963) 193–223.

[10] A.E. Bryson Jr., Optimal control – 1950 to 1985, IEEE Control Systems (1996) 26–33.

[11] A.E. Bryson, Y-C. Ho, Applied Optimal Control, Blaisdell, Waltham, MA, 1969.

[12] A.E. Bryson, S.E. Ross, Optimum rocket trajectories with aerodynamic drag, Jet Propulsion (1958).

[13] R. Bulirsch, F. Montrone, H. J. Pesch, Abort landing in the presence of a windshear as a minimax optimal control problem: Part 1: Necessary conditions, Part 2: Multiple shooting and homotopy, Technische Universität Munchen, Report No.210, 1990.

[14] R. Bulirsch, Die Mehrzielmethode zur numerischen Losung von nichtlinearen Randwertproblemen und Aufgaben der optimalen Steuerung, Report of the Carl-Cranz Gesellschaft, DLR, Oberpfaffenhofen, 1971.

[15] M.D. Canon, C.D. Cullum, E. Polak, Theory of Optimal Control and Mathematical Programming, McGraw-Hill, New York, 1970.

[16] C. Caratheodory, Variationsrechnung und partielle Differentialgleichungen erste Ordnung, Teubner, Berlin, 1935 (Translation: Calculus of Variations and Partial Differential Equations of the First Order, R. Dean, Holden-Day, San Francisco, 1967).

[17] D.A. Carlson, A. Haurie, Infinite Horizon Optimal Control, Springer, Berlin, 1987.

[18] F.H. Clarke, Optimization and Non-smooth Analysis, Wiley, New York, 1983 (Reprinted by SIAM, Philadelphia, PA, 1990).

[19] F.H. Clarke, Methods of Dynamic and Non-smooth Optimization, SIAM, Philadelphia, 1989.

[20] F.H. Clarke, Y.S. Ledyaev, R.J. Stern, P.R. Wolenski, Qualitative properties of trajectories of control systems: a survey, J. Dynamics Control Systems 1 (1995) 1–48.

[21] F.H. Clarke, R.B. Vinter, Local optimality conditions and Lipschitzian solutions to the Hamilton–Jacobi equation, SIAM J. Control Optim. 21 (1983) 856–870.

[22] J.E. Cuthrell, L.T. Biegler, On the optimization of differential–algebraic process systems, A.I.C.h.E. J. 33 (1987) 1257–1270.

[23] J.E. Dennis, M. Heinkenschloss, L.N. Vicente, Trust-region interior-point algorithms for a class of nonlinear programming problems, SIAM J. Control Optim. 36 (1998) 1750–1794.

[24] L.C.W. Dixon, M.C. Bartholomew-Biggs, Adjoint-control transformations for solving practical optimal control problems, Optim. Control Appl. Methods 2 (1981) 365–381.

[25] L. Euler, Methodus Inveniendi Lineas Curvas Maximi Minimive Propriatate Gaudientes, sive Solutio Problematis Isoperimetrici Latissimo Sensu Accepti, Bousquent, Lausannae and Genevae, 1744.

[26] G. Fraser-Andrews, A multiple-shooting technique for optimal control, J. Optim. Theory Appl. 102 (1999) 299–313.

[27] H.H. Goldstine, A History of the Calculus of Variations from the 17th to the 19th Century, Springer, New York, 1981.

[28] W.R. Hamilton, Mathematical Papers, Cunningham Memoir No. XIII, Cambridge University Press, Cambridge, 1931.

[29] R.F. Hartl, S.P. Sethi, R.G. Vickson, A survey of the maximum principles for optimal control problems with state constraints, SIAM Rev. 37 (2) (1995) 181–218.

[30] D.H. Jacobson, D.Q. Mayne, Differential Dynamic Programming, Elsevier, New York, 1970.

[31] H.J. Kelley, Gradient theory of optimal flight paths, J. Amer. Rocket Soc. 30 (1960).

[32] H.J. Kelley, R.E. Kopp, G. Moyer, A trajectory optimization technique based upon the theory of the second variation, presented at the AIAA Astrodynamics Conference, New Haven, CT, 1963.

[33] J.L. Lagrange, Théorie des Fonctions Analytiques, Paris, 1797.

[34] L.S. Lasdon, S.K. Mitter, A.D. Waren, The conjugate gradient methods for optimal control problems, IEEE Trans. Automat Control AC-12 (1967) 132–138.

[35] D. Leineweber, H.G. Bock, J.P. Schloder, J.V. Gallitzendorfer, A. Schafer, P. Jansohn, A boundary value problem approach to the optimization of chemical processes described by DAE models, University of Heidelberg, Preprint 97-14 (SFB359), 1997.

[36] P.D. Loewen, R.T. Rockafellar, New necessary conditions for the generalized problem of Bolza, SIAM J. Control Optim. 34 (1996) 1496–1511.

[37] P.D. Loewen, R.T. Rockafellar, Bolza problems with general time constraints, SIAM J. Control Optim. 35 (1997) 2050–2069.

[38] J.S. Logsdon, L.T. Biegler, Accurate solution of differential–algebraic optimization problems, Ind. Eng. Chem. Res. 28 (1989) 1628–1639.

[39] H. Maurer, Optimale Steuerprozesse mit Zustandsbeschrankungen, Habilitations-schrift, University of Wurzburg, 1976.

[40] H. Maurer, W. Gillessen, Application of multiple shooting to the numerical solution of optimal control problems with bounded state variables, Computing 15 (1975) 105–126.

[41] D.Q. Mayne, A second-order gradient method for determining optimal trajectories of nonlinear discrete-time systems, Int. J. Control 3 (1) (1966) 85–95.

[42] E.J. McShane, On multipliers for Lagrange problems, Amer. J. Math. 61 (1939) 809–819.

[43] A. Miele, Method of particular solutions for linear two-point boundary-value problems, J. Optim. Theory Appl. 2 (4) (1968).

[44] A. Miele, Gradient methods in optimal control theory, in: M. Avriel, M.J. Rijckaert, D.J. Wilde (Eds.), Optimization and Design, Prentice-Hall, Englewood Cliffs, NJ, 1973, pp. 323–343.

[45] A. Miele, R.R. Iyer, General technique for solving nonlinear two-point boundary-value problems via the method of particular solutions, J. Optim. Theory Appl. 5 (1970) 382–399.

[46] A. Miele, T. Wang, Parallel computation of two-point boundary-value problems via particular solutions, J. Optim. Theory Appl. 79 (1993) 5–29.

[47] A. Miele, K.H. Well, J.L. Tietze, Multipoint approach to the two-point boundary-value problem, J. Math. Anal. Appl. 44 (1973) 625–642.

[48] M. Morari, J.H. Lee, Model predictive control: past, present and future, Comput. Chem. Eng. 23 (1999) 667–682.

[49] K.R. Morison, R.W.H. Sargent, Optimization of multi-stage processes described by differential–algebraic equations, in: J.P. Hennart (Ed.), Numerical Analysis, Springer, Berlin, 1986.

[50] H.J. Oberle, W. Grimm, BNDSCO – a program for the numerical solution of optimal control problems, Internal Report No. 515-89/22, Institute for Flight Systems Dynamics, DLR, Oberpfaffenhofen, 1989.

[51] C.C. Pantelides, R.W.H. Sargent, V.S. Vassiliadis, Optimal control of multi-stage systems described by high-index differential–algebraic equations, in: R. Bulirsch, D. Kraft (Eds.), Computational Optimal Control, Birkhauser, Basel, 1994.

[52] E. Polak, On primal and dual methods for solving discrete optimal control problems, in: L.A. Zadeh, L.W. Neustadt, A.V. Balakrishnan (Eds.), Computing Methods in Optimization Problems, Vol. 2, Academic Press, New York, 1969, pp. 317–331.

[53] G.M. Pollard, R.W.H. Sargent, Off-line computation of optimum controls for a plate distillation column, Automatica 6 (1970) 59–76.

[54] L.S. Pontryagin, V.G. Boltyanskii, R.V. Gamkrelidze, E.F. Mischenko, The Mathematical Theory of Optimal Processes (Translation by L.W. Neustadt), Wiley, New York, 1962.

[55] E. Roxin, The existence of optimal controls, Michigan Math. J. 9 (1962) 109–119.

[56] R.W.H. Sargent, Necessary conditions for optimal control of inequality-constrained DAE systems, Centre for Process Systems Engineering Report No. B98-17, Imperial College, 1998.

[57] R.W.H. Sargent, G.R. Sullivan, The development of an efficient optimal control package, in: J. Stoer (Ed.), Proceedings of the Eighth IFIP Conference on Optimization Techniques, Part 2, Springer, Heidelberg, 1978.

[58] J. Stoer, R. Bulirsch, Introduction to Numerical Analysis, Springer, New York, 1980.

[59] H.J. Sussmann, J.C. Willems, 300 years of optimal control: from the brachystochrone to the maximum principle, IEEE Control Systems (1997) 32–44.

[60] P. Tanartkit, L.T. Biegler, A nested simultaneous approach for dynamic optimization problems, Comput. Chem. Eng. 20 (1996) 735–741.

[61] L. Tonelli, Sull' existenza del minimo in problemi di calcolo delle variationi, Ann. R. Scuola Normale Sup. Pisa, Sci. Fis. Mat. 1 (1931) 89–100.

[62] T.H. Tsang, D.M. Himmelblau, T.F. Edgar, Optimal control via collocation and nonlinear programming, Int. J. Control 21 (1975) 763–768.

[63] V.S. Vassiliadis, R.W.H. Sargent, C.C. Pantelides, Solution of a class of multi-stage dynamic optimization problems: 1, problems without path constraints, 2, problems with path constraints, I and EC Res. 33 (1994) 2111–2122, 2123–2133.

[64] R.B. Vinter, Optimal Control, Birkhauser, Boston, 1999.

[65] J. Warga, Relaxed variational problems, J. Math. Anal. Appl. 4 (1962) 111–128.

[66] K. Weierstrass, Werke, Bd. 7, Vorlesungen uber Variationsrechnung, Akademisches Verlagsgesellschaft, Leipzig, 1927.

[67] V. Zeiden, The Riccati equation for optimal control problems with mixed state-control constraints: necessity and sufficiency, SIAM J. Control Optim. 32 (1994) 1297–1321.

www.ingramcontent.com/pod-product-compliance
Lightning Source LLC
Chambersburg PA
CBHW080710220326
41598CB00033B/5367

* 9 7 8 0 4 4 4 5 0 5 9 9 6 *